7th AIC-ICMR on Sciences and Engineering 2017

Advancing Knowledge Research and Technology for Humanity

IOP Conference Series: Materials Science and Engineering
Volume 352

Banda Aceh, Indonesia
18 - 20 October 2017

Editor:

Syarizal Fonna

ISBN: 978-1-5108-6283-8
ISSN: 1757-8981

Printed from e-media with permission by:

Curran Associates, Inc.
57 Morehouse Lane
Red Hook, NY 12571

Some format issues inherent in the e-media version may also appear in this print version.

Printed by Curran Associates, Inc. (2018)

For permission requests, please contact the Institute of Physics
at the address below.

Institute of Physics
Dirac House, Temple Back
Bristol BS1 6BE UK

Phone: 44 1 17 929 7481
Fax: 44 1 17 920 0979

techtracking@iop.org

Additional copies of this publication are available from:

Curran Associates, Inc.
57 Morehouse Lane
Red Hook, NY 12571 USA
Phone: 845-758-0400
Fax: 845-758-2633
Email: curran@proceedings.com
Web: www.proceedings.com

Table of contents

Volume 352

The 7th AIC-ICMR on Sciences and Engineering 2017

18–20 October 2017, Banda Aceh, Indonesia

Accepted papers received: 16 April 2018
Published online: 11 May 2018

Preface

Papers

Mechanical and Industrial Engineering

Civil and Structural Engineering

Physical Sciences and Engineering

Chemical Sciences and Engineering

Computer Science and Information Systems

Architecture, Urban and Rural Planning

PROCEEEDING

The 7th AIC-ICMR on SCIENCES and ENGINEERING

The Annual International Conference 2017
Syiah Kuala University

"Advancing Knowledge, Research, and Technology for Humanity"

Banda Aceh, Aceh, Indonesia
October 18-20, 2017

IOP Conf. Series: Materials Science and Engineering **352** (2018) 011001 doi:10.1088/1757-899X/352/1/011001

Advisory Board

Prof. Dr. Samsul Rizal	Rector of Syiah Kuala University, Indonesia
Dr. Hizir	Vice Rector I for Academic Affairs, Syiah Kuala University, Indonesia
Prof. Dr. Husni	Vice Rector II for Financial Affairs, Syiah Kuala University, Indonesia
Dr. Alfiansyah B.C.	Vice Rector III for Alumnae and Student Affairs, Syiah Kuala University, Indonesia
Dr. Nazamuddin	Vice Rector IV for Planning, Cooperation, and Community Affairs, Syiah Kuala University, Indonesia
Prof. Dr. Hasanuddin	Head of Institute for Research and Community Services, Syiah Kuala University, Indonesia
Dr. Khairul Munadi	Syiah Kuala University, Indonesia
Dr. Syaifullah Muhammad	Syiah Kuala University, Indonesia

IOP Conf. Series: Materials Science and Engineering **352** (2018) 011001 doi:10.1088/1757-899X/352/1/011001

Editorial Board

IOP Conf. Series: Materials Science and Engineering **352** (2018) 011001 doi:10.1088/1757-899X/352/1/011001

Organizing Committee

General Chair
Dr. Ramzi Adriman, Syiah Kuala University, Indonesia

General Co-Chair
Afdhal Azmi, Syiah Kuala University, Indonesia

Treasurer and Finance Chair
Dr. Ira Devi Sara, Syiah Kuala University, Indonesia

Conference Chair
Dr. Syarizal Fonna, Syiah Kuala University, Indonesia

Conference Co-Chair
Dr. Syifaul Huzni, Syiah Kuala University, Indonesia

Publication Chair
Dr. Saiful, Syiah Kuala University, Indonesia
Dr. Yunita Idris, Syiah Kuala University, Indonesia

Correspondence Chair
Dr. Halida Yunita, Syiah Kuala University, Indonesia

OCS Administrator
Wahyu Rinaldi, Syiah Kuala University, Indonesia

Website Developer
Muhammad Chandra Gunawan, Syiah Kuala University, Indonesia
Syahabuddin, Syiah Kuala University, Indonesia

Local Arrangement Committee
Muhammad Haiqal, Syiah Kuala University, Indonesia
Purwandi Hasibuan, Syiah Kuala University, Indonesia
Zainuddin, Syiah Kuala University, Indonesia

International Scientific Committee

IOP Conf. Series: Materials Science and Engineering **352** (2018) 011002 doi:10.1088/1757-899X/352/1/011002

Welcome Speech from Rector of Syiah Kuala University

Assalamualaikum Wa Rahmatullahi Wa Barakatuh,

In the name of Allah, the Most Beneficent and the Most Merciful. May peace, mercy, and blessings of Allah be upon you.

Dear colleagues, professors, lecturers, researchers, ladies and gentlemen. On behalf of Syiah Kuala University, I would like to express my sincere gratitude and welcome you to the 7th AIC-ICMR on Science and Engineering 2017. This year, the 7th AIC-ICMR on Science and Engineering is supported and co-organized by Universiti Sains Malaysia, Hasanuddin University, Universitas Islam Sumatera Utara, BKS-PTN Barat, and IMT-GT Uninet. Thus, I would be glad to express my sincere appreciation for their collaboration and support for the conference. Moreover, I honorably welcome our keynote speakers Prof. Dato' Dr. Musa Ahmad, Vice-Chancellor at Universiti Sains Islam Malaysia (USIM, Malaysia, and Assoc. Prof. Dr. Azma Putra from Universiti Teknikal Malaysia Melaka (UTeM) Malaysia.

It is my hope that the 7th AIC-ICMR on Science and Engineering 2017 would be able to achieve its objective in providing an effective forum for academician, researchers, and practitioners to advancing knowledge, research, and technology for humanity. There has always been a gap between peoples and communities who can make effective use of technology and those who cannot. 'Humanity' (or being 'humane') can also in one usage refer to civil rights and social causes, or in other words to people treating each other with care, compassion, and dignity; respecting the common 'humanity' in the other person. Therefore, despite what we may have been falsely led to believe, matters of 'human' rights have no relation to politics or any part of the so-called political spectrum: they do not belong to any political party or faction, nor to any individual or organization; they are part of our common nature, part of the collective 'humanity' spoken for by our very existence. For that matter, all academicians, researchers, and practitioners should consider humanity things to make a peaceful world. With humanity, we change to the better environment and a better life. We believe knowledge, research and technology for all people, for minorities, people with disabilities, and to help anyone. No matter how much we can accomplish by ourselves, whether it be research or development, it is never sufficient in this world of knowledge. Therefore, the focal drive of this conference is to exchange ideas, and by participating in this exchange, it is hoped that all parties who may benefit from the conference can apply it in managing activities in their areas. It is pleasing to note that the agenda of this conference covers a wide range of interesting topics related to all theoretical and practical aspects, but not limited to Mathematical Sciences, Physics Sciences and Engineering, Chemical Sciences and Engineering, Geology, Mining and Petroleum Engineering, Mechanical and Industrial Engineering, Civil and Structural Engineering, Architecture and Planning, Electrical and Computer Engineering, Computer Sciences and Information Systems.

Last but not least, my deepest gratitude goes to the Advisory Board, Organizing Committee, International Scientific Committee, institutions, companies, and volunteer who have directly and indirectly supported the success of this seminar. The committee has organized a vibrant scientific program and is working hard to present highly respected and internationally notorious speakers to lead it. Although we try our finest to be professional, on behalf of Syiah Kuala University, please accept our sincere apologies should there be inconveniences that occur before, during, or after the event. I wish you a very productive conference with exciting and encouraging discussions and exchange of knowledge so that together we can anticipate a future of groundbreaking knowledge, research, and technology for humanities.

May God bless us all with good health to make this event a successful and enjoyable one!

Best Regards,

Prof. Dr. Ir. Samsul Rizal, M.Eng
Rector of Syiah Kuala University

IOP Conf. Series: Materials Science and Engineering **352** (2018) 011002 doi:10.1088/1757-899X/352/1/011002

Welcome Speech from General Chair

Assalamualaikum Wa Rahmatullahi Wa Barakatuh,

On behalf of the organizing committees, I would like to welcome all of you to Banda Aceh, Indonesia for the 7th AIC-ICMR on Science and Engineering 2017 as a part of the 2017 Annual International Conference (AIC) Syiah Kuala University. This conference is co-organized by Universiti Sains Malaysia (USM), Universitas Islam Sumatera Utara (UISU), Universitas Hasanuddin (UNHAS), and the Cooperation Agency of Public Universities in the Western Region of Indonesia (BKS PTN Barat) and IMT-GT Uninet. The annual event in 2017 is joint conference between the 7th Annual International Conference (AIC) Syiah Kuala University and The 6th International Conference on Multidisciplinary Research (ICMR). The AIC is an annual international conference carried out regularly by Syiah Kuala University since 2011. The ICMR is an annual event which has been jointly organized by UNSYIAH, USM, UISU and UNHAS since 2012. Therefore, we would like to express our sincere appreciation for their support to the conference.

Under the theme of the conference "*Advancing Knowledge, Research, and Technology for Humanity*", this conference features a rich program, including Syiah Kuala University Innovation Expo 2017, two keynote speeches delivered by Prof. Dato' Dr. Musa Ahmad, Vice-Chancellor at Universiti Sains Islam Malaysia (USIM, Malaysia, and Assoc. Prof. Dr. Azma Putra from Universiti Teknikal Malaysia Melaka (UTeM) Malaysia. the 7th AIC-ICMR on Science and Engineering 2017 offers a special opportunity to bring together professors, researchers and scholars around the globe, and serves as a platform to deliver innovative research results and latest trends and development in the fields of sciences and engineering.

All accepted and presented papers in this conference will have the opportunities to be published in IOP Material Science and Engineering Proceedings and indexed by Sci-Scopus, Thomson Reuters ISI, and EI Compendex. The conference has received 110 submitted papers, whereby 67 papers have been accepted by the committees for presentation and to be included in the proceedings. These papers on various topics are divided into 10 parallel sessions in the conference. To all members of the organizing committees, the international scientific committee, the reviewers, and the collaboration partners, we would like to thank all of them for their tremendous efforts to organize this conference successfully.

We look forward to having a successful conference, and we hope that all the attendees enjoy and benefit from this conference.

Best Regards,

Dr. Ramzi Adriman
General Chair

Keynote Speaker 1

Optical Sensing of Chemicals: The Principles & Challenges

Prof. Dato' Dr. Musa Ahmad
Faculty of Science & Technology,
Universiti Sains Islam Malaysia (USIM), Nilai,
Negeri Sembilan, Malaysia

Abstract

Generally, any substance surrounding us in our physical world, such as liquid, solid or gas is made of chemicals. A chemical could be pure or any mixture; occurs naturally or chemically synthesised in laboratory. Both synthetic and natural chemicals can have poisonous effects. The degree of poisoning caused by a chemical is depending on several factors, include route of entry into the body, dosage, toxicity of the chemical and biological variation. Based on this ground, it is therefore essential to sense the presence of these chemicals in the surrounding, in order to quantify their concentration.

Optical chemical sensor is an analytical tool consisting of active sensing material used in close conjunction with a device which can transforms chemical information into an analytically useful signal. Optical chemical sensors have been the focus of much research attention in recent years since it has its own merit such as low cost, robust, can be applied for in situ analysis and possibility of remote sensing with the use of optical fibre. This presentation will basically highlight the development of such analytical methods and instrumentations for quantitative quantification of chemicals or analytes particularly in food industries and environmental monitoring application.

Keywords: optical chemical sensor, optical fibre.

IOP Conf. Series: Materials Science and Engineering **352** (2018) 011003 doi:10.1088/1757-899X/352/1/011003

Keynote Speaker 1 - Profile

Prof. Dato' Dr. Musa Ahmad has been the Vice-Chancellor at Universiti Sains Islam Malaysia (USIM) since August 2014. Prior to his present position Prof. Ahmad was serving as Deputy Vice Chancellor (Academic and Inter national) at USIM since May 2010 until August 2014.

He served Universiti Kebangsaan Malaysia (UKM) for more than 20 years as an academician and administrator, the latest as Dean of Faculty of Science and Technology (December 2008 - May 2010).

Prof. Ahmad obtained his first degree in Chemistry from UKM with first class honours, and obtained his MSc (Instrumentation & Analytical Science) and PhD (Optical Chemical Sensor) degrees from the University of Manchester, England. As an academician, he has shown an excellent progress in teaching, supervision, publication, research and consultancy.

His main research interest is in the field of chemical optical sensing and optical bio-sensing. His research findings made him won more than 20 awards at R&D exhibition either at national or international levels, which include International Invention Exhibition at Geneva, Switzerland; World Exhibition of Innovation, Brussels, Belgium and INSPEX Invention in Cleaning Products/Equipment, Pittsburg, USA. Some of these products have been highlighted at local newspapers and TV stations. Today, over 200 articles of Prof. Ahmad have been published in the form of books, journal papers and conference proceedings.

IOP Conf. Series: Materials Science and Engineering **352** (2018) 011003 doi:10.1088/1757-899X/352/1/011003

Keynote Speaker 2

Towards Eco-Friendly Acoustic Absorbers

Assoc. Prof. Dr. Azma Putra
Centre for Advanced Research On Energy
Universiti Teknikal Malaysia Melaka
Malaysia

Abstract

The acoustic absorber has been widely known for its applications for noise control in industries as well as for improving acoustic quality in buildings. Sound absorption materials were first substituted by synthetic fibers in the 1970s due to public health concerns on the original asbestos-based material, which is harmful to human health. Until recent years, when global warming has became an alarming issue, studies were conducted to investigate the utilisation of sound absorber materials. According to the study on the Life Cycle Assessment (LCA) on a range of synthetic and natural absorber materials, it is found that the production of synthetic materials requires more energy and thus provide greater global warming potential. In contrast, natural plant-based material contributed less carbon footprint. In the air quality guidelines published by World Health Organisation (WHO), synthetic fibres were related to air pollutants which can potentially lead to skin and eye irritation and when inhaled can cause pulmonary diseases. The keynote talk will present the research progress on the eco-friendly natural materials as acoustic materials. The materials include the natural wasted fibres and non-fibrous type materials such as hollow structures as well as microperforated panels. The general theoretical foundation will be presented and important findings on the absorption performance of the materials will bediscussed. It will be shown that the performance of the 'new' eco-friendly absorber is comparable with that of the synthetic materials.

Keywords: eco-friendly, acoustic, absorbers

IOP Conf. Series: Materials Science and Engineering **352** (2018) 011003 doi:10.1088/1757-899X/352/1/011003

Keynote Speaker 2 - Profile

Dr. Azma Putra received his B. Eng from Bandung Institute of Technology (ITB), Indonesia in Engineering Physics. He then continued his postgraduate studies in the United Kingdom and received his M.Sc in 2004 and Ph.D in 2008 both from the Institute of Sound and Vibration Research (ISVR), University of Southampton, UK. He continued to work as a research fellow for one year in ISVR and then joined with Universiti Teknikal Malaysia Melaka (UTeM) Malaysia as a Senior Lecturer. He is now Associate Professor in Faculty of Mechanical Engineering, UTeM.

His current research interests are engineering acoustics and noise control, vibro-acoustics and structural dynamics. He has published of more than 50 articles and over 30 conference proceedings. He was awarded the best researcher in Journal Publication Award for 2013 in UTeM. Dr. Putra is a member in International Institute of Acoustics and Vibration (IIAV) UK, International of Noise Control Engineering (INCE) USA and Society of Vibration and Acoustics Malaysia (SVAM). He is now holding a post as the coordinator in Centre for Advanced Research on Energy (CARe), UTeM. Besides research activities, he also involves in consultancies with oil and gas as well as power companies regarding rotating machinery vibration. Dr. Putra is certified as Vibration Analyst Category III by the Mobius Institute Australia.

Dr. Putra is passionate to make difference in teaching and learning using technology. He creates interactive video animations and iBooks for his lectures and applies the Flipped Learning concept for his class, Mechanical Vibration. He is acknowledged as Apple Distinguished Educator (ADE) Class 2015 by Apple Inc. and he was awarded as Best Educator with E-Learning Content for 2015 in UTeM. He is a member in Society of Engineering Education Malaysia (SEEM).

IOP Conf. Series: Materials Science and Engineering **352** (2018) 011004 doi:10.1088/1757-899X/352/1/011004

Peer review statement

All papers published in this volume of *IOP Conference Series: Materials Science and Engineering* have been peer reviewed through processes administered by the proceedings Editors. Reviews were conducted by expert referees to the professional and scientific standards expected of a proceedings journal published by IOP Publishing.

IOP Conf. Series: Materials Science and Engineering **352** (2018) 012001 doi:10.1088/1757-899X/352/1/012001

Corrosion behaviour of carbon steel in aqueous solution containing Galena concentrate

Safira Fitri, Zulfiadi Zulhan, Bonita Dilasari
Metallurgical Engineering, Bandung Institute of Technology

E-mail: safiraafitri@gmail.com

Abstract. Carbon steel specimen was left in an aqueous solution containing galena concentrate for 15 days. Electrochemical Impedance Spectroscopy (EIS) method was conducted at 1st, 3rd, 5th, 7th, 9th, 11th, 13th and 15th days of immersion. Carbon steel was immersed in the lower solid region of 25 wt% and 75 wt% galena concentrate in solution. After 15 days of immersion, polarization resistance was measured using linear polarization resistance (LPR) technique. Analysis of impedance data from EIS method enables the determination of polarization resistance. The result showed that polarization resistance increases as a function of the solid percentage of galena concentrate and corrosion of carbon steel could be minimized.

1. Introduction

Corrosion of carbon steel causes major losses such as building structures, reduce process efficiency, product contamination, and may also lead to catastrophic accidents that endanger humanity. NACE International through the IMPACT program [1] has conducted the study to examine the global cost of corrosion. It was estimated to be \$2.5 trillion, which was equivalent to 3.4% of the global Gross Domestic Product (GDP) in 2013.

Isaacson [2] has studied the effect of Galena on corrosion of carbon steel. Experiments were made to observe the corrosion processes of carbon steel grinding media in a mill used to crush sulfide minerals such as galena. The results showed that the corrosion rate of carbon steel decreases with increasing amount of galena when dissolved oxygen is present. The low corrosion rate of carbon steel is due to galena acting as oxygen scavenger which decreases the dissolved oxygen content in solution. The oxidation reaction does not occur only in carbon steel but also in Galena. Galena oxidation reaction at acidic pH can be expressed by reaction (1).

$$PbS + 2H^{+} + O_2 \rightarrow Pb^{2+} + S^{0} + H_2O \quad (1)$$

The elemental sulfur (S^0) released by galena oxidation can react with water and decrease pH of the solution due to the formation of H^+. Maldonado-Zagal and Boden [3] suggested total reaction of elemental sulfur in reaction (2).

$$4S^{0} + 4H_2O \rightarrow SO_4^{2-} + 3H_2S + 2H^{+} \quad (2)$$

Hydrogen sulfide is one of the industrial aqueous corrosive solutes and corrodes carbon steel forming sulfide corrosion products [4].

Schaschi [5] also suggested a sulfur elemental reduction reaction that controls the corrosion of steel in the environment where sulfur is dissolved, following reaction (3).

$$S^0 + H_2O + 2e^- \rightarrow HS^- + OH^- \quad (3)$$

The iron sulfide film will continue to form as long as elemental sulfur is available in the solution. The formation of the iron sulfide is expressed by reaction (4) and (5).

$$Fe^{2+} + HS^- + OH^- \rightarrow FeS + H_2O \quad (4)$$

$$Fe^{2+} + S^{2-} \rightarrow FeS \quad (5)$$

A study by Miyasaka [6] concludes that the reduction of sulfur is controlled by mass transfer of oxygen in solution. If the solution contains a small amount of dissolved oxygen, then the corrosion rate will decrease. The presence of oxygen will accelerate the formation of iron oxide corrosion product. This iron oxide will react with S^{2-} to produce iron sulfide. Water content which can be expressed as a solid percentage becomes an important parameter in the corrosion process of carbon steel in galena concentrate as it affects pH, dissolved ions and oxygen, and electrolyte resistance. Corrosion of carbon steel in a different solid percentage of galena concentrate was studied using Electrochemical Impedance Spectroscopy (EIS) to find the corrosion mechanism in the interface of carbon steel and galena concentrate. The results were used as references for preparing condition of galena concentrate in contact with carbon steel to minimize corrosion damage.

2. Materials and Methods

2.1 Materials

Carbon steel specimens used for this study are 1 cm^2 in size for electrochemical measurement and 10 cm^2 in size for immersion test. Its composition (in weight %) characterized using Optical Emission Spectroscopy (OES) is : 0.05% C, 0.06% Si, 0.006% S, 0.011% P, 0.22% Mn, 0.06% Ni, 0,03% Cr and the rest was Fe. Specimens were connected to copper wire using solder and mounted in epoxy resin for the electrochemical measurements. Then, specimens were wet grounded with silicon carbide abrasive papers from P120 to P2000, followed by rinsing with alcohol prior to testing. Only one plane of specimens was exposed to the Galena concentrate.

The Galena concentrate was crushed using a ball mill and sieved to get mineral particles with a diameter smaller than 75μm. After that, the sampling process was done using a riffle to obtain a galena concentrate with a homogeneous composition for XRD and XRF analysis. Galena concentrate was composed of 61% of pyrite (FeS_2), 25.2% of anglesite ($Pb(SO_4)$), and 13.8% of galena (PbS) based on qualitative analysis XRD using MATCH! software. Then, XRF analysis is presented in Table 1. Galena concentrates were mixed with aquades to obtain 25 wt% and 75 wt% solid percentage. Aquades were added to the electrochemical cell on the 5th and 10th days of immersion to maintain the constant solid percentage of galena concentrate.

2.2 Electrochemistry

Electrochemical impedance measurement was conducted based on ASTM G-106 [7] and used three-electrode set-up. EIS measurements were performed using Potentiostat Reference 600 and software from Gamry Instrument. Impedance was measured at 1st, 3rd, 5th, 7th, 9th, 11th, 13th and 15th days of carbon steel immersion in galena concentrate. The amplitude of the sinusoidal potential signal was 10 mV concerning corrosion potential and the frequency range from 100 kHz to 1 mHz with 10 points/decade. The results of EIS measurement were analyzed to obtain electrical circuit model and electrical component value.

The aqueous and solid phase of a solution containing galena concentrate got separated after they were left for several hours without any disturbance. In this study, EIS was performed only on the solid region of 25 wt% and 75 wt% galena concentrate in solution. Schematic configuration of the electrochemical cell designed for this study is presented in Figure 1. The electrochemical cell consists of a cylindrical vessel 250 mL. There are holes in the top which are used to connect the potentiostat with the working, reference, and auxiliary electrode. The working electrode used in this study was carbon steel specimen, while the reference electrode was Ag/AgCl reference electrode and the auxiliary electrode was graphite.

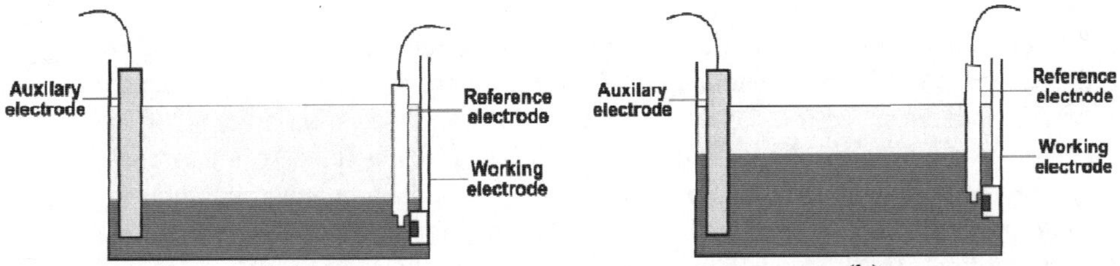

(a) (b)

Figure 1. Schematic configuration of electrochemical cell for (a) 25 wt% and (b) 75 wt%

Table 1. XRF analysis of galena concentrate.

Component	Pb	Zn	Cu	Fe	Si	S	Ca	Al
Result (wt%)	31.5	3.59	1.23	23.5	1.06	38.1	0.36	0.36

Immersion test was also conducted based on NACE TM0169/G31-12a [8]. Carbon steel coupon was wet grounded with silicon carbide abrasive papers from P120 to P2000, followed by rinsing with alcohol prior to testing. As both planes of the coupon are exposed to the solution, mounting the carbon steel coupons in epoxy resin are not required. After 15 days of immersion, carbon steel surface was characterized using X-Ray Mapping to observe corrosion products. Linear polarization resistance (LPR) method was also conducted on the test cell after 15 days of immersion to determine polarization resistance value. This measurement was prepared based on ASTM G-59 [9] using potential ±30 mV concerning corrosion potential and the scan rate of 0.167 mV/s. The polarization resistance obtained from EIS were compared to polarization resistance obtained from LPR technique to conclude the trend of polarization resistance.

3. Results and discussions

3.1 Corrosion potential and pH

Corrosion potential of specimens in the aqueous solution containing galena concentrate for an electrochemical test is shown in Figure 2. The corrosion potential is expressed concerning Ag/AgCl reference electrode. It can be seen that the corrosion potential of carbon steel in 25 wt% and 75 wt% of Galena concentrate decreases as a function of time. The decrease in corrosion potential inferred that carbon steel tends to become nobler during immersion in solid region of galena concentrate. In addition to corrosion potential measurement, pH for all the variations of galena concentrate are also measured and shown in Figure 3. It was ranged from 4.2 to 5 which indicated the test solution is in acid condition during 15 days of the test.

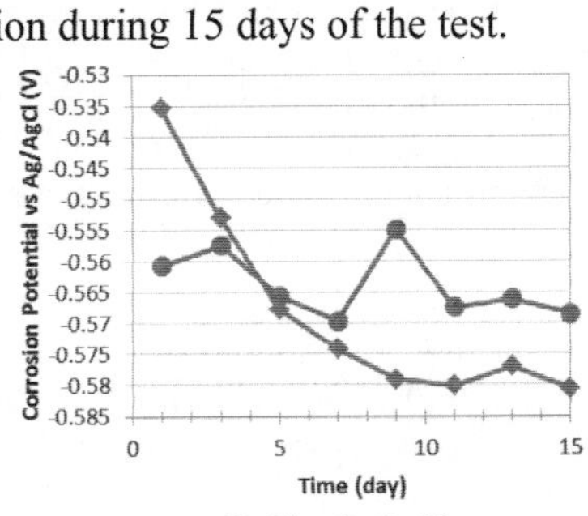

Figure 2. Corrosion potential of carbon steel vs. Ag/AgCl (V)

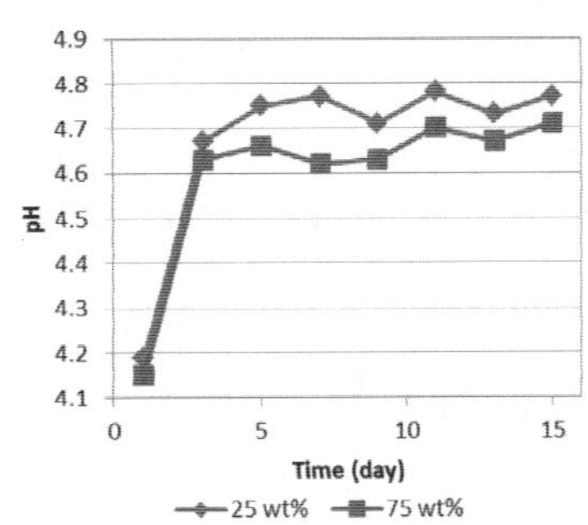

Figure 3. Electrolyte pH for all the variations of galena concentrate

IOP Conf. Series: Materials Science and Engineering **352** (2018) 012001 doi:10.1088/1757-899X/352/1/012001

3.2 Electrochemical impedance spectroscopy

3.2.1 Electrical circuit model.

Electrical circuit model used to simulate carbon steel in the aqueous solution containing galena concentrate are shown in Figure 4. The electrical circuit components are R_s indicates the resistance of the solution, R_i indicates passive layer resistance, CPE_i indicates passive layer capacitance, R_p indicates polarization resistance, CPE_{dl} indicates double layer capacitance of electrode-electrolyte, R_g indicates galena concentrate resistance, and CPE_g indicates galena concentrate capacitance. Model 1 was used to describe the carbon steel surface where the passive layer begins to form well in all parts. Model 2 with addition of CPE_g and R_g arranged in parallel which means the galena concentrate got attached on the carbon steel surface. It can also indicate the presence of some species attached to the carbon steel surface.

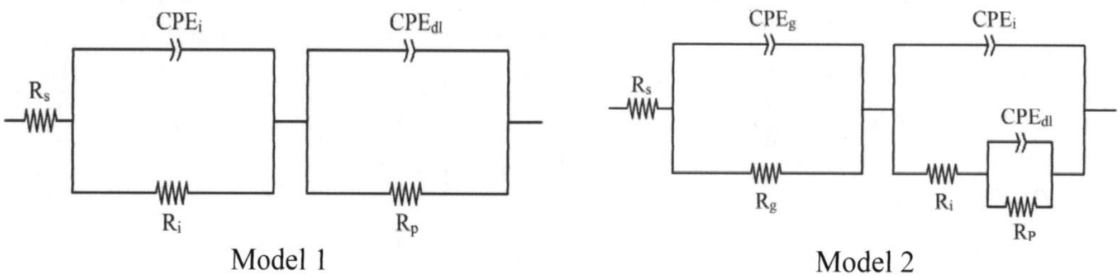

Figure 4. Electrical circuit model used to simulate EIS measurement results

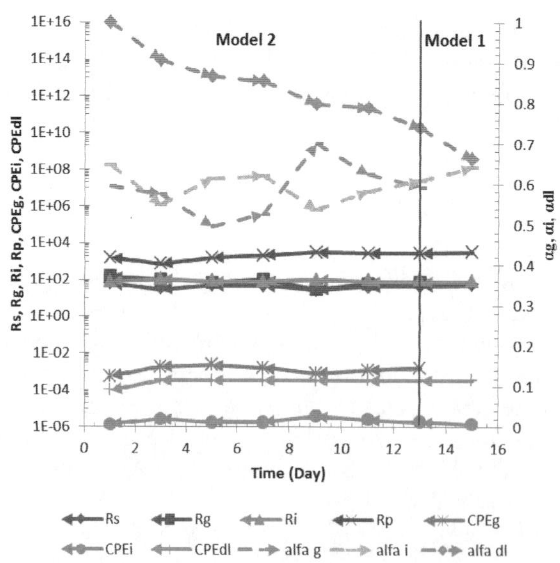

Figure 5. Electrical circuit components value for carbon steel immersed in 25 wt% of Galena concentrate on the solid region

3.2.2 Immersion of carbon steel in 25 wt% of Galena concentrate in solution.

The electrical circuit components value for carbon steel immersed in 25 wt% of Galena concentrate on the solid region are shown in Figure 5. The electrical circuit model chosen to simulate carbon steel immersed in 25 wt% of Galena concentrate on the solid part was model 2. It changed to model 1 after 13th days of immersion shows that galena concentrate was no longer attached to the carbon steel surface. It was indicated by the value of α_g and R_g which continue to decrease as a function of time. R_i tends to be stable for 15 days of immersion, while R_p increases as a function time that indicates better corrosion resistance of carbon steel. The value of capacitance tends to be constant which indicates the corrosion process on the carbon steel surface does not change as a function of time.

Figure 6. Immersion test coupon for carbon steel immersed in 25 wt% of Galena concentrate

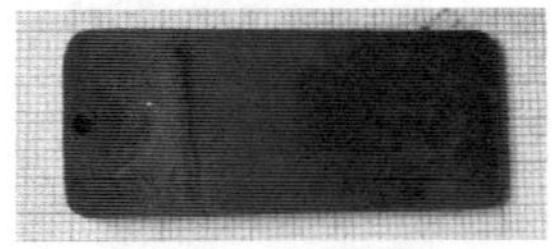

Figure 7. Immersion test coupon for carbon steel immersed in 75 wt% galena concentrate

Figure 8. X-Ray Mapping analysis of carbon steel immersed in 25 wt% of Galena concentrate

The increasing of corrosion resistance of carbon steel in galena concentrate has been studied by Isaacson [2]. He stated that corrosion rate of carbon steel in galena concentrate decreases as increasing amount of galena concentrate which is caused by the oxidation process of Galena. Galena acts as an oxygen scavenger and consumes dissolved oxygen; hence, the corrosion rate of carbon steel decreases. The increasing of corrosion resistance is by the results of the immersion test in Figure 6 which shows the absence of corrosion products on the surface of carbon steel in solid region of galena concentrate. X-Ray Mapping analysis (Figure 8) shows that Fe and O dominate carbon steel immersed in the aqueous region of galena concentrate more uniformly than the solid region. It proves that corrosion damage of carbon steel immersed in a solid region of galena concentrate was hindered.

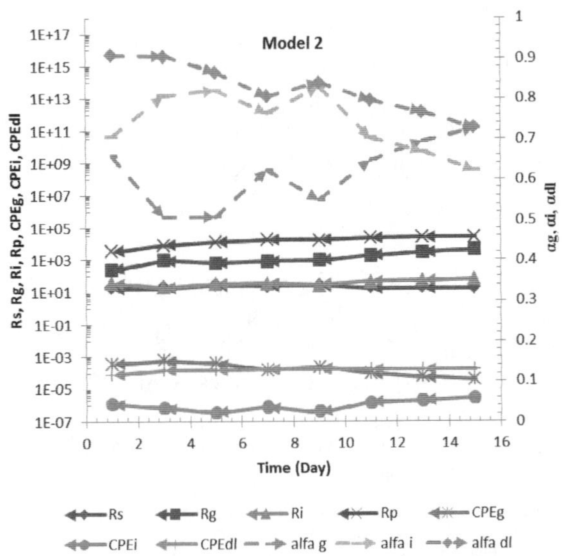

Figure 9. Electrical circuit components value for carbon steel immersed in 75 wt% of Galena concentrate on the solid region

3.2.3 Immersion of carbon steel in 75 wt% of Galena concentrate in solution.
The electrical circuit components value for carbon steel immersed in 75 wt% of Galena concentrate on the solid region are shown in Figure 9. Electrical circuit model used to simulate the carbon steel interface is model 2 with the capacitance value tends to be stable for 15 days of immersion. The increasing of R_p indicates that carbon steel in contact with the solid of galena concentrate has low corrosion rate. It can be proven from the direct observation of the carbon steel surface as shown in Figure 7 where corrosion product slightly formed on the carbon steel surface immersed in solid region. The reduction of corrosion process of carbon steel in 75 wt% galena concentrate has the same characteristics as carbon steel immersed in 25 wt% galena concentrate. It was due to galena acts as oxygen scavenger that reduces dissolved oxygen in solution so that dissolved oxygen needed for carbon steel oxidation were not enough available.

3.2.4 Comparison of polarization resistance and corrosion rate of carbon steels.
Polarization resistance indicates carbon steel resistant to corrosion. The higher value of polarization resistance means the better corrosion resistant. Carbon steel in a higher percentage of galena concentrate shows excellent corrosion resistant than carbon steel in a lower percentage of galena concentrate. The corrosion rate of carbon steel immersed in 75 wt% of Galena concentrate is lower than the 25 wt%. Both EIS and LPR method shows the same trend for the measurement of polarization resistance and corrosion rate as shown in Table 2.

Table 2. Polarization resistance (Rp) and corrosion rate of carbon steel.

	R_p EIS (Ohm)	R_p LPR (Ohm)	Corrosion Rate EIS (mm/year)	Corrosion Rate LPR (mm/year)
25 wt%	3439	3843	0.074	0.066
75 wt%	25794	26520	0.006	0.006

4. Conclusion
Carbon steel immersed in an aqueous solution containing galena concentrate can be illustrated using electrical circuit model obtained from EIS measurement. Electrical circuit model analysis results in new information about carbon steel behavior especially the corrosion mechanism. The value of electrical circuit component can also be determined including polarization resistance (R_p) value which indicates carbon steel resistant to corrosion. The higher value of polarization resistance means the better corrosion resistance, was observed in the higher solid percentage of galena concentrate. The increase of the solid percentage of galena concentrate will increase the corrosion resistant to carbon steel.

References
[1] Koch G, Varney J, Thompson N, Moghissi O, Gould M, Payer J 2016 *International Measures of Prevention, Application, and Economics of Corrosion Technologies Study* (Texas : NACE International) ed G Jacobson
[2] Isaacson A E and Huiatt J L 1985 An electrochemical study of grinding media corrosion *Annu. Meet. Assoc. Corros. Eng. NACE* 364
[3] Maldonado Z S B and Boden P J 1982 Hydrolysis of Elemental Sulphur in Water and its Effect on the Corrosion of Mild Steel *Br. Corros J* **17** 5–9
[4] Jones D A 1996 *Principles and Prevention of Corrosion 2nd Edition* (United State of America: Simon & Schuster)
[5] Schmitt G 1991 Effect of Elemental Sulfur on Corrosion in Sour Gas Systems *Corrosion* **47** 285-308
[6] Miyasaka A, Denpo K and Ogawa H 1989 Environmental Aspects of SCC of High Alloys in Sour Environments *Corrosion* **87** 5-10
[7] ASTM Comittee 1999 ASTM G106-89 Standard Practice for Verification of Algorithm and

Equipment for Electrochemical Impedance Measurements
[8] NACE International 2012 Standard Guide for Laboratory Immersion Corrosion Testing of Metals 1-10
[9] ASTM Comittee 2003 ASTM G-59 Conducting Potentiodynamic Polarization Resistance Measurements 1–4

IOP Conf. Series: Materials Science and Engineering **352** (2018) 012002 doi:10.1088/1757-899X/352/1/012002

Application of taguchi method for selection parameter bleaching treatments against mechanical and physical properties of agave cantala fiber

F Yudhanto[1,2], Jamasri[1], Heru S B Rochardjo[1]
[1] Department of Mechanical and Industrial Engineering Faculty of Engineering Universitas Gadjah Mada, Jl. Grafika No.2 Yogyakarta 55281, Indonesia
[2] Department of Mechanical Engineering, Universitas Muhammadiyah Yogyakarta, Jl. Lingkar Barat, Tamantirto, Kasihan, Bantul, Yogyakarta 55183, Indonesia

E-mail: ferriawan@umy.ac.id

Abstract. The characterized agave cantala fiber in this research came from Sumenep, Madura, Indonesia was chemically processed using sodium hydroxide (NaOH) and hydrogen peroxide (H_2O_2) solution. The treatment with both solutions is called bleaching process. Tensile strength test of single fiber was used to get mechanical properties from selecting process of the various parameter are temperature, PH and concentration of H_2O_2 with an L9 orthogonal array by Taguchi method. The results indicate that PH is most significant parameter influencing the tensile strength followed by temperature and concentration H_2O_2. The influence of bleaching treatment on tensile strength showed increasing of crystallinity index of fiber by 21%. It showed by lost of hemicellulose and lignin layers of fiber can be seen from waveforms changes of 1735 (C=O), 1627 (OH), 1319 (CH_2), 1250 (C-O) by FTIR graph. The photo SEM showed that the bleaching of fibers causes the fibers more roughly and clearly than untreated fibers.

1. Introduction

Agave Cantala (*cantala*) fiber is one of the prospective tropical plants to produce fibers from their leaves. It has high fiber strength, resistant to high salinity and is renewable after harvested every two years. These plants grow well on dry and rocky land like Sumenep, Madura, Indonesia [1]. Chemical structure of *cantala* fibers contains hemicellulose, lignin and extractive substance as amorphous regions and cellulose as a crystalline region. The chemical process effectively decreases impurities of fibers such as hemicellulose and lignin. It changes the surface topology of fibers and their crystallographic structure. Bleaching is immersing the fibers in alkali solution with reacting hydrogen peroxide (H_2O_2). It process reduces lignin content of natural fibers termed as delignification. Hydrogen peroxide as a substitute of chlorine is widely used in pulp industry for bleaching because it oxides is poisonous and can damage environment [2]. Hydrogen peroxide safer and must conserve energy than another oxide cause if react with water (H_2O) will change to form a hydronium ion (H_3O^+) and perhydroxyl ion (HO_2^-) [3]:

$$H_2O_2 + H_2O \leftrightarrow HO_2^- + H_3O^+ \quad (1)$$

The research has done by Tutus (2014) on bleaching will increase white index and mechanical strength on pulp than using chlorine [4]. Hashem et al. (2010) developed research on bleaching for cotton fibers. These process more active in alkaline medium at PH≈10.3-10.8. Taguchi is one of the designs of experiment method for selection of parameter and their level. This method is simple and efficient to find the optimum reaction from a mixture of sodium hydroxide and hydrogen peroxide at the best temperature. Optimum mixture and condition process of solution effect increase value of mechanical properties from natural fiber. The bleaching process is reaction phase liquid with the absorption of hydroxyl ion in the surface of the fiber and increasing rate of the process by adding sodium hydroxide. The following is activation of hydrogen peroxide [5]:

$$H_2O_2 + OH^- \leftrightarrow H_2O + HO_2^- \quad (2)$$

Parameters that effect of bleaching are a concentration of H_2O_2, temperature, and PH. All those parameters are very important against the quality of fibers such as brightness, tensile strength, and elongation [6].

2. Materials and methodology

2.1. Chemical material and process

Chemical treatment conducted by alkalization ($NaOH_{(liquid)}$ with purity of 48%) and bleaching ($H_2O_{2(liquid)}$ with purity of 50%). Those products obtained from Brata Chemical Yogyakarta Indonesia. Pretreatment process of fiber is immersing 5% NaOH at 100°C for 1 hour (1:50) w/w of fiber and solution. After that continue by bleaching (NaOH + H_2O_2) at PH 10, 11 and 12. The concentration of hydrogen peroxide used is 1, 2 and 3%. Time of process is 45°C, 60°C and 75°C.

2.2. Tool

The main tools to prepare this experiment are erlenmeyer glass tube 500 ml (Iwaki), Hotplate magnetic stirrer (U-Ten 78-1), measure glass 1000 ml (Iwaki), digital weigher, PH tester (Milwaukee), Thermometer.

2.3. Single fiber tensile test

A single fiber of *cantala* tensile test was conducted according to the ASTM D3379-75 [7] standard using a Tenso Lab-168 E strength tester. For every single fiber of *cantala* was glued on paper frame 20 mm gage length. The tensile test was performed using load cell 300 N at crosshead speed 5 mm/min. The supporting paper frame was cut in the middle. The dimension of single fiber tensile test sample is as presented in Figure 1.

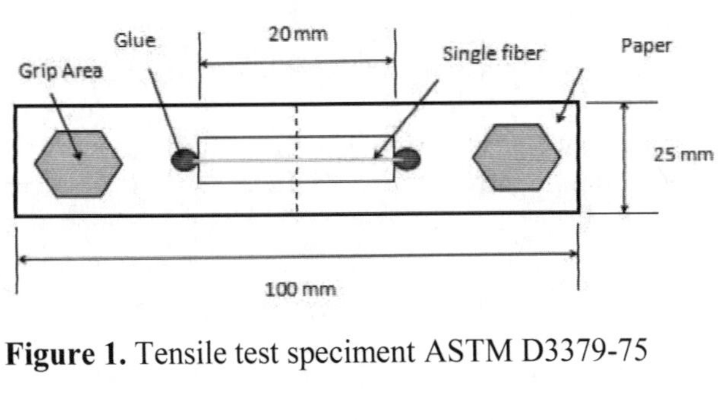

Figure 1. Tensile test speciment ASTM D3379-75

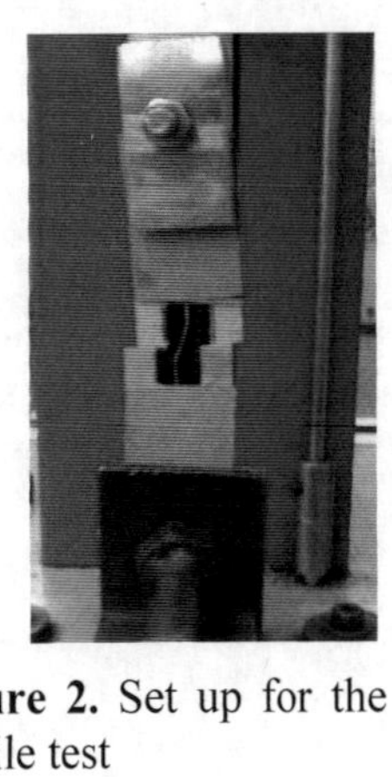

Figure 2. Set up for the fiber tensile test

The tensile strength is obtained from ratio of peak force to the cross-sectional area of plane perpendicular to the fiber axis that expressed by equation 3 [7];

$$\sigma = \frac{F}{A} \tag{3}$$

Where F is a force for failure (N), A is area of fiber diameter (mm^2)

2.4. Description of Taguchi method

Taguchi method developed with an *orthogonal array* form to design efficient experiments, analyst and minimize the number of experiments [8, 9 and 10]. The result selected is L9 orthogonal array has three control parameters are A (Temp), B (PH) and C (% H_2O_2) and their levels. The identifications show the parameter and level affect the response (tensile strength of single fiber). The result of experiments transforms into S/N ratio which are measure the deviation from characteristic quality. There are three choices of S/N ratio, i.e. *Larger is better, Nominal is best* and *Smaller is better.* This research choosing the larger is better that expressed by equation 4 [10];

$$\text{S/N ratio} = -10\log_{10}\frac{1}{n}\sum_{i=1}^{n}\frac{1}{y_i^2} \tag{4}$$

2.5. Fiber characterization

XRD (X-Ray diffraction) that tool was used to study physical properties of the crystalline material. Type of XRD is Rigaku Miniflex diffractometer 600 operate condition 40 KW, 15 mA. Sample of fiber was scanned in 2 theta ranges varying from 3° to 40°. The crystallinity index (I_C) of cellulose can be calculated using equation 5 [12];

$$I_c = \frac{(I_{002} - I_{amorph})}{I_{002}} \times 100\% \tag{5}$$

Where I_{002} is the peak intensity (maximum intensity) located at two-theta is around 22° which represent the crystalline cellulose. I_{amorph} is the valley intensity (minimum intensity) located at two-theta is around 18° which represent amorphous cellulose. FTIR (Fourier Transform Infra Red) spectroscopy is an appropriate tool to study chemical structure of *cantala*. Type of FTIR spectrometer is Paragon 1000 PC Perkin Elmer. Shape and wave number shows the bond of chemical structure in region 900 cm^{-1} to 4000 cm^{-1}. SEM (Scanning Electron Microscopy) photographs of fibers surfaces before and after bleaching treatment were taken by JEOL JSM-6510 LA instrument.

3. Results and discussion

3.1. Chemical effect of bleaching process

Bleaching process of *cantala* fiber decreases the percentage of hemicellulose, lignin and other extractives. Lowering the content of these impurities increase the percentage of cellulose, surface roughness and index whiteness of fibers [14, 16]. Chemical treatment also causes fibrillation on fiber. It reduces fiber's diameter from 200±60 μm to 60±30 μm.

Figure 3. Untreated fibers

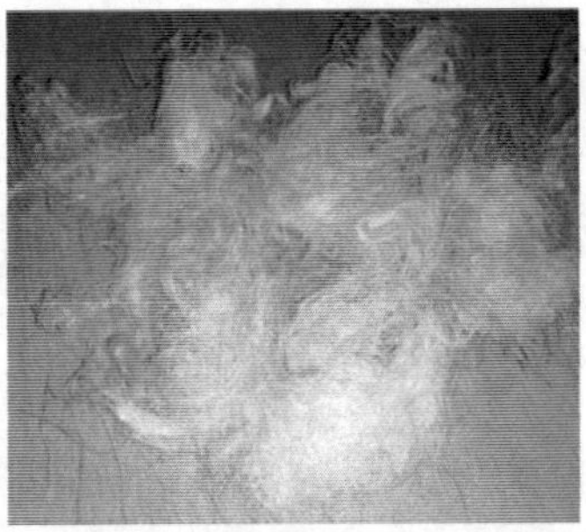

Figure 4. Bleached fibers

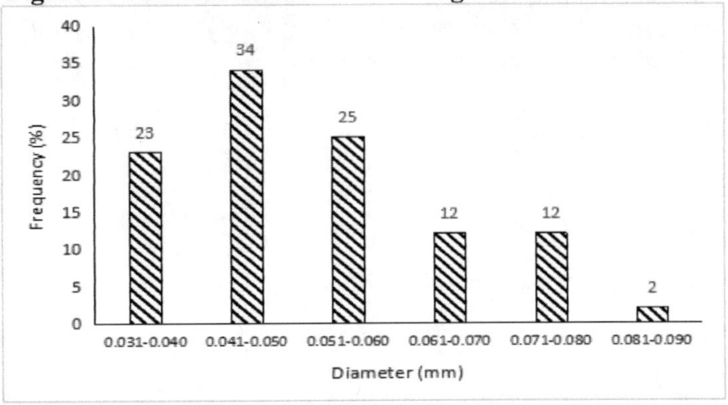

Figure 5. Size distribution of diameter *cantala* fibers after bleaching

Cantala fibers after bleaching are resulting various diameter of 30-90 μm (0.030-0.090 mm). 108 samples were grouped into intervals class as much as 6 class. Frequency distribution of average *cantala* fibers diameter are uniformly located between 30-60 μm (0.030±0.060 mm) can be seen in Figure 5.

Table 1. Experimental layout using an L-9 orthogonal array

Experiment Number	Bleaching Parameter Level A Temp (^{o}C)	B PH	C % H_2O_2	Measured Tensile Strength	Calculated S/N ratio for tensile strength
1	45	10	1	621.33	55.81
2	45	11	2	520.75	54.25
3	45	12	3	521.91	53.90
4	60	10	2	799.33	57.85
5	60	11	3	489.58	53.72
6	60	12	1	545.50	54.68
7	75	10	3	732.25	57.27
8	75	11	1	441.91	52.57
9	75	12	2	396.91	51.55

Table 1 shows the experimental results for ultimate tensile strength for the single fiber of *cantala* and the corresponding S/N ratio using Eq. (4). The mean S/N ratio for each level of the bleaching parameter is summarized and called the mean S/N ratio for ultimate tensile strength.

Table 2. Response mean S/N ratio for ultimate tensile strength parameter and significant interaction

Symbol	Bleaching parameter	Mean S/N ratio				
		Level 1	Level 2	Level 3	Max-min	Rank
A	Temperature °C	54.66	55.42	53.80	1.62	2
B	PH	56.98	53.51	53.38	3.60	1
C	% H_2O_2	54.36	54.55	54.97	0.61	3

Table 2 shows the greatest value of max-min S/N ratio is 3.60 indicate that parameter PH most influence the fiber tensile strength. Temperature is a second parameter (1.62) and third is a concentration of hydrogen peroxide (0.61). PH, temperature and concentration above the optimum parameter causing increased rate of decomposition significantly and the unutilized HOO^- anion so rapid becomes unstable reaction so may damage the fibers.

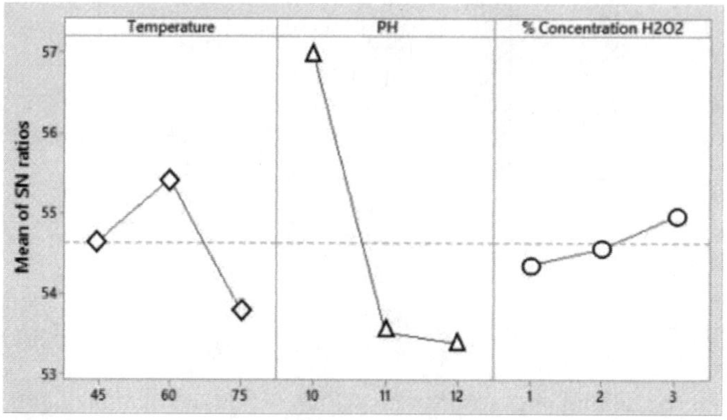

Figure 6. The mean of single-to-noise graph for ultimate tensile strength of *cantala* fiber

Figure 6 shows result mean of single-to-noise (S/N) graph for the tensile strength of cantala fiber with Taguchi method. The optimum parameters of bleaching process are temperature 60°C, PH 10 and 3% H_2O_2 (A2-B1-C3).Those process caused the rate of evolution of perhydroxyl ion is equal to the rate of consumption. The ultimate tensile strength single fiber of cantala increase from 520 MPa (raw material) into 800 MPa after bleaching with parameter A2-B1-C3.

3.2 Fiber characterization

Fiber characterization X-ray diffraction shows that choosing the optimum parameter of bleach treatment that are temperature 60°C, PH 10 and 3% H_2O_2 (v/v) can increase tensile strength and crystallinity index (Ic) of *cantala* fibers. Increase value of I_C from 64.5% to 78% indicate that non cellulosic component (lignin, hemicellulose and extractive substance) in fibers decrease significantly. This research similar to the Hay *et al.* (2012) that chemical modification of date palm fibers (DPF) can effectively increase the ultimate tensile stress and elongation [12, 13].

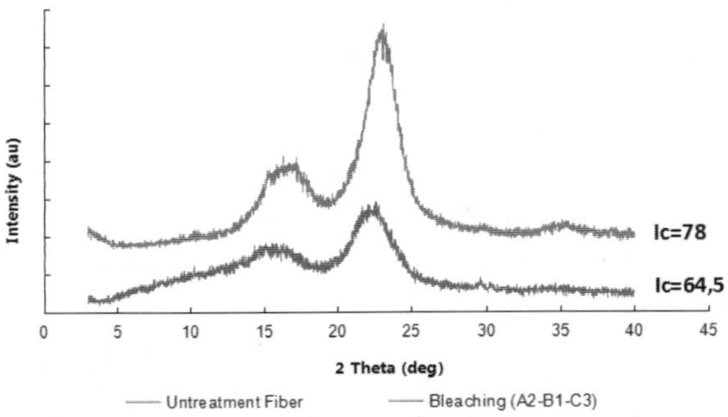

Figure 7. X-ray diffraction graph of *cantala* fibers

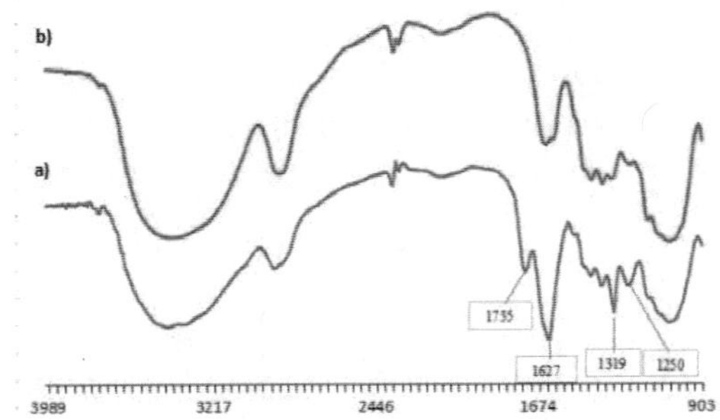

Figure 8. FTIR graph of *cantala* fibers

FTIR spectra can be seen in Figure 8, natural fibers waveform especially show four main chemical bond there are 1735 (C=O) bond of carbonyl C=O stretching of carboxylic acid or ester (lignin and hemicellulose), 1319 CH_2 Wagging (lignin), 1250 stretching of acetyl C-O (lignin), 1627 OH (water content). Change waveform numbers from sharp valley to wide peaks indicate loss of chemical bond C=O, CH_2 and C-O indicate the reduction of non cellulosic component [15]. The change of waveform at 1627 show that decrease water content in fibers causes the hydrophilic into hydrophobic properties. Hydrophilic nature is a major problem for all cellulose fibers if used reinforcement in plastics [13].

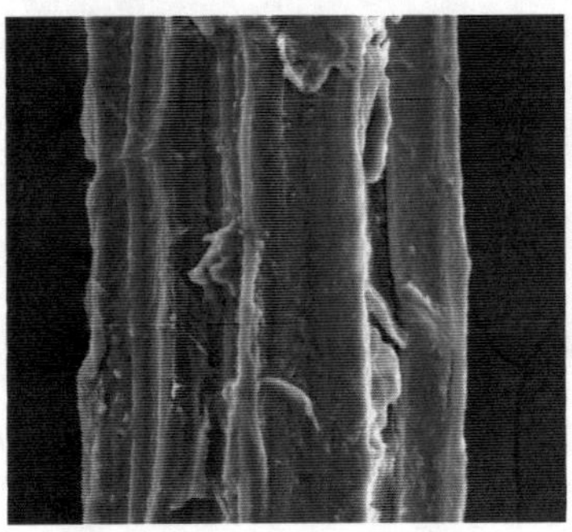

Figure 9. Untreated fiber

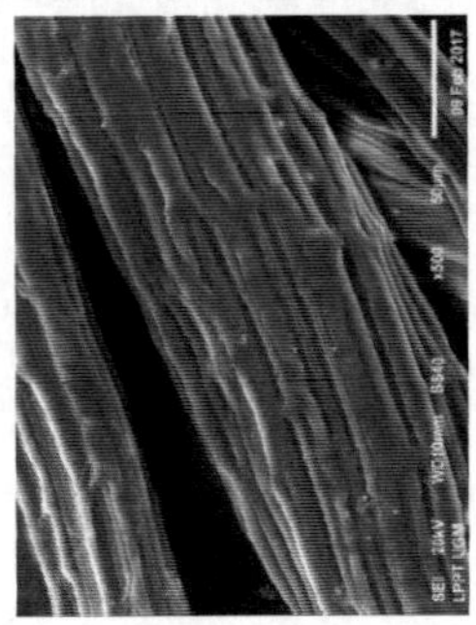

Figure10. Bleached fiber

Figure 9 show SEM images were taken with 500 x optical zoom to produce topography of *cantala* fiber untreated, the picture that the bundle of fiber does not appear because still tied and coated by non cellulosic component in fiber. These fibers after bleaching treatment can be seen in Figure 10 shows the surface of fibers more clearly, roughly and diameter of fibers became tiny.

4. Conclusion

1. In this study, Taguchi parameter design was successfully used to optimize a parameter of the bleaching process. It was found that bleaching process with temperature 60°C, PH 10 and 3% H_2O_2 (A2-B1-C3) lead to higher tensile strength.
2. Optimum bleaching with parameter A2-B1-C3 shows increase mechanical strength and physical properties indicated by increasing crystallinity index (Ic) and decreasing percentage of water content in *cantala* fibers.

References

[1] Santoso B 2015 *Perspektif* **8** 84 (in Indonesian)
[2] Bjorklund M 2008 *Bleaching of Pulp* (United State Patent, 6.777.548 B1)
[3] E S Abdel-Halim, Salem S Al-Deyab 2013 *Carbohydrate polymers* **92** 1844
[4] Tutus A 2014 *Pakistan J. Science 7* **8** 132
[5] Hashem M, El-Bisi M, Sharaf S and Refaie R 2010 *Carb.Poly* **79** 533
[6] J E G Van Dam 2002 *Coir Processing Technologies: Improvement of drying, softening, bleaching and dying coir fibre/yarn and printing coir floor covering* Tech.paper 6 (Netherland, FAO and CFC)
[7] Ronald F Gibson 1994 *Principles of Composite Material Mechanics* (Singapore, McGraw-Hill)
[8] Vishal Parashar, Shailandra S Bhadauria, Yogesh Sahu 2015 *Int.J.of Mechanical And Production Engineering* **3** 2320
[9] Vinod Kumar Vakanti and Venkateswarlu Ganta 2014 *J.Mater.Res.Technol.* **3** 35
[10] W H Yang, Y S tarng 1998 *Journal of Material Processing Technol.* **84** 122
[11] L Segal , J J Creely, A E Martin,C M Conrad 1959 *Textile Research journal* **29** 1746
[12] Abdal-Hay A, Suardana N P G, Jung D Y, Choi K-S and Lim J K 2012 *International Journal of Precision Engineering and Manufacturing* **13** 1199
[13] Bledzki A K and Gassan J 1999 *Progress Polymer Science* **24** 221
[14] Darmanto S, Heru S B, Rochardjo, Jamasri and Widyorini R 2016 *AIP conf.proc.* 1788
[15] Purwanto, Wijayanti Dwi Astuti, Harini Sosiati, Kuwat triyana 2014 *Prosiding Pertemuan Ilmiah XXVII Jateng & DIY.* 0853 (in Indonesian)
[16] Achmad W, Abdullah and Slamet P 2010 *Seminar Rekayasa Kimia dan Proses* 1411 (in Indonesian)

Quality and market chain of Aceh Cocoa Beans

Irfan, I Sulaiman, CN Ikhsan, N.Faizun
Agricultural Faculty of Syiah Kuala University, Banda Aceh, Indonesia
Aceh Food Security Agency, Banda Aceh, Indonesia

E-mail: ismail.sulaiman@unsyiah.ac.id

Abstract. After long-lasting conflict and tsunami on December 26, 2004, some international donors/NGOs supported Aceh on cocoa development. Aceh cocoa sector has experienced tremendous growth in Indonesia. This study aims to investigate quality and market chain of Aceh cocoa beans. The survey was conducted in Pidie District. A number of 21 farmers and 1 exporter were interviewed; the beans from farmer's warehouses were analyzed and compared to Indonesia National Standard (INS). The result showed that the beans were generally produced from 6 Sub-Districts: Keumala, Titeue, Glumpang Tiga, Padang Tiji, and Tangse. They were not fermented; most were exported to the USA. Based on bean count, quality was mainly included in I/A and II/B. The main quality problem was high moisture content. Presumably, the beans were bought by wholesalers with lower price although not been sufficiently dried. Other quality parameters were good: no moldy bean and contaminant, very low insect damage/hollow-/germinated beans, and tiny broken beans (quality I)

1. Introduction

Indonesia is one of the main cocoa suppliers, after Ivory Coast, and Ghana. The cocoa plant is easy to be integrated or intercropped with other plants as bananas, nuts, etc. and can be developed within the scope of organic farming.

In Aceh, cocoa was rapidly developed after the tsunami of 26 December 2004, supported by international NGOs and donors [1]. Cocoa played a special role in the economic rebuilding of the community after conflict since many potential cocoa cultivation areas located at areas where people most severely by conflict and capital requirement for cocoa plantation is relatively low compared to oil palm and rubber, making it affordable by the community. In addition the global market of cocoa in the future is expected to remain strong.

Now, cocoa is one of Aceh superior commodities. Production increases continuously, with the centers in East Aceh, East South Aceh, North Aceh, Bireuen, Pidie Jaya and Pidie Districts. A big warehouse, managed by an exporter, was established in Pidie District [5].

Unfortunately, Aceh cocoa export has not given much-added value to farmers. The quality variation affected the acceptance of consumers and long marketing chain reduced the profits of farmers. Therefore, the study of cocoa quality and marketing chain is important in order to provide greater profits for farmers. Efforts to increase the added value of cocoa plantations also can be done with the utilization of cocoa pod waste to be processed into organic fertilizer or extracted to get pectin [3].

2. Quality Standard

The quality standard can be used to produce quality classes[8]. Cocoa beans' quality based on Indonesian National Standard (INS 01-2323-2003) is presented in Table 1.

Table 1. INS of cocoa beans

Characteristic	I	II	III
Bean count/ 100 grams	AA:max.85 A: max.100	B: max.110 C: max.120	> 120
Moisture content, % max	7.5	7.5	7.5
Moldy, % max	3	4	4
Unfermented, % max	3	8	8
Insect damage, hollow-, germi-nated, % max	3	6	6
Broken, % max	3	3	3
Contaminants, % max	0	0	0

3. Research Method

The survey was conducted in Pidie District. A number of 21 respondents which consisted of 20 farmers and 1 exporter[5] were interviewed. The respondent was determined by purposive sampling technique.

Samples of cocoa beans were taken from the farmers' warehouses and analyzed in the laboratory of Agricultural Faculty of Syiah Kuala University. Quality analysis consisted of bean count, moisture content, moldy beans, fermented/not fermented beans, insect damage/hollow beans/germinated beans, broken beans, and contaminants. The results of this analysis were compared to the quality requirements of INS, so that it could be known the quality class of each sample.

In addition to the analysis of cacao beans quality parameters, it was also observed the market chain of cocoa beans in the field. The market chain here is the stages which through by the sale of cocoa beans from the producers or farmers to the exporter.

4. Results and discussion

The results of the quality analysis of Aceh cocoa beans from this research are presented in table 2.

Table 2. Quality of Aceh cocoa beans

Sub-District	Bean count	Moisture (%)	IHG (%)b	Broken (%)
Tiro	108	15	1.17	1
Glumpang Tiga	85	21	3.89	1
Keumala	88	12	1.28	0.47
Tangse	105	9	1.69	0
Titeue	88	12	2.58	0
Padang Tiji	102	9	1.63	0

a. Unfermented, moldy beans 0%, contaminants 0%
b. Insect damage/hollow-/germinated beans

4.1. Bean count

Bean count is a determination of cocoa beans quality based on the calculation of the number of beans per 100 grams of sample. The larger the bean size, the fewer the bean count, so the better the bean quality class. However, the comparison of the cocoa quality based on bean count must be done in the same moisture content, namely the moisture content of 7.5% (INS).

The field data presented in Table 2, for example, showed that the beans from Glumpang Tiga had the fewest bean count (85) but the highest water content (21%), while the beans of Keumala and Titeu had more bean count (88) but lower water content (12%). As the cocoa moisture contents were reduced or set up to 7.5% based on INS, the bean count of Glumpang Tiga was equivalent to 98 or higher that of Keumala and Titeu (92). However, since these bean counts were below 100, based on INS, they met into quality IA. The bean count of Padang Tiji was equivalent to 103 (IIB), Tangse 106 (IIB), Tiro 116 (IIC), and no sample was included in quality III of INS.

4.2. Moisture content

The moisture content affects the durability of the beans during storage [10]. To prevent the growth of micro-organisms, the moisture content of cocoa beans must be reduced by drying the cocoa beans to the limit of the maximum moisture content of 7.5%. However, if the cocoa beans dried for too long under the sun, the moisture content could be extremely low. Too dried beans could stimulate outer skin damage, so that the beans became brittle and easily broken [9].

The moisture content of the beans in this research was above INS (Table 2). It allegedly due to bad post-harvest handling, for example, the sap was not removed before the beans dried (sugar beans). On drying process, the outer skin of sugar beans will remain wet even though the inside of the beans has been dried. The cocoa beans with relatively high moisture content were also 21% due to the insufficient drying process. Supposedly the beans must be dried up to 4-5 days, however sometimes the collectors directly buy it with cheaper prices despite the drying process was still 1-2 days.

Generally, the moisture content of cocoa beans in the field was measured by using a digital tool called Aqua-Boy. The measurement was quite simply by plugging the tool into jute or sack containing cocoa beans, and then the tool automatically determines the amount of water content of the cocoa beans. The tool was also quite accurate, since there was no much difference as the result of this tool compared to that of the laboratory using the oven method.

4.3. Moldy beans

Mold can not only damage the taste and aroma of chocolate, but also has potential to produce toxic compounds that are harmful to health. The beans with high moisture are susceptible to mold attack and mold attack is considered serious if the growth has entered into pieces of beans.

However, as shown in Table 2, the moldy beans in this research were 0%. Although the moisture content did not meet the INS, allegedly the storage process in average had not lasted longer (just 5-6 days), so that the condition of the beans were still good.

4.4. Unfermented beans

Fermented cocoa beans have a brighter appearance, sharp scent, not slimy, and higher value. The unfermented beans have relatively lower quality since they have the more opaque appearance, less pungent aroma, slimy, and lower prices[10].

As shown in Table 2, the cocoa beans in this research were not fermented. These beans were exported to the American market, which based on farmer opinion did not require fermentation process. In addition, farmers generally did not understand how to do cocoa fermentation. It was also found in the field that there was no price difference which was gained by farmer between fermented and unfermented beans. The previous ones however required a longer processing time and a more intensive work compared to the last ones, so that there was so far no stimulate motivation for the farmer in order to conduct fermentation process.

4.5. Insect damage

Insect damage beans are beans those found inside the content of insects or insect body parts, or show damage due to insects that can be seen by the eye. The presence of insects could be caused by bad storage condition or by too long storage time. While the hollow beans are beans those have been empty inside, and will deflate when pressed with a finger.

As shown in Table 2, insect damage, hollow-, germinated beans (IHG) in this research ranged between 1.17%-3.89%. These met the quality I of SNI (max.3%), except Glumpang Tiga (II). Interestingly, the only contribution of IHG came from hollow beans. No insect damage and germinated beans were found. Germinated beans usually occur when the fruits are harvested too ripe.

The hollow beans occurred allegedly due to the lack of farmer's knowledge on cocoa cultivation technique. Farmers expected a lot of fruits from one cocoa plant. They did not understand that branched cocoa fruits could later cause hollow beans. Therefore the cocoa farmers still required counseling and training.

Low IHG beans were also suspected due to a relatively short storage time (5-6 days). It was a time as farmers dried or stored the cocoa beans while waiting for the collectors. The beans were then further dried by the collectors and after dry enough were stored for no longer than 1 month, while waiting for being sold to big trader or exporter. The trader kept the cocoa beans for no longer than 3 months prior to export. Before storage, the warehouse was cleaned and fumigated in order to avoid pests. As the cocoa beans were being to be exported, the analysis was carried out again in order to ensure the quality.

4.6. Broken beans

There are several causes of broken beans, such as using of sharp tools to issue the beans from the pod, as well as a too long drying process. Using of sharp tools can hurt the beans, while too dried beans are easy to be broken [9], Fortunately, as shown in Table 2, the broken beans in this research were less than 1% (quality I).

4.7. Contaminants

The presence of contaminants or foreign objects in cocoa beans can reduce the quality of cocoa beans. Usually, foreign objects such as gravel, twigs, and cigarette butts can contaminate the cocoa beans by the drying process, as the beans are dried without any mats or plastic layers under the sun next to the roadside.

INS has absolutely no tolerant to contaminants (0% of contaminants). As shown in Table 2, there were no contaminants or foreign objects found in the samples. This showed a full awareness of cocoa farmers on this quality parameter.

4.8. Market chain

The results showed that there were two types of market chain used by cocoa farmers: farmers-collectors-exporter (type I) and farmers-exporter (type II). About 70-75% of farmers sold their cocoa beans to collectors who came to their homes or who stayed in the market, the collectors sold further the beans to the exporter. Only 25-30% of farmers sold their cocoa beans directly to the exporter (Armajaro). This company has a market share of 40 percent of cocoa beans produced in Pidie.

According to Azzaino [2], the longer the market chain, the higher the marketing costs and therefore the lower profit margins gained by the farmers. The high marketing cost will also increase the prices at the consumer level or decrease the prices at the producer level.

Thus short marketing chain (type II) should be implemented in order to achieve higher marketing margin. However, cocoa farmers in the field were generally less educated or just graduated from primary school, so that they were inexperienced to deal directly with the exporter. Only a small part of the farmers had better education, ranged from junior high school to senior high school. Other reasons associated with the scale or cocoa land owned by farmers. Many cocoa plantations are still managed by the community on a small scale. Most farmers of respondents had less than 2 hectares of land, only a few farmers who own the land area of 3-4 hectare. Therefore, although inefficient, they prefer to implement Type I in the marketing chain.

IOP Conf. Series: Materials Science and Engineering **352** (2018) 012003 doi:10.1088/1757-899X/352/1/012003

5. Conclusion

Generally cocoa beans in Aceh (Pidie) were produced without fermentation process and exported to the USA. Based on bean count, quality was mainly included in I/A and II/B. The main quality problem was high moisture content. Presumably, the beans were bought with a lower price by collectors although not been sufficiently dried. The collectors or exporter had to reduce moisture content before export. Other quality parameters generally meet INS. Although less efficient, farmers with lower education level prefer to implement type I of marketing chain (farmers-collectors-exporter).

Recommendation

It needs to outreach to farmers, especially regarding of post-harvest handling, including drying and storage techniques, as well as marketing.

Reference

[1] Irfan, Syahrial H and Rusli F 2009 Support of Aceh Green livelihood program through cacao training (Banda Aceh: ESP)
[2] Azzaino Z 1981 Introduction to Agricultural trading system (Bogor: Social Sciences Department of Agricultural Economics, IPB)
[3] Irfan, Asmawati, Fahrizal, Erika, C. 2013. Physicochemical properties of pectin from cocoa pod husk extracted with Ammonium Oxalate through different extraction time and pH *7th Inter. Conf. Chemical Eng. Sci. Appl.*
[4] Jano P and Denise Mainville 2007 The Cacao Marketing Chain in Ecuador: Analysis of chain constraints to the development of markets for high-quality Cacao *IAMA Submission Conference Parma, Italy*
[5] Ngu D 2011 PT.Armajaro *Proceeding of Aceh cocoa conference* (Banda Aceh: PEKA)
[6] NSI No. 01-2323-2003 National Indonesian Standard *Quality of cocoa beans* (National Standard Authority, Jakarta)
[7] Roldan M B, Ingrid Fromm and Aido R 2013 From Producers to Export Markets: The Case of the Cocoa Value Chain in Ghana *J. Africak Development* **15** 121-38
[8] Soekarto T S 1990 *Basics of food quality control and standardization* (Bogor: IPB)
[9] Susanto F X 2011 *Cokoa: Cultivation and processing* (Yogyakarta: Kanisius)
[10] Wahyudi T, Penggabean T R and Pujiyanto. 2008 *Cocoa agribusiness management, upstream to downstream* (Jakarta: Penebar Swadaya)

IOP Conf. Series: Materials Science and Engineering **352** (2018) 012004 doi:10.1088/1757-899X/352/1/012004

Identification of corrosion product on medium carbon steel under the exposure of Banda Aceh's atmosphere

Sulaiman Thalib[1], Muharil Ikhsan[1], Syarizal Fonna[1], Syifaul Huzni[1], Syahrir Ridha[2]

[1]Department of Mechanical and Industrial Engineering, Faculty of Engineering, Syiah Kuala University, Jl. Tgk. Syech Abdur Rauf No. 7 Darussalam – Banda Aceh 23111, Indonesia.

[2]Petroleum Engineering Department, Universiti Teknologi PETRONAS, 32610, Bandar Seri Iskandar, Perak Darul Ridzuan, Malaysia.

E-mail: syarizal.fonna@unsyiah.ac.id

Abstract. This research was conducted to study the form of corrosion products of medium carbon steel under the exposure of Banda Aceh's atmosphere. The medium carbon steel samples which the size based on ASTM G 50 were exposed in open areas around the Engineering Faculty, Syiah Kuala University, Darussalam - Banda Aceh, Aceh province, Indonesia. The study was carried out from January through December 2016. The corrosion product formed on the surface of the samples was studied using X-ray Diffraction (XRD) and Scanning Electron Microscope (SEM) method. Measurements of weight loss due to atmospheric corrosion as a basis for calculating corrosion rates was referring to ASTM G1. Corrosion product found based on XRD analysis was *lepidocrocite* (FeO_2, γ-FeOOH) and *goethite* (FeO_2, α-FeOOH). The results agreed with SEM analysis that also indicates to *lepidocrocite* and *goethite*. The corrosion rate for twelve months showed that the highest rate occurs in the period of March-April that was 0.024 mpy. During twelve months exposure, the corrosion products consist of *lepidocrocite* and *goethite*. Significant changes began to occur in the eighth month, where the product of corrosion was almost entirely *goethite*.

1. Introduction

Atmospheric corrosion is a degradation of metal due to contaminating with the surrounding air and the result of pollutants contained in the liquid [1]. Corrosion or commonly called rust formed on carbon steel consists of various oxides such as hydrated oxides, oxyhydroxides, miscellaneous crystalline and other elements as a result of the reaction of iron with the environment [2]. The corrosion products need to be studied to understand how corrosion can be controlled or can be prevented.

Many investigations have been conducted on corrosion products which formed in various atmospheres using XRD and Raman spectroscopy. The corrosion products obtained are *lepidocrocite* (γ -FeOOH), *goethite* (α-FeOOH) and *magnetite* (Fe_3O_4) [3-8]. *Lepidocrocite* is usually formed at the beginning of exposure time, then with the increasing of time, this phase turns into *goethite* [4].

The corrosion product of steel is generally a compound of iron oxide, hydroxide, and oxide-hydroxides in the form of crystals [9]. The structure and size of the crystals that are formed depending on the variable condition where crystals grow. The chemical compositions of some atmospheric corrosion products are shown in table 1. However, the information of the atmospheric corrosion product for Banda Aceh city of Aceh province, Indonesia, is still limited. Hence, the purpose of this study is to identify atmospheric corrosion products formed on the surface of medium carbon steel using XRD and SEM in Banda Aceh.

Table 1. The chemical composition of atmospheric corrosion products [9].

Iron oxide	Composition	Crystal structure
Goethite	α-FeOOH	Orthorhombic
Lepidocrocite	γ-FeOOH	Orthorhombic
Akaganeite	β-FeOOH	Monoclinic
Hematite	α-Fe_2O_3	Hexagonal
Feroxyhyte	δ-FeOOh	Hexagonal
maghemite	γ-Fe_2O_3	Cubic

2. Experimental procedure

The medium carbon steel samples, that the size based on ASTM G 50, were exposed in open areas around the Engineering Faculty Syiah Kuala University, Darussalam, Banda Aceh, Aceh province, Indonesia. The calculation of atmospheric corrosion rate refers to ASTM G1. Characteristics of the atmospheric corrosion products were identified by XRD. The morphology of corrosion product was observed by SEM. The preparation of powder sample for XRD is conducted by grinding the atmospheric corrosion products using a mixer. The composition of medium carbon steel used in the study is shown in table 2.

The XRD data retrieval was performed every six month and corrosion rate calculations were done every two months from January to December. The weight losses of the specimen were measured using a digital scales with precision 0.001 gr.

Table 2. The composition of medium carbon steel used in the study

Elements	C	Si	Mn	P	S	Ni	Cr	Mo	Co	Al	Fe
Composition	0,310	0,062	0,729	0,002	0,010	0,049	0,051	0,029	0.019	0,014	6,256

3. Results and discussion

3.1. XRD analysis

The result of corrosion products identification using XRD after six-month period exposure is shown in Figure 1 (A label). The first peak of the angle 2Θ is 14.33° and corrosion product formed is lepidocrocite (FeO_2). The second peak of the angle of 2Θ is 26.97° and the corrosion products formed Fe (III) oxide hydroxide lepidocrocite (γ-FeOOH). The third highest peak which formed goethite (FeO_2) at an angle of 2Θ 36.33°.

Corrosion or commonly referred rust formed on carbon steel consists of various oxides, as a result of the reaction between iron and the environment. The corrosion product of lepidocrocite is usually formed at the beginning of exposure. With the increasing of exposure time, the lepidocrocite might turn into goethite. This is in accordance with the results obtained by other researchers with using x–rays and Raman spectroscopy [3-8, 10]. The identification using SEM verified the corrosion products formed after six months exposure was lepidocrocite as seen in figure 2(a).

The corrosion product after twelve month period of exposure was also identified using XRD. The result is shown in figure 1 (B label). Corrosion product of lepidocrocite (FeO_2) was formed at the first peak with 14.05°of 2Θ angle. The second peak with 27.01° of 2Θ angle corresponded to the corrosion products of Fe (III) oxide hydroxide lepidocrocite (γ-FeOOH). Corrosion product of goethite (α-FeOOH) was formed at the third peak with 36.31° of 2Θ angle.

The verification was also performed using SEM for the corrosion products in a period of six months to twelve months. It was obtained lepidocrocite and goethite phases. However, goethite is

IOP Conf. Series: Materials Science and Engineering **352** (2018) 012004 doi:10.1088/1757-899X/352/1/012004

dominant than lepidocrocite as shown in figure 2(b). The changes of corrosion product morphology might be affected by the presence of hydrogen on metal surfaces with the increased exposure time.

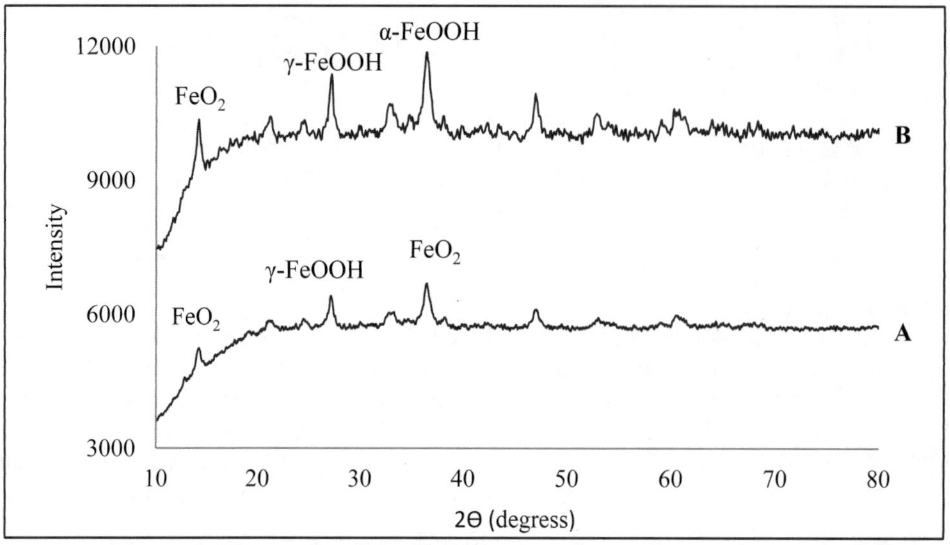

Figure 1. The results of the identification of corrosion products using XRD.

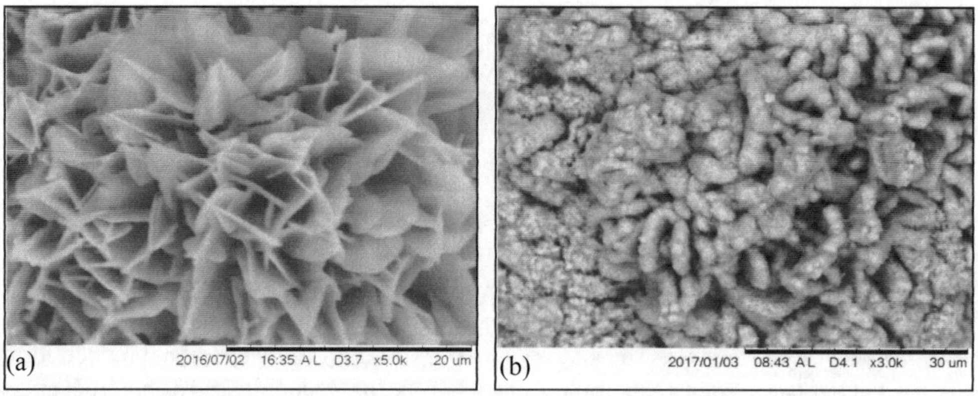

Figure 2. (a) Corrosion products of lepidocrocite that formed after six months of exposure; (b) Corrosion products of goethite that formed after twelve months of exposure.

3.2. Corrosion rate

Atmospheric corrosion rate was calculated over the last twelve months using an equation based on ASTM G1. The highest corrosion rate of 0.024 mpy occurred in the period of March-April, as shown in Figure 3(a). The lowest corrosion rate that occurred in the period July-August was 0.0125 mpy. It was suggested that the rain became a major factor for the occurrence of such a low corrosion rate. However, the corrosion rate around Engineering Faculty, Syiah Kuala University, Darussalam, Banda Aceh, Aceh province, Indonesia, is still in the outstanding categories.

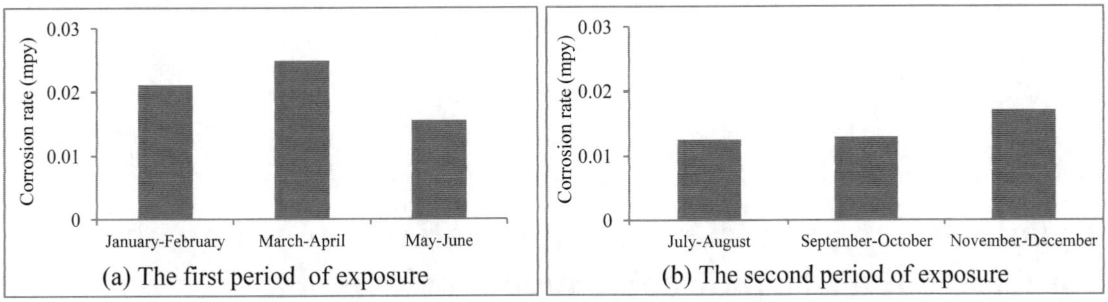

Figure 3. The corrosion rate for all periods of exposure.

4. Conclusions

The results of the identification of the corrosion products using SEM and XRD shows that in general, the corrosion products formed on medium carbon steel which exposed in the open area of Engineering Faculty, Syiah Kuala University, Banda Aceh were lepidocrocite and goethite. A chemical compound that was formed after six months exposure was FeO_2, γ-FeOOH, and FeO_2. Whereas after twelve months of exposure, the product was FeO_2, γ-FeOOH, and α-FeOOH. The highest corrosion rate in the area was 0.024 mpy during March-April. While, the lowest corrosion rate occurred in the period of July-August, i.e. 0.0125 mpy. However, the corrosion rate level in the area was still found as outstanding.

References

[1] ASM International 1992 *Corrosion* ASM Metals Handbook **13**
[2] De La Fuente D, Alcántara J, Chico B, Díaz I, Jiménez J A and Morcillo M 2016 Characterisation of rust surfaces formed on mild steel exposed to marine atmospheres using XRD and SEM/Micro-Raman techniques *Corrosion Science* **110** 253-264
[3] Antunes R A, Costa I, Araújo de Faria D L 2003 Characterization of corrosion products formed on steels in the first months of atmospheric exposure *Materials Research* **6** 403-408
[4] Misawa T, Hashimoto K and Shimodaira S 1974 The mechanism of formation of iron oxide and oxyhydroxides in aqueous solutions at room temperature *Corrosion Science* **14** 131-149
[5] Misawa T, Yamashita M, Matsuda Y, Miyuki H and Nagano H 1993 Protective Rust Layer Formed on Weathering Steel by Atmospheric Corrosion for a Quarter of a Century *Tetsu-To-Hagane/Journal of the Iron and Steel Institute of Japan* **79** 69-75
[6] Raman A, Razvan A, Kuban B and Clement K A and Graves W E 1986 Characteristics of the Rust from weathering steels in Louisiana Bridge Spans *Corrosion* **42** 447-455
[7] Sei J Oh, Cook D C, Townsend H E 1998 Characterization of Iron Oxides Commonly Formed as Corrosion Products on Steel *Hyperfine Interactions* **112** 59–66
[8] Sei J Oh, Cook D C, Carpio J J 2000 Characterization of the Corrosion Products Formed on Carbon Steel in a Marine Environment *Journal of the Korean Physical Society* **36** 106-110
[9] Cornell R M, Schwertmann U 2003 *The Iron Oxides, Structure, Properties, Occurrences and Uses*, 2nd Ed. Wiley-VCH Verlag GmbH & Co. KGaA Weinheim
[10] Marco J F, Gracia M, Gancedo J R, Martín-Luengo M A and Joseph G 2000 Characterization of the corrosion products formed on carbon steel after exposure to the open atmosphere in the Antarctic and Easter Island *Corrosion Science* **42** 753-771

A cable-driven parallel robots application: modelling and simulation of a dynamic cable model in Dymola

M F Othman [1,2], R Kurniawan [1,4], D Schramm [1], A K Ariffin [3]
[1] University of Duisburg-Essen, Lotharstrasse 1, MD 222, 47047 Duisburg, Germany
[2] Universiti Teknologi Mara (UiTM), Selangor, 40450 Shah Alam, Malaysia
[3] Universiti Kebangsaan Malaysia (UKM), Selangor, 43600 Bangi, Malaysia
[4] Universitas Syiah Kuala (Unsyiah), Darussalam, 23111 Banda Aceh, Indonesia

E-mail: othman@imech.de, aee4958@yahoo.com

Abstract. Modeling a cable model in multibody dynamics simulation tool which dynamically varies in length, mass and stiffness is a challenging task. Simulation of cable-driven parallel robots (CDPR) for instance requires a cable model that can dynamically change in length for every desired pose of the platform. Thus, in this paper, a detailed procedure for modeling and simulation of a dynamic cable model in Dymola is proposed. The approach is also applicable for other types of Modelica simulation environments. The cable is modeled using standard mechanical elements like mass, spring, damper and joint. The parameters of the cable model are based on the factsheet of the manufacturer and experimental results. Its dynamic ability is tested by applying it on a complete planar CDPR model in which the parameters are based on a prototype named CABLAR, which is developed in Chair of Mechatronics, University of Duisburg-Essen. The prototype has been developed to demonstrate an application of CDPR as a goods storage and retrieval machine. The performance of the cable model during the simulation is analyzed and discussed.

1. Introduction

Cables in engineering have a variety of usage, in electrical for instance, cables are used for power and signal transmission, for civil engineering, stranded or solid steel cables are used as structural anchors for building and bridge [1]. Meanwhile, in mechanical engineering, cables are mainly used for lifting, hauling and towing in which force is conveyed through tension by pulley and winch mechanism. For electrical and civil engineering applications, cables are basically fixed in length and the only matters are the physical and mechanical properties of the cables such as cross-sectional area, conductivity, specific mass, density, modulus of elasticity, tensile strength, etc. Modeling of cables for these two engineering applications does not need to consider the dynamic change of the cable length. Adversely, in mechanical engineering applications, the variation of cable length, mass and stiffness are necessary especially for dynamic simulation.

Researchers on cables or wire ropes as called in literatures have been carried out for many decades. Most of them have been conducted to serve for specific purposes and needs. For instance, articles from K. Spak et al. [1-3] have analyzed several configurations of stranded helical cables used for spacecraft. The inspected cables served for electrical usage which was modeled and tested in specific length and configuration. In other research by B. Xu et al. [4], an approach to model a steel wire ropes for high energy absorber apparatus is introduced. The work considered three-dimensional contacts between the wire ropes and other objects as well as kink-wave propagation. However, as far as CDPRs is concerned, the approach is not applicable for a dynamic simulation where changing of length is required for every 1 kHz to 2 kHz of the time step. The closest approach is from [5] in which it is implemented in ADAMS software. The cable model is made extensible as well as varies in mass and stiffness. Although the idea and need are alike, the modeling approach and implementation in Dymola or Modelica simulation environments are somehow different. Up to now, there is no string-like cable model that can be dragged in from its library for simulating mechanical systems such as cranes, elevators or alike.

In this paper, an approach is proposed in detail for modeling the cable model using parameters gathered from the factsheet of the cable manufacturer and experimental results. The cable properties are set based on 5mm diameter Dyneema® high-strength, high-modulus polyethylene fiber that is presently being used to actuate the CABLAR prototype. The paper is organized as follows: Section II reviews the theoretical background of cable modeling and relation between discrete system and continuous system modeling. Later on, Section III describes the flow of cable modeling systematically in Dymola. Furthermore, Section IV lists out main parameters of the CABLAR prototype and discusses the simulation results and as well as the cable model performance. Eventually, the achievement and future works of the cable model is summarized in section V.

2. Review of cable mathematical modelling

Theoretically, string-like cables can be mathematically modeled as strings in transverse vibration or rods in axial vibration using either continuous or discrete systems approach. The former approach is described by partial differential equations and the latter is described by ordinary differential equations. Since it is of the same physical system, one should expect both solving approaches would offer similar dynamical behavior if the discrete model were to have a multiple numbers of mass elements in the limit [6].

2.1. Transverse vibration formulation

Considering a boundary value problem in which both ends of the cable are fixed and applying Newton's second law, the equation of motion of the cable in transverse direction can be formulated as follows,

$$T_i \frac{y_{i+1} - y_i}{\Delta x_i} - T_{i-1} \frac{y_i - y_{i-1}}{\Delta x_{i-1}} + F_i = m_i \frac{d^2 y_i}{dt^2} \tag{1}$$

Let a system of discrete masses m_i (i = 1,2,…,n) be connected by massless strings as in figure 1, where the masses m_i are subjected to the transverse forces F_i. To derive the differential equation of motion for typical mass m_i, three adjacent masses m_{i-1}, m_i and m_{i+1} are considered. The tensions in the string segments connecting m_i to m_{i-1} and m_{i+1} are denoted by T_{i-1} and T_i, and the horizontal projections of these segments are Δx_{i-1} and Δx_i, respectively. The displacement $y_i(t)$ (i=1, 2,…, n) of the masses m_i are assumed to be small, so that these projections remain essentially unchanged during motion. Moreover, the angle between the string segments and the horizontal are sufficiently small that sine and tangent of the angles are approximately equal to one another. Rearranging the differential equation (1) becomes

$$\frac{T_i}{\Delta x_i} y_{i+1} - \left(\frac{T_i}{\Delta x_i} + \frac{T_{i-1}}{\Delta x_{i-1}} \right) y_i + \frac{T_{i-1}}{\Delta x_{i-1}} y_{i-1} + F_i = m_i \frac{d^2 y_i}{dt^2} \tag{2}$$

Since both ends are fixed, thus the transverse displacement of y_0 and y_{n+1} are set to zero. Further on, set $y_{i+1} - y_i = \Delta y_i$ and $y_i - y_{i-1} = \Delta y_{i-1}$, hence

$$T_i \frac{\Delta y_i}{\Delta x_i} - T_{i-1} \frac{\Delta y_{i-1}}{\Delta x_{i-1}} + F_i = m_i \frac{d^2 y_i}{dt^2}, \qquad i = 1, 2, \ldots, n \tag{3}$$

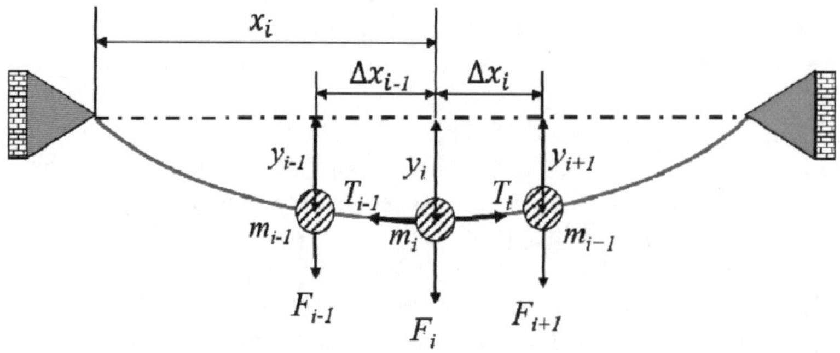

Figure 1. Mass system of a cable in transverse vibration.

Simplifying further equation (3) on it first two terms of the left-hand side, one would write it as

$$\Delta\left(T_i \frac{\Delta y_i}{\Delta x_i}\right) + F_i = m_i \frac{d^2 y_i}{dt^2}, \qquad i = 1, 2, \ldots, n \tag{4}$$

Furthermore, divide equation (4) by Δx, it would be further written as

$$\frac{\Delta}{\Delta x_i}\left(T_i \frac{\Delta y_i}{\Delta x_i}\right) + \frac{F_i}{\Delta x_i} = \frac{m_i}{\Delta x_i} \frac{d^2 y_i}{dt^2}, \qquad i = 1, 2, \ldots, n \tag{5}$$

The second term of equation (5) left-hand side is the distributed transverse force of the cable. Meanwhile, the coefficient of the right-hand side is regarded as its specific mass density at point x. Consider the number of n increase to the limit that $\Delta x \rightarrow 0$, in which one could imagine that the span between masses reduces proportionately with respect to the increase of masses, thus it could be deduced as

$$\frac{\partial}{\partial x}\left[T(x)\frac{\partial y(x,t)}{\partial x}\right] + f(x,t) = \rho(x)\frac{\partial^2 y(x,t)}{\partial t^2}, \quad 0 < x < L \tag{6}$$

In which (6) is considered as the transverse differential equation of the continuous system of the cable that for fixed boundary cases, $y(0,t)$ and $y(L,t)$ is set to zero.

3. Cable modelling in dymola

Based on a brief theoretical derivation in Section 2, the most appropriate way of modeling a cable in Dymola is by using the discrete system. Modeling a string-like cable and having a realistic dynamic behavior of it is obviously challenging in multibody simulation tools. The cable has to be modeled so that it is capable of changing in length, mass, and stiffness with respect to the change of platform pose. On top of that, the cable has to work when it is pulled and not vice versa. All these requirements have to be fulfilled for the cable model to be successfully simulated as actuators of the CDPR model. To make it comprehensible, the modeling procedure is divided into three steps as follows:

IOP Conf. Series: Materials Science and Engineering **352** (2018) 012005 doi:10.1088/1757-899X/352/1/012005

Step 1: Elementary cable model

Considering figure 2, let us start from its basic element which is called a connector. For this instance, they are denoted as `frame_a` and `frame_b`. The names are assigned by default in Dymola for which they represent as input and output connectors of a three-dimensional mechanical model respectively. Besides, they also contain information of the through and across variables of the model. Taking the figure from left to right, it is then followed by three consecutive revolute joints namely `revolute_x`, `revolute_y` and `revolute_z` which are used to allow free motion to the cable in three rotational axes, i.e. roll, pitch and yaw.

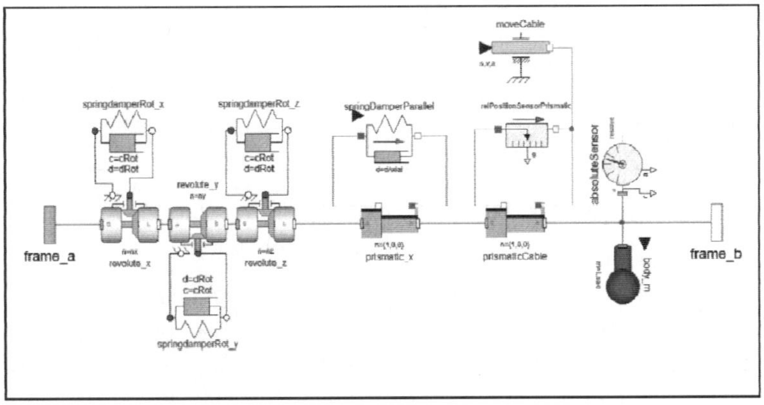

Figure 2. Elementary model of the cable.

Together with every single revolute joint is a one-dimensional rotational spring-damper so-called `springdamperRot_x`, `springdamperRot_y` and `springdamperRot_z` that are used to limit the rotational motion of every single revolute joint. Furthermore, two prismatic joints are set in the elementary cable model. The first prismatic joint is called `prismatic_x` as the horizontal direction between connectors is considered as the x-axis. This specific prismatic joint is assigned to provide elastic behavior within the cable's elementary element. It is attached in parallel with a modified one-dimensional translational spring damper named as `springDamperParallel`. The blue triangle next to it is a real input connector used to link the model to the external input from the upper-class model. The second prismatic joint is called `prismaticCable`. This prismatic joint serves as a length-changer for the elementary cable model. Connected parallel to it is two default models that are called `relPositionSensorPrismatic` and `moveCable` in which they serve as a sensor to measure the relative length with respect to its initial length, l_0 and actuator to the prismatic joint. Furthermore, in order to link the elementary cable with gravitational influences, a mass element, i.e. `body_m` is considered. The chosen mass element or rigid body consists of mass and inertia tensor as input parameters. To make it rational, the mass and inertia tensor ought to vary with regard to the change of cable length and this should happen simultaneously within the cable. Therefore, a similar procedure applied to `springDamperParallel` element in `prismatic_x` is also carried out on the mass element, a real input is attached to it as to allow variable mass inputs to be transferred from the upper-class model.

Step 2: Intermediate cable element

Now, Step 2 explicitly shows the second layer of the cable model named as an intermediary or middle layer. It is used to link the upper cable model to its elementary model.

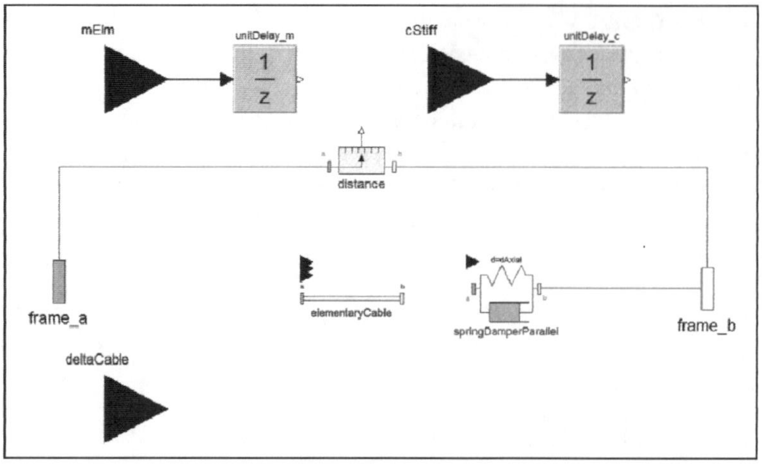

Figure 3. Intermediate model of the cable.

Considering figure 3 from left to right, the intermediate model starts with `frame_a` as the input connector. Then, it is followed by `elementaryCable` which is the outlook of the lower class model of Step 1. There are three real inputs attached on top of its input frame that allow it to connect with the upper-class model. They are used to inherit variable inputs of cable length, mass, and stiffness for the elementary cable model. Next to it is a three-dimensional `springDamperParallel` which is used to provide elasticity behavior on the overall cable model. There is also a real input attached to it for inheriting the variable stiffness. To make it understandable and easy to model, the stiffness of the cable is formulated as

$$k_{c_{Total}} = \frac{EA}{l_0 + \Delta l} \tag{7}$$

in which E is Young's modulus of the cable, A is the cross-section area of the cable and Δl is the change of cable length. Therein, the stiffness of each cable element is derived further as

$$k_{c_{element}} = \frac{k_{c_{Total}}}{nElm + 1} \tag{8}$$

where $nElm$ is a pre-defined number of cable element. The addition of one in the denominator of equation (8) signifies the additional spring stiffness for the three-dimensional parallel spring damper element. Furthermore, as to relate to the change of cable length, the mass of cable is formulated as

$$m_{c_{Total}} = \rho V = \rho A l = \rho A(l_0 + \Delta l) \tag{9}$$

where V is the cable volume, which can be calculated from its diameter, d and length, l whereas its density, ρ is taken from the manufacturer factsheet. Hence, for single element, the mass is further deduced as

$$m_{c_{element}} = \frac{m_{c_{Total}}}{nElm} \tag{10}$$

In the case of cable density is unknown, instead of using equation (9), one could also use the specific mass of the cable or cable mass per unit length, m_{spec} as shown in equation (11).

$$m_{c_{element}} = \frac{m_{spec} l}{nElm} = \frac{m_{spec}(l_0 + \Delta l)}{nElm} \tag{11}$$

Considering figure 3 further, there are three real inputs used to connect the current cable model to its upper class named as `deltaCable`, `mElm` and `cStiff`. For initialization and discretization of the cable model, `unitDelay_m` and `unitDelay_c` are used to predefine the initial cable mass and stiffness. The initial of both variables are defined as in equation (7) and equation (9) by excluding the change of cable length, Δl. Finally, the intermediate model is wrapped up by an output connector `frame_b`.

Step 3: Top cable element

The top layer of the cable model consists of a user interface as indicated in figure 4. Here, all the important parameters of the cable are predefined. As could be seen in the figure, the cable parameters are separated into three parts. The first part is a general part in which user has to define the number of cable element, *nElm* and initial length of the cable, l_0.

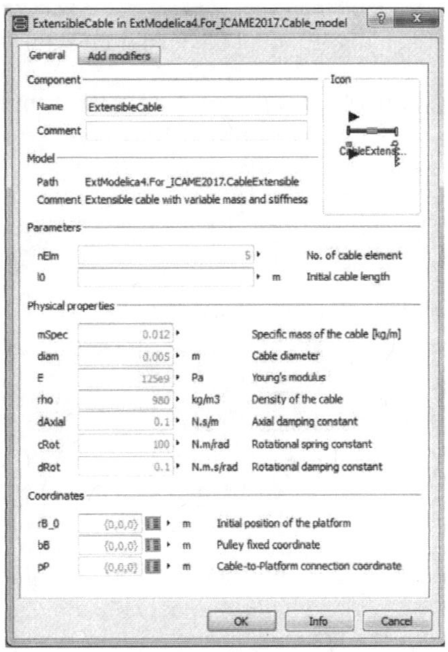

Figure 4. Parameter dialog of the cable model.

Besides, ones could choose any finite number of cable elements that suit their need. Note that, by having more elements in line, the cable model would react closer to real cable and possess sagging and oscillation effects. This fact has been explained in Section II, where discrete and continuous systems have been correlated. However, too many elements would trade off the computing time accordingly. Later comes the initial length, `l0` of the cable in which for CDPR case, the length has to be pre-calculated by inverse kinematics. Then, it is followed by a list of common physical properties of the cable that have been gathered from the factsheet and experiment such as its specific mass per unit length, `mSpec`, diameter, `diam`, Young's modulus, `E`, density, `rho`, and as well as the damping properties. Finally, yet importantly, it is followed by main reference coordinates of CDPR such as the initial platform position, `rB_0`, fixed coordinate of the pulley, `bB`, and the cable-to-platform connection point, `pP`. For a clearer view, placement of the major coordinates is shown in figure 5. The first two coordinates have to be set with reference to its inertial frame, *O* whereas the third coordinate has to be set with respect to the local coordinate of the platform, *P*. The green, red and blue arrows respectively represent the direction of the coordinate vectors.

IOP Conf. Series: Materials Science and Engineering **352** (2018) 012005 doi:10.1088/1757-899X/352/1/012005

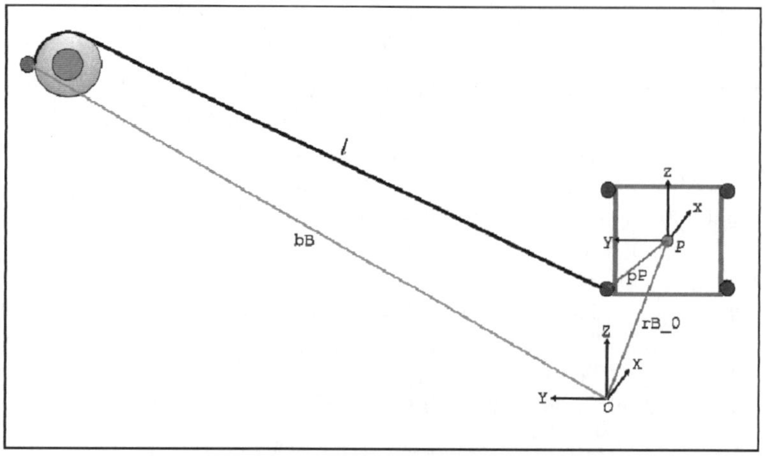

Figure 5. Layout of CDPR major coordinates.

To make it understandable, the flow of input and output variables of the cable model is summarized in figure 6. Respectively, note that the flow of modeling process is opposite to the simulation flow.

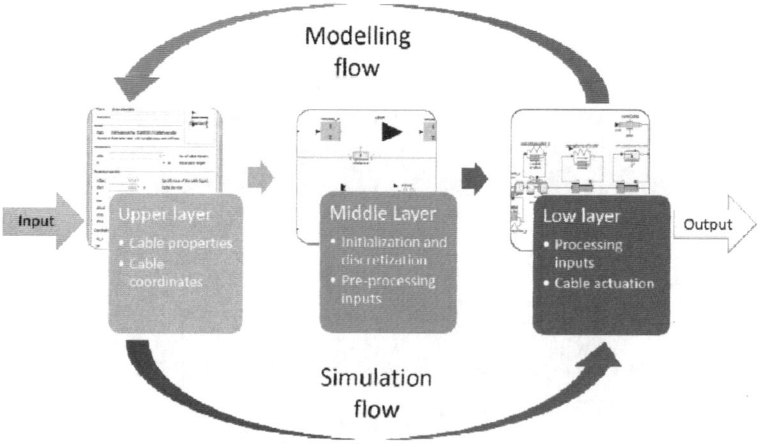

Figure 6. Process flow of cable modeling.

4. Results and discussion

4.1. Simulation setup

Simulation of the cable model is carried out on a complete CDPR model in which the parameters are based on the CABLAR prototype developed in Chair of Mechatronics, Universität Duisburg-Essen. The prototype is a planar eight-cable CDPR, which is meant for goods storage and retrieval operation. The main coordinates of the prototype are tabulated as follows

Table 1. Pulley fixed coordinates

Pulley fixed coordinates, bB_i	x(m)	y(m)	z(m)
bB_1	0.4000	-5.0791	0.5335
bB_2	0.4000	-5.1289	0.5269
bB_3	0.4000	-5.0992	4.7472
bB_4	0.4000	-5.1114	4.7549
bB_5	0.4000	5.0873	0.5142
bB_6	0.4000	5.1144	0.5314
bB_7	0.4000	5.1136	4.7678
bB_8	0.4000	5.1240	4.7688

Table 2. Cable-to-platform connection points

Cable-to-platform points, pP_i	x(m)	y(m)	z(m)
pP_1	0.0420	-0.4530	0.2350
pP_2	-0.0420	-0.4530	0.2350
pP_3	0.1375	-0.4530	-0.2350
pP_4	-0.1375	-0.4530	-0.2350
pP_5	0.0420	0.4530	0.2350
pP_6	-0.0420	0.4530	0.2350
pP_7	0.1375	0.4530	-0.2350
pP_8	-0.1375	0.4530	-0.2350

Table 2 and 3 are the input parameters for all eight cables of the CDPR model that set the initial direction vector of the cables in terms of position and orientation. Note that, the pP_i is the coordinates of the cable-to-platform connection points with reference to the local coordinate system of the platform. The following tabulated figures in table 3 are a list of major properties of Dyneema cable applied in the cable model.

Table 3. Major properties of the cable model

Cable Properties	Figures
Modulus of elasticity, E	109 - 132 GPa
Density, ρ	970 - 980 kg/m^3
Area of cross-section, A	19.635x10^{-6} m^2
Damping constant, c	10 N.s/m
Diameter, d	0.005 m

Table 4. Trajectory coordinates of the platform

Trajectory coordinates, P_i	x(m)	y(m)	z(m)
P_1	0.0000	-2.5000	1.3600
P_2	0.0000	0.8050	2.0000
P_3	0.0000	2.9800	1.3900
P_4	0.0000	-2.0000	2.0000
P_5	0.0000	0.8050	2.0000
P_6	0.0000	0.2030	1.4000
P_7	0.0000	-2.0000	2.0000
P_8	0.0000	1.1650	2.0000
P_9	0.0000	0.2300	1.3800
P_{10}	0.0000	-2.2000	1.7000
P_{11}	0.0000	1.1650	2.0050
P_{12}	0.0000	2.0000	1.4100
P_{13}	0.0000	0.0000	0.7650

For simplification, the damping properties of the cable have been considered as constant throughout the simulation even though based on the experiment, the axial damping constant is varied below the given figure. Therefore, the given damping constant has been considered as the maximum damping constant for simulating the CABLAR CDPR model. Furthermore, it is also discovered that the damping properties changed dynamically with respect to cable tension and length. Physical experiment on the cable has shown that an increase of cable tension proportionally reduces the damping constant whereas an increase of cable length correspondingly increases the damping constant. To test the cable model to it limit, a set of trajectory coordinates as listed in table 4 are used for the simulation. The complete cable robot model is simulated for 70 seconds on a system with specification given as in table 5.

Table 5. System Specification

Specification details	
Windows edition:	Windows 7 Enterprise
Processor:	Intel® Core™ 2 Quad Q6600 @ 2.4 GHz
Installed memory (RAM):	4.00 GB
System type:	64 Bit Operating System

4.2. Validation of cable model

For validation, static analysis is conducted on the cable by a pure bending test. The sagging of the cable model with regard to a variation of cable length is simulated and compared with the results of cable sagging from the experiment. The experiment is conducted such that the cable is let to fall freely due to gravity at minimum cable tension of 6N. The same setup is model in Dymola. The setup of the experiment is as shown in figure 5. The results of the experiment and simulation are tabulated as in table 6. The vertical displacement of the mid-mass element of the cable model is cable model is compared with the experimental results of a free fall sagging test.

Table 6. System specification

Cable length (m)	Sagging (mm)		Absolute error (mm)
	Experiment	Simulation	
0.90	1.677	1.673	0.004
1.15	2.889	2.879	0.010
1.40	4.470	4.469	0.001
1.65	6.436	6.431	0.005
1.90	8.800	8.753	0.047
2.15	11.576	11.553	0.023
2.40	14.775	14.657	0.118
2.65	18.407	18.401	0.006
3.00	24.237	24.134	0.103

4.3. Simulation results

Based on the trajectory coordinates listed in table 4, the desired and simulation trajectory of the CDPR is sub-plotted as depicted in figure 7. The axes are plotted separately so that the behavior of the platform with respect to the desired trajectory for every single Cartesian coordinate can be clearly visualized. Note that, since it is a planar type CDPR, the movements are obviously significant on y and z-axes. As can be observed, the trajectory of the platform for y and z-axes are successfully controlled except for x-axis.

IOP Conf. Series: Materials Science and Engineering **352** (2018) 012005 doi:10.1088/1757-899X/352/1/012005

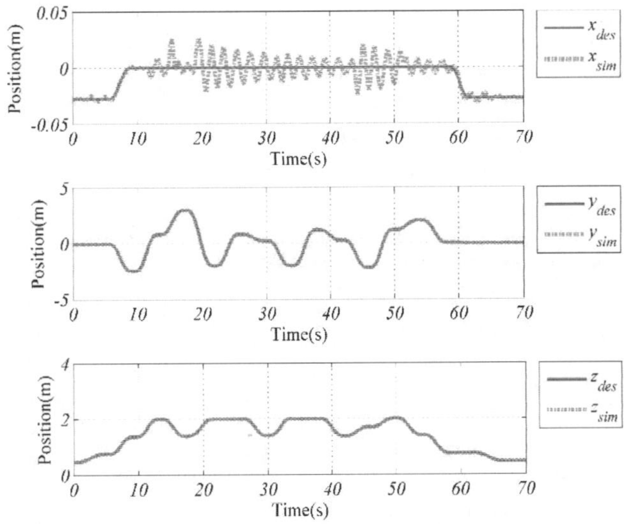

Figure 7. Comparison between the desired and simulation position of the platform.

Through the generated animation, it is clearly viewed that the platform is wobbling in x-axis for every coordinate-to-coordinate transition. The wobbling range is within 2 centimeters and similar behavior is observed on the prototype. Up to now, it can only be described qualitatively as there is no specific onboard sensor available in the market to measure the real-time pose of the platform. Furthermore, this kind of behavior on the platform happened when the stiffness and mass of the cables are made to vary with respect to the change of their length.

On the other hand, the orientation of the platform consists of its roll, pitch and yaw are shown in figure 8. The range of orientation overshoots due to coordinate-to-coordinate transition within 2 degrees.

To show the achievable variation of cable length throughout the simulation, instead of showing all the eight cables variation, which may cause cumbersome, four selected cables have been chosen and plotted in figure 9. Taking cable 1 (c1) in solid red line, for example, the length and stiffness are inversed proportional to each other. As the cable length grows longer, the stiffness of cable drops correspondingly. This successfully shows the relation between cable length and its stiffness as in equation (7). Meanwhile, the mass of the cable is proportionally correlated with the change of cable length. This correlation has been described in equation (9). This kind of correlation is important for multibody dynamic analysis as it displays a true behavior of the developed model specifically for a nonlinear element such as string-like cable.

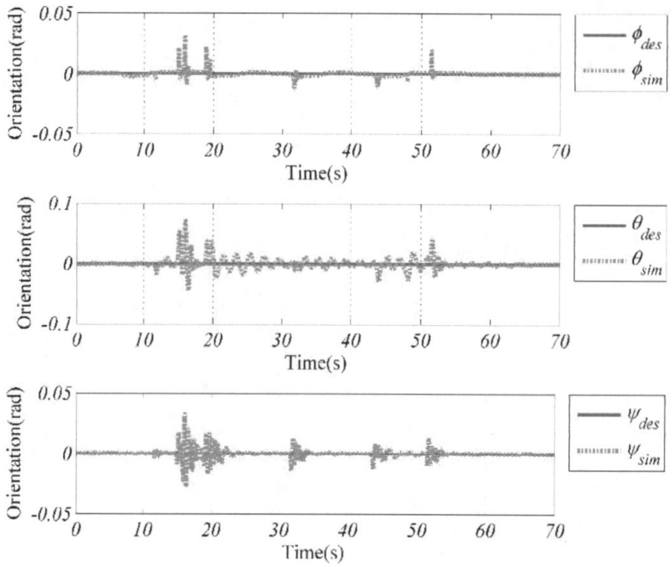

Figure 8. Comparison between the desired and simulation orientation of the platform.

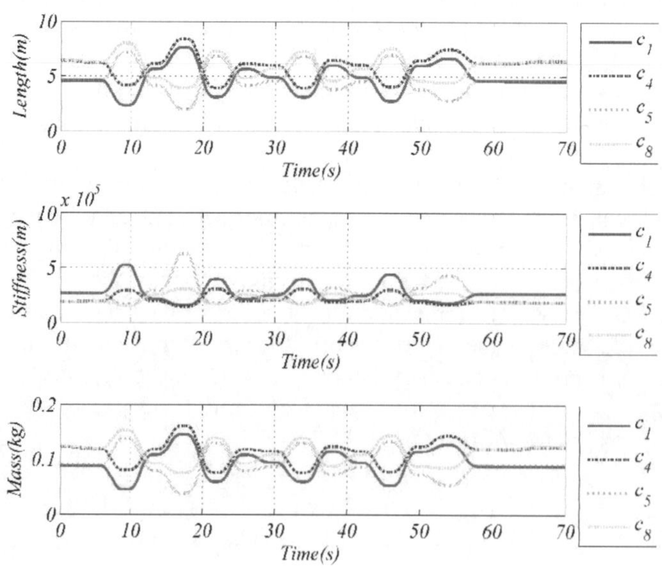

Figure 9. Variation of cable length, stiffness, and mass with respect to the trajectory planning of the platform

5. Conclusion and future works

This paper introduced in detail the procedure of developing a dynamic cable model, which is useful for performing a dynamic analysis of CDPR. The cable model was dynamically varied in stiffness and mass with respect to the change of its length. Despite complex trajectory, the CDPR platform was well tracked and controlled, which somehow due to correct approach of modeling the cable. Qualitatively, the behavior of the platform based on coordinate-to-coordinate transition has shown the close behavior of the CABLAR prototype. Besides that, the cable model could be used to simulate different type of string-like cable applications such as crane and elevator systems. Furthermore, the cable modeling approach could be applied to other Modelica simulation environments such as OpenModelica, MapleSim, SimulationX and many more. For future works, the subjected cable will be examined further using an improved experimental setup for discovering its true dynamic properties such as stiffness and damping as they showed very strong correlation. From the experimental data, a function that relates its dynamic properties with the change of length and tension will be modeled and deems to tackle hysteresis problem of the cable as well as give more realistic behavior to overall CDPR and string-like cable applications.

Acknowledgements

The authors would like to express high appreciation to all members of Mechatronics Chair, University of Duisburg-Essen for their vigorous supports and assistance to this piece of work. This indebtedness also goes to Universiti Teknologi MARA and Ministry of Higher Education Malaysia for their financial support.

References

[1] Spak K S, Agnes G S, and Inman D J 2015 Modeling vibration response and damping of cables and cabled structures *Journal of Sound and Vibration* **336** 240-256

[2] Spak K S, Agnes G S, and Inman D 2013 Towards modeling of cable-harnessed structures: cable damping experimentsin *54th AIAA/ASME/ASCE/AHS/ASC Structures, Structural Dynamics, and Materials Conference*

[3] Spak K, Agnes G, and Inman D 2013 Cable modeling and internal damping developments *Applied Mechanics Reviews* **65** p 010801

[4] Xu B, Dong P, Zhang J and Zhang S 2014 A three-dimensional approach to model steel wire ropes used in high energy absorber apparatus *International Journal of Solids and Structures* **51** 4280-4293

[5] Elliott A S 2001 Efficient modeling of extensible cables and pulley systems in ADAMS *The 1st MSC. ADAMS European User Conference*

[6] Meirovitch L 1975 *Elements of vibration analysis*

The development of furrower model blade to paddlewheel aerator for improving aeration efficiency

Samsul Bahri[1], Radite Praeko Agus Setiawan[2], Wawan Hermawan[2] and Muhammad Zairin Junior[3]
[1] Department of Mechanical Engineering, Lhokseumawe State Polytechnic, Lhokseumawe, Indonesia
[2] Department of Mechanical and Biosystem Engineering, Bogor Agricultural University, Bogor, Indonesia
[3] Department of Aquaculture, Bogor Agricultural University, Bogor, Indonesia

E-mail: samsul@pnl.ac.id

Abstract. The successful of intensive aquaculture is strongly influenced by the ability of the farmers to overcome the deterioration of water quality. The problem is low dissolved oxygen through aeration process. The aerator device which widely used in pond farming is paddle wheel aerator because it is the best aerator in aeration mechanism and usable driven power. However, this aerator still has a low performance of aeration, so that the cost of aerator operational for aquaculture is still high. Up to now, the effort to improve the performance of aeration was made by two-dimensional blade design. Obviously, it does not provide the optimum result due to the power requirements for aeration is directly proportional to the increase of aeration rate. The aim of this research is to develop three-dimensional model furrowed blades. Design of Furrower model blades was 1.6 cm diameter hole, 45° of vertical angle blade position and 30° of the horizontal position. The optimum performance furrowed model blades operated on the submerged blade 9 cm with 567.54 Watt of electrical power consumption and 4.322 m^3 of splash coverage volume. The standard efficiency aeration is 2.72 kg O_2 kWh^{-1}. The furrowed model blades can improve the aeration efficiency of paddlewheel aerator.

1. Introduction

Aeration is a mechanism of adding some amount of oxygen into water to provide sufficient amount of oxygen. Aeration is carried out by increasing water and air contact using aerator device. One type of aerator device which widely used in pond farming is paddle wheel aerator [1]. Paddlewheel aerator is considered as the most appropriate aerator device due to aeration mechanism and wide usable driven power [2]. Standardized aeration efficiency is directly proportional to the standard oxygen transfer rate and inversely proportional to the power consumption.

Aeration rate is influenced by water and air surface contact, differential oxygen concentration, film surface coefficient and turbulence [3]. Geometry, size and wheel speed affect aeration performance [4]. Largest size tends to have higher aeration which coincides followed by higher driven power needs due to higher drag force. This condition causes the certain problem in utilizing paddle wheel aerator as it causes increasing operational cost including electrical and fuel consumption. Electric power consumption is an important parameter in utilization paddlewheel aerator because effected on operating costs. Also, to provide the value of the energy consumption needs in the factual, the measurement of electrical power is also a reference for the farmers in selecting the aerator [5].

Up to now, the effort to improve the performance of aeration was made by two-dimensional blade design. Obviously, it does not provide optimal result because of the power requirements for aeration is directly proportional to the increase of aeration rate. Therefore, the aim of this research is to develop the model of a three-dimensional furrowed model blade and operational condition for improving aeration efficiency.

2. Materials and Methods

2.1. Materials

2.1.1. Furrower model blades.

Furrower model blades are based on the geometry of the arch blade wheel optimization results that have been done before [6], the hole diameter was 16 mm, 45° of vertical angle blade position and 30° of horizontal position as shown in Figure 1. The blade consists of eight pieces with the outer width was 174 mm, the length 176 mm and 300 mm for the radius of curvature of the blade.

Figure 1. Rim and furrowed model blades

2.1.2. Paddlewheel test unit.

Structural design paddlewheel consists of a frame, reduction rotation transmission systems, paddlewheel blades and laying the instrument. The frame using the steel angle with the dimensions, shape, and strength customized with support and load other components. The motor used is 1 phase AC electric motor with a power of 1 HP at 1440 rpm rotational speed.

2.2. Procedure

2.2.1. Treatments variations.

The tests were conducted in the freshwater pool with dimensions of length 350 cm, 200 cm width with a water depth of 40 cm. Submerged variations blade is done by changing the height position paddlewheel support. The tests were conducted at a rotational speed of 154 rpm with the blade submerged 6, 9, and 12 cm.

2.2.2. Torque measurement.
Torque measurement was done by using a strain gauge mounted on the wheel shaft. The sensor was connected to the strain amplifier (DAS-406B DC Strain Amp) through the slip ring, and the bridge box was recorded by using data loggers (minilab 1008) and stored in the computer (Figure 2). Measurement data in the form of a voltage (mVolt) was converted to strain (µst) and torque measurement values (N·m) with calibration values that have been done before.

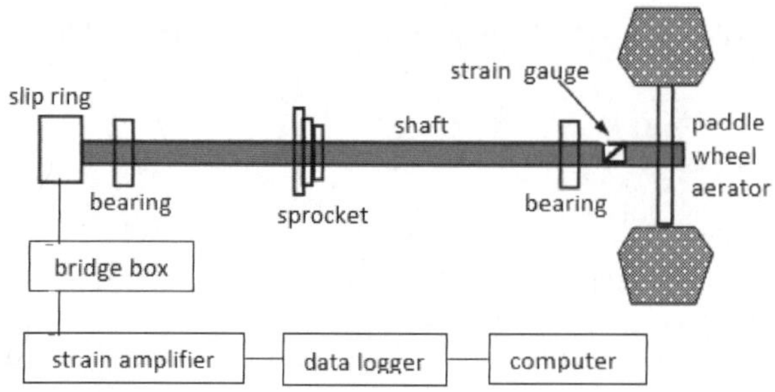

Figure 2. Rim and furrowed model blades

2.2.3. Power measurement.
Paddlewheel power measurement was done by measuring the electrical power consumption of electric motors using Ammeter (DO2A) which was connected to an electrical outlet. Reading the power measurement value (Watt) was done by using a digital video camera recording on display Ammeter. Rated power was taken on the average value that often was showed from the reading video playback recording for each treatment testing.

2.2.4. Splash coverage volume measurement.
Splash coverage volume was done by taking the recorded images used a digital camera from the front side and side the wheel at the time of testing. Then, the digital image was processed by using a CAD program to create segments (grid), which the splash coverage volume was the number of multiple results of segments area at the front side water splash with the side of water splash.

2.2.5. Standard aeration efficiency (SAE) measurement.
The SAE test was conducted according to ASCE standard [7]. The test was done in pool size 4x5x1.25 m and water depth 1 m. Paddlewheel was used to stir sodium sulfite with water until it reached 0% saturation or below 1 mg l^{-1}. Measurements of DO and water temperature using Oxygen meter. Measurements of DO increases begin when DO water started to increase until saturation is 80%.

3. Results and Discussion
The water splash produced visible from the front and side the wheel of the submerged blade 12 cm as shown in figure 3.

Figure 3. The image of water splash and calculation segment of splash coverage volume.

Torque occurred, electrical power consumption and splash coverage volume produced at different blade submerged shown in table 1.

Table 1. Performance test results at different blade submerged.

Blade submerged (cm)	Torque (N·m)	Power consumption (Watt)	Splash coverage volume (m^3)
6	45.85	552.19	3.754
9	47.13	567.54	4.322
12	48.28	581.42	4.095

The increasing of blade submerged resulted in greater torque and increasing the power consumption, while the splash coverage volume had optimum value at the blade 9 cm.

The increase in torque due to changes blades submerged of 6 cm to 9 cm and 12 cm respectively 2.79 % and 5.30%, while the electrical power consumption were 2.78% and 5.29%. This is caused by the increase of water mass that driven by the paddlewheel, thus increasing the drag force of the blade wheel as revealed by Kang (2004). While the difference between the increase in torque with the electric power is due to lose in the reduction system.

Splash coverage volume maximum occurred in submerged blade 9 cm was 4.322 m^3 (up 15.13%) with 567.54 Watt power consumption, whereas in 12 cm increased by 9.08%. Submerged blade 6 cm have little mass of water to be splashed, whereas in 12 cm of displaced water mass was too large, thus slowing rotation due to the great drag force of blade. The coverage volume is obtained approaching splash coverage volume of Taiwan paddlewheel models (two-wheel) 4.356 m^3 with 852 Watt power consumption.

Based on the ratio of the electric power consumption, showed that furrowed model blades have a good effect on reducing the drag force of the blades. This is as mentioned by Munson et al. (2006) that the drag coefficient is determined by the geometry and dimensions of an object. While the increase in volume is obtained by the spread of water splash due to furrowed blades. Reducing of water collision momentum due to the tilt angle of the blade overcome by a rotational speed of paddlewheel was 154 rpm compared to Taiwan paddlewheel models was 120 rpm. This is as stated by Park et al. (2014) and Liu and Peymani (2015) that the power consumption and the flow rate increased directly proportional to the depth of submerged blades and increasing paddlewheel rotation.

The standard aeration test at pool volume 20 m^3 and water temperature 28 °C, obtained saturation dissolved oxygen 9.07 mg l^{-1} and 588 Watt of electric power used, so obtained the relation of natural logarithm (Ln) deficit DO as shown in figure 4.

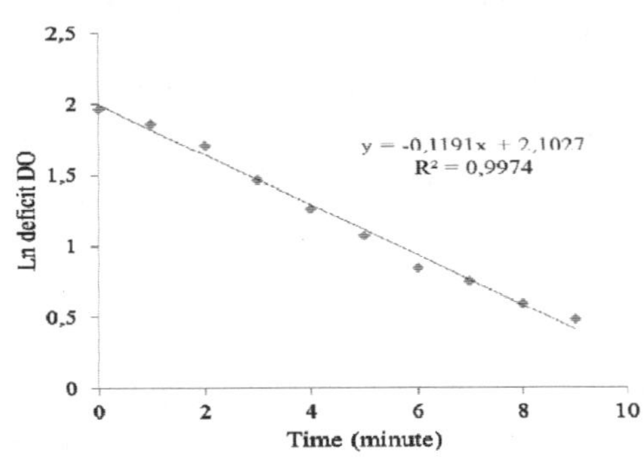

Figure 4. Graph for calculation of aerator *SOTR* test

The standard oxygen transfer rate (SOTR) was 1.60287 kg O_2 h^{-1}, and the standard aeration efficiency (SAE) was 2.72 kg O_2 kWh^{-1}. This achievement was better than the previous, the average SAE was 2.2 kg O_2 kWh^{-1} [3]. The increase of SAE is obtained due to the decrease in drag force of blades use the angle blades so reduces the consumption of electrical power. However, the splash of water remains widely as a result of angle blades, so the water spread. It has shown the efficiency of standard aeration proportional to the rate of oxygen transfer and inversely to the consumption of electric power as mentioned by Lekang (2007). The performance was better than the standard aeration efficiency of the standard model paddlewheel of Taiwan and the previous development wheel.

4. Conclusions

The paddlewheel of Furrower model blade consumed less electrical power but produced greater splash coverage volume. Design of Furrower model blades was 1.6 cm diameter hole, 45° of vertical angle blade position and 30° of the horizontal position. The standard oxygen transfer rate and standard aeration efficiency were produced 1.60 kg O_2 h^{-1} and 2.72 kg O_2 kWh^{-1}. The furrowed model blade was able to improve the aeration efficiency of the paddlewheel.

References

[1] Laksitanonta S, Sahdev S and Gajendra S 2003 A review of aerators and aeration practices in Thai aquaculture *AMA Agr. Mech. ASIA AF* **34** 64-71

[2] Romaire R P and Merry GE 2007 Effect of paddlewheel aeration on water quality in crawfish pond *Appl. Aquac.* **19** 61-75

[3] Boyd C E 1998 Pond water aeration systems *Aquac. Eng.* **18** 9-40

[4] Moulick S, Mal BvC, and Bandyopadhyay S 2002 Prediction of aeration performance of paddlewheel aerators *Aquac. Eng.* **25** 217-237

[5] Kumar A, Moulick S and Mal BC 2013 Selection of aerators for intensive aquacultural pond *Aquac. Eng.* **56** 71-78

[6] Bahri S, Setiawan R P A, Hermawan W and Junior M Z 2015 Simulation on blade geometry and operational condition toward torque requirement and drag force in paddle wheel aerator *IJSER* **6** 812-816

[7] [ASCE] American Society of Civil Engineers 1992 *Measurement of oxygen transfer in clean water* (New York: ANSI/ASCE) 2-91

[8] Kang Y H, Lee M O, Choi S D and Sin Y S 2004 2-D Hydrodynamic model simulating paddlewheel-driven circulation in rectangular shrimp culture ponds *Aquaculture* **231** 163-179

[9] Munson B R, Young D F and Okiishi T H 2006 *Fundamentals of fluid mechanics* (USA: John Wiley & Sons) chapter 9 pp 518-537

[10] Park J, Heikes D, Recsetar M and Ror L A 2014 Performance evaluation and engineering considerations for a modular-and culvert-based paddlewheel circulator for split-pond system *Aquac Eng* **61** 1-8

[11] Liu Y and Peymani F Y 2015 Development and computational validation of an improved analytic performance model of the hydroelectric paddle wheel *DGAE Journal* **30** 58-79

[12] Lekang O I 2007 *Aquaculture engineering* Blackwell Publishing chapter 8 p 97-103

IOP Conf. Series: Materials Science and Engineering **352** (2018) 012007 doi:10.1088/1757-899X/352/1/012007

Determination of UAV pre-flight Checklist for flight test purpose using qualitative failure analysis

Hendarko, T Indriyanto, Syardianto and F A Maulana
Faculty of Mechanical and Aerospace Engineering, Institut Teknologi Bandung, Jalan Ganesha 10, Bandung, Indonesia

E-mail: hendarko@ae.itb.ac.id

Abstract. Safety aspects are of paramount importance in flight, especially in flight test phase. Before performing any flight tests of either manned or unmanned aircraft, one should include pre-flight checklists as a required safety document in the flight test plan. This paper reports on the development of a new approach for determination of pre-flight checklists for UAV flight test based on aircraft's failure analysis. The Lapan's LSA (Light Surveillance Aircraft) is used as a study case, assuming this aircraft has been transformed into the unmanned version. Failure analysis is performed on LSA using *fault tree analysis* (FTA) method. Analysis is focused on propulsion system and flight control system, which fail of these systems will lead to catastrophic events. Pre-flight checklist of the UAV is then constructed based on the basic causes obtained from failure analysis.

1. Introduction

The use of Unmanned Aerial Vehicle (UAV), either for military or civil applications, has been increasing rapidly in the last decades. Acceleration of technology developments, especially in electronics, has driven massive developments and productions of UAVs with a wide range of capabilities and relatively low cost. As a consequence of the increasing market, many enterprises which are designing, developing and manufacturing UAVs with wide variation of size, configuration, and mission have been significantly increasing as well.

In the process of designing, developing, and manufacturing UAV, flight testing will become an important phase which has to be performed. In practice, incidents and/or accidents occasionally occur in flight tests. The application and use of checklists during flight tests helps eliminate the errors that can be made in an intense high workload flight test environment. These checklists should be developed and presented in the flight test plan [1].

Meanwhile, failure analysis is a process of collecting and analysing data to determine the cause of a failure, often with the goal of determining corrective actions or liability. Failure analysis has been used in the development of new UAVs and for the improvement of existing ones, either for a specific subsystem such as flight control system [2] or for the whole UAV system [3][4].

In order to minimize the probability of incidents/accidents occur in flight test, a more detail pre-flight checklists should be determined. In this research, a new approach for determination of pre-flight checklist is developed by using qualitative failure analysis of the UAV.

IOP Conf. Series: Materials Science and Engineering **352** (2018) 012007 doi:10.1088/1757-899X/352/1/012007

2. Method of Analysis

The UAV evaluated in this study is the Lapan's LSA (Light Surveillance Aircraft) manufactured by Stemme AG, a German aircraft manufacturer. In this analysis, the aircraft is assumed to have been transformed into the unmanned version. Failure analysis performed on the UAV in this study uses a deductive approach, namely Fault Tree Analysis (FTA) method.

2.1. Lapan's LSA (Light Surveillance Aircraft)

The National Institute of Aeronautics and Space (Lapan) has an aircraft named LSA (Light Surveillance Aircraft). LSA is actually the Stemme ASP S15-1 aircraft. It is a two-seat single-engined, all composite construction, powered sailplane with the engine mounted in the center fuselage. The cockpit has room for two in side-by-side configuration. It has a shoulder wing, a conventional T-tail and a retractable nose wheel landing gear.

The aircraft received type certificate from European Aviation Safety Agency (EASA) in October 2013 [5]. The general specifications and 3-view drawing of the aircraft are shown in table 1 and figure 1, respectively [6].

Table 1. General geometry and performance characteristics of LSA.

Parameters	SI Unit	British Unit
Length	8.52 m	27 ft 11 in
Wingspan	18 m	59 ft 1 in
Height	2.45 m	8 ft 0 in
Wing area	17.4 m2	187 sq ft
MTOW	1,100 kg	2,425 lb
Powerplant: 1x Rotax914F2	84.5 kW	113.3 hp
Propellers diameter: 3-bladed Muhlbauer Type MTV-7-A/170/051	1.7 m	5 ft 7 in
Stall speed	107 km/h	58 kts
Take-off distance	833 m	2832 ft
Maximum rate of climb	3.0 m	591 ft/min
Maximum cruise speed	234 km/h	126 kts
Maximum endurance	6.22 h	6.22 h
Maximum range	1119.6 km	604.5 nm

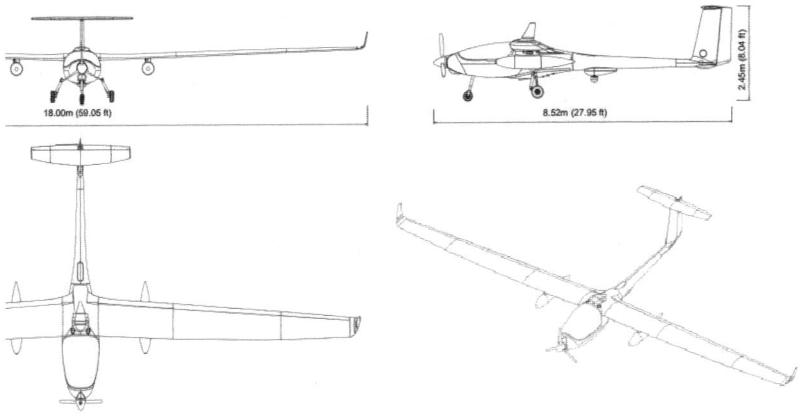

Figure 1. 3-view drawing of LSA [6].

2.2. Fault Tree Analysis of LSA

Fault Tree Analysis method is selected to be used in conducting failure analysis on the LSA. Fault Tree Analysis (FTA) is a deductive, failure-based approach. As a deductive approach, FTA starts with an undesired event, such as failure of the main engine, and then determines (deduces) its causes using a systematic, backward-stepping process. In determining the causes, a fault tree (FT) is constructed as a logical illustration of the events and their relationships that are necessary and sufficient to result in the undesired event, or top event. The FT is a qualitative model that provides extremely useful information on the causes of the undesired event. The FT can also be quantified to provide useful information on the probability of the top event occurring and the importance of all the causes and events modeled in the FT [7].

3. Failure Analysis and Pre-flight Checklists Determination of LSA

Failure analysis conducted on the LSA is focused on propulsion system and flight control system, which fail of these systems will lead to catastrophic events. In the case of LSA, the propulsion system is equipped with an engine and a propeller, which both are connected with drive section, to provide thrust to the UAV platform, as shown in figure 2.

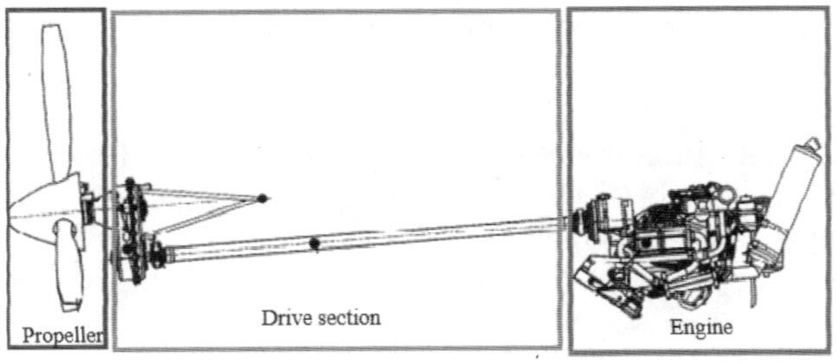

Figure 2. The propulsion system of LSA [6].

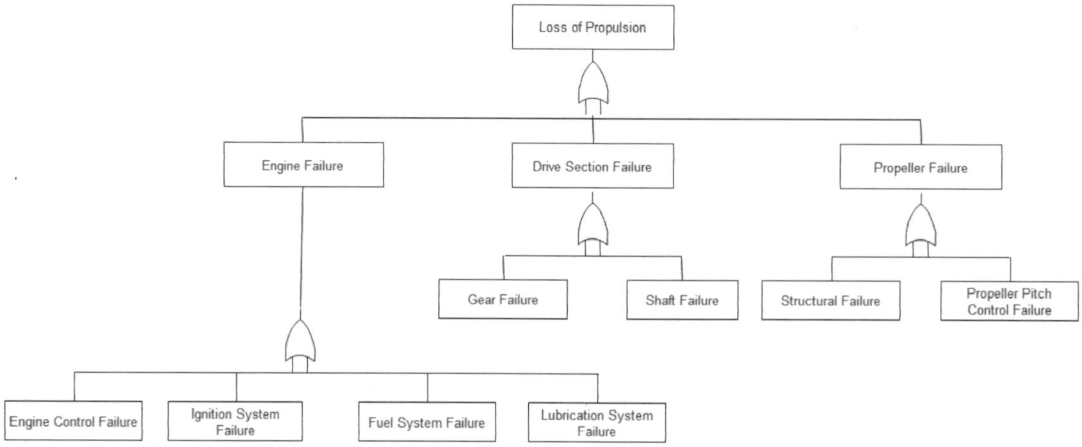

Figure 3. FT of the propulsion system.

It is considered Loss of Propulsion as a top event, which may be caused by either Engine Failure, Propeller Failure, or Drive Section Failure. The engine comprises four important subsystems, i.e. engine control system, ignition system, fuel system, and lubrication system. Failure of the engine is considered because of failure of either one of them. Meanwhile, the Propeller Failure may be caused by either structural damage of the propeller or failure of propeller pitch control. The drive section, which transfers power from the engine to propeller, can fail because of either the gear failure or the shaft failure. The Fault Tree of the propulsion system is shown in figure 3 and the Fault Tree of the ignition system, fuel system, and lubrication system are shown in figure 4 to figure 6, respectively.

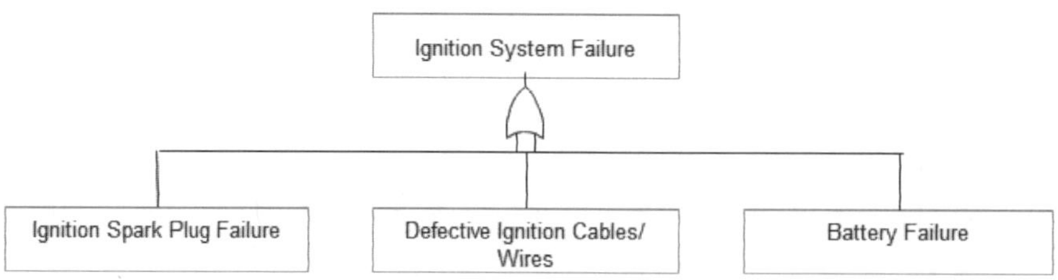

Figure 4. FT of the ignition system.

Ignition System failure results from three derivative failures, i.e. spark plug failure that results in an ignition process does not occur, cable/wire failure that results in generator not working properly, and the battery as a power supply does not work well.

Fuel system failure may occur due to five derivative failures, i.e. failures/defects of the filter, fuel tank, pipes/hoses, pumps, and sensors. The defective filter may prevent the filter to screen out dirt and rust particles from the fuel, which in turn can defect other components or degrade the system performance. Defective fuel tanks such as leakage may cause running out fuel and hence engine stop. Blocked pipes/hoses usually affect the failure to start the engine. Failure of the fuel pump, commonly using an electric pump, gives rise to unconveyability of fuel to the engine. Failure of fuel sensors (pressure and temperature) can cause the fuel system not working properly.

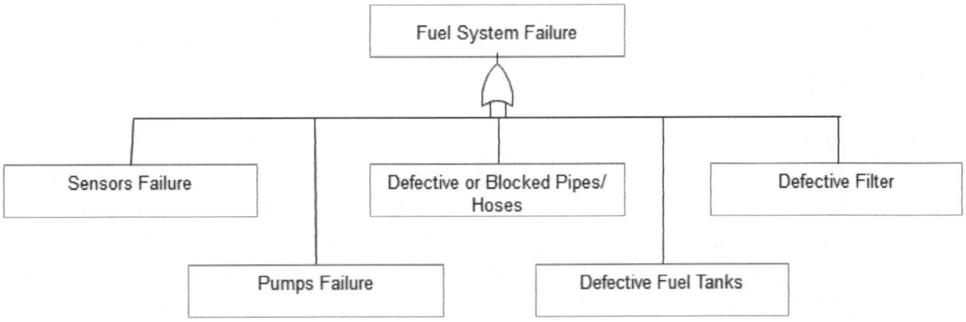

Figure 5. FT of the fuel system.

The failure of the lubrication system has three derivative failures. Failure of this system is almost same as the failure of the fuel system, consisting of sensors failure, pumps failure, and defective or blocked pipes/hoses. These three failures can make the engine not get enough oil and cause the engine to be damaged.

Flight control system of LSA is mainly comprised the control surfaces (elevator, ailerons, rudder, trim tab, flaps, and airbrakes), the actuators which deflect the control surfaces through the linkages, and Flight Control Computer. Failure of either one of them will cause a failure of flight control system. Meanwhile, control surface failure results from structural damage or jammed linkages. Beside hardware or software failure, failure of sensors (air data sensors or deflection angle sensors) can lead to a failure of flight control computer. Figure 7 illustrates the Fault Tree of Flight Control System of LSA.

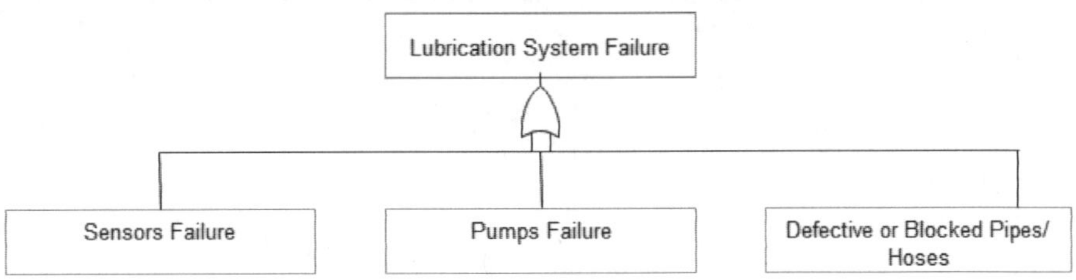

Figure 6. FT of the lubrication system.

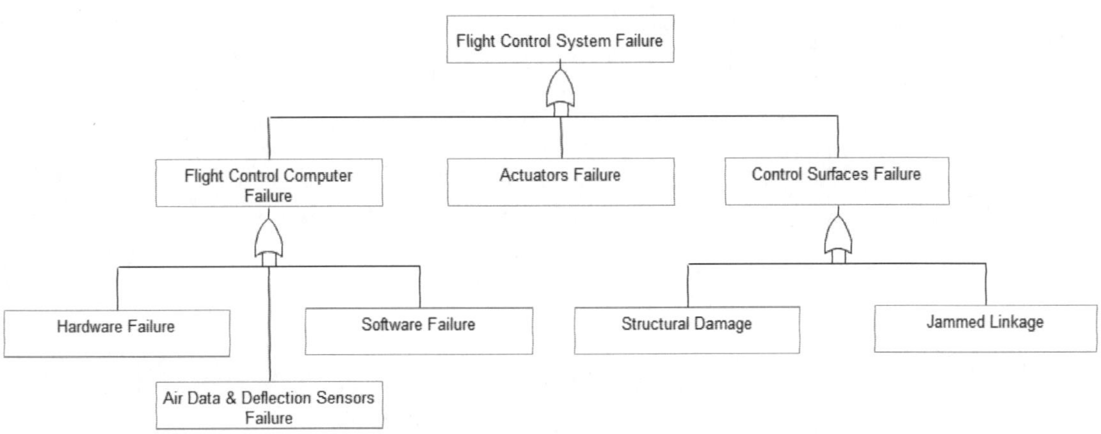

Figure 7. FT of flight control system.

Based on the basic causes obtained from fault tree analysis of the LSA, one can determine pre-flight checklists to mitigate the likelihood of incident/accident occurrence. The developed checklist resulted from qualitative failure analysis of the LSA is shown in Appendix.

4. Concluding Remarks

In this report, qualitative failure analysis of a UAV has been performed using the Lapan's LSA as a study case. Pre-flight checklist of the aircraft is then determined from the failure analysis. This concludes development of a new approach for determination of pre-flight checklist of UAV, i.e. based on qualitative failure analysis. In the future research, the developed checklist can be tested and implemented to the LSA aircraft.

References

[1] Stoliker F N 2005 Introduction to flight test engineering *RTO AGARDograph 300 Flight Test Techniques Series* **14**

[2] Okafor E G and Eze I H 2016 Failure analysis of UAV flight control system using markov analysis *NIJOTECH* **35** 167-173

[3] Murtha J F Evidence theory and fault tree analysis to cost-effectively improve reliability in small UAV design

[4] Juliana B and Goes L C S 2007 Failure analysis methods in Unmanned Aerial Vehicle (UAV) applications *Proc. 19th Int, Congress of Mechanical Engineering (Brasilia)*

[5] European Aviation Safety Agency (EASA) 2013 *Type Certificate Data Sheet, EASA TC* No.A.612

[6] Stemme AG 2013 *Flight Manual for the Aircraft ASP S15-1* Doc.-No.: K400-911.960

[7] Stamatelatos M, Vesely W, Dugan J, Fragola J, Minarick J and Railsback J 2002 *Fault Tree Handbook with Aerospace Applications* version 1.1

IOP Conf. Series: Materials Science and Engineering **352** (2018) 012007 doi:10.1088/1757-899X/352/1/012007

Appendix. Pre-flight checklist of the LSA resulted from qualitative failure analysis

Systems	Yes	No
engine (general)		
breather inlet clear		
engine exhaust clear		
all connections (bolts, nuts, etc) properly tightened		
fuel system		
amounts of fuel, oil, and coolant in proper level		
fuel lines connected and no leakage or blockage		
fuel filters clean and no defect		
fuel tanks have no cracks and leaks		
fuel sensors are well-functioned		
fuel pumps are well-functioned		
lubrication system		
amounts of oil in proper level		
oil sensors are well-functioned		
oil pumps are well-functioned		
oil tank has no cracks and leaks		
oil lines connected and no leakage or blockage		
ignition and electrical system		
spark plugs cleaned, undamaged, and properly torqued		
all wiring connected, undefect, and secured		
battery capacity is sufficient (min. 16Ah)		
propeller		
propeller blades free for cracks and damage		
propeller hub free for cracks, rusts and damage		
propeller pitch control well-functioned		
spinner has no cracks and mounted correctly		
drive connection		
shaft has no cracks and damage		
gear has no cracks and damage		
cowling has no cracks and mounted correctly		
flight control system		
control surfaces free from cracks and damage		
control rods and cables free from cracks, rusts and damage		
actuators well-functioned		
control surfaces movement has no jam and resistance		
no foreign object/contamination in pitot-static tube		
air data sensor well-functioned		
FCC well-functioned		
datalink well-functioned		
GPS signal available		
weight and c.g position within the limits		

IOP Conf. Series: Materials Science and Engineering **352** (2018) 012008 doi:10.1088/1757-899X/352/1/012008

Corrosion resistance of new beta type titanium alloy, Ti-29Nb-13Ta-4.6Zr in artificial saliva solution

Gunawarman[1], D D Giatmana[1], Ilhamdi[1], J Affi[1], S Fonna[2], M Niinomi[3] and M Nakai[3]

[1]Mechanical Engineering Dept., Andalas University, Padang 25163, Indonesia
[2]Mechanical Engineering Dept, Syiah Kuala University, Banda Aceh, Indonesia
[3]Formerly with Institute for Materials Research, Tohoku Univ, Sendai 980-8577, Japan

E-mail: gunawarman@ft.unand.ac.id

Abstract. The corrosion resistance of Ti-29Nb-13Ta-4.6Zr (TNTZ) and Ti-6Al-4V alloys in oral cavity environment were studied by investigating its corrosion rate in artificial saliva solution. Corrosion measurement was conducted in 600 ml solution of Fusayama-Meyer artificial saliva containing 0.4g NaCl, 0.4g KCl, 0.795g $CaCl_2.2H_2O$, 0.69g NaH_2PO_4, and 1 g urea using a potentiostat controlled by a personal computer. The solution was maintained at pH 5.2 and controlled the temperature of 37°C to imitate oral cavity condition. After corrosion test, specimen surfaces were examined by SEM and EDX. The results show that the average corrosion rate of TNTZ and Ti-6Al-4V is $4{,}5\times10^{-9}$ mmy^{-1} and $6{,}4\times10^{-8}$ mmy^{-1}, respectively, indicating that the corrosion resistance of TNTZ is slightly better than Ti-6Al-4V. This is suggested mainly due to the formation of multiple layers of Ti, Nb and Zr oxides in the surface of TNTZ. However, the formation of micro-pitting corrosion is more severe in TNTZ as compared to that of Ti-6Al-4V. The intense pitting corrosion in TNTZ is found strongly corresponded to its high impurities content and wide elemental segregation. It is recommended, therefore, a longer homogenizing process is required in TNTZ for reducing pitting corrosion attack. However, the details of corrosion mechanism are needed to be explored further.

1. Introduction

Titanium alloys, in particular Ti-6Al-4V, has been used as the most attractive dental materials due to their excellent combination of mechanical properties, corrosion resistance and biocompatibility [1-4]. However, the release of V and Al ions into the surrounded cell tissue by passive film dissolution [5] and wear corrosion [6], induce possible toxic effects in the human body [7]. Moreover, Young's modulus of Ti-6Al-4V is still much greater compared to that of the cortical bone's. This can be understood as these alloys are originally used for structural applications which need high modulus. Beta (β) type titanium alloys such as Ti-15Mo-5Zr-3Al and Ti-13Nb-13Zr with low Young's modulus and greater strength have been, therefore, developed for biomedical applications [4,5]. A new beta type titanium alloy composed of non-toxic and non-allergic elements like Nb, Ta, and Zr, Ti-29Nb-13Ta-4.6Zr (TNTZ) [6], has been developed by Niinomi and co-workers in order to achieve much lower Young's modulus and excellent mechanical performances. This alloy can have a wide range of mechanical properties by performing heat treatment or thermo-mechanical treatments on this alloy [6-8]. This alloy was found to have an excellent corrosion resistance in air and body fluids [9-13]. This may indicate that TNTZ also has potential to be used for orthodontic appliances. Some fundamental works are still necessary to carry out in order to know the behavior of this alloy in oral cavity environment for such application.

Previous work on the corrosion test of TNTZ in 0.5% HCl and Ringer's solution indicated that the corrosion rate of TNTZ varies according to the heat treatment and thermo-mechanical treatment where the cold-rolled of TNTZ has the highest corrosion rate among others [9]. The corrosion resistance of TNTZ, however, is slightly greater than those of forged Ti-15Mo-5Zr-3Al and Ti-6Al-4V ELI. While, measurement of corrosion rate in a modified artificial saliva using weight loss method indicates that the weight loss of TNTZ is zero up to exposure time of 480h [12]. At the same time, weight loss of two conventional alloys for dental application; stainless steel (SS) and cp-Ti can be easily obtained, that is 0.01g and 0.03 g, respectively. In order to understand the corrosion resistance of TNTZ in a artificial saliva, the corrosion rate is then measured by using a potensiostat in a Fusayama-Meyer artificial saliva medium at pH and temperature near oral cavity condition.

The corrosion rate of the most popular titanium, Ti-6Al-4V was also determined for comparison. The corrosion rates of these alloys are then compared to the corrosion rate of some conventional dental materials that are available in the literature.

2. Experimental Detail

As received samples in this study were rolled plates of TNTZ and bars of Ti-6Al-4V. Both materials were provided by Prof. Niinomi (Institute for Material Research, IMR, Japan). Some square plate specimens of TNTZ with a size of 10×10 mm and 3 mm thickness, and round Ti-6Al-4V with a diameter of 10 mm and 3 mm thickness were machined from the as-received materials (figure 1a). All sample surfaces were grinded gradually with emery papers from mesh size of #100 to #2000, and then polished and also cleaned prior to corrosion test to obtain smooth surface specimens. The surface of specimens was then fully covered with epoxy resin except for top surface for exposing to the artificial saliva solution (figure 1b).

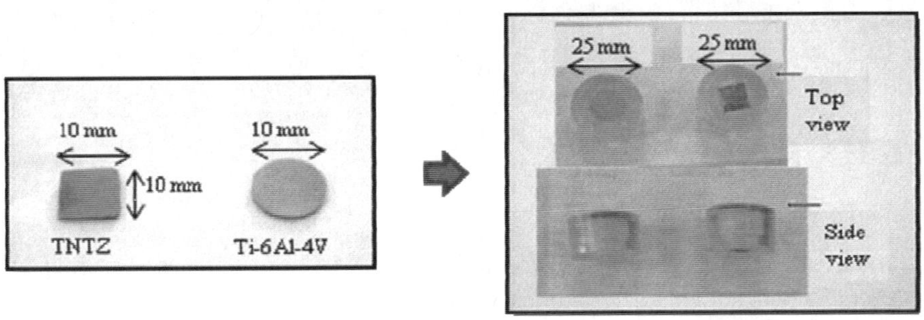

(a) (b)

Figure 1. Corrosion test samples (a) blank and (b) mounted specimens

Corrosion test was conducted in 600 ml solution of Fusayama Meyer artificial saliva solution containing 0.4g NaCl, 0.4g KCl, 0.795g $CaCl_2.2H_2O$, 0.69g NaH_2PO_4, and 1 g urea [14] which was provided by Faculty of Biomedical Engineering and Health Science, University Technology of Malaysia. This solution was maintained at pH 5.2 and 37°C during corrosion test using a potentiostat model (VERSA studio-200) controlled by a personal computer. Potentiodynamic polarization studies were carried out for 3h at a scan rate of 1 mV/s to obtain Tafel slot for each sample. Corrosion potential (E_{corr}) and corrosion current density (I_{corr}) of each sample were then calculated from the Tafel slot. Corrosion rate (CR) of each sample was then determined using following formula (ASTM standards G01):

$$CR = K_1 \frac{I_{corr} EW}{\rho} \quad (1)$$

where, CR is corrosion rate in mmpy, K_1 = 3,27 x 10^{-3} mm g/μA Cm yr, I_{corr} = corrosion current

IOP Conf. Series: Materials Science and Engineering **352** (2018) 012008 doi:10.1088/1757-899X/352/1/012008

density at E_{corr}, EW = equivalent weight of material, and ρ = density of material (g/cm^3).

Specimen surfaces were then observed by Scanning Electron Microscope (SEM) and Energy Dispersive Spectroscopy (EDS/EDX) and X-ray Photoelectron Spectroscopy (XPS). SEM examination was conducted at an acceleration voltage of 15 kV using Hitachi SEM S-340N. EDX was conducted at the same acceleration using EDX Horiba attached to the SEM machine. Both examinations were conducted at Andalas University, Padang, Indonesia.

3. Result and Discussion

Typical SEM micrograph of TNTZ and Ti-6Al-4V prior to corrosion process is shown in figure 2. The chemical composition of the alloy using EDX is tabulated in table 1. It can be seen that the presence of porosities and inclusions in both materials as indicated by black spots in the micrographs. The number of such impurities is more severe in TNTZ as compared to that of Ti-6Al-4V. This is due to the different preparation of both materials. TNTZ is still a new type material that has been manufactured for laboratory testing only. While, Ti-6Al-4V is a commercial one. The EDX result in Table 1 indicates that TNTZ, as expected, contains alloying elements of Nb (26.5%), Ta (13.5%) and Zr (3.9%) with a chemical composition close to the standard composition (29%Nb-13%Ta-4.6%Zr). While, Ti-6Al-4V contains alloying elements of Al (6.3%) and V(3.9%) which is also close to the standard one (6%Al and 4%V). However, the contents of other elements in TNTZ is much higher than that of Ti-6Al-4V. This verifies that TNTZ has much higher impurities than Ti-6Al-4V (Fig.2).

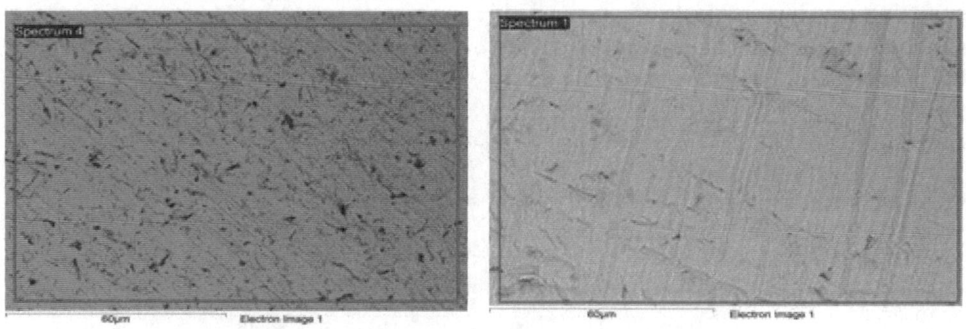

(a) (b)

Figure 2. Surface condition of (a) TNTZ and (b) Ti-6Al-4V prior to corrosion process

Table 1. Chemical composition of the alloy by EDX

TNTZ			Ti-6Al-4V		
Element	%mass	%atom	Element	%mass	%atom
Ti	40.4	37.8	Ti	86.9	77.1
Zr	3.9	1.9	Al	6.3	9.9
Nb	26.5	12.8	V	3.9	3.2
Ta	13.5	3.3	Others	2.9	9.8
Others	15.7	44.2			

Tafel curve of all samples obtained from the potensiostat test is shown in figure 3. While, the calculation result of the Tafel curve is tabulated in table 2. It can be seen that I_{corr} of TNTZ is much smaller than that of Ti-6Al-4V. As for the average E_{corr} of TNTZ is also smaller that of Ti-6Al-4V. However, the E_{corr} of Ti-6Al-4V is more stable than that of TNTZ. Since corrosion rate (CR) is proportional to I_{corr} value, the corrosion rate of TNTZ is much lower than that of Ti-6Al-4V. Averaged corrosion rate of TNTZ is 4,48x10^{-9} mmy^{-1}, while Ti6Al4V is 6,37x10^{-8} mmy^{-1}. This value indicates that the corrosion rate of TNTZ is much smaller than the conventional dental wires of NiTi and stainless steel at pH 5 [7]. In orthopedic application, the low corrosion rate is essential as it minimizes the release of ion and particle to human digestion system, which in turns suppress the toxicity and allergic problems of elemental deposit inside human body [4].

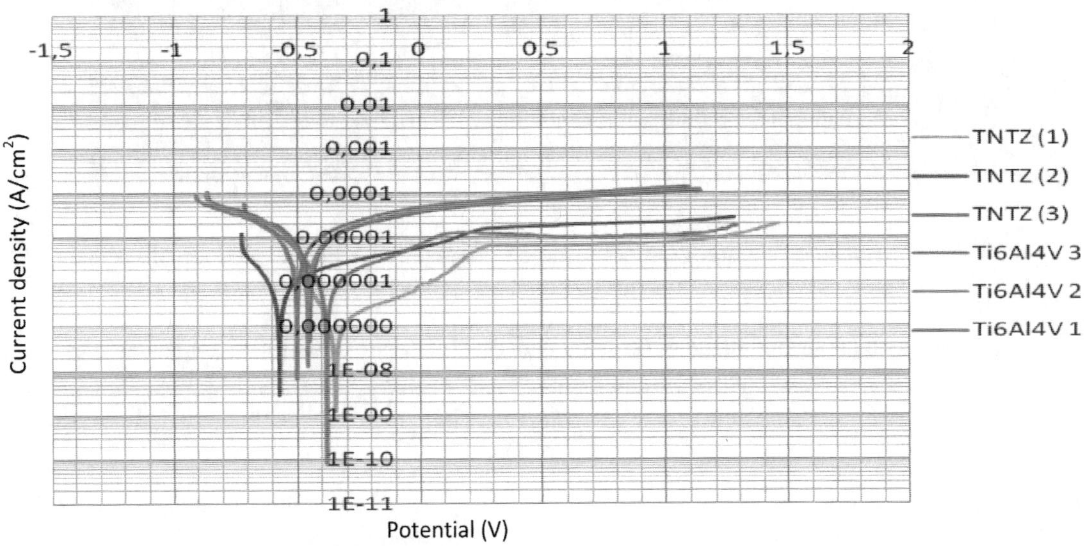

Figure 3. Tafel curve TNTZ and Ti-6Al-4V specimens

Table 2. Result of potensiostat test on TNTZ and Ti6Al4V

Specimen		I_{corr} (A)	E_{corr} (mV)	CR (mmy^{-1})
TNTZ	1	1,57 x10^{-7}	-350	1,36 x10^{-9}
	2	6,99 x10^{-7}	-570	6,00 x10^{-9}
	3	6,91 x10^{-7}	-380	6,07 x10^{-9}
	Avrg	5,16 x10^{-7}	-433	4,48 x10^{-9}
Ti-6Al-4V	1	7,29 x10^{-6}	-406	6,33 x10^{-8}
	2	7,51 x10^{-6}	-440	6,52 x10^{-8}
	3	7,20 x10^{-6}	-490	6,25 x10^{-8}
	Avrg	7,33 x10^{-6}	-463	6,37 x10^{-8}

SEM observation result on the sample surfaces shows black spot in TNTZ and Ti-6Al-4V samples (figure 4). Elemental distribution in the black spots is tabulated in table 3. The high content of oxygen in the black spots indicates corrosion product. It can be seen clearly in Fig. 4 that corrosion area in TNTZ is wider than that in Ti-6Al-4V. The average content of oxygen in TNTZ is 46.7% that is much higher than that of Ti-6Al-4V (32.7%). This indicates that the formation higher amount oxides in TNTZ rather than Ti-6Al-4V. Moreover, the high content of Nb and Ta in the surface of the alloy may indicate that the oxide in TNTZ surface is not only titanium oxide, but also Niobium oxide and Tantalum oxide. In the case of Ti-6Al-4V, the oxide is predominantly titanium oxide as indicated by the very high content of Titanium in the corrosion area of Ti-6Al-4V. Elemental distribution in table 3 also confirms the corrosion attack in TNTZ is contributed directly by the artificial saliva solution. Almost all artificial saliva solution elements can be detected with a high content of chloride that is accumulation from NaCl, KCl, and $CaCl_2.2H_2O$. However, the reason of the saliva elements undetected in all spots of Ti-Al-4V is not clear. A further investigation is needed to clarify this phenomenon, which will be reported separately.

IOP Conf. Series: Materials Science and Engineering **352** (2018) 012008 doi:10.1088/1757-899X/352/1/012008

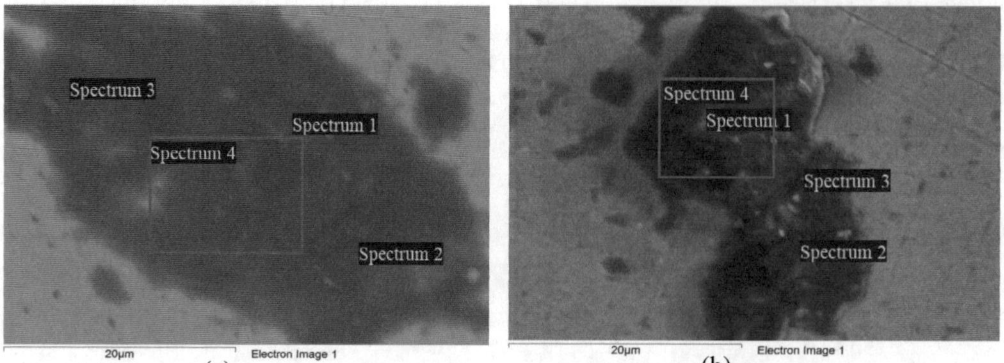

(a) (b)

Figure 4. Typical SEM micrograph of sample surfaces (a) TNTZ and (b) Ti-6Al-4V

Table 3. Elemental distribution near corrosion area of TNTZ and Ti-6Al-4V sample surfaces

Sample	Elements	Spectrum 1	Spectrum 2	Spectrum 3	Spectrum 4	Avrg	Range
TNTZ	Oxygen (O)	70,9	27,7	36,4	50,0	46,7	43,3
	Titanium (Ti)	7,2	36,0	24,6	23,2	23,3	28,8
	Niobium (Nb)	6,3	22,7	18,8	16,8	15,7	16,4
	Tantalum (Ta)		11,5	15,7		11,7	4,2
	Chloride (Cl)	9,6	1,4	3,0	5,9	5,6	8,3
	Kalium (K)	9,6	0,5	0,9	1,9	4,6	9,2
	Calcium (Ca)	2,6	0,3	0,7	2,2	1,6	2,3
Ti-6Al-4V	Oxygen (O)	58,8	3,9	24,8	23,9	32,7	55,0
	Titanium (Ti)	82,5	85,6	67,0	66,4	67,5	19,2
	Aluminium (Al)	0,8	10,6	7,3	7,6	6,8	9,8
	Vanadium (V)	0,6		0,8	2,1	1,3	1,5
	Natrium (Na)	18,3					-
	Chloride (Cl)	17,1					-
	Kalium (K)	4,5					-

SEM micrographs on the cross-section area at the surrounding of the corroded area and its elemental distribution are shown in figure 5 and table 4, respectively. Figure 5 shows pitting corrosion (or can be called as micro-pitting regard to the size of the pitting area) in both alloys, where, the micro-pitting in TNTZ is much deeper and swallower than that in Ti-6Al-4V. The oxygen content in the pitting area of TNTZ is also much higher than that of Ti-6Al-4V (table 4). This confirms that the formation of pitting corrosion in TNTZ is much more intense rather than in Ti-6Al-4V. Easier pitting corrosion attack in TNTZ is strongly suggested due to high element segregation in TNTZ. The range of alloying elements distribution in TNTZ is much wider than that of Ti-6Al-4V (table 3 and 4). The content of alloying elements (Nb, Ta, and Zr) is not detected in all spot of measurements. Moreover, high impurities content mentioned above is also contributed to induce more severe pitting corrosion attack in TNTZ (figure 2). In the case of Ti-6Al-4V, each alloying element is detected in all spectrum as can be seen in Table 4. Moreover, the range of composition of each alloying element in Ti-6Al-4V is tighter than that of TNTZ. This seems to provide a uniform protection on the surface of Ti-6Al-4V. In order to improve pitting corrosion resistance of TNTZ, an intense homogenizing process is necessary to apply in TNTZ prior to dental product application. It is well-known that homogenizing process is commonly applied to reduce element segregation in all alloys. Applying multi thermomechanical treatment is also improve element distribution in TNTZ [9].

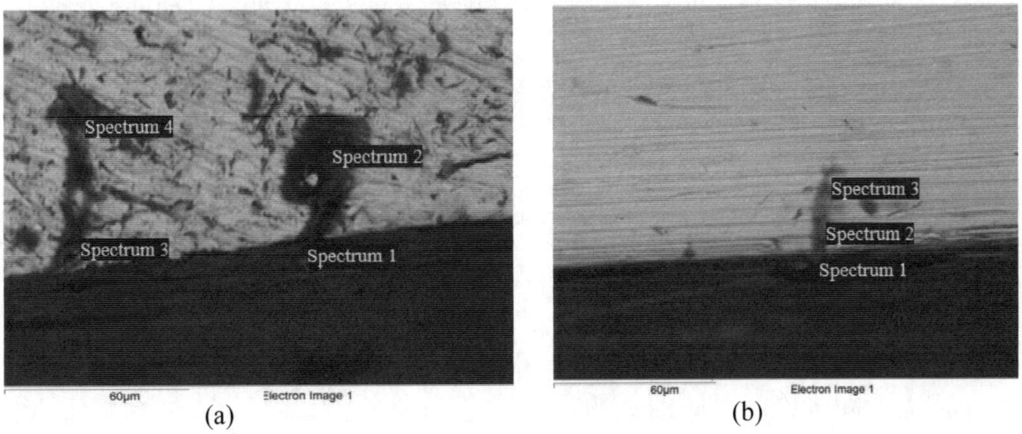

Figure 5. SEM Micrograph of cross-section sample (a) TNTZ and (b) Ti-6Al-4V

Additional examinations on the surface of the sample are required to understand the corrosion resistance of TNTZ, e.g. by examining typical binding energy spectra of TNTZ using X-ray Photoelectron Spectroscopy (XPS) in order to determine the oxide compounds in the surface. The previous study showed the formation of multilayer (bi-layer) oxides in the surface of TNTZ [11]. The high intensity of oxygen and Titanium in TNTZ as compared to those of Ti-6Al-4V may indicate a high activation energy for uniform corrosion in TNTZ, and thus (uniform) corrosion resistance of TNTZ becomes better than that of Ti-6Al-4V.

Table 4. Elemental distribution near micro-pitting area of TNTZ and Ti-6Al-4V

Sample	Elements	Spectrum 1	Spectrum 2	Spectrum 3	Spectrum 4	Avrg	Range
TNTZ	Oxygen (O)	88,7	96,6	72,6	48,4	71,3	48,2
	Titanium (Ti)	7,4	3,5	12,1	35,2	18,4	31,8
	Zirconium (Zr)	3,2		4,8	3,7	3,6	1,6
	Niobium (Nb)	0,8		10,5	12,7	8,2	12,0
Ti-6Al-4V	Oxygen (O)	2,1	41,8	35,1		27,1	39,7
	Titanium (Ti)	82,5	45,5	54,8		58,0	37,0
	Aluminium (Al)	2,2	2,1	3,7		2,6	1,6
	Vanadium (V)	5,8	1,6	1,9		3,5	4,2
	Chloride (Cl)	7,4	9,3	4,5		6,6	4,8

4. Conclusion

In order to understand the corrosion resistance of a new developed β type titanium alloy Ti-29Nb-13Ta-4.6Zr (TNTZ) in oral cavity environment, the corrosion rate of this alloy in a modified artificial saliva was investigated by using potensiostat . The corrosion rate of the most popular titanium, Ti-6Al-4V was also determined for comparison. The following results are obtained:

1. Corrosion resistance of TNTZ in the modified artificial saliva solution is greater than that of Ti-6Al-4V. It is suggested mainly due to the formation of multi-layer of Ti, Nb and Zr oxides, as compared to single-layer of Ti oxide in the surface of Ti-6Al-V.
2. Micro-pitting is observed on the surface of both alloys, but micro-pitting in TNTZ is wider,

deeper and swallower than that in Ti-6Al-4V due mainly to a higher alloying element segregation in TNTZ as compared to that of Ti-6Al-4V. Moreover, high impurities of TNTZ is also contributed to the pitting corrosion attack.

3. In order to improve corrosion resistance, in particular pitting corrosion attack, an intense homogenizing process is recommended to apply in TNTZ for reducing the pitting corrosion.

Acknowledgments

The first author thanks to Engineering Faculty of Andalas University for supporting this work under Andalas University GB Grant no. 102/UN.16/HKRGB/LPPM/2016. Thank you very much for Dr. Hendra Hermawan and members of Biomaterial Laboratory of University Technology Malaysia for providing the artificial saliva solution and helping Potensiostat test.

References

[1] Williams D F 1977 Titanium as a metal for implantation, part 1; physical properties *J. Med. Eng. Technol.* **6** 195-198 and 202-203

[2] Schmalz, Gotrfried, Arenholt-Bindslev and Dorthe 2009 *Biocompatibility of dental materials* Berlin : Springer)

[3] Leyens C and Peters M *2003 Titanium and titanium alloys: Fundamentals and applications* (Weinheim: Wiley-VCH)

[4] Niinomi M 2002 *Mat. Mater. Trans. A* **33**A 477–486

[5] Okazaki Y and Gotoh E 2005 Comparison of metal release from various metallic biomaterials in vitro *Biomaterials* **26** 11–21

[6] Khan M A, Williams R L and Williams D F 1999 The corrosion behaviour of Ti–6Al–4V, Ti–6Al–7Nb and Ti–13Nb–13Zr in protein solutions *Biomaterials* 20 765–772

[7] Lopez M F, Jimenez J A and Gutierrez A 2003 Corrosion study of surface-modified vanadium-free titanium alloys *Electrochim. Acta* **48** 1395–1401

[8] Sakaguchi N, Niinomi M, Akahori T, Takeda J and Toda H 2005 *Mater. Sci. Eng.* C **25** 370

[9] Akahori T, Niimoni M, Fukui H and Suzuki A 2004 *Materials Transactions* **45** pp 1540 to 1548

[10] Diomidis N, Mischler S, More N S, Roy M and Paul S N 2011 *Wear* **271** 1093– 1102

[11] Karthega M, Raman V and Rajendran N 2007 *Acta Biomaterialia* **3** 1019–1023

[12] Gunawarman, et al 2012 *Proc. of JSME chapter Indonesia Seminar* (Jakarta: SNTTM XII)

[13] A. M. Barcelosa, et al, Materials Research, 16:1 (2013) 50-64.

[14] R.A. Subari, et al, Australian Journal of Basic and Applied Sciences 7(9), (2013) 87-96

IOP Conf. Series: Materials Science and Engineering **352** (2018) 012009 doi:10.1088/1757-899X/352/1/012009

Effect of roughness on stiction

Zahrul Fuadi[1], Hassan Zahouani[2], Toshiyuki Takagi[3], Hiroyuki Miki[4]

[1]Mechanical Engineering Department, Syiah Kuala University
Jl. Tgk. Syech Abdurrauf No. 7 Darussalam Banda Aceh 23111, Indonesia
[2]Laboratory of Tribology and System Dynamics
Ecole Central de Lyon, 69311 Ecully Cedex, France
[3]Institute of Fluid Science, Tohoku University
2-1-1 Katahira, Aoba-ku Sendai 980-8577, Japan
[4]Frontier Research Institute for Interdisciplinary Science, Tohoku University
Aramaki-aza Aoba, Aoba-ku Sendai 980-8678, Japan

E-mail: zahrul.fuadi@unsyiah.ac.id

Abstract. In this paper, the viscoelastic material was used to investigate the effect of roughness on stiction. The material is chosen because it is highly deformable so that contact during friction can be fully elastic. The soft surfaces were prepared by casting the silicon material on metal surfaces having smooth and unidirectional grooved texture. Two tests were conducted, indentation and friction, to find out the effect of roughness on parameters of normal contact stiffness, friction force and the difference between static and kinetic friction coefficient, μs-μk. As the results, it is found that all parameters are related to the surface roughness. Smoother surface tends to have a higher value of normal contact stiffness and higher value of friction force thus resulting in a larger difference between the static and kinetic coefficient of friction. Since the value of μs-μk is commonly related to the stick-slip motion, the smoother surface tends to have a larger propensity of stiction. It is shown by the result that the texture can reduce the stiction because it reduces the value of normal contact stiffness, resulting in a lower value of μs-μk.

1. Introduction

Dry friction controls two different phenomena in nature, i.e. the resistance against the beginning of a motion from the state of rest and the resistance against an existing motion [1,2]. The first phenomenon is known as stick and the latter is known as slip. Coulomb describes this as static friction and kinetic friction.

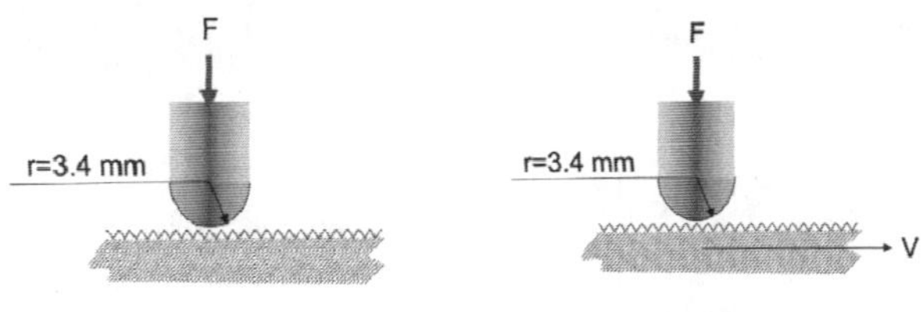

Figure 1. Schematic diagram of the tests; indentation (a) and friction (b)

In a continuous motion, both phenomenon could take place successively, resulting in stick-slip motion, or also known as stiction. This motion is undesirable in the mechanical system because it has negative effects on the system's performances such as positioning errors, fatigue failure, severe wear, and sound emission. Stiction is caused by various factors such as material properties, forces, and surface irregularities. In literature, the stiction is commonly related to various parameters such as difference of static and kinetic coefficient of friction [3], negative slope of friction-velocity relationship [4], and the parameter of contact stiffness [5]. Various experimental models have also be employed in the investigation of stiction phenomena such as a caliper-slider model [6].

One of the most important factors in stiction is the surface roughness. In this paper, soft materials were used to demonstrate the effect of surface roughness on stiction. Soft material is chosen because it is highly elastic so that plastic deformation effect during surface interaction can be omitted. The roughness on the soft surface was prepared by casting the soft material on a metal surface having a smooth and unidirectional grooved texture. Thus, there are two different surfaces; smooth and textured.The static (indentation) and friction tests were conducted to find out the parameters of normal contact stiffness, friction force, and the difference between static and kinetic friction coefficient, μs-μk. The objective of this study is to find out the effect of surface roughness on stiction propensity and to relate it to the parameters of normal contact stiffness and the difference between the static and kinetic coefficient of friction, μs-μk.

2. Methodology

Static (indentation) tests and friction tests were conducted in this analysis. The schematic diagram of the tests is given in figure 1. In the indentation test (figure 1 (a)) the force applied (normal) was 100 μN. The indenter had a diameter of 3.4 mm and made of stainless steel. The similar test setup was used in the friction test. In the friction test, after the normal force of 100 μN is applied, the indenter moves translationally for a traveling length of 2000 μm. Thus, sliding takes place and the friction force can be measured as well as the difference between static and kinetic friction coefficient, μs-μk.

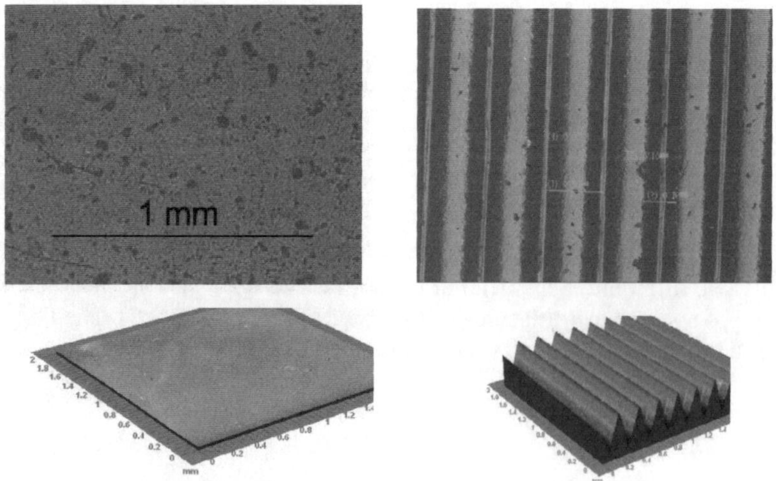

(a) (b)

Figure 2. Surface roughness; Smooth (Ra = 0.03μm) (a) and Textured (b) surfaces

Two different roughness were prepared for the surface specimens. The specimens were marked as "smooth" and "textured". The detail of the specimens was given in figure 2. Figure 2(a), smooth specimen, was achieved by casting the silicon on the metal surface having a surface roughness Ra of

0.03 μm. The textured specimen was achieved by casting the silicon on a metal surface having unidirectional grooves.

Figure 1 shows the schematic diagram of the tests. The indentation test (figure 1(a)) has two steps; 1) force application and 2) relaxation. As a result, the relationship between normal displacement and normal force can be achieved (figure 3(a)). Using the result, two quantitative parameters can be calculated i.e. normal contact stiffness, Kn, and the adhesion force between indenter and surface, N.
In the friction test (figure 1(b)), there are also two steps; 1) force application and 2) translational displacement (figure 3(b)). As a result, the parameter of friction force, N, and the difference between the static and kinetic coefficient of friction (μs-μk) can be achieved.

3. Results and discussions

The results of indentation tests were given in figure 4. Figure 4(a) indicates the hysteresis loop of the normal force as it is applied and relieved for the smooth and textured surfaces. The vertical axis indicates the normal force of 100 mN applied for both conditions. The horizontal axis indicates the displacement, i.e. the elastic deformation of the surface occurred as the effect of normal force application. As indicated by the figure, in the case of a smooth surface (the one on the left) the displacement started from 100 μm and reached the peak at 222 μm. Thus the overall displacement for the smooth surface is 122 μm. In the case of textured surface, the displacement started from 212 μm and reached a maximum at 329 μm. Thus, the overall displacement of the textured surface was 117 μm. This indicated that in both conditions, the material properties are similar. For the relatively similar normal force, the displacement is also relatively similar (differ by only 5 μm).

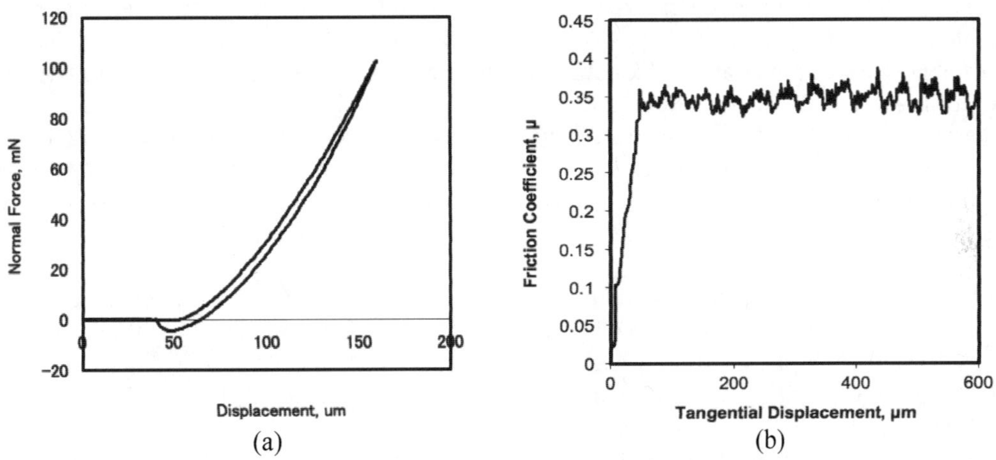

Figure 3. Relationship between normal displacement and normal force; smooth (a), and grooved (b)

Referring to the method given in figure 3(a), the normal contact stiffness value, Kn, for both conditions can be calculated. For the case of a smooth surface, the value of Kn is 650 N/m. For the case of textured surface, the value of Kn is 738 N/m. Thus, the normal contact stiffness of the textured surface is lower than that of smooth surface. This results indicated that there is a quantitative relationship between the surface parameter and the normal contact stiffness. Therefore, in the future, it is interesting to find out the relationship between surface parameters and the normal contact stiffness.

Figure 4(b) shows the detail image of the adhesion force between the ball and the surface after the normal force is relieved. In the case of smooth surface (the left one), the adhesion force is 9 mN and for that of the textured surface, it is 2 mN. This figure indicates that the textured surface has a lower value of adhesion force. This could be caused by the reduction of the surface contact area. In smooth surface, the contact occurs at almost entire ball's surface. This area is reduced by the presence of grooves on the surface. The reduce of contact area caused by the presence of groves, the texture, resulting in the reduction of the normal contact stiffness. According to Hertzian contact theory, one of the parameters affecting the contact stiffness is real contact area. Thus, the smoother the surface, the larger the contact area.

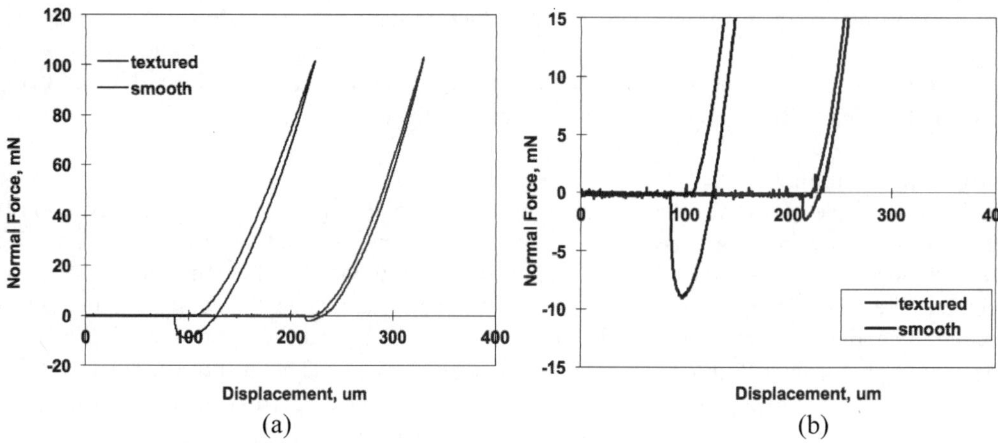

Figure 4. Indentation test result (a), and adhesion force (b)

So far, the presence of grooves on the surface has resulted in the reduction of normal contact stiffness in which this reduction can be attributed to the reduction of real contact area according to available contact theory. Furthermore, it is interesting to find out the effect of grooves on the value of the difference between the static and kinetic coefficient of friction, μs-μk. This value has been related to the occurrence of a stick-slip motion in the mechanical system [3].

The result of friction test is given in figure 5. It can be seen that the coefficient of friction of smooth surface is larger than that of the textured surface. In both cases, the normal force applied is 100 mN. In the case of smooth surface, the coefficient of friction is 0.35, but in the case of the textured surface, it is 0.15. The value of μs-μk for the smooth surface is 0.04 and that for the textured surface is 0.02, averagely.

From the results it is clear that smoother surface tends to have a larger value of μs-μk. Since the value of normal contact stiffness for the smoother surface is also larger than that of the textured one, the larger value of μs-μk can also be related to the contact stiffness. Thus, a higher value of contact stiffness tends to have higher value of μs-μk. Since the value of μs-μk is related to the stick-slip motion, the smoother surface tends to have a larger propensity of stiction.

It is shown by the result that rougher surface could reduce the stiction because rougher surface tends to have a lower value of μs-μk. Here, it is important to notice that a low value of μs-μk is caused by the reduction of normal contact stiffness which is caused by the reduction of the real contact area. Therefore, stiction can be directly related to the parameter of contact stiffness. The relation between the contact stiffness value and stick-slip motion has also been highlighted in ref [6].

Therefore, since the stiction is undesirable in mechanical systems, it is possible to minimize the stiction by decreasing the value of normal contact stiffness. This study has clearly shown that introducing the texture to a surface is an effective method to minimise the stiction. Unlike roughness that is highly random, texture can be designed according to the purpose. In this case, the texture has not only decreased the value of μs-μk but also the friction force drastically. In another case, it is also possible to design a textured surface that will only decrease the value of μs-μk but will have a relatively similar friction force.

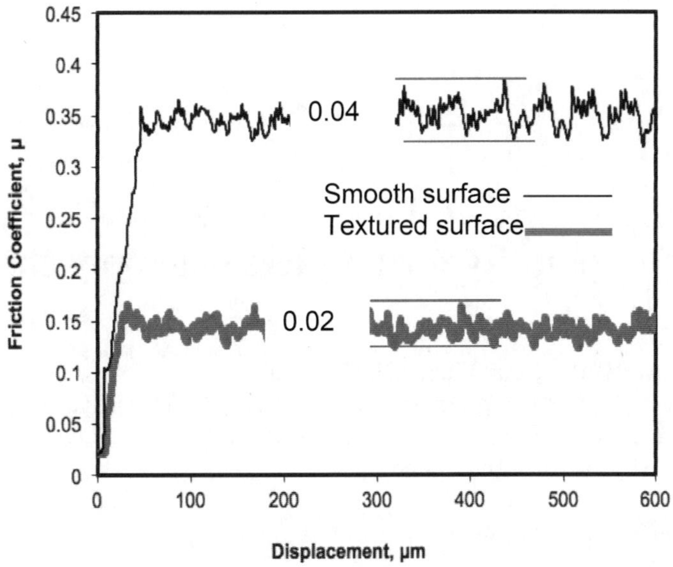

Figure 5. Friction test result

4. Conclusion

The effect of the surface parameter to the stiction propensity has been investigated by using the viscoelastic material. The result can be summarized as follows:

1. The normal contact stiffness of the textured surface is lower than that of the smooth surface. This effect is caused by the reduction of real contact area due to the presence of the texture.
2. The value of the difference between the static and kinetic coefficient of friction, μs-μk, for the textured surface is lower than that of the smooth surface. Thus, introducing texture to a contact interface could reduce the stiction propensity.

References

[1] Popp K 1992 Some model problems showing stick-slip motion and chaos *Friction-induced Vibration, Chatter, Squeal and Chaos* Ibrahim R A and Soom A ed ASME 1-12

[2] Wei D, Ruan J, Zhu W and Kang Z 2016 Properties of stability, bifurcation, and chaos of the tangential motion disk brake *Journal of Sound and Vibration* **375** 353-65

[3] Nakano K 2006 Two dimensionless parameters controlling the occurrence of stick-slip motion in a 1-DOF system with Coulomb friction *Tribology Letters* **24** 91-98

[4] Cabboi A and Woodhouse J 2017 Validation of a constitutive law for friction-induced vibration under different wear conditions Wear. In Press August 2017

[5] Fuadi Z, Maegawa S, Nakano K and Adachi K 2010 Map of low-frequency stick–slip of a creep groan. Proceedings of the Institution of Mechanical Engineers *Part J: Journal of Engineering Tribology* **224** 1235-1246

[6] Fuadi Z, Adachi K, Ikeda H, Naito H and Kato K 2009 Effect of contact stiffness on creep-groan occurrence on a simple caliper-slider experimental model *Tribology Letters* **33** 169-78

IOP Conf. Series: Materials Science and Engineering **352** (2018) 012010 doi:10.1088/1757-899X/352/1/012010

Numeric model to predict the location of market demand and economic order quantity for retailers of supply chain

Edy Fradinata[1] , Zurnila Marli Kesuma[2]
[1]Industrial Engineering Department, Syiah Kuala University, Banda Aceh, Indonesia.
[2]Statistics Study Program, Faculty of Mathematics and Natural Science, Syiah Kuala University, Banda Aceh, Indonesia.

E-mail: edinata69@gmail.com

Abstract. Polynomials and Spline regression are the numeric model where they used to obtain the performance of methods, distance relationship models for cement retailers in Banda Aceh, predicts the market area for retailers and the economic order quantity (EOQ). These numeric models have their difference accuracy for measuring the mean square error (MSE). The distance relationships between retailers are to identify the density of retailers in the town. The dataset is collected from the sales of cement retailer with a global positioning system (GPS). The sales dataset is plotted of its characteristic to obtain the goodness of fitted quadratic, cubic, and fourth polynomial methods. On the real sales dataset, polynomials are used the behavior relationship *x*-abscissa and *y*-ordinate to obtain the models. This research obtains some advantages such as; the four models from the methods are useful for predicting the market area for the retailer in the competitiveness, the comparison of the performance of the methods, the distance of the relationship between retailers, and at last the inventory policy based on economic order quantity. The results, the high-density retail relationship areas indicate that the growing population with the construction project. The spline is better than quadratic, cubic, and four polynomials in predicting the points indicating of small MSE. The inventory policy usages the periodic review policy type.

1. Introduction

The retail is transactions between the business store in the particular area to the consumer where it involves the money, service, and goods, and it is called retailer. The retailer usually uses the prediction of customer need-based on the historical data. There are many methods used to predict the demand based on the area clusters by retailers [1]. The regression of the market area for the retailer is significant in the competitiveness of business. The prediction will determine the level of profit for the single business company. The prediction methods could be used to predict the certain point of sales dataset in the retailer in the supply chain [2].

Before tsunami 2004, The Lafarge Cement Indonesia-Aceh (LCI) is a single cement producer and distributes them to the distributors for all over Aceh province. The cement product is called semen andalas. When the tsunami came, many infrastructures include LCI Aceh were destroyed by the wave of a tsunami. This situation pushed the mitigation condition for the cement need in Aceh Province. The emergency cement needs to build the infrastructure in fast many rebuilding infrastructures in Aceh province where it was needed as soon as possible. The local government took place the new regulations that allow Padang cement substitute the market area in Aceh province while waiting the

IOP Conf. Series: Materials Science and Engineering **352** (2018) 012010 doi:10.1088/1757-899X/352/1/012010

LCI-Aceh recovered the plant. Nowadays, there are two competitor cement products in retailers in Aceh province.

Serambi Indonesia (2017) supposed that the demand of cement fluctuated base on the infrastructure market demand need. The distributors predict the market based on the retailer's requirement. Nowadays, the infrastructure project of the subsidiary house is rapidly increased to build the community houses. Cement demand is higher than the LCI-Aceh production. The situation creates the distributors highlight the purchase order (PO) from the beginner order of wholesalers and retailers. This condition influences the housing contractors to get the appropriate amount of cement to their project [3].

Moreover, the cement production is influenced by the determinants of cement demand from the infrastructure buildings environmental. This creates the challenges to the producer of cement and their supply chain to fulfill the market demand in Aceh province, especially in Banda Aceh and Aceh Besar districts. The cement demand has the trend of fluctuating increase characteristic. Based on these situations the local business can be the entrepreneurship of distributors and retailers. Hence, the retailers have high competitiveness in the region of their place due to many new growth retailers and the distance relationships among the retailers are a very closed each other. It influences the retailers to get the more revenue.

There are some prediction models to predict the location of the relationship retailer places. In the beginning, it is better to see in overall view of the place for the retailer geographic positions. This study used the GPS to track the point on the real geographic dataset. Then, it was overlaid on the google map to measure the distances. Then, it produced the waypoint before the data plotted on the x-abscissa and y-ordinate graph.

The contribution of this paper is to explore some advantages. They are to produce the performance of numerical spline and polynomials models, estimate the relationship area for the retailers, predict the demand area of customer demand from a retailer in the region of less than 1 km. At last, measure the factor economic order quantity in the case of this characteristic to determine the inventory policy on the retailer level.

2. The Methods

2.1. Data collection

The data are collected by the open-source internet of google map. This method can measure the distance between the points in the driving direction. This method is a very quick result to obtain the retailer's distance the relationship between them and from customer area to the retailers. The raw data were collected by using the GPS Garmin tool before they were overlaid on the paper map. However, this method applies in manually to obtain the distance in the engine map point's record. Furthermore, this method is very simple and be the basic approach to calculate in the paper map sales region. This process has the similar characteristic procedure when using the geographic information system (GIS) method capability to measure the distance of the track point on the paper map. As our objectives here is to obtain the efficiency but still have a precision of it where it is compared to the manual ruler based.

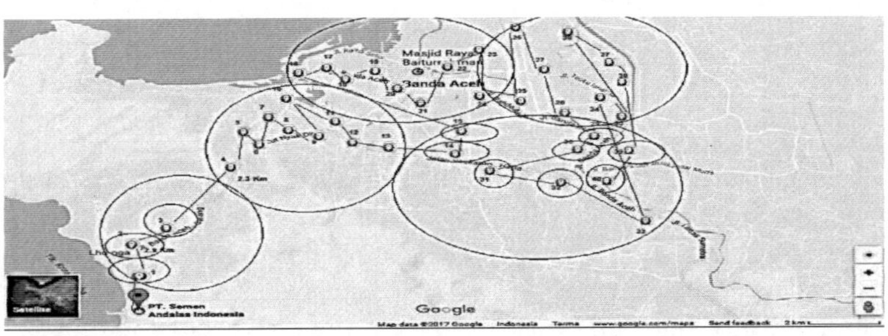

Figure 1. The spread of cement retailers and their customer's area.

IOP Conf. Series: Materials Science and Engineering **352** (2018) 012010 doi:10.1088/1757-899X/352/1/012010

The area for this dataset is a relatively new area to obtain the quality of the dataset. The data are collected from 100 cement retailers that were collected tracking the points manually. Figure 1 shows that the analysis of the retailer area relationship requires the data of the retailer's distance each other and the area of representing one sales region from the retailer. Then the data overlay the distance on the google map and their customer regions around Banda Aceh.

2.2. Spline and Polynomial models

The Spline and polynomial methods are applied due to these methods can work more accuracy than the linear method for the un-linear dataset. The spline can overlay the points with the very small deviation between the real dataset, and the polynomial can do the better than the linear model in the un-linear and dispersion dataset.

Polynomial regression is indeterminate of variables where it involves the coefficients or mathematical functions.

$$Y = \beta_0 + \beta_1 x + \beta_2\, x^2 + \cdots + \beta_r x^r + e \tag{1}$$

Where $\beta_0, \beta_1, \ldots, \beta_r$ are regression coefficients where it would estimate the value. If the dataset consists of n pairs $(x_i, y_i), i = 1, \ldots, n$ then the least squares estimators of $\beta_0, \ldots, \beta_r$ called them $\beta_0, \ldots, \beta_r$ are the values to minimize.

$$\textstyle\sum_{i=1}^{n} (Y_i - B_0 - B_1 x_1 - B_2 x_i^2 - \cdots - B_r x_i^r)^2 \tag{2}$$

To determine these estimators, the partial derivatives concerning $B_0 \ldots B_r$ of the previous sum of squares, and set these equal to 0 to determine the minimizing value on doing so, and then rearranging the answer from the equations, The least square estimators should be $\beta_0, \beta_1, \ldots, \beta_r$ satisfy the following set of $r + 1$ linear equations is named the normal of formula.

The polynomial equation with the single variable is $2x^3 - 2x^2 + 2x - 5$ and with two variables are $2xy^3 - 2xy^2 + 2x - 5$. The polynomial equations represent in many real areas of science, chemistry, physic, economic, statistic, mechanical design, and mathematics. Many problems can be solved to apply with the advanced mathematical function such as algebraic, calculus, geometric, and polynomial. These usually use many variables to identify the real cases [4].

Spline regression is the numerical analysis that is the unique piecewise polynomial. The spline polynomials usually produce the minimal error when it is applied in the best fit on the real dataset. It can avoid the problem of oscillation occur between the points in the interpolating case.

$$S^n(x) = \textstyle\sum_{j=0}^{n} a_j x^j + \sum_{t=1}^{n} d_t (x - \varepsilon_t)_+^n \tag{3}$$

Where, the symbol of a parameter coefficient of independent variables, x is the independent variable, ε is knotted the point, d is the parameter coefficient of the knot, n is the degree of the spline and under the restriction, in Equation (2) as follow:

$$(x - \varepsilon_1)_+^n = \begin{cases} 0, & x \leq \varepsilon_1 \\ x_1 - \varepsilon_1 & x \geq \varepsilon_1 \end{cases} \tag{4}$$

In this logic, the knot points and the ε_1 are known in advance. The parameters of spline a_j and d_t. Spline is linear in this parameter [5].

The goodness of fit for sales data is the fundamental to obtain the best models in statistical methods to realize the behavior relationship between the data and the predicted dataset. It is quite important to understand the underlying behavior market dataset in the entire market environment

Furthermore, the equation from the goodness of fit can be determined by the adding explanatory variables Whether these variables are meaningless to be accurate in the fitting where measured by R^2 statistic [6].

Some problem is using the polynomial equation to predict the sales-distance relationship among them. The reasons are the weak theoretical support, the allowance of negative sales, and non-monotone behavior, unclear elasticity, intractable structural form.

However, some factors are offered by the shop where they have different attributes, such as offer the diversification products, more services, expensive goods, or other factors. The other location of the shop may offer differences factor where it would make affecting sales factor that offers to customers.

This study collects the data from the distance of the cement product in their retailers and consumers. In such case, the data are available for sales from retailers and the factors that distinguish retail outlets, and the predictive model can be drawn as

$$S_i = f(d, m, y) \tag{5}$$

Where S_i shows the sales of the store i in the region, d is a vector of distance variables. This means that the distance between the store i and another store, m is a vector of amnesty variables such as size, hours, quality or price merchandise, and the atmosphere of the area surrounding the retail outlet, and y is a vector of data on the disposal income available to the customer in the area. The income variable comes from the retails outlet from different areas.

The distance in the demand equation is an indicator of time and other costs of traveling to and from the retailer and customer to the retailer. Part of the time cost is the same whether the trip is across the street. If the customer within 1 km and other 5 km, it means that it is indeed closer than the majority is 10 km or more away.

The normal equations (2) continue the calculation to fit a polynomial to determine the necessary degree of a polynomial by a scatter diagram.

$$\sum_{i=1}^{n} Y_i = B_0 n + B_1 \sum_{i=1}^{n} x_i + B_2 \sum_{i=1}^{n} x_i^2 + \cdots + B_r \sum_{i=1}^{n} x_i^r \tag{6}$$

$$\sum_{i=1}^{n} x_i Y_i = B_0 \sum_{i=1}^{n} x_i + B_1 \sum_{i=1}^{n} x_i^2 + B_2 \sum_{i=1}^{n} x_i^3 + \cdots + B_r \sum_{i=1}^{n} x_i^{r+1} \tag{7}$$

$$\sum_{i=1}^{n} x_i^2 Y_i = B_0 \sum_{i=1}^{n} x_i^2 + B_1 \sum_{i=1}^{n} x_i^3 + .. + B_r \sum_{i=1}^{n} x_i^{r+2} \tag{8}$$

$$\sum_{i=1}^{n} x_i^r Y_i = B_0 \sum_{i=1}^{n} x_i^r + B_1 \sum_{i=1}^{n} x_i^{r+1} + .. + B_r \sum_{i=1}^{n} x_i^{2r} \tag{9}$$

The explanation of the polynomial can be emphasized to the lowest possible degree that appears to describe the spread of data. To obtain the degree of n where n pairs (x_i, Y_i), i=1, …, n, it would be hard to fit. The polynomial solution model is noted in Equation (2). Cubic spline interpolation is a good global interpolation; furthermore, it is easy to separate. This can be explored with the function of the spline curve with the computed from [7].

$$f'_{i,i+1}(x) = \frac{k_i}{6}\left[\frac{3(x-x_{i-1})^2}{x_i - x_{i+1}} - (x_i - x_{i+1})\right] - \frac{k_{i+1}}{6}\left[\frac{3(x-x_i)^2}{x_i - x_{i+1}}\right] - k_{i+1}\frac{x-x_i}{x_i - x_{i+1}} \tag{10}$$

$$f'_{i,i+1}(x) = k_i \frac{x - x_{i+1}}{x_i - x_{i+1}} - k_{i+1}\frac{x - x_i}{x_i - x_{i+1}} \tag{11}$$

There are various measures of forecast errors. One of them is a *mean squared error (MSE).* The MSE is the square of quantity the unit term of an equal to the original value, which is the difference between the actual value and an estimated value, it can be defined in Equation 1.

IOP Conf. Series: Materials Science and Engineering **352** (2018) 012010 doi:10.1088/1757-899X/352/1/012010

$$MSE = \frac{\sum_{i=1}^{n}(Y_t-\hat{Y}_t)^2}{n} \tag{12}$$

Where:

Y_t is actual value for a period t,
$\hat{Y}_t$, is a period value of a forecast t, and
n is a period number

Defining market area, the relationship between the sales and distance in a quantitative manner is called the trade areas that are created on the density of sales. It is expected the retail to attract from different geographic areas. It depends on the purpose of analysis. It can be divided into the primary, secondary and tertiary market areas. Primary is the consumer very closest to the retailer and followed by the secondary and tertiary market area. The closest distance of retailer would have the high density of the population in the area. The criteria of the market area can also determine by the AHP method [8].

2.3. Determinant of demand

Demand forecasting is the process of predicting demand need in the future from the sale or historical dataset. This study took the sample from some retailers to obtain the information of the demand in the sales area of their locations. Most of the retailers used the similar system to predict the demand from belong customers and how they keep the safety stock on the certain level in their store or warehouse.

2.4. Determine the economic order quantity

One of the main objectives of this study is to determine the policy through the economic order quantity at the retailer level. In the real case, the retailers have a different system when they keep the inventory in the store and the policy of the retailers. Nevertheless, most of them have the assumption the safety stock on the level of 10% to 25%. The probability of the demand has the fluctuation in each month where it correlates to the stock inventory system in the retailer [9].

In real life situations, the review of inventory is mandatory to do regular intervals to manage the buffer in the store. The certain amount of quantity must be ordered. The review of safety stock is in the policy reviews the situation when the inventory falls under reorder point, where the store or warehouse places an order. In theoretical, this model is named the *s*, *S* policy. Where *s* is the reorder point and *S* is maximum order size. The notation *s*, *S* means that the inventory fall below *s* then the system places an order to bring inventory to *S*. The EOQ formula can be seen in the Equation (13).

$$h = \frac{z\,x\,b}{t} \tag{13}$$

Where z is the safety factor is chosen from the table of probability service level, *b* is periodic time per year. The stock out during lead time is precisely $1 - \alpha$, it means that reorder level in the condition of (14).

$$\Pr\left[D \geq L\,x\,\bar{D} + z\,x\,sd\,x\,\sqrt{L}\right] = 1 - \alpha \tag{14}$$

Where D is demand during lead time, L is lead time, $\bar{D}$= average demand, *sd* is standard deviation and α is service level.

$$Q = \sqrt{\frac{2\,x\,k\,x\,\bar{D}}{h}} \tag{15}$$

Where k is ordering cost, $\bar{D}$ is average demand, h is the holding cost. In addition, the safety stock position is in the Equation of (16).

$$z\ x\ sd\ x\ \sqrt{L} \quad (16)$$

The expected level of inventory immediately after receiving the order is in the Equation (17).

$$Q + z\ x\ sd\ x\ \sqrt{L} \quad (17)$$

Thus, the average inventory level is the average of these two values with Equation (18)

$$\frac{Q}{2} + z\ x\ sd\ x\ \sqrt{L} \quad (18)$$

In the periodic review policy, the base stock level is formed by two variable; they are average demand during an interval of r + L days, where is shown in Equation (19)

$$(r+L)\ x\ \bar{D} \quad (19)$$

In this equation, r is the length of the review period, where the orders are placed every r periods. While the safety stock can be illustrated in Equation (20)

$$\frac{Q}{2} + z\ x\ sd\ x\ \sqrt{r+L} \quad (20)$$

Hence, the expected level of inventory after receiving an order is equal to (21). [10].

$$r\ x\ \bar{D} + z\ x\ sd\ x\ \sqrt{r+L} \quad (21)$$

Moreover, the safety stock is equal to (22).

$$z\ x\ sd\ x\ \sqrt{r+L} \quad (22)$$

$$\frac{r\ x\ \bar{D}}{2} r\ x\ sd\ x\ \sqrt{r+L} \quad (23)$$

The average inventory level is in the (23) has the value where the inventory keeps on the demand average from the customer need.

3. Result and Discussion

3.1. Spline and Polynomial models

Figure 2 is the underlying sales-distance relationship between the retailers in the area for 100 points. The spread of this dataset, it shows that the retailers are mostly building around the high density of population.

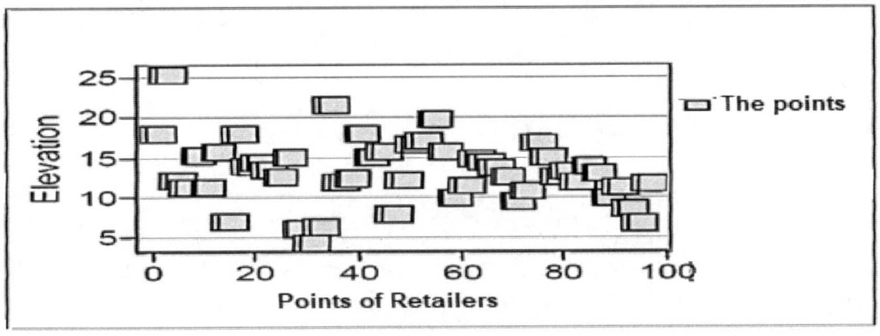

Figure 2. Underlying sales-distance relationship using waypoints of Garmin GPS.

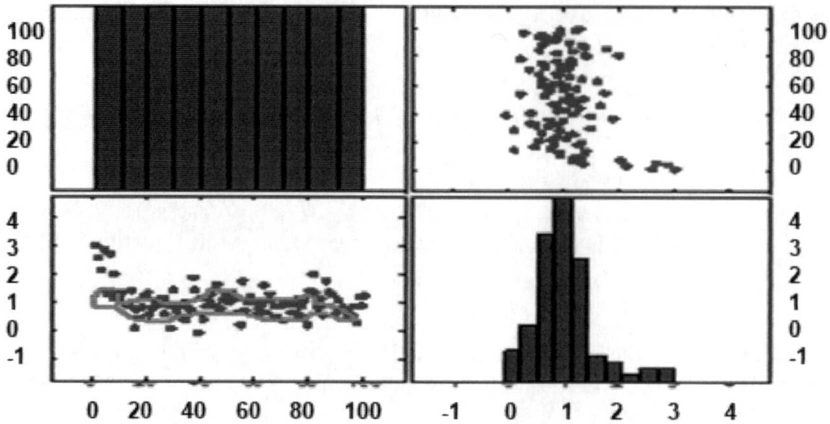

Figure 3. Plot matrix from tracking data point from GPS dataset.

This study develops the spatial dimension to use the Garmin GPS and spatial GIS software to obtain the result of retailer points [11]. The spatial dimension dataset obtains with the Garmin GPS type PSMAP-64s to track the points in the area of cement retailers. Moreover, the dataset from the GPS type was overlaid on the google map to view the points to measures the weight distance between the retailer points with the spline, quadratic, cubic, and four-degree polynomials methods.

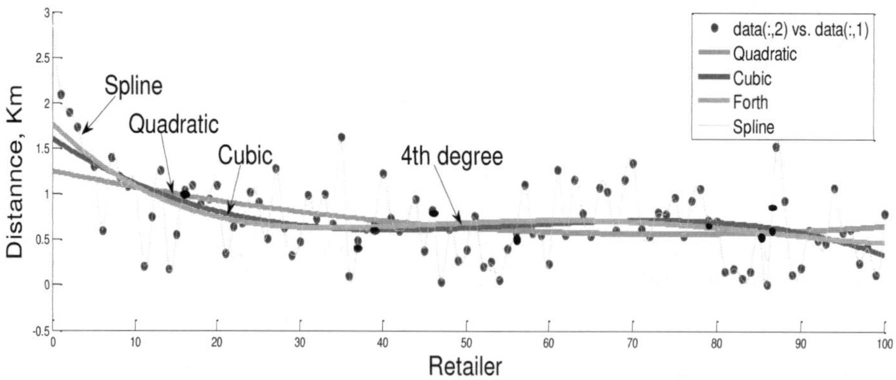

Figure 4. Distance retailer sales relationship in the area with the spline and polynomials.

The spline model fits on the exactly on the raw dataset while other models such as quadratic (24), cubic (25), and four degrees (26) from a distance models between retailers can be calculated:

$$y = 0.00013\,X^{2} - 0.019\,X + 1.2 \quad (24)$$

$$y = -\,6.7e\text{-}06\,X^{3} + 0.0011\,X^{2} - 0.06\,X + 1.6 \quad (25)$$

$$y = 1.1e\text{-}07\,X^{4} - 2.8e\text{-}05X^{3} + 0.0025X^{2} - 0.092X + 1.8 \quad (26)$$

The model of *X*-abscissa for the quadratic, cubic and four degrees have their values itself, and this means that the model has different characteristic when they are transformed into the plot fitting models.

IOP Conf. Series: Materials Science and Engineering **352** (2018) 012010 doi:10.1088/1757-899X/352/1/012010

Table 1. The comparison of MSE in four methods.

MSE	Spline	Polynomial		
		Quadratic	Cubic	Fourth
MSE	0.0000	0.2428	0.1974	0.1744

Moreover, each model should have the different mean squared error. The comparison of mean square error for spline and polynomial methods can be shown in the following table 1. The table indicates that the spline has the best MSE compared to other polynomials, such as in quadratic, cubic and fourth polynomial. The spline does the best work to explain the variations of the dataset. The line goes down at the end indicates that the curve is formed from the closer distribution of the dataset. This line helps to understand the estimate a behavior relationship of data. Getting the excellent model means that the understanding not just one situation, but an entire market. Furthermore, the well-known result from econometrics that the goodness of fit from any equation can be adding more explanatory variables, even if these variables are meaningless.

The interesting for polynomial equations to model the distance sales relationship at the end of the line usually goes down. The concept that good fit is a good thing. The go down to the line depending on the degree of the polynomial pattern and this shows that the polynomial has a weakness or problems to the sales distance, there are some factors where the consumer want to choose the distance to shop.

1. There is no data to identify the behavior of the customer when they choose the shop to buy. Moreover, they compare the low price to other distance.
2. Some consumers want to buy goods in the difference distance due to the service and quality of the goods. It is hard to identify the loyal consumers.
3. The closest distance is just the cost advantage for the customers whether it is still a consideration for the consumer. It means that the distance has a value of time and the costs when going to the retailer for consumers.
4. Cannibalism is the variation of a new product with the low price. This situation can absorb many consumers. These situations make convenience for the customer to get the new series of products but contribute the new problem for retailers.

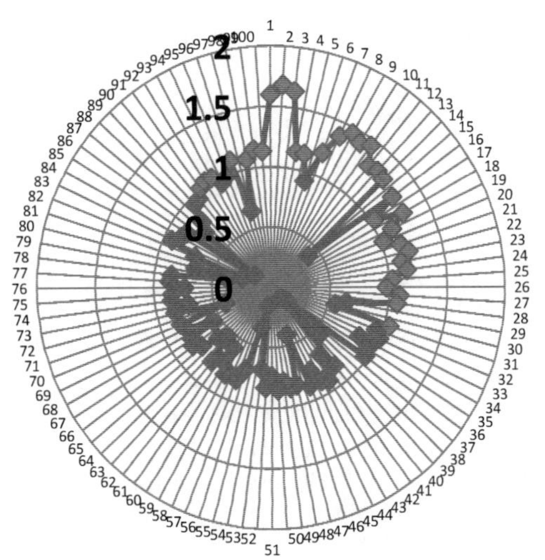

Figure 5. The distribution of the closer dataset from the fourth-degree polynomial.

The distance identifies just for predicting the area of potential customers due to no guaranty that the closest distance between consumers to the retailer will get many buyers. Even though mathematical models are used to predict the distance of retailer to the customer, it just only the hopes of predicting. The minimum and the maximum area of predicting may be a solution to minimize the error of predicting.

Figure 5 shows the minimum distance truncation where it is many residential properties in the area of one km or more of the retail establishment. The overestimate occur when the minimum retail distance of zero of the total sales of the store with a primarily residential customer base, or allow a handful of observations to bias the estimated parameters of the model. The distance sales relationship is the fundamental of customer demand.

The point of retailers who have the closest distance each other where it less than 1 km drawn in the first range. Moreover, the points which outside the first range are the long distance with the low density of population.

3.2 Stock inventory model for retailer

Understanding the stock control in the retailers is mandatory important to maintain the customer need. The study took some samples of the inventory system from the fields. Each retailer has the different policy of stock inventory to fulfill the demand.

It is commonly they keep the safety stock at a low level of percentage in the store for rolling reorder of stock. They also refuse the system that the order in big batches because it is not necessary. If high inventory is savings in the store, it would be high risk for inventory cost in holding cost. Most of the retailer controls the level of inventory in every two weeks with one month's order cost. Some time the retailer order out of schedule to fulfill many customer requirements. This happens when the contractor has many jobs for building the government infrastructure and the house of community. This system is more appropriately used the periodic review policy. Let say the probability condition, if there are 1000 sacks sales per month, means that the retailer will keep the same amount about 1000 sacks. This condition will calculate the average demand is 500 sacks include the safety stock is 100 sacks. Means when the stock fall under the safety stock, and even at zero points the retailer start to reorder min around 150 until 250 sacks per week while the lead time is one week. This means that the average level of demand reduces to be around 125 sack per weeks with the safety stock is 25 sacks per week. The retailer can control the store more economic in a week or 2 weeks than they must order and keep in big baths. This is a significant reduction in stock and cost without any effect on customer service. The probability demand must be quite considered by the retailer to obtain revenue.

In the probability situation, the scheduling of order is not appropriate to do so, but based on the short demand requirement control can obtain the advantage to the retailer.

3.3 EOQ retailer model

Local retailers in this area have the characteristic of the probabilistic EOQ system. Even though, in practise, there are no certain rules how to determine the order quantity in the inventory system in the supply chain.

There are some policies to determine the condition of stocks, periodic review policy and continues order policy at the across of the supply chain. In the periodic review policy, the system of the inventory model considers on bath size, Q, where to keep the model equal to the cost of holding stock and the cost of ordering it. The retailer just keeps the based stock condition when the stock falls into the safety stock, and they start to reorder. Generally, in this area, they keep the safety stock around 10%-25% to reorder of stock. The result of the EOQ result can be seen in figure 6.

IOP Conf. Series: Materials Science and Engineering **352** (2018) 012010 doi:10.1088/1757-899X/352/1/012010

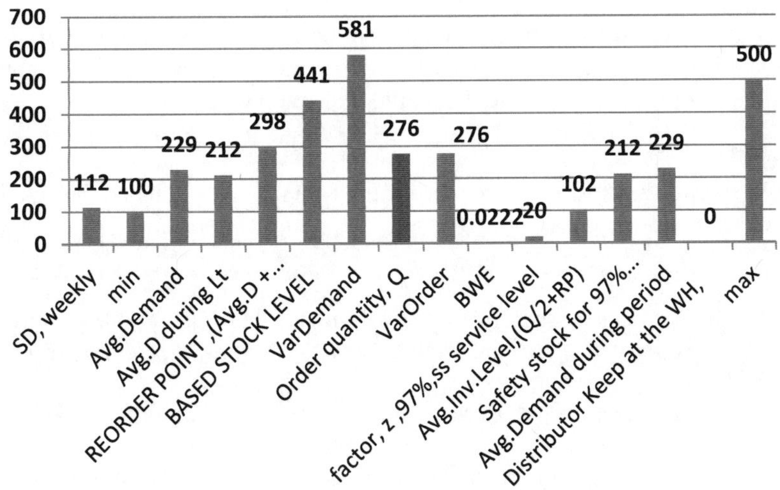

Figure 6. The stock inventory system in the retailer.

The figure 6 illustrates that the demand is 500 sacks per 2 weeks but have the variation demand in probabilistic with the max 581 sacks, and average demand is 229 sacks, from 500 sacks of demand. Average during lead-time is 212; reorder point is 298, the base stock level is 441. Order quantity (EOQ) is 276, variance order 276 sacks; the starting of the bullwhip effect is 0.0222. The calculation uses the z factor 97% service level (z = 1.88), and the safety stock for this level is 212 (it is around 45%),. Average inventory level 102, the average demand during the period is 229 sacks [12].

The safety stock indicates the high because influenced by the variance demand of high fluctuation. Then the retailer keeps more the safety stock to anticipate the unpredicted demand from consumers.

4. Conclusion

This study is the initial identification of the retailer's competitiveness based on the distance relationship. The models are the distance relationship for cement retailer in the Banda Aceh province of Indonesia. It concludes that the retailers have the strong competitiveness in the closest distance between retailers, means that the locations have many populations in the area. The mathematical models of spline and polynomials can be used to compare the accuracy deviation with the MSE of the retailer based on the distance. These models can be applied for identification; the new location for the new retailer's business with the assumption is the closest relationships have the high density of population. Other assumptions can be determined to obtain the good location-based on infrastructure activities. The MSE from the four methods indicates that spline is outperformed than the others where it is followed by the fourth, cubic and quadratic polynomial. The characteristic of models influences the inventory policy for the retailer. This policy influence the economic order quantity of retailer where from this study has the value of 276 sacks with the safety stock in 97% service level (z = 1.88) has the value of 212 (42%) to keep the inventory for the unpredicted high demand. The policy uses the periodic review policy. The demand fluctuates with the maximum high range 581 sacks. However, this study concludes that the closer of retailers indicate that the area has the high population and many infrastructure projects in that area. The locations are promising the good condition of investment to the business of retailer. Recently, the retailers could get the more revenue on the subsidized housing project in the area. Moreover, the retailers must manage the inventory policy management on the right track to obtain the maximum revenue from their supply chain; otherwise, it will be lost the opportunity to the retailers to get the more revenue. The additional conclusion that the marketing strategy must be applied refers to the situation to make more advantage to the retailer.

Future Work

The suggestion for the future work is trying to develop more models of inventory in the cannibalism retailers, identify the behavior of the customer when they choose the shop to buy, identify the variety of services quality, investment strategic and the marketing strategic, etc.

Acknowledgement

The authors would like to thank the research assistants to track the points of the retailer in the area with the Garmin GPS.

References

[1] E.Fradinata, S. Suthummanon, and W. Suntiamorntut, "Forecasting Determinant of Cement Demand in Indonesia with Artificial Neural Network," *Journal of Asian Scientific Research,* vol. 5, p. 373, 2015.
[2] I. N. Pujawan, "The effect of lot sizing rules on order variability," *European Journal of Operational Research,* vol. 159, pp. 617-635, 2004.
[3] http://aceh.tribunnews.com/2017/08/26/suplai-semen-andalas-mulai-lancar, colleced on 23 sept 2017
[4] B. Santosa, *Data mining teknik pemanfaatan data untuk keperluan bisnis* vol. 978, 2007.
[5] W. Gnad, "Regression by spline functions," *Empirical Economics,* vol. 2, pp. 69-77, 1977.
[6] E. Fradinata, S. Suthummanon, N. Sirivongpaisal, and W. Suntiamorntuthq, "ANN, ARIMA and MA timeseries model for forecasting in cement manufacturing industry: Case study at lafarge cement Indonesia—Aceh," in *Advanced Informatics: Concept, Theory and Application (ICAICTA), 2014 International Conference of,* 2014, pp. 39-44.
[7] V. F. Jones, "A polynomial invariant for knots via von Neumann algebras," *Bulletin of the American Mathematical Society,* vol. 12, pp. 103-111, 1985.
[8] U. Ciptomulyono, "Fuzzy goal programming approach for deriving priority weights in the analytical hierarchy process (AHP) method," *Journal of Applied Sciences Research,* vol. 4, pp. 171-177, 2008.
[9] E. Fradinata, "Stock Market Prediction with Neural Network Method," in *ASEAN/Asian Academic Society International Conference Proceeding Series*, 2012.
[10] D. Simchi-Levi, E. Simchi-Levi, and P. Kaminsky, *Designing and managing the supply chain: Concepts, strategies, and cases*: McGraw-Hill New York, 1999.
[11] E. Murakami and D. P. Wagner, "Can using global positioning system (GPS) improve trip reporting?," *Transportation research part c: emerging technologies,* vol. 7, pp. 149-165, 1999.
[12] E. Fradinata, S. Suthummanon, and W. Suntiamorntut, "Reducing the bullwhip effect from signal demand of hybrid artificial neural network models of supply chain in Indonesia," *International journal of advanced and applied sciences,* vol. 4, pp. 64-75, 2017.

Investigation of turbine ventilator performance after added wind cup for room exhaust air applications

D Harun, Zulfadhli, H Akhyar
Department of Mechanical Engineering, Syiah Kuala University
Jln. Syech Abdurrauf No.7 Darussalam – Banda Aceh, 23111, Indonesia

E-mail: darwinmtir@unsyiah.ac.id, akhyar@unsyiah.ac.id

Abstract. The turbine ventilator is a wind turbine with a vertical axis that has a combined function of the wind turbine and a suction fan. In this study, the turbine ventilator modified by adding a wind cup on the top (cap) turbine ventilator. The purpose of this experiment is to investigated the effect of the addition of wind cup on the turbine ventilator. Turbine ventilator used is type v30 and wind cup with diameter 77 mm. The experiment was conducted using a triangular pentagon model space chamber which was cut off to place the ventilator turbine ventilation cup with a volume of 0.983 m^3 (equivalent to 1 mm^3). The results of this study indicate that at an average wind speed of 1.8 m/s, the rotation of the turbine produced without a wind cup is 60.6 rpm while with the addition of a wind cup in the turbine ventilator is 69 rpm. The average increase of rotation turbine after added win cup is 8.4 rpm and the efficiency improvement of turbine ventilator is 1.7 %.

1. Introduction

Turbine ventilator is widely used in the roof of factories as additional ventilation. Installation of turbine ventilator is beneficial to improve room air quality, furthermore to reduce the usage of air conditioning (AC). The turbine ventilator is one of equipment to dispose of exhaust airflow through negative pressure down-stream end in the air by using turbine rotating. The efficiency of turbine ventilators has been investigated by varied of three sizes with no inner vanes, such as 6", 14" and 20". One Ventilator with 0.5 m has inner vanes [1]. The result shows the better ventilators with inner vanes. Turbine ventilator is very helpful to improve ventilation for building and factory ventilation.

The prototype of rooftop turbine ventilator hybrid by wind and photovoltaic energy has been developed and investigated. The rotation speed recommended is 1500 rpm best performance to add the inner fan at turbine ventilator [2]. The influence of inclined roof can extend safety limits of turbine ventilator by reducing wind force applied to the ventilator [3]. Turbine ventilator was used to cool down the photovoltaic (PV) cell. The result can be seen the performance of combination between wind-driven and turbine ventilator can increase the output power PV cell up to 46.54 % [4]. The testing of flow rates conducted to four commercial turbine ventilators have been studied. Different wind speed affected by rotational speed and ventilation flow rates, it is has tested by comparing open column and two vent hats. It was found that the throat diameter, blade configuration and construction material of the ventilator impact on the ventilator's capacity to induce flow [5].

This experiment aims to investigate the performance (such as wind velocity, temperature, humidity, and rotation) of turbine ventilator before and after added wind cup.

2. Materials and Methodology

The parameters that influencing the performance of the turbine ventilator are the rotation produced from the ventilator turbine (*n*), the velocity of free stream fluid ($U\infty$), the fluid density (ρ), and the diameter of the funnel pipe of turbine ventilator (*D*). Data collected in some locations and the experiment scheme as shown in figure 1. This study uses a fan with three variations of distance, resulting in different wind speeds. The fan distance is 1, 2, and 3 meters from the turbine ventilator that has been added to the wind cup. Temperature and humidity data of the chamber are recorded in the area indicated by the red dot in figure 1. The rotation data is obtained by measuring the turbine ventilator of the cause by the fan with the distance variation (red dot point).

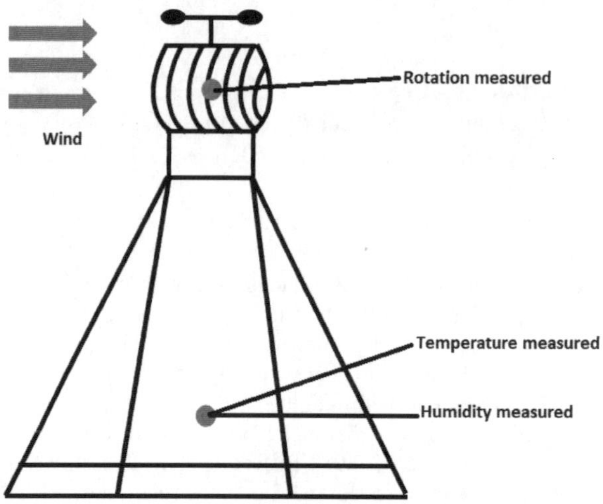

Figure 1. Experimental scheme and location measured of static pressure at the tested model.

The efficiency of turbine ventilator can be calculated by equation (1) [6].

$$\eta = \frac{n}{v} x100\% \quad (1)$$

where the efficiencies of turbine ventilator (η), rotation of turbine ventilator, *rpm* (*n*), wind velocity, *m/s* (v).

In this study, the turbine ventilator is modified by adding a wind cup on the top of the turbine ventilator that connected by a shaft. The wind cup can increase the speed rotation of turbine ventilator which serves to absorb the existing hot air in the room. The scheme of turbine ventilator combines with wind cup that modified, and the tested chamber was modeled was developed in this experiment. The model tested chamber consisted of wall glass, aluminum frame and floor chamber of the aluminum plate. A fan used for wind speed that was varied with three-speed variations in this experiment.

The testing stages are carried out by preparing the tested room model, the temperature and humidity of the test model chamber were measured and recorded. Install the turbine ventilator with wind cup in the test model chamber. The effect of modification turbine ventilator with adding wind cup is analyzed by a previous comparative study (standard turbine ventilator). The experiment apparatus after modified as shown in figure 2, where wind cup (1), horizontal shaft or with three cups (2), turbine ventilator cap (3), turbine shaft/SS (4), Teflon SS bearings (5), and Neck of turbine ventilator (6).

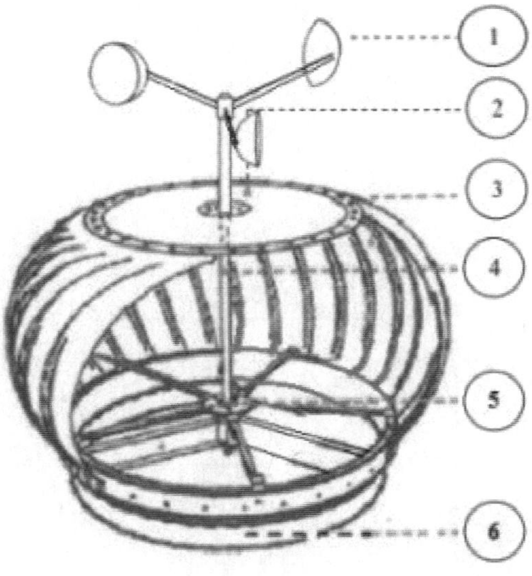

(a)

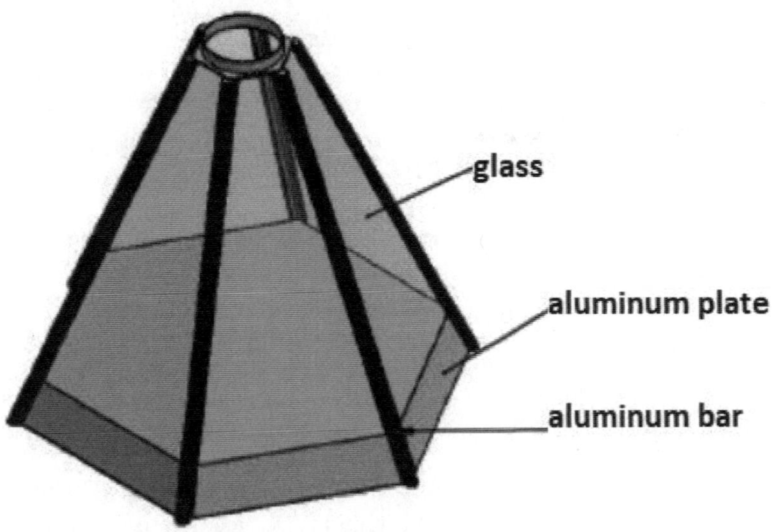

(b)

Figure 2. (a) turbine ventilator modified by adding wind cup, (b) model of the tested chamber.

3. Results and Discussions

Based on the three variations of wind speed that generated by the fan, the results of the measurements obtained in this study are shown in table 1. The results show the rotation of turbine ventilator after modified by wind cup after that three variations of fan distance and then comparing turbine ventilator without wind cup added was conducted in this experiment. Three fan distances are 2, 1.5 and 1 meters for each turbine ventilator.

Table 1. The turbine ventilator with and without wind cup varied by the distance of fan.

No.	Fan Distance (m)	Wind velocity (Km/h)	Temperature (°C)	Humidity (%)	Rotation (rpm)	Turbine ventilator (With and without wind cup)
1	2	1.8	30.3	55.3	60.6	None
2	2	1.8	29.9	62.6	69	Wind cup
3	1.5	2.1	30.2	57.6	67.6	None
4	1.5	2.1	29.6	64.3	84.3	Wind cup
5	1	3.5	30	58.6	91.3	None
6	1	3.5	29.3	64.6	108	Wind cup

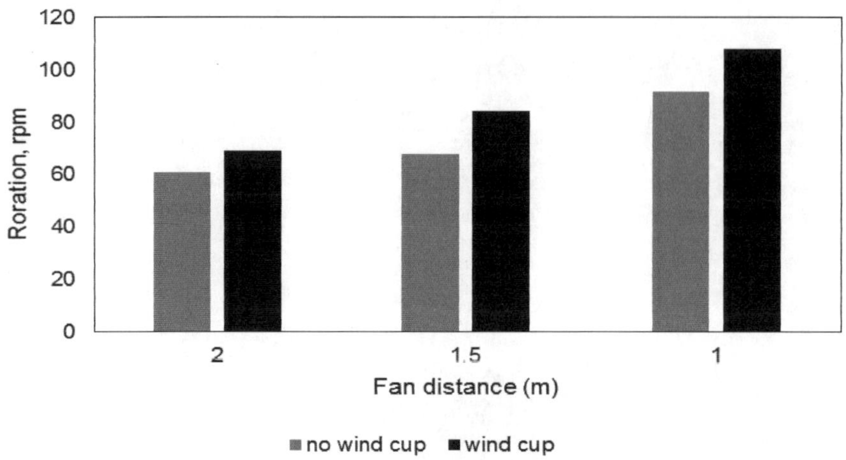

Figure 3. The rotations of turbine ventilator before and after adding wind cup.

Figure 3 shows the rotation of turbine ventilator comparing with the modified turbine ventilator with added wind cup. Fan distance at 2 meters, the rotation index of turbine ventilator modified by wind cup is increasing when compare with turbine ventilator standard. The increasing rotation value is 8.4 rpm, and average wind speed was constant at 1.8 m/s.

The rotation number of turbine ventilator modified increases when adding the wind cup; it shows a growing from 67.6 to 84.3 rpm when fan position is brought close to the turbine ventilator at 1.5 meters. The average wind speed is 2.1 m/s. Subsequently, when fan distance at 1 meter, the rotation of turbine ventilator standard is 91.3 rpm, it increases after modified by wind cup at 108 rpm. The average of wind speed is 3.5 m/s.

Temperature range from 29.3 to 30.3 °C, it is recorded from the model of tested chamber. The temperature declines from 0.4 to 0.7 °C and humidity reduced from 7.3 to 6 % when fan distance brought closer. The temperature ambient of Banda Aceh-Indonesia is usually increased in the morning and reached the maximum at noon, and then slightly decreased in the afternoon [6,7]. The range of ambient temperature about 29 °C until the maximum at 33 °C. The data collected from drying apparatus on the investigation for drying bilimbi and cocoa.

The turbine ventilator standard and modified illustrate the performance of wind speed increase when the fan distance brought closer to the turbine. The increasing of wind speed value causes the raising rotation of both turbine ventilator (standard and modified). So as an overall it can be concluded that the efficiency of turbine ventilator standard is 12.6 % according to equation (1), while the turbine ventilator after modified by win cup can increase the efficiency until 14.3 %.

4. Conclusions

The addition of wind cups on the turbine ventilator can be taken several conclusions:

1. The efficiency of turbine ventilator is 1.7 % after adding the wind cup.
2. The turbine ventilator with added wind cup can significantly increase the wind speed and the rotation.
3. The turbine ventilator modified can slightly reduce room temperature and humidity.
4. With the addition of wind cups on the turbine, the ventilator can increase the efficiency of the turbine spin of the ventilator.

Acknowledgment

This publication was made possible by a Grand from the 2017 PNBP of Syiah Kuala University. The authors would like to thank Sayed Rizky Al-Idrus for his support through this research.

References

[1] Chi-ming L 2013 Experiments on the ventilation efficiency of turbine ventilators used for building and factory ventilation *Energy and Buildings* **35** 927–932

[2] Chi-Ming L 2006 Prototype development of the rooftop turbine ventilator powered by hybrid wind and photovoltaic energy *Energy and Buildings* **38** 174–180

[3] Shao Ting J L, Noor A A 2011 Effect of inclined roof on the airflow associated with a wind driven turbine ventilator *Energy and Buildings* **43** 358–365

[4] Peyvand V, Masoud R, Amin P, Mohammad M M 2014 Using a wind-driven ventilator to enhance a photovoltaic cell power generation *Energy and Buildings* **73** 115–119

[5] Naghman K, Yuehong S, Saffa B R, Colin B 2008 Performance testing and comparison of turbine ventilators *Renewable Energy* **33** 2441–2447

[6] Darwin H, Muhammad I M, Hasan A, Husaini 2016 Experimental investigation on open sun-drying and solar drying system of bilimbi *6th International Annual Engineering Seminar (InAES)* 271-275

[7] Darwin H, Hasan A, Razali T 2016 Performance investigation and development of solar dryer tunnel type apparatus of Cocoa" *Proc. of International Conference on Engineering and Science for Research and Development (ICESReD* (Banda Aceh, Indonesia)

Assessment of modularity architecture for recovery process of electric vehicle in supporting sustainable design

D K Baroroh and D Alfiah
Department of Mechanical and Industrial Engineering, Universitas Gadjah Mada Grafika 2, Yogyakarta, Indonesia

E-mail: dawi.karomati.b@mail.ugm.ac.id

Abstract. The electric vehicle is one of the innovations to reduce the pollution of the vehicle. Nevertheless, it still has a problem, especially for disposal stage. In supporting product design and development strategy, which is the idea of sustainable design or problem solving of disposal stage, assessment of modularity architecture from electric vehicle in recovery process needs to be done. This research used Design Structure Matrix (DSM) approach to deciding interaction of components and assessment of modularity architecture using the calculation of value from 3 variables, namely Module Independence (MI), Module Similarity (MS), and Modularity for End of Life Stage (MEOL). The result of this research shows that existing design of electric vehicles has the architectural design which has a high value of modularity for recovery process on disposal stage. Accordingly, so it can be reused and recycled in component level or module without disassembly process to support the product that is environmentally friendly (sustainable design) and able reduce disassembly cost.

1. Introduction

The rising discussion of issues related to the concept of environmentally-sound industry encourages companies to improve environmental performance [1]. Moreover, the government provides strict regulations related to environmental issues so that the manufacturing industry is also required to comply with environmental regulations and produce environmentally friendly products [2–5]. Electric vehicles are made for purposes to reduce environment pollution by replacing the use of fossil energy into electrical energy. It does not produce exhausted gas and does have an impact on reducing environmental damage, but when viewed from the use of materials and recovery process, it does not necessarily have a good impact on the environment when the product has entered the disposal stage (cannot be used). Products that have complex architectures when they enter the disposal stage will have difficulty during disassembly processes, individual parts of the product will be difficult to reuse and may increase the likelihood of damage to parts [6]. Damaged and non-reusable products will then become waste which will have a negative impact on the environment. The complex architecture of the product recovery process will lead to difficulties in the disassembly stage, individual parts of the product will be difficult to reuse and may increase the likelihood of damage to parts [5,7]. Modular product design becomes one of the solutions to overcome difficulties during the disassembly process. The modular product can be the most critical factor in determining the preparation of product architecture [8,9]. Most of the design methods for modularity of product architecture from previous studies only consider the physical structure and function of the product architecture to improve the effectiveness of the manufacturing process. The backward approach is to sequence the disassembly

IOP Conf. Series: Materials Science and Engineering **352** (2018) 012012 doi:10.1088/1757-899X/352/1/012012

sequence of the disk drive product and then reassemble the modules which are considered optimal (Disassembly Sequence Structure Matrix) [10]. Other studies have added considerations to the purchasing function [11], from the life cycle aspect, manufacturing costs and interactions between components using Liason Network [12]. However, the use of such methods still has limitations in its ability to identify complex system elements such as in electric vehicle products. Approaches to the modular design of product architecture have been widely implemented, but the most thing to consider is only regarding function or physical relationship of product architecture [13,14].

2. Method

The object of this study used the product of electric vehicle X which has entered the disposal and recovery stage, shown in figure 1. This research was done by following the steps as follows:

a. Data collection. Data about components or parts of products, materials used, lifespan, and functions of each component.
b. Product Architecture Analysis. The linkages between the components to one another are represented using the design structure matrix (DSM) [15,16].
c. Identify Modular Drivers. This grouping stage is based on information on (a) the complexity of the relationship (categorized) according to table 1. Its value is calculated using (1) and table 2). (b) the material similarity (value follows (2)), and (c) the lifespan of the component (its value refers to (3)).

Figure 1. Product Design of Electric Vehicle X

Table 1. Definition of Interface Type [17]

Interface Type	Definition
Attachment (A)	Structural connections between two components that require a type of connector
Spatial (S)	Geometrical and locational constraints of a components
Transfer (T)	Flow of materials or power between components
Control & Communication (C)	Relationship of signal or information flow is communicated or controlled by another component
Field (F)	Interaction between two components in which one component can generate heat, vibration, or magnetic field
Power (P)	Electrical connection between two components unlike communications and control interfaces

$$\omega MCS = \begin{matrix} [\omega_i \text{ x } (\# \text{ of } 1's \text{ x } 1)] \\ + [\omega_i \text{ x } (\# \text{ of } 2's \text{ x } 2)] \\ + [\omega_i \text{ x } (\# \text{ of } 3's \text{ x } 3)] \\ + [\omega_i \text{ x } (\# \text{ of } 4's \text{ x } 4)] \\ + [\omega_i \text{ x } (\# \text{ of } 5's \text{ x } 5)] \\ + [\omega_i \text{ x } (\# \text{ of } 6's \text{ x } 6)] \end{matrix} \quad (1)$$

with ω_i is weight value for each type of interface based on table below.

Table 2. Weight Value of Interface Type [17]

Interface	Weight	Interface	Weight	Interface	Weight	Interface	Weight
A	2.974	SC	21.435	ATC	64.025	PCF	76.142
S	1.795	SF	22.843	ATF	66.137	TCP	75.598
P	6.832	SP	17.253	ATP	57.752	TFP	77.710
C	8.923	TC	36.735	ACF	64.568	ASTC	92.546
T	9.445	TF	38.143	ACP	56.184	ASTF	95.362
F	9.627	PF	32.916	AFP	58.296	ASTP	84.182
AS	9.538	CF	37.098	STC	60.488	ATCF	123.872
AT	24.838	PC	31.508	STF	62.600	ATCP	112.692
AC	23.793	AST	42.642	STP	54.215	ACFP	113.418
AF	25.201	ASC	41.074	SCF	61.032	ASTCF	163.815
AP	19.611	ASF	43.186	SCP	52.647	ASTCP	149.841
ST	22.480	ASP	34.801	SFP	54.759	ASTCFP	237.568

$$\omega(v_i, v_j) = \begin{cases} 1 & \text{if the same material is used in ith and jth components} \\ 0.5 & \text{if similar materials are used} \\ 0.3 & \text{else} \end{cases} \quad (2)$$

$$\omega(v_i, v_j) = \frac{\text{min lifespan }(v_i, v_j)}{\text{max lifespan }(v_i, v_j)} \quad (3)$$

d. Identification of Modular Architecture. After obtaining modules based on a modular driver then identify existing modular architecture of the electric vehicle X.
e. Evaluating Modular Architecture. Perform measurement Module Independence (MI) and Module Similarity (MS). Evaluation category of Modularity Architecture (MA) is based on MI and MS values, which is calculated using (4)-(7) and the category of MA shown in fig. 2.

$$MI = \frac{\sum_{i=1}^{N} \text{Intra-module interface complexity}_i}{\text{Total interface complexity}} \quad (4)$$

$$MS = \frac{\sum_{i=1}^{N} \text{Max}\,(S_{i,j})}{N} \quad (5)$$

$$S_{i,j} = \left[\frac{S_{i,1}}{n}, \frac{S_{i,2}}{n}, \dots, \frac{S_{i,j}}{n}\right], \sum_{j=1}^{n} S_{i,j} \quad (6)$$

$$MA = \begin{cases} A, & \text{if } MS \geq \theta_1, MI \geq \theta_2 \\ B, & \text{if } MS \geq \theta_1, MI < \theta_2 \\ C, & \text{if } MS < \theta_1, MI \geq \theta_2 \\ D, & \text{if } MS < \theta_1, MI < \theta_2 \end{cases} \quad (7)$$

Value of Θ_1 and Θ_2 as the constant of the decision limits of the product architecture category is 0.5.

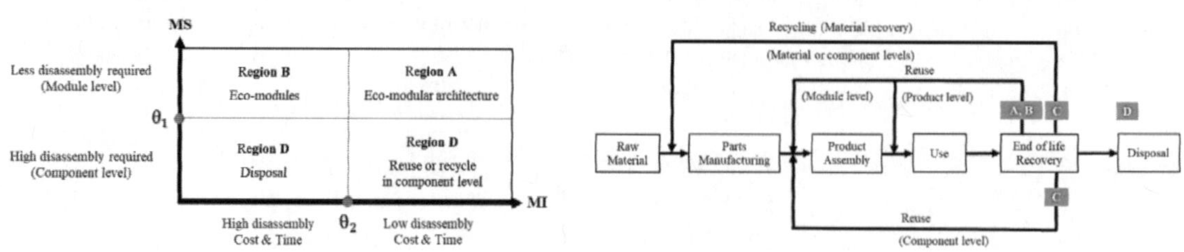

(a) (b)

Figure 2. Category of Modularity Architecture (MA) [6]

IOP Conf. Series: Materials Science and Engineering **352** (2018) 012012 doi:10.1088/1757-899X/352/1/012012

3. Results and Discussions

DSM is used to represent inter-relationship between components. Based on DSM built adjacency matrix, then generate the relationship between module and component. In the electric vehicle models analyzed, there are 102 components (shown in table 3) grouped into 9 modules or clusters. Its Namely (a) Frame, (b) Front Left and Front Right Wheel, (c) Brake System, (d) Steering, (e) Seat, (f) Drive Train, (g) Electronic System, (h) Rear Left and Right Back, and (i) Body.

Table 3. Component List of Electric Vehicle X

No.	Part Name	No.	Part Name	No.	Part Name	No.	Part Name
1	Frame	27	Tire right	53	Adaptor sprocket diff	79	Single row ball bearing right
2	a-arm upper left	28	Shaft	54	Chain	80	Rim wheel left
3	a-arm lower left	29	Disk brake left	55	Shaft	81	Rim wheel right
4	a-arm upper right	30	Brake caliper left	56	a-arm upper left	82	Tire left
5	a-arm lower right	31	Disk brake right	57	a-arm lower left	83	Tire right
6	Bushing left	32	Brake caliper right	58	a-arm upper right	84	Disk brake left
7	Bushing right	33	Pedals gas	59	a-arm lower right	85	Brake caliper left
8	Push rod left	34	Pedals brake	60	Bushing left	86	Disk brake right
9	Push rod right	35	Master rem	61	Bushing right	87	Brake caliper right
10	Rocker arm left	36	Wire brake	62	Spacer left	88	Battery pack
11	Rocker arm right	37	Rack	63	Spacer right	89	Shutdown circuit
12	Absorber left	38	Pinion	64	Push rod left	90	Sensor
13	Spring left	39	Housing rack & pinion	65	Push rod right	91	Controller motor
14	Absorber right	40	Tie rod left	66	Rocker arm left	92	Vehicle control
15	Spring right	41	Tie rod right	67	Rocker arm right	93	Dash board
16	Knuckle left	42	Bushing	68	Absorber left	94	DC to DC converter volt
17	Knuckle right	43	Steering rod	69	Spring left	95	MCB
18	Bushing knuckle left	44	Rolling bearing	70	Absorber right	96	Contactor
19	Bushing knuckle right	45	Housing bearing	71	Spring right	97	Telemetric antenna receiver
20	Spacer left	46	Steering wheel	72	Tie rod left	98	Telemetric antenna deliver
21	Spacer right	47	Seat	73	Tie rod right	99	Front body
22	Rolling bearing left	48	Motor brushless	74	Bushing left	100	Rear body
23	Rolling bearing right	49	Adaptor sprocket diff	75	Bushing right	101	Left side
24	Rim wheel left	50	Sprocket up	76	Knuckle left	102	Right side
25	Rim wheel right	51	Sprocket low	77	Knuckle right		
26	Tire left	52	Differential gears	78	Single row ball bearing left		

Table 4. Interface Matrix Steering

Part	41	42	43	44	45
41		S(2)			
42	S(2)				
43				S	
44			S		S
45				S	

Table 5. Adjacency Matrix Interface Complexity Steering

Part	41	42	43	44	45
41		3.590			
42	S(2)				
43				1.795	
44			1.795		1.795
45				1.795	

Table 6. Adjacency Matrix Similarity Material Steering

Part	41	42	43	44	45
41		1.00			
42	1.00				
43				0.50	
44			0.50		0.30
45				0.30	

Table 7. Adjacency Matrix Similarity Lifespan Steering

Part	41	42	43	44	45
41		0.188			
42	0.188				
43				0.188	
44			0.188		0.250
45				0.250	

For example, the interface form of the matrix in the steering module can be seen in table 4, while the adjacency matrix of the steering module for complexity, material similarity and lifespan can be seen in tables 5, 6 and 7 respectively.

Table 8. Modularity Architecture on Existing Design

Architecture	MI	MS Material	MS Lifespan	M_{EOL}
Existing Design	0.685	0.718	0.696	0.700

Based on table 8, the value is then plotted into four regions based on the Module Independence (MI) and Module Similarity (MS) values as shown in fig. 3. The decision point of modular architecture category is determined based on two thresholds (Θ_1, Θ_2) = (0.5, 0.5). The threshold value is flexible depending on the design strategy including customer demand for product or module recovery, the quality of the recovery product, or the cost of recovery and profit [15]. Therefore, based on fig. 3, the existing design of electric vehicle X has been classified Eco-Modular architecture, it has seen from the MI and MS values of the material and lifespan that is above 0.5.

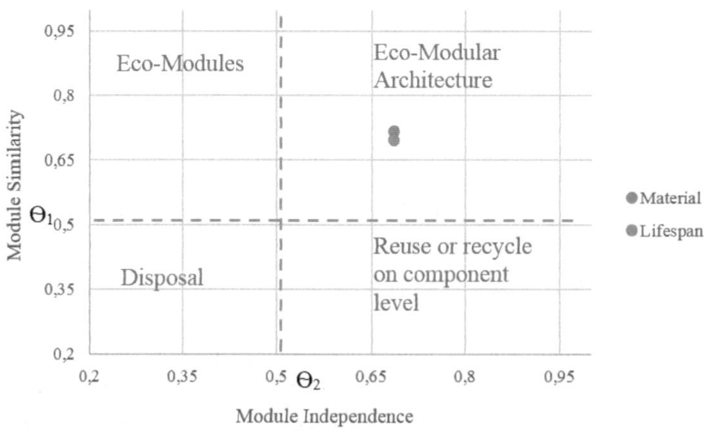

Figure 3. Category of Modularity Product Architecture

Based on the classification of architectural type of existing design of electric, vehicle X consists of 9 modules in one product. It can be said that the design can be recycled and reused without further disassembly process at component level or module. The value of modularity product architecture at the end of life or M_{EOL} stage also shows that the architectural design of electric vehicle X product is relatively high at 0.7. This value is obtained because each value of MI, $MS_{Material}$ and $MS_{Lifespan}$ is above 0.5. The highest value contribution is obtained from the value of $MS_{Material}$ and $MS_{Lifespan}$ whose equation contains the variable N or the number of modules in the product. The number of modules in the product in the calculation of Modular Similarity both material and lifespan becomes the divisor in the calculation to enable an effect on the high value of M_{EOL}.

4. Conclusion

The modular design of the architecture in electric vehicles X has been classified as an eco-modular architecture with an M_{EOL} value of 0.7 so that for recovery process in the disposal stage of the product the electric vehicle can be reused and recycled without further disassembly processes at the component level or modules so as to support sustainable design, and able to reduce the cost of disassembly.

Acknowledgment

The research was funded by the Grant Department of Mechanical and Industrial Engineering, Faculty of Engineering Universitas Gadjah Mada No. 858 / H1. 17 / TMI / LK / 2017.

References

[1] Zhu Q, Sarkis J and Lai K 2007 Green supply chain management: pressures, practices and performance within the Chinese automobile industry *J. Clean. Prod.* **15** 1041–52
[2] Ramani K, Ramanujan D, Bernstein W Z, Zhao F, Sutherland J, Handwerker C, Choi J-K, Kim H and Thurston D 2010 Integrated Sustainable Life Cycle Design: A Review *J. Mech. Des.* **132** 91004
[3] Ilgin M A and Gupta S M 2010 Environmentally conscious manufacturing and product recovery (ECMPRO): A review of the state of the art *J. Environ. Manage.* **91** 563–91
[4] Jeong M G, Morrison J R and Suh H W 2015 Approximate Life Cycle Assessment via Case-Based Reasoning for Eco-Design *IEEE Trans. Autom. Sci. Eng.* **12** 716–28
[5] Ljungberg L Y 2007 Materials selection and design for development of sustainable products *Mater. Des.* **28** 466–79
[6] Kim S, Baek J W, Moon S K and Jeon S M 2015 A New Approach for Product Design by Integrating Assembly and Disassembly Sequence Structure Planning *Proc. 18th Asia Pacific Symp. Intell. Evol. Syst. Vol. 1* **1** 247-257
[7] Dahmus J B, Gonzalez-Zugasti J P and Otto K N 2001 Modular product architecture *Des. Stud.* **22** 409–24
[8] Kristianto Y and Helo P 2015 Reprint of "Product architecture modularity implications for operations economy of green supply chains" *Transp. Res. Part E Logist. Transp. Rev.* **74** 63–80
[9] Ulrich K T and Eppinger S D 2012 *Product Design and Development* **384**
[10] Smith S, Smith G and Chen W H 2012 Disassembly sequence structure graphs: An optimal approach for multiple-target selective disassembly sequence planning *Adv. Eng. Informatics* **26** 306–16
[11] Fernández I and Kekäle T 2005 The influence of modularity and industry clockspeed on reverse logistics strategy: Implications for the purchasing function *J. Purch. Supply Manag.* **11** 193–205
[12] Shoval S, Qiao L, Efatmaneshnik M and Ryan M 2016 Dynamic Modular Architecture for Product Lifecycle *Procedia CIRP* **48** 271–6
[13] Yu T-L, Yassine A a. and Goldberg D E 2007 An information theoretic method for developing modular architectures using genetic algorithms *Res. Eng. Des.* **18** 91–109
[14] Du X, Jiao J and Tseng M M 2000 Architecture of product family for mass customization *Proceedings of the 2000 IEEE International Conference on Management of Innovation and Technology* **1** 437–43
[15] Kim S and Moon S K 2016 Eco-modular product architecture identification and assessment for product recovery *J. Intell. Manuf.* 1–21
[16] Helmer R, Yassine A and Meier C 2010 Systematic module and interface definition using component design structure matrix *J. Eng. Des.* **21** 647–75
[17] Asikoglu O 2012 *A New Method For Evaluating Design Dependencies In Product ArchitectureS*

An integrated production-inventory model for food products adopting a general raw material procurement policy

G Fauza[1], H Prasetyo[2] and B S Amanto[1]
[1] Dept. of Food Technology, Universitas Sebelas Maret, Surakarta, Indonesia
[2]Pusat Studi Logistik dan Optimisasi Industri (PUSLOGIN), Dept. of Industrial Engineering, Universitas Muhammadiyah Surakarta, Indonesia

Abstract. Studies on an integrated production-inventory model for deteriorating items have been done extensively. Most of the studies define deterioration as physical depletion of some inventories over time. This definition may not represent the deterioration characteristics of food products. The quality of food production decreases over time while the quantity remains the same. Further, in the existing models, the raw material is replenished several times (or at least once) within one production cycle. In food industries, however, a food company, for several reasons (e.g., the seasonal raw materials, discounted price, etc.) sometimes will get more benefit if it orders raw materials in a large quantity. Considering this fact, this research, therefore, is aimed at developing a more representative inventory model by (i) considering the quality losses in food and (ii) adopting a general raw material procurement policy. A mathematical model is established to represent the proposed policy in which the total profit of the system is the objective function. To evaluate the performance of the model, a numerical test was conducted. The numerical test indicates that the developed model has better performance, i.e., the total profit is 2.3% higher compared to the existing model.

1. Introduction

Inventory plays an important role in industry since 40% of companies' total asset is invested in the form of inventory [1]. Since EOQ was introduced by Harris in 1915, it has been developed extensively. Currently, business becomes more competitive. Thus, a conventional approach where a company only focuses on managing its own inventory is no longer sufficient. The joint economic lot size (JELS) model is firstly introduced by [2], where a company establishes corporations with its customers/retailers. The producer and retailer manage inventory collaboratively due to cost reduction of the joint system. Reference [3] introduced a joint economic-lot-size for purchaser and vendor based on a lot for lot basis. Later, [4] advanced it by developing one setup multiple delivery JELS that allowed the retailer to receive finished goods in a smaller batch during the production cycle. Unlike the lot for lot basis, the single setup multiple deliver strategy successfully reduced inventory cost in retailer. However, the study still disregarded the material procurement plan in inventory system. Reference [5] included raw materials procurement plan in the single vendor single buyer (SVSB) system while [6] generalized the model by applying a general procurement policy. Extensive reviews regarding the integrated production-inventory model can be seen in [7] and [8]

The aforementioned studies, however, do not pay sufficient attention to quality losses in food products. Multi-echelon models considering quality losses in raw materials and finished goods is proposed by [9] and [10]. In the models, the quality of raw materials follows a kinetic model function while the quality of finished goods is perceived by customers based on product's expiration dates. For the latter, since the customers perceive the quality of aging products as declining, the customers' willingness to pay (WTP) for the product is also decreasing and influence the demand rate. To maintain the demand, retailers reduce the products' price and promote the discounted products [11]. Representing this phenomenon, [9] and [10] developed a shelf-life based price function where the price of finished goods is varied during the selling period. The inventory models, however, still limit the frequency of raw material replenishment. There must be at least one lot of raw materials is replenished during one production cycle. This restriction reduces the flexibility of the model in representing some real cases. In food industries, manufacturers sometimes deal with a situation where, for some reasons, they may need to purchase raw materials in a bigger quantity. Considering the possibility of ordering one lot of raw materials to fulfil more than one production cycle (i.e. implementing a more general procurement policy) will give companies more alternatives. Therefore, this study relaxes the procurement policy in the current research by applying a more general policy in a SVSB system dealing with food inventory issues.

The remaining part of this paper is organized as follows. Section 2 defines the objective problem as well as its assumptions and notations to develop the proposed model. The model development is presented in Section 3 while the numerical test and the result are discussed in Section 4. Finally, Section 5 summarizes this study.

2. Problem definition, assumptions, and notations

This study extends [9] by applying a more general raw material procurement policy into the SVSB system. The integrated system deals with problems of managing food inventories, i.e., the quality of raw material losses during storage time and the finished goods suffering from perceived value decrease. The raw material enters warehouse in maximum quality condition (Q_{max}) and, by the time, its quality degrades until the minimum, but it is still acceptable to be used (Q_{min}). Different from [9], the proposed model considers options either to order raw materials more frequent ($m \geq 1$) or less frequent ($m \leq 1$) during the production cycle (m denotes the number of ordering frequencies). However, regardless of which option is chosen, the ordering cycle should be shorter than the usable period of the raw materials. This is important to ensure that no shortage occurs due to spoilage/damage materials. Therefore, in this study, a constraint is added to prevent the ordering cycle of raw material longer than its usable period.

The proposed model is developed based on the following assumptions:

a. The production rate and demand rate are constant;
b. One unit raw material is required to produce one finished unit goods;
c. Once the customers' willingness to pay (WTP) decreases, the product's price declines linearly to maintain the demand;
d. All delivering batches enter the retailer' warehouse before the customers' WTP decline;
e. The ordering cycle of the raw material is shorter than its usable period;
f. Lead time is neglected, and no shortage is allowed.

Further, the notation below is used to symbolize the model.

For the manufacturer

P	:	production rate (unit product/unit time)
D	:	demand rate from retailer (unit product /unit time)
k	:	deterioration rate (unit quality/ unit time)
Q_{max}	:	maximum quality level of raw material (unit quality)
Q_{min}	:	minimum quality level of raw material (unit quality)
$Q(t)$	:	remaining quality level of raw material at time t (unit quality)
c_{raw}	:	purchasing cost of raw material ($/unit product)
A_{raw}	:	delivering cost of raw material ($/shipment)
q_{raw}	:	ordering size of raw material (unit product)

H_{raw}	:	holding cost of raw material (\$/unit product/ unit time)
c_{loss}	:	cost of quality loss (\$/unit quality/unit time)
$L(m,T)$	:	total cost of quality loss (\$/unit time)
$TC_{\text{raw}}\ (m,T)$	:	total cost of raw material system (\$/unit time)
c_{mfc}	:	manufacturing cost (\$/unit product)
A_{mfc}	:	production setup cost (\$/setup)
H_{mfc}	:	holding cost of finished goods (\$/unit product/unit time)
$\bar{I}_{\text{mfc}}$	:	average inventory of finished goods at the manufacturer (unit product)
$TC_{\text{mfc}}(m,T)$	:	total cost of finished goods system (\$/unit time)
$TP_{\text{mfc}}(m,n,T)$	:	total profit of the manufacturer (\$/unit time)

For the retailer

c_{ret}	:	product selling price from manufacturer to retailer (\$/unit product)
H_{ret}	:	holding the cost of finished goods at the retailer (\$/unit product/ unit time)
A_{ret}	:	transportation cost for delivering product to the retailer (\$/shipment)
τ_{sl}	:	product shelf-life (unit time)
τ_{start}	:	the time when customer's WTP starts decreasing (unit time)
$p_{\max}$	:	maximum product price (\$/unit product)
$p(t)$	:	price of the product at age t (\$/unit product)
E_i	:	age of batch i when entering retailer (unit time)
$TP_{\text{ret}}(n,T)$	:	total profit of the retailer (\$/unit time)
$TP(m,n,T)$	:	total profit of the integrated supply chain system (\$/unit time)

Decision variables

m	:	the number of raw material procurement cycles
n	:	the number of finished goods deliveries
T	:	the length of production cycle

3. Model development

The proposed model is developed in two phases. The first is developing the quality loss model of raw material which implements a general procurement policy.

3.1. Modelling the quality loss in raw material using a general procurement policy

Reference [9] has developed a mathematical model to represent the quality loss in raw material during the storage time as shown by equation (1) and (2). However, the model is restricted only for systems that adopt one procurement policy, i.e., $m \geq 1$. In the food industry, however, companies sometimes are benefitted from purchasing raw materials in huge lot size. Hence, one raw material replenishment may cover more than one production cycle. The illustration of such situation can be seen in figure 1 where $m=1/2$ or one replenishment to supply two production cycles. Also, in [9], there is no guaranty that the length of ordering cycle is always shorter than product's usable time. This is contradictive with the assumption that no shortage is allowed. As can be seen in figure 1, intuitively, the shortage will occur if product's usable period (t_Q) is shorter than the length of procurement cycle (t_{raw}). This study, therefore, solves this problem by adding a constraint ensuring that t_{raw} is always shorter than t_Q as shown by equation (3).

$$L(m,T) = c_{\text{loss}} \frac{mP}{T} \int_0^{\frac{DT}{mP}} \Delta Q(t)dt \tag{1}$$

Subject to

$$m \geqslant 1 \tag{2}$$

$$t_Q > \frac{DT}{mP} \tag{3}$$

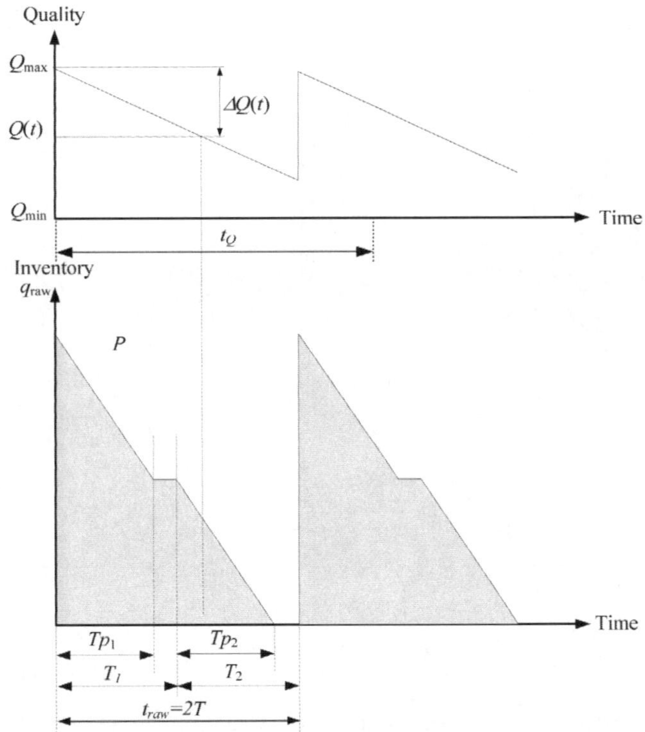

Figure 1. Quality loss in inventory using a general raw material procurement $m = ½$ or $t_{raw}=2T$

As mentioned previously, in this study, a general policy will be implemented in the model. Instead of restricting the number of ordering cycle ($m \geq 1$) such as in [9], the model considers the possibility of ordering raw material for covering more than one production cycle ($m<1$). As can be seen in figure 1, one replenishment cycle of raw material is used to fulfill the demand for raw material in two production cycles. Hence, the value of m is equal to ½. During the first production run length (Tp_1), the quantity of raw material is depleted by raw material demand d_{raw} which is equal to the production rate or P and remains the same during the off-production period (T_1-Tp_1). Further, at the next cycle, it is continuously depleted until the end of Tp_2. Following the steps in [9], the total relevant cost per unit time due to the quality loss in the raw material is represented by equation (4)-(6). This formula complements [9] for the case m less than one.

$$L(m,T) = c_{\text{loss}} \frac{mP}{T} \sum_{i=1}^{1/m} \int_{(i-1)T}^{(i-1)T+DT/P} \Delta Q(t)dt \tag{4}$$

Subject to

$$m < 1 \tag{5}$$

$$t_Q > \left(\frac{1}{m} - 1\right)T + \frac{DT}{P} \tag{6}$$

3.2. Modelling the average inventory in the integrated production-inventory system

The next phase of this section is to find a formula to calculate the average inventory in the integrated system. Figure 2 represents the inventory status of either raw materials or finished goods in a single manufacturer (vendor) single retailer (buyer) environment (or usually called SVSB).

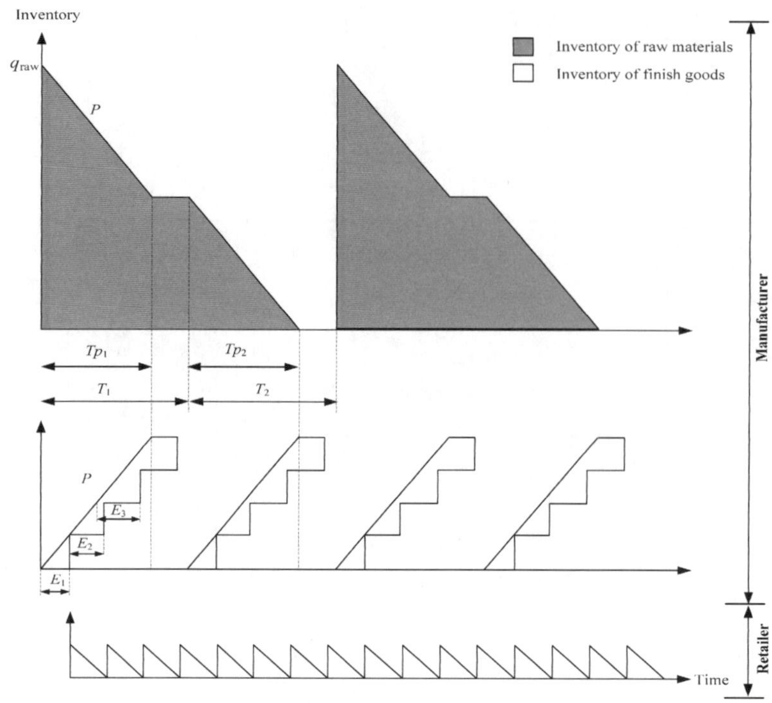

Figure 2. The integrated system using a general raw material procurement policy (m=1/2)

As figure 2 shows, this model relaxes the restricted rule in [9] by allowing m=1/2 (one replenishment of raw material to fulfill two production cycles, i.e., $t_{raw}=2T$). The relaxing influences the formula for calculating the average inventory of raw material. Equation (7) is the equation used in [9], complemented by equation (8) as a result of the general procurement policy.

$$I_{raw} = \frac{D^2T}{2mP}, \text{ for } m \geq 1 \tag{7}$$

$$I_{raw} = mT\left[\frac{D^2}{2mP} + \sum_{i=1}^{1/m}(i-1)\frac{D^2T}{P}\right], \text{ for } m < 1 \tag{8}$$

The inventory cost of carrying raw materials in the manufacturer's system is obtained by multiplying the average inventory in equation (7) or (8) with raw material inventory cost (H_{raw}). Hence, the total cost of procuring and holding raw materials in the manufacturer's system is presented by equation (9).

$$TC_{raw}(m,T) = C_{raw}D + A_{raw}\frac{m}{T} + H_{raw}I_{raw} + c_{loss}L(m,T) \tag{9}$$

Further, following [4], the equation for calculating the average of finished goods is represented by equation (10).

$$I_{mfc} = \frac{DT}{2n}\left[\frac{D}{P}(2-n) + (n-1)\right] \tag{10}$$

Similar to the previous step, the inventory cost of carrying finished goods is obtained by multiplying equation 10 with the cost for holding the finished goods. Hence, the total cost of processing and holding finished goods in the manufacturer's system is presented by equation (11) while equation (12) represents the total profit obtained from using the general raw material procurement policy.

IOP Conf. Series: Materials Science and Engineering **352** (2018) 012013 doi:10.1088/1757-899X/352/1/012013

$$TC_{mfc}(n,T) = C_{mfg}D + \frac{A_{mfc}}{T} + H_{mfc}I_{mfc} \tag{11}$$

$$TP_{mfc} = c_{ret}D - TC_{mfg}(m,n,T) - TC_{ret}(n,T) \tag{12}$$

The last stage is to formulate the mathematical model that accommodates the shelf-life based pricing. The formula for calculating the age of each batch (E_i) as well as the formula for calculating the total profit in retailer or $TP_{ret}(n,T)$ is similar to [9]. However, since it is assumed in this study that all batches should enter the retailers' warehouse before the customers' WTP declines, $E_i < \tau_{start}$ is added as a constraint to ensure that the age of all receiving batches is younger than τ_{start} . Equations (13)-(16) represent the total profit of the integrated system.
Maximise:

$$TP(m,n,T) = TP_{ret}(n,T) + TP_{mfc}(m,n,T) \tag{13}$$

Subject to:

$$P \geq D; \tag{14}$$

$$E_i < \tau_{start};\ \text{for } i = 1, 2, \ldots, n \tag{15}$$

$$T > 0; \tag{16}$$

4. Numerical example

In this numerical example, model in [9] is used as the benchmark model. Hence, equation (10) in [9] is compared to equation (13) with parameters shown by table 1.

Table 1. Parameters in numerical test

Manufacturer	P	D	k	c_{raw}	A_{raw}	H_{raw}	c_{loss}	c_{mfc}	A_{mfc}	H_{mfc}
Value	19,200	4,800	0.01	5	1000	1	10	5	600	6
Retailer	c_{ret}	H_{ret}	A_{ret}	O_{ret}	τ_{sl}	τ_{start}	p_{max}			
Value	35	7	25	50	0.208	0.167	50			

To determine the controlling variables (m,n,T), a genetic algorithm tool in Matlab 2009 is utilized. The result can be seen in table 2.

Table 2. Total Cost, Revenue and Total Profit of the SVSB System

	Decision Variables		
	m	n	T
Benchmark Model	1	4	0.1921
	Manufacturer	Retailer	Total
Total Cost	130,060	146,050	276,110
Revenue	144,000	239,970	383,970
Total Profit	13,940	93,920	107,860
	Decision Variables		
	m	n	T
Proposed Model	0.2500	3	0.1825
	Manufacturer	Retailer	Total
Total Cost	127,610	145,980	273,590
Revenue	144,000	240,000	384,000
Total Profit	16,390	94,020	110,410

As can be seen from table 2, decision variables obtained from the benchmark model are m=1, n=4 and T=0.1921. It means that during the period of 0.1921, the manufacturer procures raw material once and deliver the finished goods four times to retailers. This method leads to \$107,860 of the total profit.

On the other hand, the proposed model yields \$110,410 of the total profit, in which general raw material procurement is applied. The manufacturer of the latter strategy is allowed to order raw material in a larger quantity to fulfill the demand of four production cycles. In this case, the proposed strategy performs better regarding the total profit (2.3% higher). Further, still from table 2, the total cost in the manufacturer system is reduced from \$130,060 to \$127, 610 leading to 17.6% increase in the total profit of the manufacturer applying the proposed strategy. Meanwhile, in the retailer system, the total cost declines slightly leading to 0.1 % increase in the total profit of retailers. Although in overall applying the general procurement strategy is advantageous for all players in the SVSB system, the manufacturer seems benefitted most.

5. Conclusion

In this study, an integrated inventory model for food product applying general raw material procurement has been developed. Unlike the previous model, the procurement policy in this model is more flexible. The manufacturer has more alternatives/strategies, i.e., establishing one, two or more raw material replenishments during one production cycle or ordering one raw material replenishment to support more than one production cycle. Using the given parameter, the developed model has a better performance, i.e., the total profit of the integrated system is higher than that of the current model. The numerical test shows that reducing the ordering frequencies and the length of production cycle potentially diminish the manufacturing and retailer costs. Applying the proposed strategy increases the total profit of either the manufacturer or retailers. However, the numerical test result also reveals that the manufacturer obtains more benefits from this strategy. Therefore, to distribute the advantages among players, a profit-sharing scenario between the manufacturer and retailers is worth to invcstigate in future research.

References

[1] Tersine R J 1994 *Principles of inventory and materials management* (New Jersey: Prentice Hall)

[2] Goyal S K 1977 An Integrated Inventory Model for a Single Product System *Op. Res. Quart.* **28** 539-45

[3] Banerjee A 1986 A joint economic-lot-size model for purchaser and vendor *Dec. Sci.* **17** 292-311

[4] Kim S L and Ha D 2003 A JIT lot-splitting model for supply chain management: Enhancing buyer–supplier linkage *Int. J. of Prod. Econ.* **86** 1-10

[5] Banerjee A and Kim S L 1995 An integrated JIT inventory model *Int. J. of Op. & Prod. Man.* **15** 237-44

[6] Lee W 2005 A joint economic lot size model for raw material ordering, manufacturing setup, and finished goods delivering *Om.* **33** 163-74

[7] Ben-Daya M, Darwish M and Ertogral K 2008 The joint economic lot sizing problem: Review and extensions *Euro. J. of Op. R.* **185** 726-42

[8] Glock C H 2012 The joint economic lot size problem: A review *Int. J. of Prod. Econ.* **135** 671-86

[9] Fauza G, Amer Y, Lee S H and Prasetyo H 2015 An inventory model of production-inventory policy for food products considering quality loss in raw materials *IEEE Int. Conf. on Industrial Engineering and Engineering Management (IEEM)* 910-914

[10] Fauza G, Amer Y, Lee S H and Prasetyo H 2016 An integrated single-vendor multi-buyer production-inventory policy for food products incorporating quality degradation *Int. J. of Prod. Econ* **182** 409-17

[11] Tsiros M and Heilman C M 2005 The effect of expiration dates and perceived risk on purchasing behavior in grocery store perishable categories *J. of Marketing* **69** 114-29

Solving Capacitated Closed Vehicle Routing Problem with Time Windows (CCVRPTW) using BRKGA with local search

H Prasetyo[1], M A Alfatsani[2], G Fauza[3]
[1] Pusat Studi Logistik dan Optimisasi Industri (PUSLOGIN), Dept. of Industrial Engineering, Universitas Muhammadiyah Surakarta, Indonesia
[2] Dept. of Industrial Engineering, Universitas Muhammadiyah Surakarta, Indonesia
[3] Dept. of Food Technology, Universitas Sebelas Maret, Surakarta, Indonesia

E-mail: hari.prasetyo@ums.ac.id

Abstract. The main issue in vehicle routing problem (VRP) is finding the shortest route of product distribution from the depot to outlets to minimize total cost of distribution. Capacitated Closed Vehicle Routing Problem with Time Windows (CCVRPTW) is one of the variants of VRP that accommodates vehicle capacity and distribution period. Since the main problem of CCVRPTW is considered a non-polynomial hard (NP-hard) problem, it requires an efficient and effective algorithm to solve the problem. This study was aimed to develop Biased Random Key Genetic Algorithm (BRKGA) that is combined with local search to solve the problem of CCVRPTW. The algorithm design was then coded by MATLAB. Using numerical test, optimum algorithm parameters were set and compared with the heuristic method and Standard BRKGA to solve a case study on soft drink distribution. Results showed that BRKGA combined with local search resulted in lower total distribution cost compared with the heuristic method. Moreover, the developed algorithm was found to be successful in increasing the performance of Standard BRKGA.

1. Introduction

The main problem in Vehicle Routing Problem (VRP) is to find distribution routes between depot and outlets so that all outlets are covered with the most minimum distance, and delivery cost [1], with the vehicle, is limited to visit an outlet only once [2]. VRP is aimed to distribute products to outlets that are spread in many locations to optimize distribution distance [3] and distribution cost [4]. It is assumed that VRP is an open route, with the vehicle goes from and returns to the same depot [5]. Examples of VRP cases in the industrial world is the case of soft drink distribution [6], rice distribution [7], distribution of vegetables from places of higher altitudes [8], waste collection and disposal [9], etc.

Reference [6] has solved the problem of VRP for the case of soft drink distribution that used to have less optimum routes, resulting in high cost for companies to pay overtime for workers. A similar case in the distribution of computer spare parts by [4] and distribution of subsidized rice [7]. Some vehicle routing problems accommodate limited vehicle capacity and numbers or known as Capacitated

Vehicle Routing Problem (CVRP). Reference [10] dealt with CVRP for newspaper distribution, while [11] addressed CVRP for soft drink distribution. Another characteristic of VRP was time limitation (Time Windows), where product distribution was based on the time set by the companies. Vehicle Routing Problem with Time Windows (VRPTW) had successfully been applied in the study of [12] in the case of instant noodles delivery, where the decision on routes was highly important to ensure optimum delivery time so that the products sent to customers on time. Another example is the distribution of blood sample, where time is crucial considering the nature of blood that rapidly degrades. Thus time frame had to be decided carefully in distribution [3]. The route where vehicles leaving from and returning to the depot is called a closed route, while if the vehicles start from a specific place to distribute products to several outlets before ending up in the depot, it is called an open VRP route (with the assumption that vehicles are not owned by the company). In practices, both capacity and time constraints may simultaneously exist. Therefore, this research is concerned with Capacitated Closed Vehicle Routing Problem with Time Windows (CCVRPTW). There have been some studies dealing with CCVRPTW, such as [13], [14] and [15]. This research contributes in proposing a biased random-key genetic algorithm approach for solving CCVRPTW efficiently.

In general, VRP is difficult to solve with analytical methods, since functions cannot be integrated. Thus solutions are difficult to find or even cannot be found [16]. Even though the solution suggested could yield optimum results, but it took a long time to compute. Reference [6] solved the case of soft drink distribution using the heuristic method, which is called Nearest Neighbor. Several researchers have used heuristic methods for instance [17], [11] and [18]. However, the heuristic method does not consider whether the solution could bring optimum results. Therefore, a computing model was designed to suit the research, which is a metaheuristic method. Metaheuristic could provide a more optimum solution with shorter computing time compared to an analytical/numerical method and heuristic method. Thus, this method has been used more frequently in solving complicated problems like in this study. Several famous metaheuristic methods are Simulated Annealing, Ant Colony Optimization, and Genetic Algorithm [19]. Simulated Annealing has been used by [13], while Ant Colony Optimization was utilized by [20] to solve scheduling problem and Genetic Algorithm was applied by [8]. Reference [21] used the Tabu Search method to develop optimum distribution route for bread industry. In this study, the metaheuristic method that was used was Biased Random Key Genetic Algorithm (BRKGA) that takes the concept of human evolution. Introduced in 2002, BKRGA has successfully solved many optimization problems due to its flexibility and good performance. Reference [22] presented a review of the application of this algorithm. In this study, to improve the performance of BKRGA, the algorithm was modified and combined with local search. Therefore, several experiments were performed to proof that BRKGA combined with local search was better than both Standard BKRGA and heuristic method.

2. Methods

2.1 Problems and assumptions

The problem addressed in this study is the same problem addressed by [23] who discussed VRP problems in the soft drink industry. The distributor is aimed to find optimum route so that products are distributed from a depot to 45 outlets. Moreover, the vehicle has a limited capacity that affects the order limit from every outlet. The company has set the maximum distribution time for each delivery.

In this study, data were collected to support the main results. Those data were for example location of outlets, distance and travel time between each outlet and depot, order from each outlet, vehicle capacity, distribution time, vehicle setup time, service time at every outlet, as well as loading and unloading time at every outlet. Several assumptions were also needed in the study such as an order from each outlet had been known based on order record and outlet's order was assumed constant in every distribution batch. The vehicle that was used for the distribution was a truck with the capacity of 130 crates. The delivery time limit was assumed to be equal and constant in every sub route, which was 480 min, including 15 min for vehicle setup at every sub-route and 19 min service, loading and unloading time at every outlet. Distribution time allowance of 20% was used for unforeseen circumstances that might happen during distribution, for example, if the vehicle broke down or if there was a traffic jam.

2.2 Solution approach
Biased Random Key Genetic Algorithm (BRKGA) is a variant of Random Key Genetic Algorithm (RKGA) that was introduced by [24] to perform optimization, and it was first proposed in [25] and [26]. BRKGA consisted of a population (p) that is represented in a vector of random real numbers within the interval of 0 to 1 and it has been used from generation to generation. Numbers form a chromosome that represents an outlet to be decoded. Decoding a population is done by translating every gene at every chromosome into outlets. Therefore, the order of distribution route can be formed based on the limitations in capacity and time, so that distribution cost can be calculated. In figure 1, chromosomes with the genes 0.27, 0.65, 0.74, 0.53, and 0.81 represent every outlet to be decoded. The order of a route is developed by sorting genes from the smallest to the largest, so that the sorted genes become 0.27, 0.53, 0.65, 0.74, and 0.81 resulting in the ordered route of 1-4-2-3-5. At the next iteration, new genes are developed by encoding them from the outlet with minimum fitness and next translated into the best chromosomes.

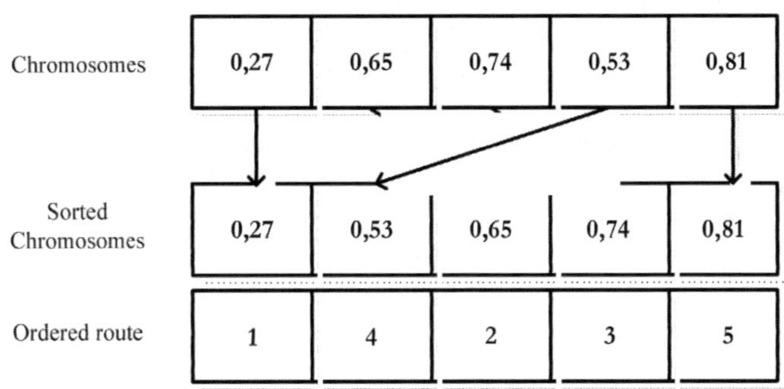

Figure 1. Decoding chromosomes

For g generation, it consists of the elite category (pe) and non-elite category (p-pe) in the initial population (p). One part of the generation (g+1) is formed from the copy of the elite category (pe) in the generation g to ensure that the best solution stays within the population. Next, the mutant (pm) is added to the population. The mutant comes from the random numbers similar to the initial population. The final part of the generation (g+1) is the result of a crossover (p-pe-pm) between the elite category (pe) with the non-elite category (p-pe) with probability values shown in table 1 [16]. RKGA selects two parents from all individuals in the population to be paired [3] (Grasas et al., 2014), while BRKGA selects two parents from each chromosome of the elite category and the non-elite category in the population to be paired [27]. As an illustration, parent 1 (pe) has the genes 0.27, 0.65, 0.74, 0.82, 0.39 and parent 2 (p-pe) has the genes 0.78, 0.71, 0.18, 0.53, 0.81. Random numbers with the interval of 0 to 1 along the parents' chromosomes are generated into 0.32, 0.69, 0.42, 0.87, and 0.76. If random numbers are lower than 0.7, the next generations will inherit the genes of the elite parents; if the numbers are higher than 0.7, the next generations will inherit the genes of the non-elite parents. Based on the crossover, the new generations will have chromosomes with the genes 0.27, 0.65, 0.74, 0.53, and 0.81. The similar process is applied to crossover generation (g+2+3+…).

Table 1. brkga recommended parameters (Goncalves & Resende, 2011)

Parameter	Description	Recommended value
p	size of population	p = ax, where $1 \le a \in \mathbb{R}$ is a constant and x is the length of the chromosome
pe	size of elite population	$0.10p \le p_e \le 0.25p$
pm	size of mutant population	$0.10p \le p_m \le 0.30p$
ρ_e	elite allele inheritance probability	$0.5 \le \rho_e \le 0.8$

Local Search Procedure (LSP) by two methods i.e. swapping two genes randomly and incrementing a gene, can further improve BRKGA. To illustrate for swapping method, the chromosome with the genes 0.27, 0.65, 0.74, 0.53, and 0.81 results in the route order 1-4-2-3-5. The random search of neighboring genes results in the genes 0.27 and 0.53 as shown in figure 2. The local search found new chromosomes by swapping these genes resulting in new chromosome 0.53, 0.65, 0.74, 0.27, and 0.81 and distribution route 4-1-2-3-5. For incrementing method, suppose gene 0.65 is chosen randomly. The local search procedure is performed by searching 45 values near 0.65.

In their report, [22] stated that BRKGA combined with local search had been applied by [28] and [29] with different methods, but similar local search concept. BKRGA was reported to give more optimum results compared to other methods. Therefore, in this study, several methods were compared, namely heuristic method, Standard BKRGA, and BKRGA combined with local search to identify if BKRGA could provide more optimum results.

3. Results and Discussion

3.1 Setting of parameters for BKRGA combined with local search

BKRGA program has been implemented in MATLAB version 7.11.0.584 (R2010b), 64 bit (win64) and runs on a notebook Intel® Core™ i5-2450M @ 250 GHz with the capacity of 4 GB RAM. The setting of BKRGA parameters was done through extensive experiment. With population size of 45, a number of set parameters are studies, i.e. *pe* (0.1, 0.15, 0.2, and 0.25), *pm* (0.1, 0.2, and 0.3) and ρ_e (0.5, 0.6, 0.7, and 0.8). Parameter setting configuration resulted in 48 parameters, every parameter was computed with 50 samples of fitness values with the iteration of 700 repetitions, and average fitness value was obtained in every parameter. The selected parameter was the parameter with the lowest average fitness values among all parameters. The lowest fitness value was obtained from the first parameter configuration with the setting as follows: the elite probability of 0.1, the elite generation probability of 0.5, and the mutant probability of 0.1, as shown in table 3, with the value of Rp 4,054.02. Figure 2 shows the convergent point of the computation of BKRGA with the local search using the best setting parameters.

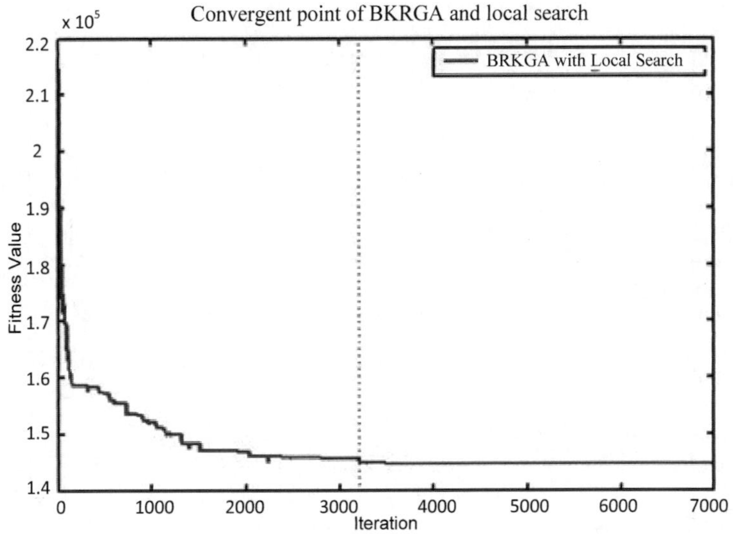

Figure 2. Convergent point of BKRGA with local search

After finding the best setting parameters, the algorithm was run to get final minimum cost. The computation was performed with the best setting parameters and repeated with 7000 iterations, resulting in the minimum cost of IDR 144,760.00.

3.2 Performance of the proposed BRKGA over standard BRKGA

The performance of the proposed BKRGA was measured by comparing standard BKRGA, and BKRGA combined with local search. Every computation resulted in the lowest fitness value with a specific distribution route. Figure 3 showed the graph of the convergent point of minimum fitness values of Standard BKRGA and BKRGA combined with the local search using the best setting parameters. It was shown that standard BKRGA reached its convergent point at the 2375th iteration, as shown by the vertical line on Figure 3, with the total cost of IDR 152,057.60. Meanwhile, BKRGA with local search could find the minimum cost of IDR 144,760.50. Therefore, there was the budget saving of IDR 7,297.10 that was gained, which was approximately 4.8% of the total distribution cost.

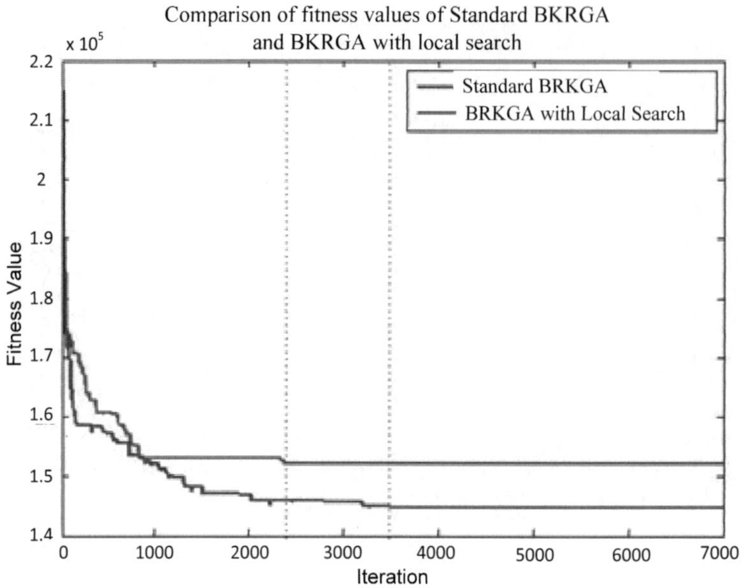

Figure 3. Comparison of BKRGA with local search and Standard BKRGA

3.3 Comparison between BKRGA with heuristic method

Other than comparing BKRGA with local search and Standard BKRGA, the main objective of this study was to compare BKRGA with local search with the heuristic method. This study compared the two methods using similar data used by Sembiring (2008). After the analysis was performed on the BKRGA algorithm, results show that BKRGA won over the heuristic method regarding the obtained distribution cost. Total distribution cost used by the heuristic method was IDR 236,500.00, while total distribution cost for BKRGA with local search was IDR 144,760.50. Thus, there was a saving of cost of IDR 91,739.50, which represented 38.8% of the total cost. Regarding the computation time the heuristic method performed faster than the proposed BRKGA. However, the computation time of fewer than one minutes for the proposed BRKGA is worthy when noting that the obtained result was a significant saving.

4. Conclusions

This article offers the solution for CCVRPTW case using BKRGA with local search approach. BKRGA configuration was very efficient in solving the problems of CCVRPTW: limitation in vehicle capacity and time allocation. In industry, one of the applications of this method was in soft drink distribution. Cutting of order of distribution route was based on the limitations in capacity and time. Thus the order of distribution route obtained was feasible. The algorithm of BKRGA with local search was able to improve the performance of Standard BKRGA by 4.8% regarding the distribution cost. The proposed algorithms could save the cost, as compared to the heuristic method. The saved amount was IDR 91,740.00, which represented 38.8% of the total distribution cost.

The results of this study cannot always be applied on all probabilistic situation since consumer demand can change anytime. Thus it needs further study for different cases of probabilistic. Moreover, several assumptions used in this study limit it from being used for all other cases of CCVRPTW.

References

[1] Arvianto A, Setiawan A H, and Saptadi S 2014 Model vehicle routing problem dengan karakteristik rute majemuk, multiple time windows, multiple products dan heterogeneous fleet untuk depot tunggal. *J. Teknik Industri* **16** 85–96 (in Indonesian)

[2] Fajarwati I A and Anggraeni W 2012 Penerapan algoritma differential evolution untuk penyelesaian permasalahan vehicle routing problem with delivery and pick-up *J. Teknik ITS* **1** A391-96 (in Indonesian)

[3] Grasas A, Ramalhinho H, Pessoa L S, Resende M G C, Caballe I and Barba N 2014 On the improvement of blood sample collection at clinical laboratories *BMC Health Services Research* **14** 1-19

[4] Pradhana F E, Sugiharti E and Kharis M 2012 Penerapan algoritma Tabu Search untuk menyelesaikan Vehicle Routing Problem *Jurusan Matematika Universitas Negeri Semarang* **1** 15-20 (in Indonesian)

[5] Fleszar K, Osman I H and Hindi, Khalil S 2009 A variable neighbourhood search algorithm for the open vehicle routing problem *Euro. J. of Op. Res.* **195** 803–09

[6] Amri M, Rahman, A, and Yuniarti, R 2014 Penyelesaian Vehicle Routing Problem dengan menggunakan metode Nearest Neighbor (Studi Kasus: MTP Nganjuk Distributor PT. Coca Cola) *J. Rekayasa dan Manajemen Sistem Industri* **2** 36–45 in Indonesian

[7] Putri, F B, Mahmudy, W F and Ratnawati, D E 2014 Penerapan Algoritma Genetik untuk Vehicle Routing Problem with Time Windows (VRPTW) pada Kasus Optimasi Distribusi Beras Bersubsidi *J. Mahasiswa PTIIK Universitas Brawijaya* **5** 1-9 (in Indonesian)

[8] Slamet A S, Siregar H H and Kustiyo A 2014 Vehicle Routing Problem (VRP) dengan Algoritma Genetika pada pendistribusian sayuran dataran tinggi *J. Teknologi Industri Pertanian* **24** 1-10 (in Indonesian)

[9] Fitria L, Susanty S and Suprayogi 2009 Penentuan rute truk pengumpulan dan pengangkutan sampah di Bandung. *J. Teknik Industri* **11** 51-60 (in Indonesian)

[10] Faiz E and Fahmi F E 2013 Studi komparasi penyelesaian Capacitated Vehicle Routing Problem (CVRP) dengan Metode Saving Matrix dan Generalized Assignment *J. Mahasiswa Matematika* **1** 276-279 (in Indonesian)

[11] Purnomo A 2010 Penentuan rute pengiriman dan biaya transportasi dengan menggunakan Metode Clark and Wright Saving Heuristic (Studi kasus PT The Botol Sosro Bandung) *J. Logistik Politeknik Pos Indonesia* **1** 97-117 (in Indonesian)

[12] Saputri M W, Mahmudy W F, Ratnawati D E 2015 Optimasi Vehicle Routing Problem with Time Window (VRPTW) menggunakan algoritma genetika pada distribusi barang *J. Mahasiswa PTIIK Universitas Brawijaya* **5** 12 (in Indonesian)

[13] Lukitasar W, Suyanto and Dayawati R N 2011 Capacitated Vehicle Routing Problem Time Windows (CVRPTW) dengan menggunakan algoritma Improved Ant Colony System (IACS) dan algoritma Simulated Annealing (SA) *Telkom University (*in Indonesian)

[14] Lydia P R, Suyanto, Dayanti R N 2011 Capacitated Vehicle Routing Problem Time Windows (CVRPTW) menggunakan Algoritma Harmony Search *Telkom University (*in Indonesian)

[15] Widyaningrum, N O 2011 Capacitated Vehicle Routing Problem with Time Windows (CVRPTW) menggunakan Differential Evolution dan Algoritma Genetika Capacitated Vehicle Routing Problem with Time Windows (CVRPTW) using Differential Evolution and Genetic Algorithm *Telkom University* (in Indonesian)

[16] Gonçalves J F and Resende M G C 2011 A biased random-key genetic algorithm for job-shop scheduling *AT&T Labs Res. Tech. Report* **46** 253–71

[17] Arinalhaq F, Imran A, and Fitria L 2013 Penentuan rute kendaraan pengangkutan sampah dengan menggunakan metode Nearest Neighbour (studi kasus PD kebersihan kota Bandung) *J. Online Institut Teknologi Nasional* **1** 22–32 in Indonesian

[18] Octora L, Imran A and Susanty S 2014 Pembentukan rute distribusi menggunakan algoritma Clarke & Wright Savings dan algoritma Sequential Insertion *J. Online Intitut Teknologi Nasional* **2** 1-11 in Indonesian

[19] Resende M G C 2012 Biased Random-Key Genetic Algorithms with applications in telecommunications *Top* **20** 130–53

[20] Liliani and Alfian A 2014 Usulan pendajwalan produksi dengan algoritma Ant Colony (studi kasus PT Shima Prima Utama Pelembang) *Simposium Nasional RAPI XIII FT UMS* I 9-1

[21] Pavela V and Purwanto I N 2009 Penyelesaian Vehicle Routing Problem dengan menggunakan Algoritma Nearest Neighbor dan Tabu Search (Studi Kasus di PT Nippon Indosari Corpindo) *Jurusan Matematika Universitas Brawijaya* 1 244–47

[22] Prasetyo H, Fauza G, Amer Y and Lee S H 2015 Survey on applications of biased-random key genetic algorithms for solving optimization problems *IEEE Int. Conf. on Industrial Engineering and Engineering Management 2015 6-9 Dec. Singapore* 863–70

[23] Sembiring, A C 2008 Penentuan rute distribusi produk yang optimal dengan menggunakan algoritma heuristik pada PT. Coca-Cola Bottling Indonesia Medan (Tugas Akhir Strata 1 :Universitas Sumatera Utara)

[24] Bean J C 1994 Genetic algorithms and random keys for sequencing and optimization ORSA *J. on Comp.* **6** 154-61

[25] Ericsson M, Resende M G C and Pardalos P M 2002 A genetic algorithm for the weight setting problem in OSPF routing *J. Comb. Opt* **6** 299–333

[26] Gonçalves J F and Almeida J R D 2002 A hybrid genetic algorithm for assembly line balancing *J. of Heuristics* **8** 629-42

[27] Fernando J and Mauricio G 2011 A parallel multi-population genetic algorithm for a constrained two-dimensional orthogonal packing problem *J. Of Comb. Opt* **22** 180–201

[28] Silva R M A, Resende M G C, Pardalos P M and Faco J L 2013 Biased random key genetic algorithm for linearly constrained global optimization *Proceedings of The 15th Annual Conference Companion on Genetic and Evolutionary Computation* 79-80

[29] Silva R M A, Rensende M G C and Pardalos P M 2014 Finding multiple roots of a box constrained system of nonlinear equations with a biased random key genetic algorithm *J. of Glob. Opt.* **60** 289-306

IOP Conf. Series: Materials Science and Engineering **352** (2018) 012015 doi:10.1088/1757-899X/352/1/012015

Mass variation effect of teki grass (cyperus rotundus) composite against tensile strength and density

Muhammad Rafiq Yanhar, A Haris Nasution
Department of Mechanical Engineering Islamic University of North Sumatera, Indonesia

Abstract. The primary purpose of this study is to determine the tensile strength using ASTM D638 - 02a type IVB and density of *teki* grass (Cyperus rotundus) composite. The production process is carried out by mass variation of 2 gr, 3 gr, and 4 gr. Hand lay-up method with three repetitions is applied. Teki grass is chosen because it is easy to find and has some advantages biodegradable, harmless to health, available in large quantities, and cost-efficient. The test result showed the largest tensile strength is 21,61 MPa at 2-gram mass fiber. Fiber addition to 3 gram and 4-gram cause tensile strength decreases to 18,51 MPa and 11,65 MPa. It happens because the fibers are random and spread in all directions, so many fibers are undirectional with the tensile force. Beside that fibers addition made matrix volume reduced and a bond between fiber and matrix decreases, finally make fiber unable to hold the tensile force properly. It is recommended to use another type of ASTM D638 - 02a which has a larger narrow section like type I (13 mm) and type III (19mm) so specimens are not broken when removed from the mold, and there isn't any decrease in tensile strength.Density test showed that fiber mass does not significantly affect the density.

1. Introduction

Until now, industries remain to use synthetic fiber as their composite reinforcement materials. Synthetic fiber is specially employed in manufacturing industries of various products such as aircraft, boat hulls, wind turbine blades, car body, among others [1]. In 2009 the consumption of fiberglass around the world had reached 4 to 5 million tons per year and was expected in 2017 to reach 8.5 million tons per year [2].

Unluckily the fiberglass has some negative impacts, such as high cost, can not decompose naturally, limited, and is harmful to health [3]. Moreover, fiberglass can increase the risk of lung cancer. Since the size of fiberglass is varied, the small particles of this fiber will be inhaled if the mouth is not covered by masker and runs into the blood; meanwhile, the larger particles can irritate to skin, eyes, nose and throat. Therefore, researchers have currently attempted to find an alternative by replacing the synthetic fiber with natural fiber which has several advantages: easy to find, biodegradable, harmless to health, large availability in nature, and low cost [3].

In this study, the natural fiber used is teki grass (Cyperus rotundus) because it is not fully utilized and often found in the open fields. The teki grass is very adaptive and therefore it consists of weeds that are very difficult to eradicate. It has tubers which are capable of reaching one meter in depth, so such tubers are saved from tillage (30 cm). This grass is discovered all over the world, grows well when sufficient water is available, tolerant towards inundation, and able to withstand drought conditions. In addition, the matrix for binding the fiber is the polyester resin.

2. Research Methods

2.1 Equipments and Materials

During this research there are some equipments and materials involved, such as, print specimen with the same size as suggested by ASTM D638 - 02a type IVB [4], a tensile testing machine servo pulser, polyester resin and its hardener, teki grass (Cyperus rotundus), wax, glass coatings, gloves, measuring cups, scissors , digital scales, clippers, and other supporting equipments.

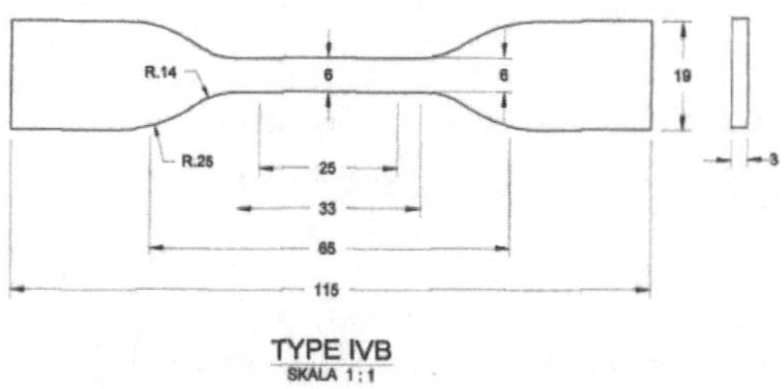

Figure 1. Tensile test specimens of ASTM D638 - 02a type IVB

2.2 Research Stages

1. The teki grass is pulled out from the ground, cleaned with water, soaked for one hour in the 5% solution of NaOH to remove the sap (lignin) and the impurities that can reduce the bond quality between the matrix and the fiber.
2. Drying the teki in the shiny sun for five days would remove its liquid.
3. After the teki is dried, it is weighed and cut into the desired size as determined in the manufacture of the specimen.
4. The metal mold is smeared with wax to make easy to remove the specimen from the mold when it becomes hardened. Glass is placed in the bottom part of the mold and is coated also smeared with wax.
5. The ratio of polyester resin and the hardener is in the range of 100: 1 and this mixture is poured into the mold.
6. Teki grass is laid down on the specimen in the fiber weight of 2 gr, 3 gr, and 4 gr.
7. Then the polyester resin is poured on the surface of the fiber.
8. Once the mixture began to thicken, resin and the glass placed on the top of the mold and pressed with ballast to eliminate the voids (air bubbles) which are trapped and to flatten the surface of the specimen.
9. Allow the specimen to become hardened for 12 hours; open the mold after 12 hours and the specimen is completely formed and ready for a tensile test.
10. The same stages are carried out to make density specimen.

3. Results and Discussion

3.1 Tensile Test

After the specimens are removed from the mold, the next step is to hold the tensile test in which each composition is carried out three times. The test results of the tensile strength of composite specimens can be seen in table 1:

Table 1. Composite Tensile Strength

Fiber Mass	Specimen (n)	Tensile Strength (MPa)	Average Tensile Strength (MPa)
2 gr	Specimen 1	21,69	
	Specimen 2	22, 44	21,61
	Specimen 3	21,52	
3 gr	Specimen 1	19,07	
	Specimen 2	18,36	18,51
	Specimen 3	18,11	
4 gr	Specimen 1	12,12	
	Specimen 2	11,06	11,65
	Specimen 3	11,78	

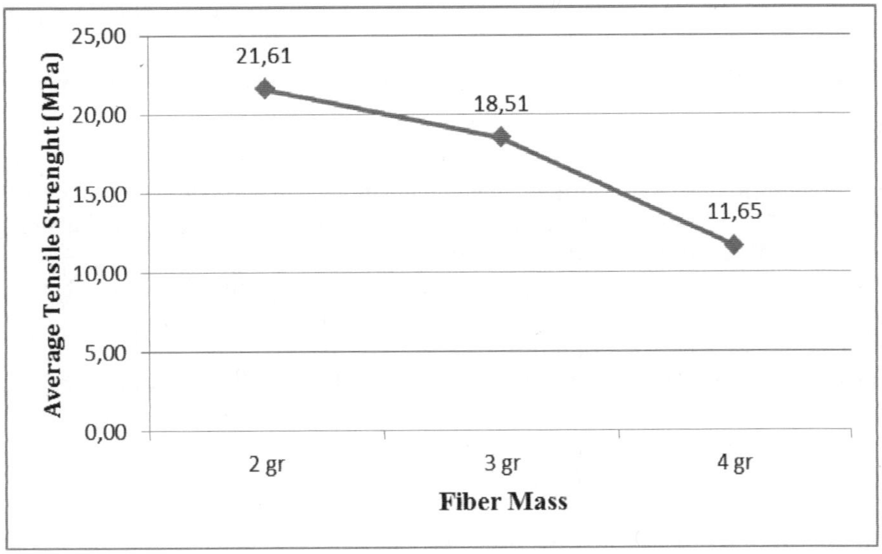

The test result showed the largest tensile strength of the average is on the weight of 2 grams of 21.61 MPa. The addition of fiber weight to 3 gr and 4 gr makes the composite tensile strength decreasing to 18,51 MPa and 11,65 MPa. This happens because the fibers used are random fibers that spread in all directions so that many fibers are not unidirectional with the tensile force. Besides that, fibers addition made matrix volume reduced and the bond between fiber and matrix decreases, finally making fiber unable to hold the tensile force well.

From the observation, there are several broken specimens when removed from the mold that may occur because the specimen dimension is too small, especially in the narrow section (6 mm). So on unbroken specimens, it is estimated that there are small cracks that cause a decrease in the tensile strength. Therefore it is recommended to use another type of ASTM D638 - 02a which has a larger narrow section like type I (13 mm) and type III (19 mm).

3.2 Density Test

Soon after the specimens are removed from the mold, the following step is density test, which each composition is carried out three times with the fiber mass 2 gr, 3 gr, and 4 gr. The result showed that the lowest average density is 1,491 gr/cm^3, while the fiber mass addition to 3 gr and 4 gr only slightly raises the density to 1,586 gr/cm^3 and 1,679 gr/cm^3. It means the fiber mass addition does not significantly affect the density.

Table 2. Composite density

Fiber Mass	Specimen (n)	Mass (gr)	Volume (cm^3)	Density (gr/cm^3)	Average Density (gr/cm^3)
2 gr	Specimen 1	11,89	8	1,486	
	Specimen 2	11,93	8	1,491	1,491
	Specimen 3	11,97	8	1,496	
3 gr	Specimen 1	12,54	8	1,567	
	Specimen 2	12,73	8	1,591	1,586
	Specimen 3	12,8	8	1,6	
4 gr	Specimen 1	13,24	8	1,655	
	Specimen 2	13,61	8	1,701	1,679
	Specimen 3	13,47	8	1,683	

4. Conclusion

The largest tensile strength of 2 grams weight is 21.61 MPa.

1. Fiber addition to 3 gr and 4 gr make the composite tensile strength decreasing to 18,51 MPa and 11,65 MPa. It happens because the fibers are random and spread in all directions, so many fibers are unidirectional with the tensile force.
2. Fibers addition make matrix volume reduced and the bond between fiber and matrix decreases, finally making fiber unable to hold the tensile force properly.
3. Fiber mass addition does not significantly affect the density.
4. It is recommended to use another type of ASTM D638 - 02a which has a larger narrow section like type I (13 mm) and type III (19 mm) so specimens are not broken when removed from the mold, and there isn't any decrease in tensile strength.

References

[1] Owens Corning 2010 *Top Ten Composite Apps.*
[2] Composites World 2009 *The making of glass fiber* CompositesWorld.htm
[3] Yudo H and Kiryanto 2010 Analisa teknis rekayasa serat Enceng Gondok sebagai bahan pembuatan komposit ditinjau dari kekuatan tarik *Jurnal Ilmiah* (in Indonesian)
[4] ASTM 2002 *Annual Book of ASTM Standard* West Conshohocken

IOP Conf. Series: Materials Science and Engineering **352** (2018) 012016 doi:10.1088/1757-899X/352/1/012016

Investigating the road surface effect to the fatigue life of an automotive coil spring

T E Putra and Husaini
Laboratory of Computational Mechanics, Department of Mechanical Engineering, Syiah Kuala University, Darussalam 23111, Banda Aceh, Indonesia

E-mail: edi@unsyiah.ac.id

Abstract. This work aims to estimate the life of a coil spring considering road surface profiles. Strain signals were measured by installing a strain gage at the highest stress location of the coil spring and then driving the vehicle on country and village roads. The village road gave high amplitudes containing spikes when the tire touched a curb, bump or pothole. These conditions contributed to a higher loading rate to the car component, contributing to shorter useful fatigue life, which was only 140 reversals of blocks. Driving on the village road resulted in a 6-times decrease in the useful fatigue life of the component in comparison to the country road. In conclusion, the village road caused stronger vibrations to the component because it has a rough surface; meanwhile, the country road provided lower vibrations because the road was smooth.

1. Introduction

The vehicle body weight and its loads provide an initial force to the components of the suspension system. The force rises if the vehicle is driven on rough road surfaces at a higher speed. After experiencing shocks, there is a probability of a high displacement of tires from the main car body and transmission of vibrations to the car. Under such circumstances, the coil springs show forced vibrations. High shocks affect the coil springs significantly, more than the other components in the car. Hence, the coil springs are seen to play an important role in the car failure as they have to tolerate the many vibrations due to the roughness of the road surfaces; which increases the fatigue-dependent mechanical failure, as the car component gets exposed to cyclic loads.

Components of automotive suspension must be changed with a traveling distance of 73,500 km, or every five years [1]. The fault of 13.18 % of 24.2 million vehicle tests was recorded [2]. Thus, the current study aims to compare the fatigue life of an automotive coil spring while being driven on country and village roads.

2. Experimental setup

In this study, the research acquired the strain signals after installing a strain gage in the highest stress location in the coil spring of the vehicle and then driving the vehicle on the public road surfaces. The finite element analysis [3-5] was used for detecting the high-stress area in the component and the strain gage was installed in that area. The SAE5160 carbon steel was selected as the material, which is generally applied in many automotive industries for the fabrication of the coil springs [6]. Table 1 describes the material properties.

Table 1. Mechanical properties of carbon steel of the SAE5160.

Properties	Values
Ultimate tensile strength (MPa)	1,584
Young modulus (GPa)	207
Yield strength (MPa)	1,487
Coefficient of the fatigue strength (MPa)	2,063
Exponent of the fatigue strength	-0.08
Exponent of the fatigue ductility	-1.05
Coefficient of the fatigue ductility	9.56
Exponent of the cyclic strain hardening	0.05
Coefficient of the cyclic strength (MPa)	1,940
Poisson ratio	0.27

The strain gage was linked to a data logging device with the help of a connector. The device was utilized for recording the signals which were acquired by the sensor. The sampling frequency value for acquiring strain signals must be higher than 400 Hz. This prevents the loss of the essential components of the signals [7]. After placing the equipment, the vehicle was then driven on the country and the village roads at speed ranging between of 30 - 40 km/h and 20 - 40 km/h, respectively.

Mean stress is an important factor in the fatigue analysis [8,9], and it is seen to affect the low and the high cycle fatigue regimes [10,11]. Hence, Smith, Watson and Topper suggested a model, known as the SWT parameter [12], which is defined mathematically as below:

$$\sigma_{\max}\varepsilon = \frac{(\sigma'_f)^2}{E}(2N_f)^{2b} + \sigma'_f\varepsilon'_f(2N_f)^{b+c} \tag{1}$$

with $\sigma_{\max}$ is the maximum stress, ε is the strain, σ'_f is the coefficient of the fatigue strength, E is the Young modulus, N_f is the number of cycles, b is the exponent of the fatigue strength, ε'_f is the coefficient of the fatigue ductility and c is the exponent of the fatigue ductility.

Total fatigue damage is determined by adding normal and shear energy for each peak-valley in a loading, Thus, fatigue damage is influenced by variable amplitude and frequency cycles. The fatigue damage for each loading cycle D_i is defined as below:

$$D_i = \frac{1}{N_f} \tag{2}$$

Thereafter, the Palmgren-Miner rule [13,14] is used for estimating the total fatigue damage D, which is expressed as:

$$D = \sum\left(\frac{n_i}{N_f}\right) \tag{3}$$

where n_i is the number of cycles. The fatigue damage ranges between 0 to 1, with 0 indicates no damage (i.e., infinite cycles towards failure), while 1 indicates a failure (i.e., 1 cycle towards failure) [15].

The reason to conduct fatigue tests with variable amplitude loading (VAL) in the current work was to predict the life of the coil spring based on the strain signals measured on the country and the village roads. The VAL fatigue tests were conducted as it was impossible to assess the fatigue lives of the complex loadings using any hypothesis [16-19]. A VAL fatigue test is principally carried out like a constant amplitude loading (CAL) fatigue test at different loading levels, with a given load being continuously repeated until a failure is obtained. The only difference is that in cases of CAL, the amplitude remains unchanged. Thus, VAL fatigue tests are more complicated, and often, more time-consuming and expensive than CAL fatigue tests [20].

All the samples for the fatigue testing were provided as per ASTM E606-92 [21], norms as described in figure 1. The geometry and the dimensions of the samples and the appropriate grips used in the fatigue testing device have been specified. The samples were fabricated with the help of the computer numerical controls and grinding machines. Thereafter, they were polished [22] for ensuring the removal of the machining marks during testing. As the fatigue cracks develop on irregular road surfaces, the sample condition plays a vital role during the fatigue testing. It can decrease the stress concentration which further prevents the initiation of cracks at those specific sites.

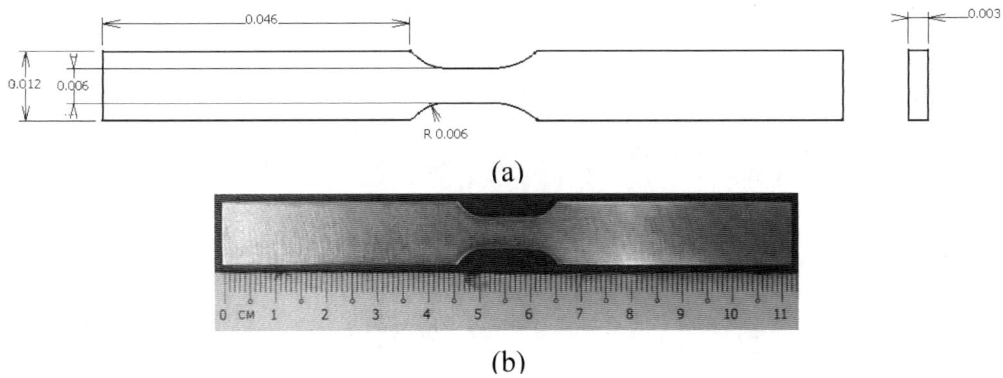

Figure 1. A sample for the fatigue test: (a) sample geometry and dimensions in millimeters, (b) photograph of the sample.

The strain signals have been converted into stresses with the help of the Ramberg-Osgood equation [23], described below:

$$\varepsilon = \frac{\sigma}{E} + K^{'}\left(\frac{\sigma}{E}\right)^{n^{'}} \tag{4}$$

with σ is the stress, K' is the coefficient of the cyclic strength and n' is the exponent of the cyclic strain hardening. Thereafter, the stresses have been converted into loads which were used as an input value in the testing device which help with the expression below:

$$\sigma = \frac{P}{A_0} \tag{5}$$

with P is the force and A_0 is the initial unloaded cross-sectional area.

The reversed VAL fatigue tests were carried out, using a 100 Hz frequency value [24,25] at a room temperature of 20 °C [21]. The researchers did not note any effect of a high frequency on the crack initiation [11,24]. The frequency effect on results is generally low for frequencies up to 1,000 Hz [25].

3. Results and discussion

The signals generated by the sensor were considered a VAL as the coil spring subjected to multiple amplitudes. 30-second long signals for each of the 15,000 data points were recorded and described in figures 2-3 for the country and the village roads, respectively, and they were said to be sufficient for evaluating fatigue life. The country road gave lower amplitude range, distributed between 40.3 με and 214.3 με. A higher amplitude was provided in the village road, distributed between -141.9 με and 179 με. The village road provided a higher amplitude than the country road because its surface was rough.

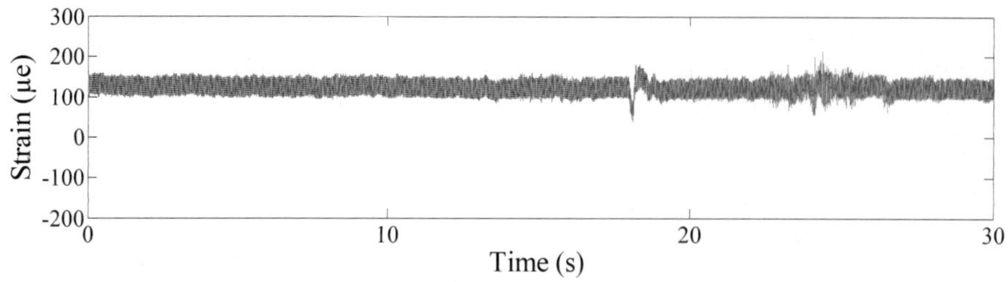

Figure 2. Time history of the country strain signal.

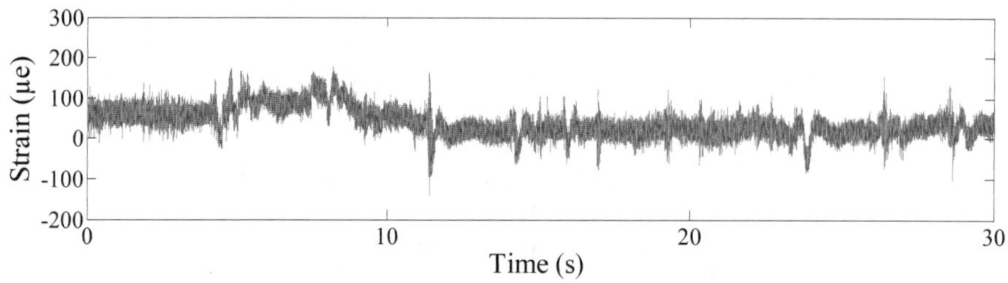

Figure 3. Time history of the village strain signal.

The signals were tensile giving the mean value of 122.6 με and 39.2 με for the country and the village roads, respectively. Thus, the SWT model was suitable for this type of loading [26]. As a structure consists of few loads at the single stress or strain levels, the fatigue damage can be estimated based on the number of cycles towards failure. The number of cycles for the country and the village strain signals was 3,021 and 3,013. Since the village strain signal provided higher strain amplitude leading to higher stress at the coil spring, the strain signal was found to have higher fatigue damage, reaching 7.14 x 10^{-4} damage per block. The country strain signal showed fatigue damage of 1.15 x 10^{-4} damage per block.

Based on the fatigue testing results, the country strain signal needed 868 reversals of blocks until failure in a period of more than 2,169 minutes. The village strain signal gave a shorter time for the testing because it consisted of higher amplitudes. About 350 minutes were required to conduct the fatigue test, involving 140 reversals of blocks. The village road resulted in 83.9 % decrease in the useful life of the coil spring in comparison to the country road. Car driving on rough surfaces can increase the fatigue damage of various car components, in particular, the suspension parts, which can decrease their useful lives.

4. Conclusion

This work estimates the useful life of a coil spring subjected to the strains measured on the country and the village roads. The village road provided higher strain amplitude that contributes to higher fatigue damage, reaching 7.14 x 10^{-4} damage per block. Based on the fatigue testing results, the village useful life was seen to be 6-times shorter in comparison to the country useful life. Some of the non-stationary vibrations include bumps, holes, turns, speed changes, accelerations, brakes and uneven tar patch on the road. These factors lead to high strains and cause high fatigue damage.

References

[1] Roman L, Florea A and Cofaru I I 2014 *Fas. Manag. Technol. Eng.* **1** 289
[2] Hamed M, Tesfa B, Gu F and Ball A D 2014 Vehicle suspension performance analysis based on full vehicle model for condition monitoring development *Proc.VETOMAC X* **23** 495-505
[3] Putra T E, Abdullah S, Schramm D, Nuawi M Z and Bruckmann T 2015 *Mech. Sys. Sig. Proc.* **60-61** 485
[4] Putra T E, Abdullah S, Schramm D, Nuawi M Z and Bruckmann T 2017 *Mech. Sys. Sig. Proc.* **90** 1
[5] Putra T E, Abdullah S, Schramm D, Nuawi M Z and Bruckmann T 2017 *Mech. Sys. Sig. Proc.* **94** 432
[6] Prawoto Y, Ikeda M, Manville S K and Nishikawa A 2008 *Eng. Fail. Analy.* **15** 1155
[7] Ilic S 2006 *Methodology of evaluation of in-service load applied to the output shafts of automatic transmissions* Ph.D. Thesis The University of New South Wales
[8] Klimkeit B, Nadot Y, Castagnet S, Nadot-Martin C, Dumas C, Bergamo S, Sonsino C M and Büter A 2011 *Int. J. Fatigue* **33** 766
[9] Ghafoori E, Motavalli M, Nussbaumer A, Herwig A, Prinz G S and Fontana M 2015 *Eng. Struct.* **84** 29
[10] Tao G and Xia Z 2007 *Int. J. Fatigue* **29** 2180
[11] Kovacs S, Beck T and Singheiser L 2013 *Int. J. Fatigue* **49** 90
[12] Smith K N, Watson P and Topper T H 1970 *J. Materials* **5** 767
[13] Palmgren A 1924 *Zeitschrift VDI* **68** 339
[14] Miner M A 1945 *J. Appl. Mech.* **67** A159
[15] Upadhyaya Y S and Sridhara B K 2012 *Materials and Design* **35** 220
[16] Sonsino C M 2007 *Int. J. Fatigue* **29** 1080
[17] Aykan M and Çelik M 2009 *Mech. Sys. Sig. Proc.* **23** 897
[18] Carvalho A L M, Martin J P and Voorlwad H J C 2010 *Procedia Eng.* **2** 2037
[19] Niesłony A and Böhm M 2013 *Int. J. Fatigue* **52** 49
[20] Rognin F, Abdi F, Kunc V, Lee M and Nikbin K 2009 *Procedia Eng.* **1** 55
[21] ASTM E606-92 1998 *Standard Practice for Strain-Controlled Fatigue Testing* (West Conshohocken: ASTM International)
[22] ASTM E112-96 2004 *Standard Test Methods for Determining Average Grain Size* (West Conshohocken: ASTM International)
[23] Ramberg W and Osgood W R 1943 *Nat. Advisory Committee for Aeronautics*
[24] Hosoi A, Takamura K, Sato N and Kawada H 2011 *Int. J. Fatigue* **33** 781
[25] Schaumann P and Steppeler S 2013 *Procedia Eng.* **66** 88
[26] Ince A and Glinka G 2011 *Fatigue & Frac. Eng. Mat. Str.* **34** 854

IOP Conf. Series: Materials Science and Engineering **352** (2018) 012017 doi:10.1088/1757-899X/352/1/012017

Stress and strain analysis from dynamic loads of mechanical hand using finite element method

Iskandar Hasanuddin[1], Husaini[1], M. Syahril Anwar[1], Sandy Yudha B.Z.[2] and Hasan Akhyar[3]

[1]Department of Mechanical and Industrial Engineering, Universitas Syiah Kuala Jl. Tgk. Syech Abdurrauf No. 7 Darussalam – Banda Aceh 23111, Indonesia

[2]Department of Mechatronics Engineering, Politeknik Aceh Jl. Politeknik Aceh, Pango Raya, Banda Aceh 23116, Indonesia

[3]Department of Mechanical Engineering, Universitas Gajah Mada Jl. Grafika No. 2 Sleman - Yogyakarta 55284, Indonesia

E-mail: iskandarhasanuddin@unsyiah.ac.id

Abstract. This research discusses the distribution of stress and strain due to the dynamic loads of mechanical hand. Thestress and strain that occur on mechanical hand are the main concern for comparing the value of finite element analysis (FEA) and calculating for its material properties. The stress and strain analysis is done with a loading condition. The given loading condition is dynamic. The loading input condition in the simulation of using hydraulic hand dynamometer is from the grip strength measurement of ten samples.The form of the given loading to the mechanical hand is the increment value with a maximum of 708 N/m^2 within 1 minute. The amount of maximum stress (von Mises) simulation is 1.731x10^{5}Pa, and the amount of maximum strain is 7.441 x 10^{-7}. The amount of maximum reaction force is 5.864 x 10^{-2}N, while the amount of maximum displacement that occurs on the distal part is 1.223 x 10 m. Based on the analysis, the maximum stress and strain were found both to occurat the extension part. The result of this study has shown that the stress and strain still occurfar below from the yield strength and the shear strength from the material AISI 1010. It can be concluded that the mechanical hand is durablefor the given loading and can hold an object with a minimum diameter of 45 mm.

1. Introduction

The hand is a multifunctional part of the body. It can do multiple things at the same time. Clenching the fingers can form a grip. Humans have hands which consist of four fingers and a thumb each. The main functions of the hand are to hold, to grab, and to grasp. In certain conditions, the hand cannot touch the target object directly, for instance when holding a beaker glass containing certain chemicals, lest the chemicals splash on the hand. In that case, a device is required to hold or handle the chemicals.

Nowadays, the development of such device has been greatly studied. Ganipineni Sai Krishna designed and analyzed the mechanical hand for industrial application for grasping oddly shaped objects. The result is that the mechanical hand can grasp the objects well [1]. Ivan Virgala analyzed mechanical hand modeling, but his model is only able to form a grip [2]. Sandy Yudha Barry Zaqy designed a mechanical hand and analyzed the strength in a static start-up loading condition with a load of 1000 N, and the result is that the load applied to the finger does not exceed the strength of the elastic material [3]. However, there has not been any analysis regarding dynamic-implicit loading condition, and the load is given merely an assumption.

Due to the problem above, this research continues the dynamic-implicit loading condition using finite element analysis. The purpose of this research is to analyze the stress and strain forces caused by the dynamic load using finite element analysis.

2. Literature Review

2.1 Mechanical Hand
The mechanical hand is modeled using the software Autodesk Inventor Professional 2015. The movements of the mechanical hand mimic the movements of the human hand [3]. The mechanical hand has five fingers; the index finger, the middle finger, the ring finger, the little finger, and the thumb.

Stress is the quantity describing the ratioof the pull force that is applied to the cross-section area of an object. The stress that occurs affects the mechanical properties of the material of the object. Stress occurs due to pressure, pull [4].Stress is defined as:

$$\sigma = \frac{F}{A} \quad (1)$$

where F is the force or the pull load applied, which is measured in the unit Newton (N), and A is the cross-section area before the load is applied (m^2). Stress is measured in megapascal MPa (SI) (where 1 MPa = 10^6 N/m^2).

2.2 Strain Analysis
The strain is defined as the ratio of the added length of an object to the original length. The word strain relates to the relative change in the dimension or shape of the object where a pressure is applied, defined as:

$$\varepsilon = \frac{l-l_0}{l_0} = \frac{\Delta l}{l_0} \quad (2)$$

Where:ε = strain, Δ = length ofobject (m), l_0 = original length (m) dan Δl = length difference (m) [4].

2.3 Finite Element Analysis
FEA is a numerical study that can be used to solve problems like stress analysis of structure and deformation [5], because exact calculation cannot be applied in complicated cases. Using this method, static and dynamic analysis can also be done. The simulation of finite element analysis is done using the software ABAQUS™.

3. Research Method

3.1 Research Location
This research is done in the Computer Aided Mechanical Laboratory and the Work System Design and Ergonomics Laboratory, Faculty of Engineering, Syiah Kuala University and collaborated with Mechanical Engineering Department,Gajah MadaUniversity

3.2 Research Equipments
To find out the load that will be applied, a measurement is made on the strength of the grip using the device Hydraulic Hand Dynamometer, because the data is needed for the application to FEA.

3.3 Testing Material
The material used is AISI 1010, with the mechanical properties as shown in table 1.

Table 1. Mechanical Properties of Material AlSI 1010 [6, 7, 8]

Modulus Young	(E)	210 Gpa
Poison'sratio	(ν)	0,30
Density	(ρ)	8,030 Kg/m^3
Yield Strength	(σ_y)	305 MPa
Tensile Strength	(σ_B)	365 MPa
Shear Strength	(τ)	260 MPa

3.4 Testing Procedure
In this research, the components in the mechanical hand that will be analyzed with the finite element analysis are the connecting pins in the mechanical hand, which are assumed to be the joints of the human fingers, as shown in figure 1.

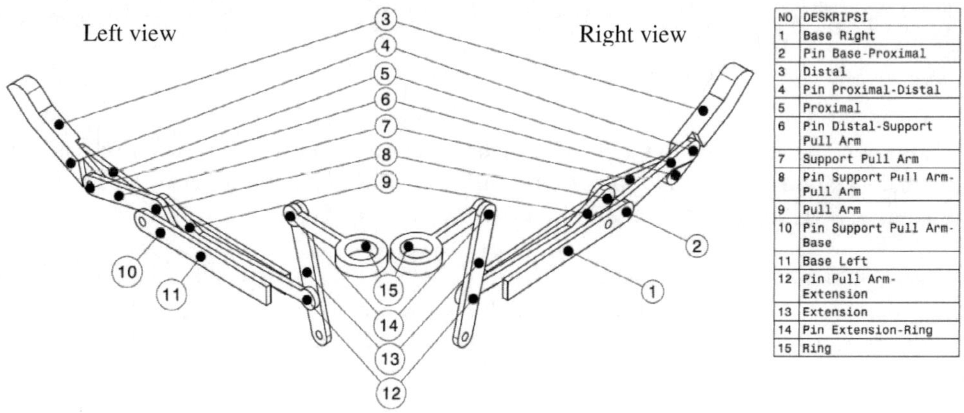

Figure 1. Index finger component to be analyzed using stress and strain analysis [3]

The load given is based on the increment value to time. The minimum increment value is 20 to 708 N/m^2 in 1 minute. The area observed can be seen in figure 2. The position of the loading can be seen in figure 3.

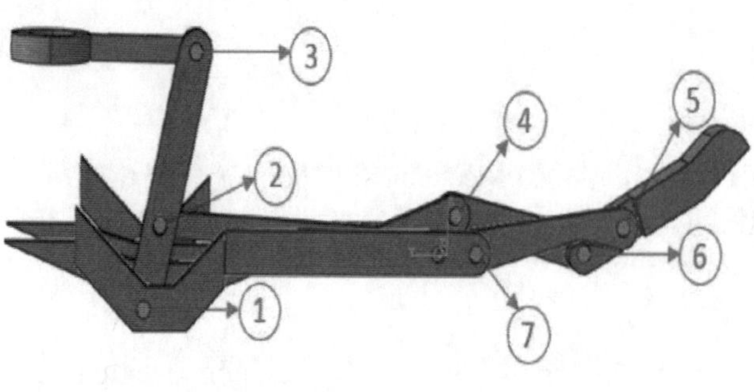

Figure 2. Observed pins

The boundary condition for thedynamic-implicit loading position is chosen at the base. The type chosen is *ENCASTRE* (U1 =U2 =U3 =UR1 =UR2 =UR3 =0).

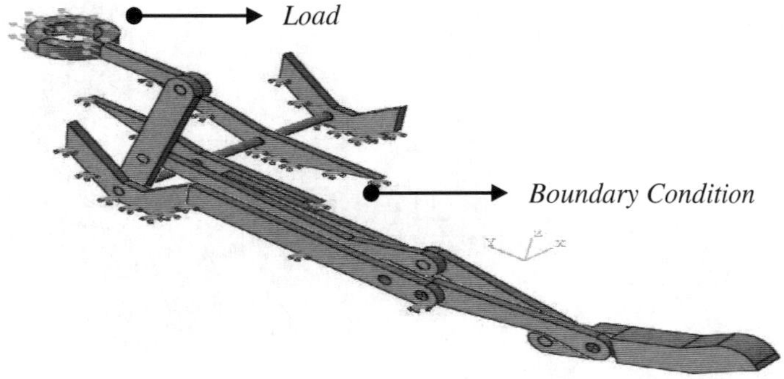

Figure 3. Loading position and boundary condition

Themeshingstep is done with separately or make dependent, with the seed part at 0.004, and the mesh type is a hexahedron. The number of nodes is found to be 10826 and the number of elements 5752.

4. Result and Discussion

4.1 Simulation Result

Below is a depiction of the simulation result of the finger of the mechanical hand. Figure 4 is the result of stress (*von Mises*), with the maximum value at node 358 in the extension part. A contour of red coloring is seen with the value of1,731 x 10^5Pa.

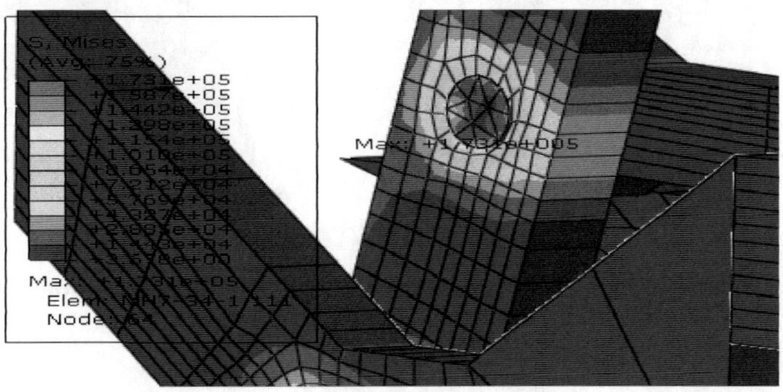

Figure 4. Stress contour occurring at the extension part

Meanwhile for the maximum displacement at node 35 is 1,223 x 10^{-1}m as shown in figure 5, with the brightest red contour on the distal part.

IOP Conf. Series: Materials Science and Engineering **352** (2018) 012017 doi:10.1088/1757-899X/352/1/012017

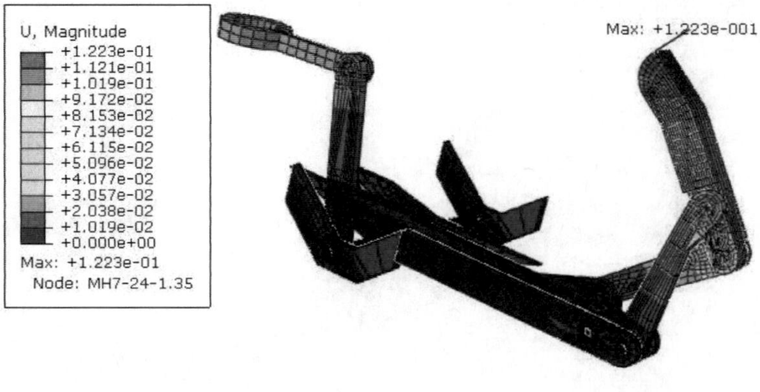

Figure 5. Displacement contour occurring at mechanical hand finger

Figure 6 below shows the contour of the maximum reaction force at node 329 at the base area of the mechanical hand, seen with red contour at 5,864 x 10^{-2} N.

Figure 6. Contour of reaction force at mechanical hand finger

Figure 7 shows the contour of the maximum strain at node 135 at the extension part at 7,441 x 10^{-7}.

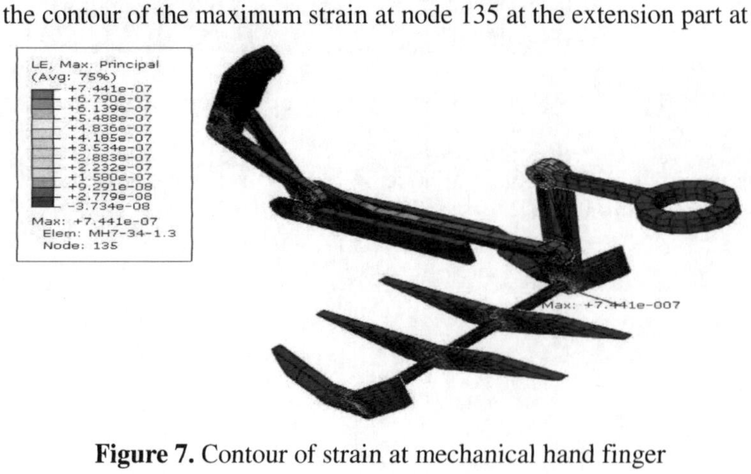

Figure 7. Contour of strain at mechanical hand finger

Considering all the parts, maximum stress (von Mises) occurs on the extension part at 1,731 x 10^5 Pa, and maximum strain occurs at the extension part at 7,441 x 10^{-7}. Considering pin 1 through 7, maximum stress (von Mises) occurs to pin 5, which is the connecting part between the proximal and the distal part. A Maximum strain occurs on pin 2, which is the connecting part of the extension part and the pull arm. Maximum stress displacement occurs on pin 1, which is the connecting part of the extension part and the base.

The yield strength of the material AISI 1010 is 305 MPa, maximum stress*(von Mises)*occurs on the mechanical hand finger is 1,731x10^5 Pa, maximum stress displacement is 260 MPa, stress displacement that occurs on the mechanical hand finger is 2,021 x 10^4 Pa. This means that the design of the mechanical hand is safe for the given load.

5. Conclusion

Based on the stress and strain analysis on the mechanical hand finger, with the dynamic load using the finite element analysis, it can be concluded that:

- Maximum stress value *(von Mises)*is 1,731x10^5 Pa, which occurs on the extension part.
- Maximum strain value is 7,441x10^{-7}, which occurs on the extension part.
- The yield strength of the material AISI 1010 is 305 MPa,with a stress displacement of 260 MPaand maximum stress displacement of 2,021x10^4 Pa. Therefore, it can be concluded that the mechanical hand is safe for the given load.
- The corner of the finger can move ±45 mm from the distal part of the base of the hand.

References

[1] GanipineneiS K, and Venkata R N 2014 Grasp and manipulation of five-fingered hand robot in unstructured environments *International Journal of Engineering Research and Applications* 497-102
[2] Ivan V and Michal K 2013 Kinematic analysis of humanoid robot hand *American Journal of Mechanical Engineering*1 443-446
[3] Sandy Y B Z 2016 Perancangan Pembuatan dan Analisa Tegangan pada Prototipe Tangan Mekanik dengan Pendekatan Kinematik Model (Tesis Teknik Mesin, Universitas Syiah Kuala, Darussalam Banda Aceh) (in Indonesian)
[4] M Souisa 2011 Analisa Modulus Elastisitas dan Angka Poisson Ratio Bahan Dengan Uji Tarik *Jurnal Ilmu Matematika dan Terapan* 51-6 (in Indonesian)
[5] Rao S S 2005 *The Finite Element Method in Engineering*, *4th edition* (Elsevier)
[6] Ridha M, Husaini and Husni 2013 *Material Teknik* (Syiah Kuala University Press.)
[7] http://www.azom.com/article.aspx?ArticleI.(Accessed 25 February2017)
[8] https://icme.hpc.msstate.edu/mediawiki/index.php/1010_Steel. (Accessed 05 February 2017)

IOP Conf. Series: Materials Science and Engineering **352** (2018) 012018 doi:10.1088/1757-899X/352/1/012018

Pitting corrosion as a mixed system: coupled deterministic-probabilistic simulation of pit growth

Israr B M Ibrahim[1,2], S Fonna[2] and R Pidaparti[1]
[1]College of Engineering, University of Georgia, Athens, GA, US, 30622
[2]Corrosion and Computational Research Group (CCRG), Dept. of Mechanical and Industrial Engineering, Syiah Kuala University, Jl. Tgk. Syech Abdur Rauf No. 7 Darussalam, Banda Aceh, Indonesia 23111

E-mail: rmparti@uga.edu

Abstract. Stochastic behavior of pitting corrosion poses a unique challenge in its computational analysis. However, it also stems from electrochemical activity causing general corrosion. In this paper, a framework for corrosion pit growth simulation based on the coupling of the Cellular Automaton (CA) and Boundary Element Methods (BEM) is presented. The framework assumes that pitting corrosion is controlled by electrochemical activity inside the pit cavity. The BEM provides the prediction of electrochemical activity given the geometrical data and polarization curves, while the CA is used to simulate the evolution of pit shapes based on electrochemical activity provided by BEM. To demonstrate the methodology, a sample case of local corrosion cells formed in pitting corrosion with varied dimensions and polarization functions is considered. Results show certain shapes tend to grow in certain types of environments. Some pit shapes appear to pose a higher risk by being potentially significant stress raisers or potentially increasing the rate of corrosion under the surface. Furthermore, these pits are comparable to commonly observed pit shapes in general corrosion environments.

1. Introduction

Pitting corrosion is a destructive process that still poses a major problem in engineering. It is known to occur on stainless steel, and aluminum allows surface that has a protective layer on its surface. Failure to maintain this layer leads to a localized process of corrosion that digs into the depth of metals, and forms pits (and hence it is called pitting corrosion). These pits act as stress amplifier and lead to major failure [1-3]. Several case studies can be found describing the risk of pitting corrosion to engineering components such as boiler tube [4-5], steam turbine [6] and the major problem in aerospace engineering [7].

Pits can assume various shapes and densities. Shapes and densities are two parameters that are used to measure the severity of pitting corrosion [8]. The shapes and densities consequently also affect the stress distribution on metallic surfaces [9-13]. Pits can take various shapes depending on the environment and metal undergoing corrosion. Figure 1 shows some commonly observed pit shapes redrawn from [8].

Pitting corrosion occurs electrochemically, like in most metallic corrosion. The mathematical model of electrochemical reactions in corrosion is typically governed by Laplace equation. This model has been discussed extensively in numerous publications (among them are [14 – 17]). Boundary Element Method (BEM) is a popular choice for numerical method applied for corrosion model since it can be used to compute solution only on the boundary (surface). Since corrosion is a surface phenomenon, this method saves a great deal of computational effort.

However, corrosion itself is a material-degrading process, and this becomes the main focus in pitting corrosion study as it seeks the shape evolution, that is pit cavity growth. A technique that is widely used for this problem is moving mesh method. Mai et al [18] have recently made a remarkable contribution to the modeling of pit cavity growth. In their study, pitting corrosion is seen from the multi-phase physics perspective. The two phases (the metal and corrosive substance) dynamically diffuse and the moving boundary can be conveniently described by the flux of diffusing ions (current density). The diverse shape of pit cavity in this model is induced by the irregularity of metallic grain influencing diffusion. In this study, moving mesh was incorporated into the model along with stochastic approach for pit cavity growth.

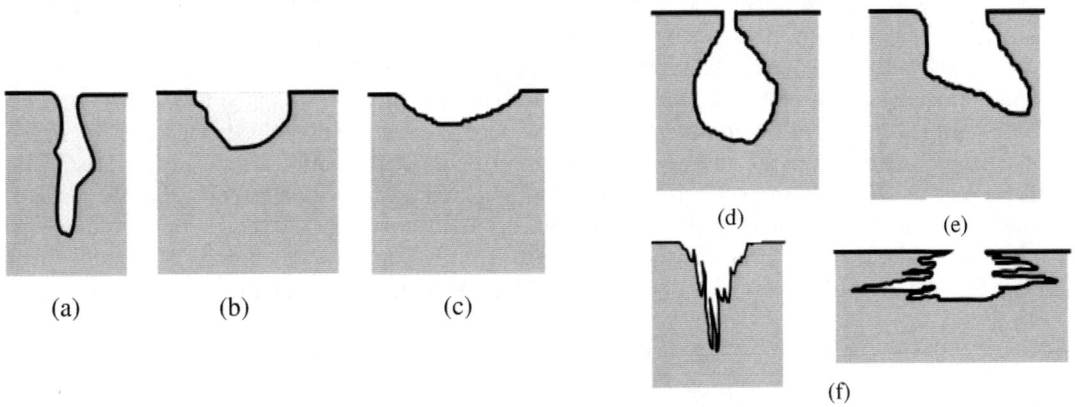

Figure 1. Some examples of common experimentally observed pitting shape, adapted from [3], (a) Narrow and deep, (b) Elliptical, (c) wide and shallow, (d) subsurface, (e) undercutting, (f) shapes influenced by microstructural orientation

Several previous works on pitting corrosion with the stochastic approach are available, including probabilistic model [19-20], Markov-based model [21] and Monte Carlo approach [22]. However, in this study, the electrochemical basis of corrosion was incorporated along with stochastic morphogenesis of pitting cavity. A well-known tool for the stochastic model is Cellular Automata (CA). CA is a method that has been used extensively for modeling non-linear system such as reaction diffusion [23], biological morphogenesis [24], the inflammatory response by immune cells [25-26]. These phenomena typically show stochastic impression at a smaller scale, but as the scale increases, they show apparent patterns. CA is widely used to model this behavior for its practicality in implementation. Several CA model for studying pitting corrosion is available as well. Pidaparti et al developed a CA that models the rate of spread a pit once initiated [27], including the interaction between pits in multi-pit growth [28]. However, this does not account for morphogenesis of pit cavity (growth in the depth direction). Li et al. [29] and di Caprio et al. [30] presented models on pit cavity growth in two- and three-dimensions. Their models are based on complete stochastics CA rules. This study incorporates established an electrochemical model of corrosion (as explained previously) into stochastic CA rules.

In this study, a framework for corrosion pit growth simulation based on the coupling of the Cellular Automaton and Boundary Element Method was developed. The framework assumes that pitting corrosion is controlled by electrochemical activity inside the pit cavity. The BEM provides a prediction of electrochemical activity given the geometrical data and polarization curves, while the CA is used to simulate the evolution of pit shapes based on electrochemical activity provided by BEM. An example case is presented to demonstrate the applicability of the developed framework.

IOP Conf. Series: Materials Science and Engineering **352** (2018) 012018 doi:10.1088/1757-899X/352/1/012018

2. Methods

The model in this study assumes that pitting is controlled by electrochemical activity inside the pit cavity, which is modeled using the BEM, while the CA is used to simulate the evolution of pit shapes based on electrochemical activity provided by BEM. Each of the aspects of BEM and CA is described below along with detailed information regarding how the coupling is carried out to achieve the pitting growth simulation.

2.1 Electrochemical Corrosion

In this study, it is assumed that electrochemical corrosion is the main driving force in the pitting corrosion process, and thus pitting growth. Pitting corrosion is assumed to be a particular electrochemical corrosion problem. This investigation is based on the electrostatic potential distribution that occurs on the metallic surface once electrochemical reactions of corrosion are established, and the growth of pits is stable. Because electrochemical reactions play the main role, the environmental condition is incorporated in the model as a polarization of electrochemical potential.

Pitting corrosion occurs on a metal that forms a passive layer that prevents the metal surface from interacting chemically with its environment. Once an area of passive layer breaks down, a portion of the metallic surface is exposed to the environment, and electrochemical reactions occur in this area. This area can be seen as a local corrosion cell. The surrounding area of the corrosion cell is highly passive; hence the corrosion cell that forms pitting can be seen as similar to a cathodic protection system where the pit area acts as the anode. Hence in this study, pitting corrosion is modeled as an electrochemical system governed by Laplace's equations (Figure 2) with three boundary conditions:

$$\phi = f_1(i) \quad \text{at } \Gamma_1 \qquad (1)$$

$$\phi = f_2(i) \quad \text{at } \Gamma_2 \qquad (2)$$

$$i = 0 \quad \text{at } \Gamma_3 \qquad (3)$$

Where ϕ is electrostatic potential (V), f_1 and f_2 are experimentally determined functions that account for polarization phenomenon, and i is current density (m/A^2). The current density follows the definition of flux of field ϕ,

$$i = \kappa \frac{\partial \phi}{\partial \mathbf{n}} \qquad (4)$$

where κ is conductivity (1/Ω.m), and $\boldsymbol{n}$ is the normal vector.

The Laplace's equation is solved using BEM by following a scheme explained in [31]. By using BEM, the surface of a model is discretized into elements. In this study, only the two-dimensional model is considered. Hence constant line elements are used. However, the CA discretizes the domain into grids of uniform square cells. Thus, to couple the computational power of the two methods, an interface is necessary to be developed which will be explained later.

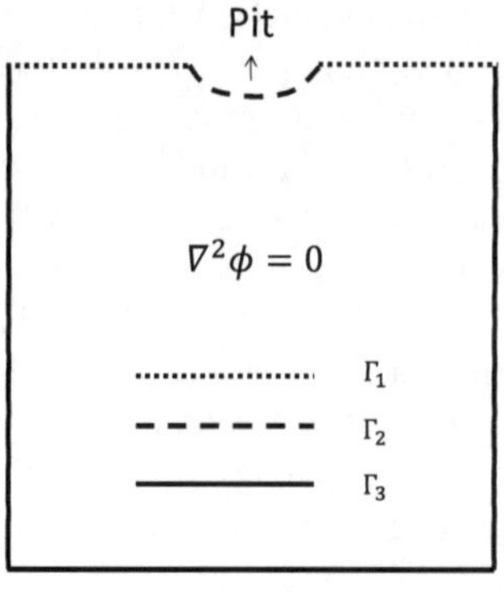

Figure 2. A cross-section of a metallic body undergoing pitting corrosion modeled as a domain governed by Laplace's equation

2.2. *The Cellular Automata*

Cellular Automata (CA) is a model that consists of a lattice of discrete, identical finite-state machines that evolves by a set of transition rules. The CA models dynamics of the system regarding discrete space and time. Hence, a cell in CA evolves only by transition rules to itself and cells in the vicinity (neighborhood). Each cell has its state, which is the binary state in early development, but can be represented in real numbers as well allow modeling of physical phenomena [23] such as diffusion. Several neighborhood schemes are available, but Moore and Neumann neighborhood [23] are typically used. The CA implementation in this study is explained below.

The CA grid in this study consists of rectangular or square cells with uniform size arranged to form the model. Hence, to accommodate these two approaches of discretization, a scheme is needed to interface the results provided by the two methods. In this study, the BEM line elements are used directly to construct the CA grid by taking the dimension of line elements as the dimensions of the squares that compose the CA grid. Hence, the BEM line elements are generated with uniform size. To generate the BEM line elements from the CA grid, the dimension of the uniform squares that build up the CA grid are taken as the dimension of the line elements. Figure 3 illustrates this approach in synchronizing both discretizations of the domain.

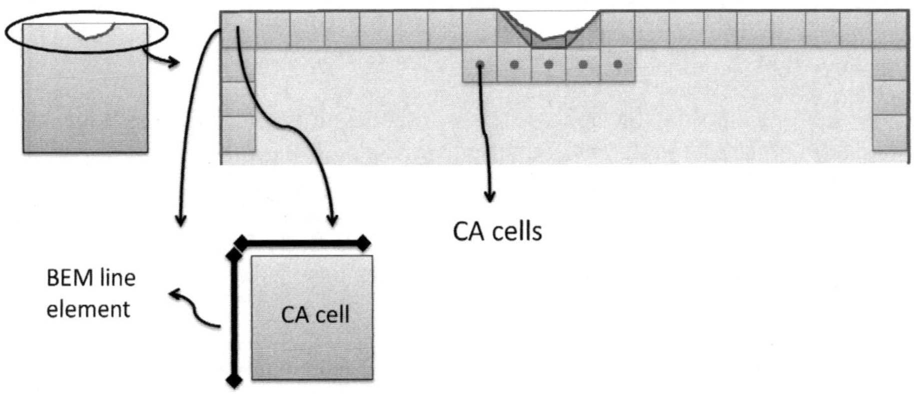

Figure 3. Coupling strategy adopted in CA and BEM

In this study, it is assumed that pit shape evolution is influenced by the electrochemical activity of the local corrosion cell. Hence, the potential distribution over the surface of the local corrosion cell is assumed to play the main role. However, the pit initiation and the path of pit growth are uncertain. In this case, the CA is used to compute the probability of initiation of new pits and the way it grows, based on the condition of the local corrosion cell. In this study, only potential distribution is considered as a major factor in determining the probability of initiation of pits and its growth.

To apply this mechanism, each cell in CA is assigned with three states representing the cathode, anode and the passive state. These can be encoded into integers. The corresponding potential value of a BEM line element is used to calculate the probability of the cell to change state. The influence of a neighboring cell is represented by summing the corresponding potential values of a cell with its neighboring cell's corresponding potential value.

The cathode cells with lower sum-of-potentials have higher chances of changing into anodes. This corresponds to viewing the pitting corrosion as a cathodic protection system, where a minimally protected area of a cathode can still undergo self-corrosion. The passive-state cells represent the subsurface body of the metal. Since corrosion occurs only on the surface, these cells have a higher chance of turning into cathodes or anodes, depending on which neighboring cells exerts the most influence on these cells. The influence is again represented by summing up all the potential values of the neighboring cells. The above-mentioned rules are implemented as below.

1. The anode cells will always be removed (corroded and converted into corrosion product) after each time iteration. If there is any non-electrode cell that shares a boundary with anode cells, then the states of anode cell are transferred into the cell.
2. The cathode and passive-state cells randomly change state into the anode. The chance is calculated as follow,

Cathode cells
$$S_{i+1} = \frac{\sum_{k,l=1}^{N} S_{i_{k,l}}}{m \times N_{cathode} \times E_{corr}} \tag{5}$$

Passive-state cells
$$S_{i+1} = \frac{m \times E_{corr}}{\sum_{k,l=1}^{N} S_{i_{k,l}}} \tag{6}$$

where,

S = Potential value from BEM simulation (V)
E_{corr} = Corrosion potential of cathode or anode (V)
m = number of electrode neighbors
$N\_cathode$ = number of cathode cells
k, l = index of cell and its neighbors

The cathode area is assumed to be highly passive. Once the pit grows, the cathode area reduces and increases the chance of the area to start corroding and grow pits. The neighborhood scheme used in this study is the Moore neighborhood. In this study, the surface is allowed to have multiple pits in this way so that it will consist of original pit growing in its original location and new pits growing and surrounding it. It is to be noted that equation (5) and (6) governs the evolution of the state of the discretization grid (as a "cathode", "passive"), and hence dimensionless.

2.3. Coupling the CA and BEM

The simulation is done iteratively. The BEM computation of potential distribution is carried out for a given domain at each iteration. The computation results (the potential distribution) are then delivered to CA as the basis to compute the rules. Between each execution of BEM and CA, an interface is needed to deliver the data into and from both discretizations. In this study, the discretization of BEM directly corresponds to the CA grid, and thus the line elements of BEM are uniform in size since CA consists of uniform square cells. The overall modeling methodology including the coupling of CA and BEM procedure is shown in Figure 4.

IOP Conf. Series: Materials Science and Engineering **352** (2018) 012018 doi:10.1088/1757-899X/352/1/012018

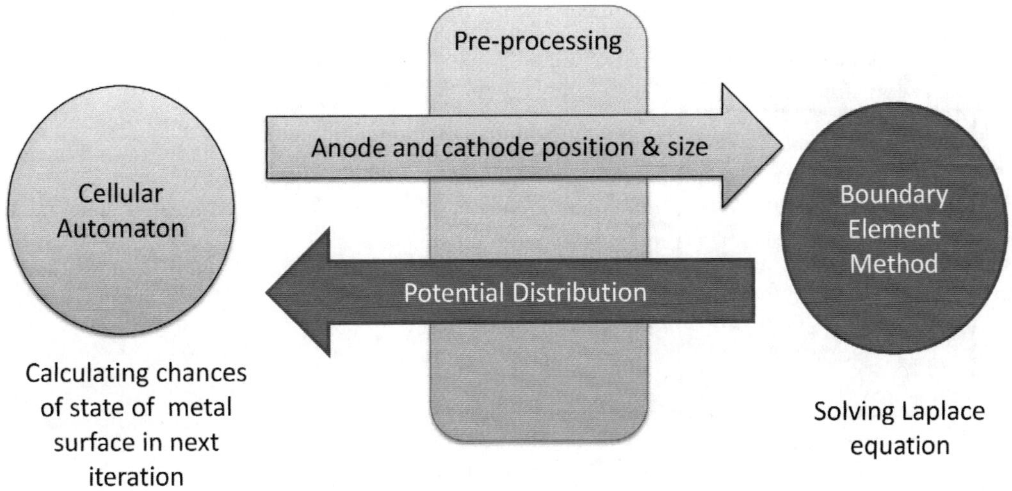

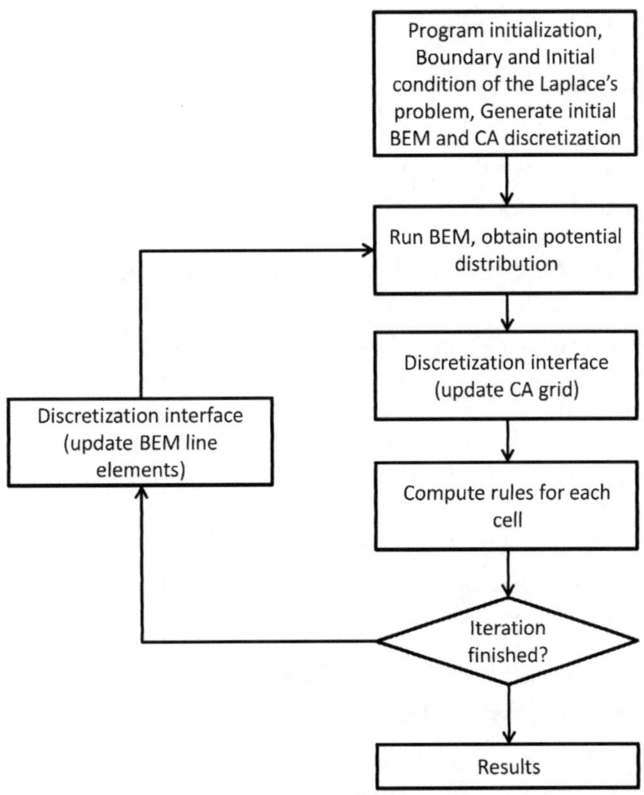

Figure 4. Procedure of the methodology for pitting growth simulation

3. Results and Discussion
A simple case is considered to demonstrate the proposed framework for pitting growth simulation. This case only considers a small localized site on which pitting was already initiated. When a pit initiates, it forms a region of cathode and anode. The anode area is smaller than the cathode area. It is assumed that the current exchange between anode and cathode takes place uniformly in and on the system. Thus, the corrosion cell formed can be modeled as a two-dimensional rectangular area as shown in Figure 5.

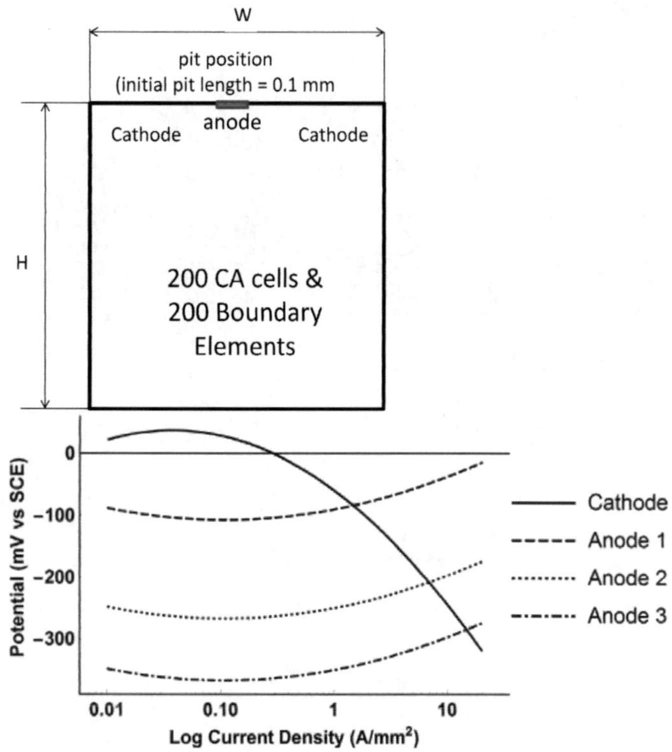

Figure 5. Left: Model of pitting corrosion simulation, unit is in mm. Right: Plot of polarization function 1, 2 and 3.

The simulation was done with different polarization functions that represent the corrosion cell behavior in specific environments, and different widths, W and depths, D, of the region. The width and depth of the region determine the area that currents will have to travel once a corrosion cell is formed. The dimensions are varied into 5x5, 10x10 and 20x20 mm for width x depth. The boundary condition of initial iteration is as in Equation 1,2 and three where the location of the initial pit is assigned as boundary Γ_2. The length of the initial pit is 0.1 mm. The polarization functions used as boundary condition Equation 2 and 3 are as shown in equation (7), (8) and (9), and plotted in Figure 5.

1
$$\begin{aligned} \text{Cathodic: } \phi_c &= -48\ \text{Log}^2(i) - 137\ \text{Log}\ (i) - 60 \\ \text{Anodic: } \phi_a &= 18\ \text{Log}^2(i) + 35\ \text{Log}\ (i) - 90 \end{aligned} \tag{7}$$

2
$$\begin{aligned} \text{Cathodic: } \phi_c &= -48\ \text{Log}^2(i) - 137\ \text{Log}\ (i) - 60 \\ \text{Anodic: } \phi_a &= 18\ \text{Log}^2(i) + 35\ \text{Log}\ (i) - 250 \end{aligned} \tag{8}$$

3
$$\begin{aligned} \text{Cathodic: } \phi_c &= -48\ \text{Log}^2(i) - 137\ \text{Log}\ (i) - 60 \\ \text{Anodic: } \phi_a &= 18\ \text{Log}^2(i) + 35\ \text{Log}\ (i) - 350 \end{aligned} \tag{9}$$

Where ϕ_c is cathode potential (V), ϕ_a is anode potential (V), and i is current density (A/m^2). Polarization function 2 (Equation (8)) is an approximation of polarization of corrosion potential measured in [17]. The rest of the polarization functions are modifications of polarization function 2 to show the effect of variations of polarization in the example case. Figure 5 also shows a plot of the polarization functions. The cathodic polarization function is kept the same while the anode polarization is varied, so it has a larger difference from that of the cathode, thus making the corrosion cell more corrosive. It can also be seen that polarization curves I have lower corrosion current than the rest. As for conductivity κ, it is treated a unity. Conductivity reflects environmental parameter. However, polarization curves (Equation (7), (8) and (9)) have incorporated environmental conditions in this model.

A series of simulation was done for various dimension and polarization curves. Each case consists of 5 trials of simulation. The number of pitting cases and the ratio of width over depth values obtained from the simulation are averaged. Results of the simulations showed that each polarization curve leads to different pit cavity properties. Figure 6 shows the width-to-depth ratio of pit cavity formed by all polarization curves with different corrosion cells. It shows that polarization curve 1 tends to lead to a higher width-to-depth ratio of pits despite corrosion cell size, while the smallest ratio is generated by polarization curve III. This means that pits in environments represented by polarization I lead to a wider opening of pits, while polarization curve III produces pits with a narrower opening. This provides a commonly used measure of pit width-depth ratio, but does not tell us anything about its shape. However, it shows a consistent geometrical trend in the simulation.

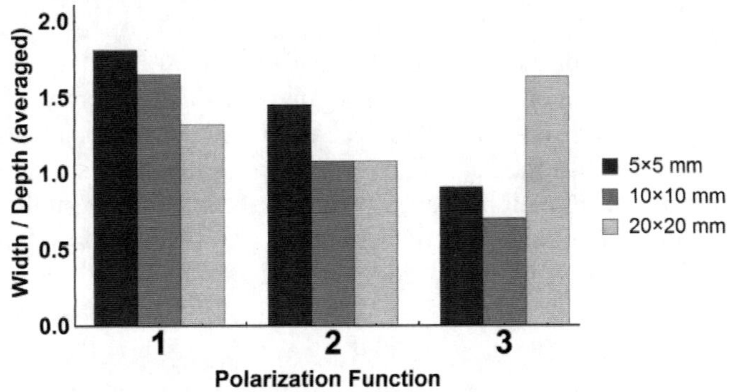

Figure 6. The width-to-depth ratio of pits formed in simulation after t=10 iterations.

Figure 7 shows samples of pit cavity shapes with a corrosion cell size of 5x5 mm in all three polarization curves. It shows the original pit shape after the 15th iteration at the center of the top of the rectangular model and new pits are formed on its right and left. It also shows that Polarization curve 1 tends to produce a conical pit, while polarization curves 2 and three show increasing irregularity of pit cavity shapes.

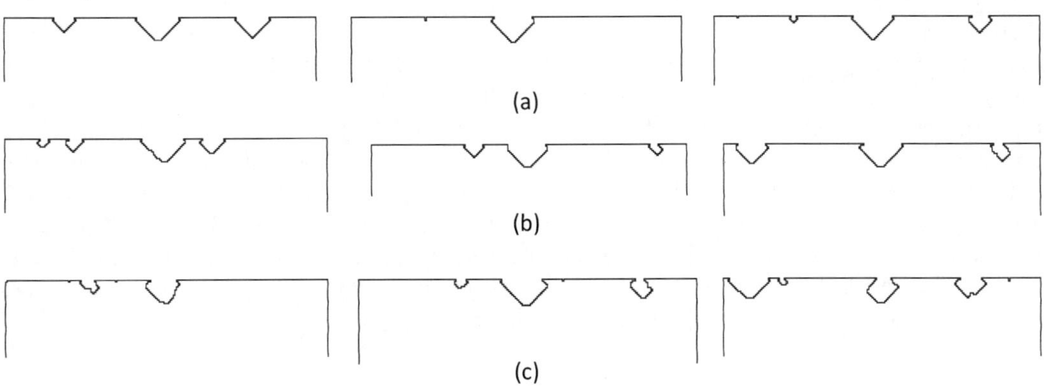

Figure 7. Three samples of pit shape produced at 15th iteration under polarization function (a) 1, (b) 2 and (c) 3

Pit cavity shapes generated by polarization one curve show consistent regular conical shapes despite repeated iterations and trials. However, when the corrosion cell size was increased, polarization 1 produced a more irregular conical pit shape, as shown in Figure 8. This result suggests that the corrosion cell size generated by pitting initiation can also influence pit growth.

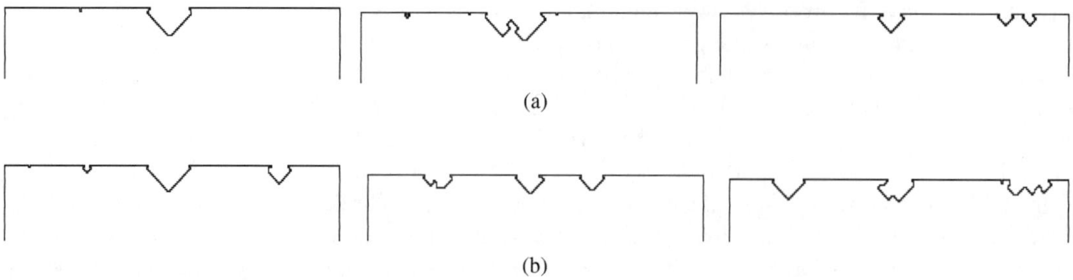

Figure 8. Three samples of pit shape produced at 15th iteration under polarization 1 with corrosion cell size of (a) 10x10 mm and (b) 20x20 mm.

Polarization curves 2 and 3 produced pit shape that tends to grow under the surface. Figure 9 shows a few samples of these pit cavity shapes. These pit shapes may behave differently than the regular conical or elliptical pit shapes since they provide room to accumulate active substance inside the pit and thus lead to the easier generation of subsurface pits. Furthermore, irregular shapes that result from branching inside the subsurface pit may generate higher stress concentrations inside the pit than the regular shapes. Assuming that pitting corrosion initiation is unavoidable, our results show that pits produced by polarization I are safer than ones that are produced by polarization III.

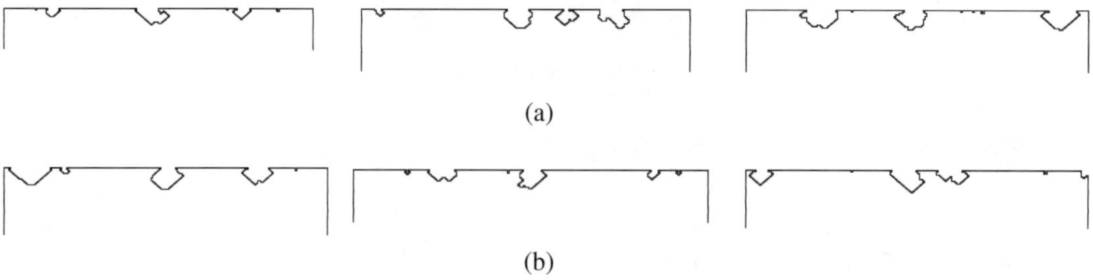

Figure 9. Three samples of pit shape produced at 15th iteration under polarization (a) 2 and (b) 3.

Figure 10 shows only the original pit and its growth from the first iteration to the 15th. It shows typical pits that are produced by in this study. In Figure 10(a), the growth (under polarization 1) shows a consistency of conical shape from the first iteration. This resembles commonly observed conical or elliptical pit shapes, such as illustrated in Figure 1(b) or (c). In Figure 10(b), the pit grew under polarization 3. The pit was initially conical in shape, but then the tendency changed toward subsurface growth. Figure 10(c) shows pit growth (under polarization 3) that was initially elliptical but then grew to a subsurface pit with a narrow opening. This pit grew to resemble the subsurface pit in Figure 1(d). Furthermore, some pits dug into the subsurface of the metal preferentially in a particular direction such as shown in Figure 10(d) and Figure 1(e). Hence, it has been shown that pits can assume different shapes based on the behavior of the local corrosion cell formed by pitting corrosion initiation.

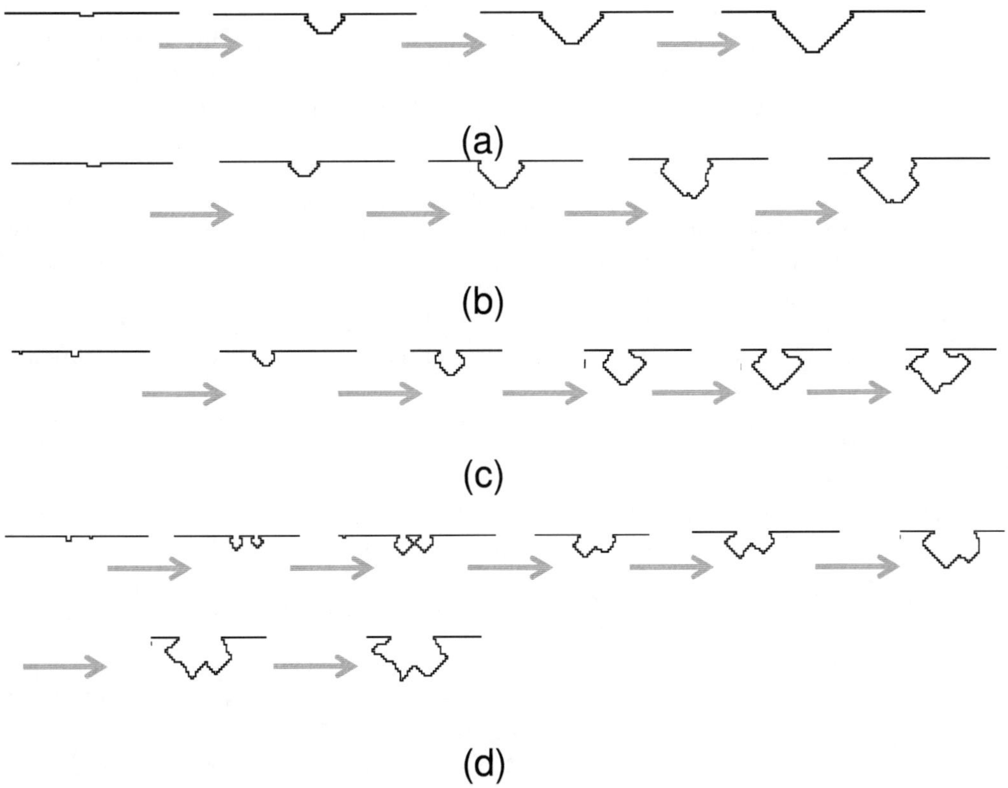

Figure 10. Growth of original pits under (a) polarization 1 with corrosion cell dimensions of 5x5 mm (b) polarization 3 with corrosion cell dimensions of 5x5 mm and (c) 10x10 mm, and (d) 20x20 mm

4. Concluding Remarks

In this study, pit cavity growth was treated as a stochastic event in which deterministic electrostatic model act as the probability distribution. The main finding in this study is that the more corrosive pit cavity increases the chance to produce subsurface pit (depicted in Figure 1(d)). Simulation results suggest that corrosion system with polarization functions 1 (the least corrosive among other two functions) is more likely to produce pit with a high width-to-depth ratio, or similar to an "elliptical" pit (Figure 1(b)). The corrosion system with polarization functions 3 (the most corrosive among the three) is more likely to develop pits with a lower width-to-depth ratio. As shown in Figure 11(c) and (d), these pits are similar to subsurface pits, where cavity continue to grow underneath the surface. These conclusions may not be generalized to every case of pitting corrosion. The experiment designed to measure correlation pit morphologies and potential distribution for a specific case (the type of materials and solutions) is needed to improve and validate the model. This is a planned future study.

Acknowledgements

The first author thanks the Fulbright Foundation for supporting his studies at VCU. RP thanks the US National Science Foundation for partially supporting the research through a grant DMR-0505039.

References

[1] Szklarska-Smialowska Z 2005 *Pitting and Crevice Corrosion* NACE International

[2] Rokhlin S I, Kim J-Y Nagy H, Zoofan B 1999 Effect of pitting corrosion on fatigue crack initiation and fatigue life *Eng. Fracture Mechanics* **62** 425–444
[3] Sankaran K K, Perez R and Jata K 2001 Effects of pitting corrosion on the fatigue behavior of aluminum alloy 7075-T6: modeling and experimental studies *Mater. Sci. Eng. A* **297** 223–229
[4] Heyes A M 2001 Oxygen pitting failure of a bagasse boiler tube *Engineering Failure Analysis* **8** 123–131
[5] Turnbull A and Zhou S 2004 Pit to crack transition in stress corrosion cracking of a steam turbine disc steel *Corrosion Science* **46** 1239–64
[6] Pantazopoulos G, Vazdirvanidis A, Tsinopoulos G 2011 Failure analysis of a hard-drawn water tube leakage caused by the synergistic actions of pitting corrosion and stress–corrosion cracking *Engineering Failure Analysis* **18** 649–57
[7] Hoeppner D W and Arriscorreta Carlos A 2012 Exfoliation corrosion and pitting corrosion and their role in fatigue predictive modeling: State-of-the-art review *International Journal of Aerospace Engineering* **2012**
[8] Frankel G S 2003 Corrosion: Fundamentals, Testing, and Protection *ASM Handbook* ASM International **13**A 236–41
[9] Pidaparti R M and Patel Ronak R. 2008 Correlation between corrosion pits and stresses in Al alloys *Materials Letters* **62** 4497-99
[10] Pidaparti R M, Rao S A 2008 Analysis of pits induced stresses due to metal corrosion *Corrosion Science* **50** 1932–38
[11] Turnbull A and L Crocker L Wright 2010 New insight into the pit-to-crack transition from finite element analysis of the stress and strain distribution around a corrosion pit *Corrosion Science* **52** 1492–98
[12] Cerit M, Genel K, and Eksi S 2009 Numerical investigation on stress concentration of corrosion pit, Engineering Failure Analysis, Volume **16** 2467–72
[13] Acuña N, González-Sánchez J, Kú-Basulto G and Domínguez L 2006 Analysis of the stress intensity factor around corrosion pits developed on structures subjected to mixed loading *Scripta Materialia* **55** 363–366
[14] Brebbia C A and Magureanu R 1987 The boundary element method for electromagnetic problems *Engineering Analysis* **4** 178–185
[15] Aoki S and Amaya K 1997 Optimization of cathodic protection system by BEM *Engineering Analysis with Boundary Elements* **19** 147–156
[16] Fonna S, Huzni S, Ridha M and Ariffin A K 2013 Inverse analysis using particle swarm optimization for detecting corrosion profile of rebar in concrete structure *Engineering Analysis with Boundary Elements* **37** 585-593
[17] Fonna S, Ibrahim I M, Ridha M, Huzni S, and Ariffin A K 2016 Simulation of the Ill-Posed problem of reinforced concrete corrosion detection using boundary element method *International Journal of Corrosion* **2016**
[18] Mai W, Soghrati S and Buchheit R G 2016 A phase field model for simulating the pitting corrosion *Corrosion Science* **110** 157-166
[19] Ahammed M and Melchers R E 1995 Probabilistic analysis of pipelines subjected to pitting corrosion leaks *Engineering Structures* **17**74–80
[20] Yuan X-X, Mao D, Pandey M D 2009 A Bayesian approach to modeling and predicting pitting flaws in steam generator tubes *Reliability Engineering & System Safety* **94** 1838–1847
[21] Valor A, Caleyo F, Alfonso L, Velázquez J C, And Hallen J M 2013 Markov chain models for the stochastic modeling of pitting corrosion *Mathematical Problems In Engineering*
[22] Caleyo F, Velázquez J C, Valor A and Hallen J M 2009 Probability distribution of pitting corrosion depth and rate in underground pipelines: A Monte Carlo study *Corrosion Science* **51** 1925–1934
[23] Schiff J L 2007 *Cellular Automata, A Discrete View Of The World* (Wiley Interscience)

[24] Ermentrout G B and Edelstein-Keshet L 1993 Cellular automata approaches to biological modeling *J. Theor. Biol* **160** 97–133
[25] Ibrahim I, Oruganti S V and Pidaparti R 2017 Simulation of healing threshold in strain-induced inflammation through a discrete informatics model *IEEE J. Biomed. Heal. Informatics* 1–1
[26] Aghasafari P, Ibrahim I B M and Pidaparti R 2017 Strain-induced inflammation in pulmonary alveolar tissue due to mechanical ventilation *Biomech. Model Mechanobiol* **16**
[27] Pidaparti R M, Palakal M J and Fang L 2004 Corrosion pit growth model using cellular automata *AIAA Journal* **42** 2562-2569
[28] Pidaparti R M, Fang Long and Palakal M J 2008 Computational simulation of multi-pit corrosion process in materials *Computational Materials Science* **41** 255–265
[29] Lei Li, Xiaogang Li, Chaofang Dong and Yizhong Huang 2009 Computational simulation of metastable pitting of stainless steel *Electrochimica Acta* **54** 6389–6395
[30] Di Caprio D, Vautrin-UI C, Stafiej J, Saunier J, Chausse A, Feron D and Badiali J P 2010 Morphology of corroded surfaces: Contribution of cellular automaton *Modelling Corrosion Science* **53** 418-425
[31] Aoki S and Kishimoto K 1991 Prediction of galvanic corrosion rates by the boundary element method *Mathematical and Computer Modelling* **15** 11–22

Hemicellulose and lignin removal on typha fiber by alkali treatment

Ikramullah, Samsul Rizal, Sulaiman Thalib, Syifaul Huzni[1]
Department of Mechanical Engineering, Syiah Kuala University, Banda Aceh, Indonesia

E-mail: syifaul@gmail.com

Abstract. One of the methods commonly utilized to alter the surfaces of natural fibers for improving the interface compatibility among fiber and polymer matrix is by alkali treatment. Several natural fibers have been experimented with alkali treatments such as abaca, borassus and kenaf. There is a relatively few of literature that reports the FTIR investigation of Typha fibers. The purpose of this study is to determine the effect of alkali treatment on Typha fiber. Two of three bundle fibers are immersed in a 5% NaOH solution for one and two hours. The chemical structure of alkali-treated and untreated fibers are both being analyzed by Fourier Transform Infrared Spectroscopy (FTIR) instrument. The emergence of peak at 1155.36 cm^{-1} in strong intensity denotes the C-O-C asymmetric stretching in cellulose compound. The lignin composition of the fiber is typified by the stretching band of C-O group at 1247 cm^{-1}. Meanwhile, the peak at 1735.03 cm^{-1} wavenumber is allegedly C=O stretching evidencing the existence of hemicelluloses and pectin. The peaks which are suspected to be hemicellulose, lignin and pectin are no longer visible in alkali treated Typha fiber. Giving alkali treatment to Typha fiber has been successfully removed impurities (hemicelluloses and lignin), as approved by the FTIR analysis. This will lead to a better contact and bonding mechanism between fiber and polymer matrix.

1. Introduction

The advantages gained from using natural fibers as a composite polymer reinforcement are low cost, low density, low energy consumption, renewable and biodegradable [1]. However, natural fibers are less compatible with the polymer matrix due to moderate adhesion between natural fibers and matrix which make natural fibers do not stick well to hydrophobic polymers. On the other hand, natural fibers possessing disadvantages upon composite manufacturing process, i.e. poor moisture resistance, limited processing temperatures, the tendency of fibers to agglomerate into bundles, unequal matrix distribution, and poor wettability upon matrix which resulting weak interface adhesion [2].

Several chemical and physical modifications were tried to reduce the hydrophilicity of natural fiber, both modification aimed to increase fiber's wettability on polymers matrix and improving bonding mechanism between natural fiber and matrix. Physical and chemical modifications have their own different advantages and mechanisms to optimize the characteristics of the composite interface. Electric discharge, generally one of the physical modification methods that modify the surface of natural fibers and structural properties by inserting the bonds to the surface of the fiber, with the change of the fiber surface, bonding mechanism of natural fiber and matrix will increase [3].

[1]syifaul@gmail.com

Meanwhile, chemical modification on natural fibers permanently transforms the properties of fiber by cross-linking cell walls [4].

One of the methods commonly utilized to alter the surfaces of natural fibers for improving the interface among fiber and polymer matrix by alkali treatment. The alkaline treatment is conducted by immersing the fiber in the water with NaOH solution. The advantage of alkali treatment in natural fibers is to make some of the hemicelluloses, lignin, pectin and wax disappear which cover the surface of the natural fiber [5]. The alkali treatment was capable of changing the smooth surface of the natural fibers to be rough, the roughness of the natural fiber surface resulting in better mechanical bonding. Furthermore, alkali treatment improves the bonding of natural fiber and matrix. The mechanical interlocking and bonding mechanism are the two factors responsible for the strength of the composite interface [6]. The alkali-treated natural fibers have been proved to optimize fiber and matrix interface bond which make composites being stronger compared to natural fibers that did not undergo alkali treatment processes. However, it should be noted the overuse alkali concentrations will damage the fibers [7].

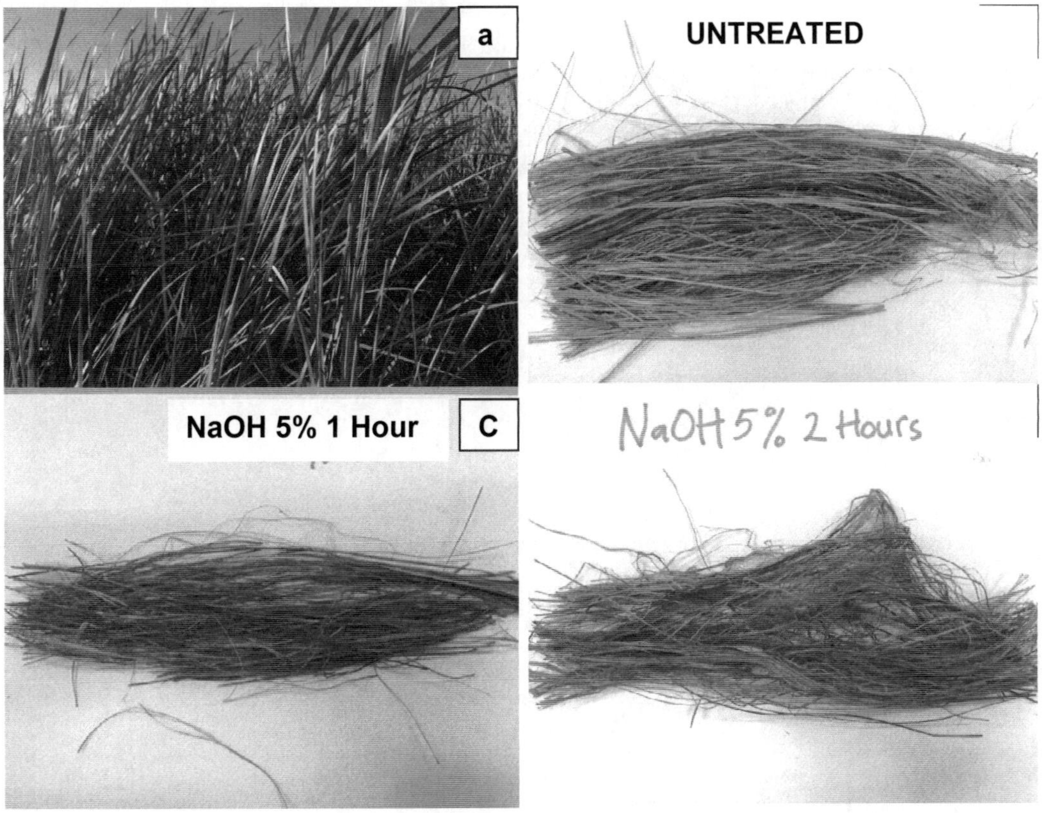

Figure 1. (a) Typha plant ; (b) untreated Typha fiber; (c) Alkali-treated Typha fiber 5 % 1 hour; (d) Alkali-treated Typha fiber 5 % 2 hour

The treatment with alkali has been experimented on various numbers of natural fibers such as abaca, borassus and kenaf [8]. There is a relatively few of literature that reports the FTIR investigation of Typha fibers. Typha plant is commonly found in swamps and is considered being parasitic and it is not widely used. The purpose of this study is to determine the effect of alkali treatment on Typha fiber.

2. Material and method

2.1. Materials

The Typha plants with 1.5-2 meters length were collected from swamps around Syiah Kuala districts, Banda Aceh. Typha trunks which are the base of the plant were manually extracted to obtain fibers, its trunk was hit by timber to ease the stripping of fibers. The fiber then stripped from its stalks by hand, afterward, the fibers were soaked in 5% NaOH solution for one and two hours, then fiber washed by water to quit the effect of alkali on fiber.

2.2. Fourier Transform Infrared Spectroscopy

The chemical structure of alkali-treated and untreated fibers were both analyzed by Shimadzu IRPrestige-21 Fourier Transform Infrared Spectroscopy (FTIR) device. All of the fiber samples were recorded in the 400 to 4000 cm^{-1} wavelength to determine the chemical composition of fibers, such as cellulose, lignin, hemicellulose and other chemical compositions that may be present in Typha fibers.

3. Result and discussion

The effect of alkali treatment on Typha fiber is marked by the disappearance of hemicelluloses, lignin and pectin. It has been observed from the FTIR spectrum that the peaks that are assigned to be hemicelluloses, lignin and pectin have been dismissed from Typha fiber. Two hours of treatment appears to be ineffective because it can damage all contents of the fiber including cellulose, in which it is an important element that will withstand the load when the fiber becomes a reinforcement on the composite. Therefore, the proposed practical procedure is by retting Typha fibers in NaOH solution for an hour as it docs not damage all components of the fiber.

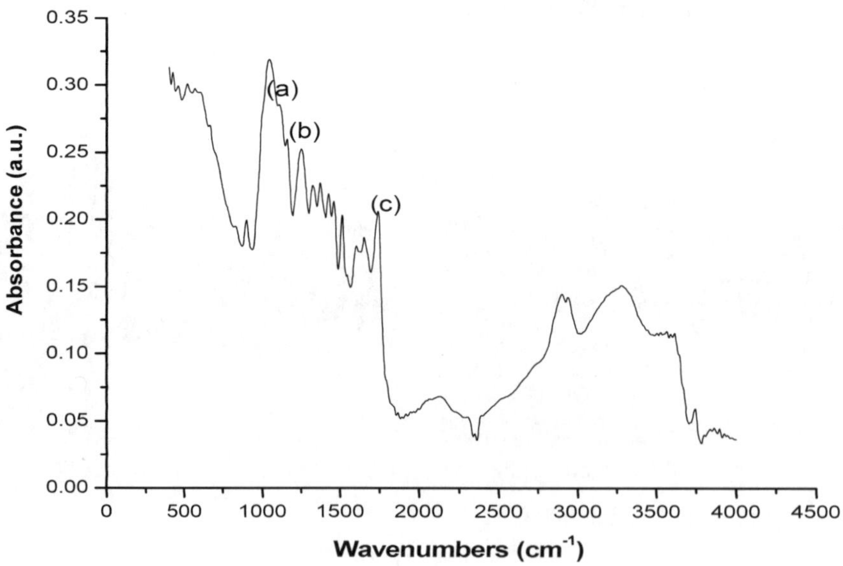

Figure 2. FTIR spectrum of untreated Typha fiber.

Each peak of the absorption bands that resemble various chemical groups as summarized from [8] are presented in table 1. The characteristic of cellulose, lignin and hemicelluloses can still be found in untreated Typha fiber as shown in figure 2. The presence of hemicelluloses can be observed by the appearance of a peak at a wavenumber of 896.90 cm^{-1} that indicates C-OH functional group representing the relationship of β-glycoside, which is a bond between monosaccharide in cellulose and hemicelluloses. As can be seen from figure 2, the emergence of the peak (a) at 1155.36 cm^{-1} in strong intensity denotes the C-O-C asymmetric stretching in cellulose compound. This peak usually appears at a spectrum range of 1070-1240 cm^{-1}. The stretching band of C-O group at 1247 cm-1(b), typifies the lignin composition in the fiber. Meanwhile, the peak (c) at 1735.03 cm^{-1} wavenumber is allegedly C=O stretching evidencing the existence of hemicelluloses and pectin. The medium intensity of the C=O strain delineates the carboxyl and acetyl groups of high-concentrated hemicelluloses content.

Table 1. FTIR Characteristic bands of Typha Fiber.

Wavenumbers (cm^{-1})	**Assignment and Remark**	**Possible Assignment**
896,90	C-OH	The relationship of β-glycosidic is a bond between monosaccharide in cellulose and hemicelluloses
1155,36	C-O-C asymmetric stretch	Cellulose
1247,94	C-O stretching	Lignin
1735,03	C=O stretching	Hemicelluloses and pectin

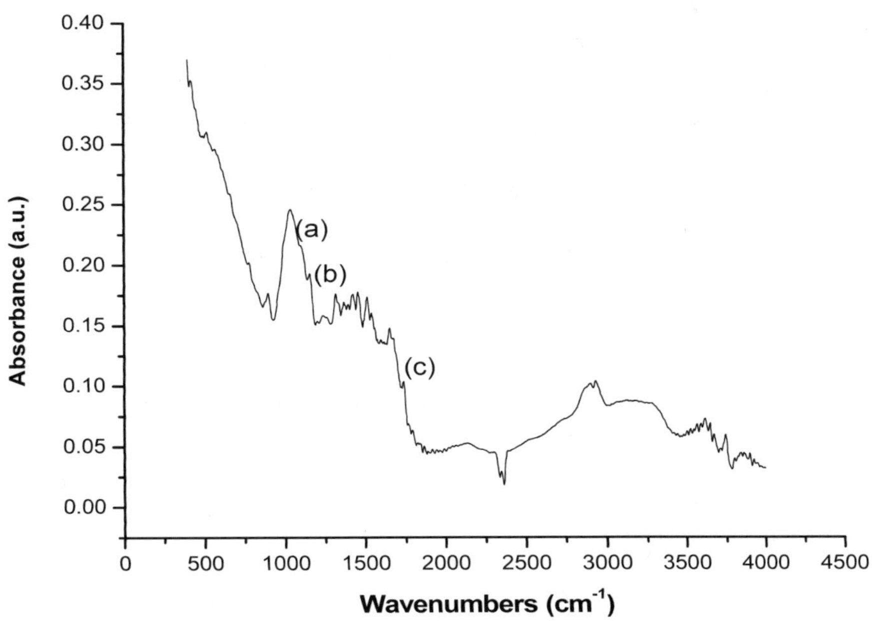

Figure 3. FTIR spectrum of treated Typha fiber for 1 hour.

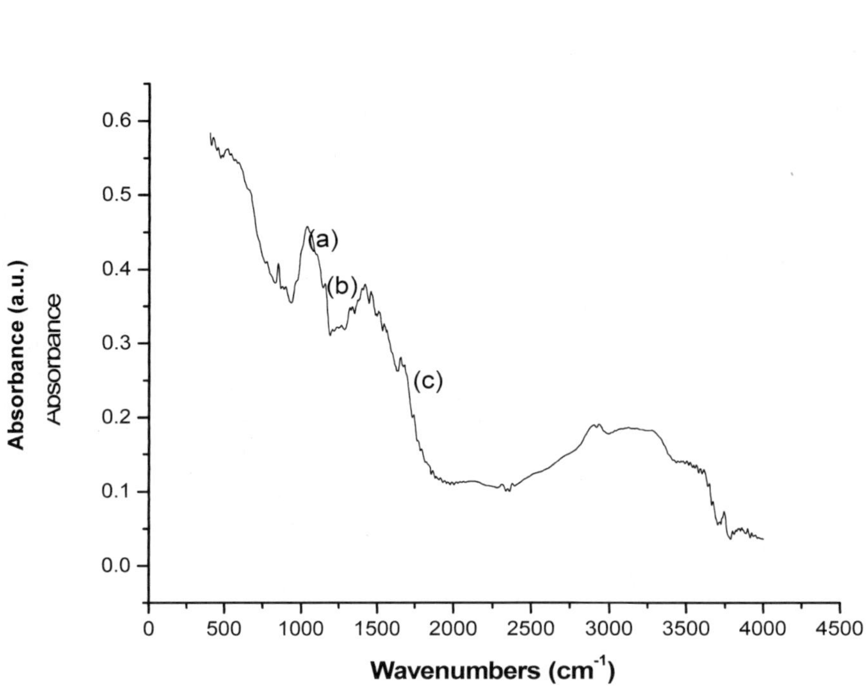

Figure 4. FTIR spectrum of treated Typha fiber for 2 hours.

The FTIR spectrum of alkali-treated Typha fibers for one hour and two hours are presented in figure 3 and 4. The peak (a) at 1155.36cm^{-1} wavenumber which assumed to be cellulose is still clearly visible even after the treatment. Meanwhile, the peak (b) at a wavenumber of 1247 cm^{-1} which indicates the stretching of C-H in lignin is no longer detectable. Similarly, the peak (c) at 1735.03 cm^{-1} which allegedly hemicellulose and pectin have disappeared from Typha fiber. This indicates that hemicelluloses, lignin and pectin have been successfully removed from the fiber by alkali treatment.

From the results of FTIR spectroscopy analysis, it is confirmed that the alkaline treatment changes the three major chemical composition components of Typha fiber, namely; cellulose, hemicelluloses and lignin. The hemicelluloses content of Typha fiber is affected by the duration of alkaline treatment, because hemicelluloses are more sensitive to NaOH than cellulose or lignin. This can be assumed that increasing the duration of the alkaline treatment will eliminate the hemicelluloses content. However, longer duration of alkaline treatment, will decrease the level of cellulose, and increase the lignin content. The possibility of this occurrence is because the longer immersion time in alkaline will cause damage to cellular molecular structures [7].

4. Conclusion

The experimental results show that the untreated Typha fiber still has a lot of fiber contamination. The characteristic of cellulose, lignin and hemicelluloses can still be found in untreated Typha fiber. Giving alkali treatment to Typha fiber has been successfully removed impurities (hemicelluloses and lignin), as approved by the FTIR analysis. This will lead to a better contact and bonding mechanism between fiber and polymer matrix.

Acknowledgments
Acknowledgements awarded to KEMENRISTEKDIKTI, which has funded this research through the scheme "Pendidikan Magister Menuju Doktor Untuk Sarjana Unggul (PMDSU)" 2017

References

[1] Malkapuram R, Kumar V and Negi Y S 2008 Recent development in natural fiber reinforced polypropylene composites *Journal of Reinforced Plastics and Composites* **28** 1169-1189
[2] Azwa Z, Yousif B, Manalo A and Karunasena W 2013 A review on the degradability of polymeric composites based on natural fibres *Materials & Design* **47** 424–42
[3] Zhou Y, Fan M and Chen L 2016 Interface and bonding mechanisms of plant fibre composites: An overview *Composites Part B: Engineering* **101** 31–45
[4] Xie Y, Hill C A, Xiao Z, Militz H and Mai C 2010 Silane coupling agents used for natural fiber/polymer composites: A review *Composites Part A: Applied Science and Manufacturing* **41** 806–19
[5] Mohanty A, Khan M A and Hinrichsen G 2000 Surface modification of jute and its influence on performance of biodegradable jute-fabric/Biopol composites *Composites Science and Technology* **60** 1115–24
[6] Ouajai S and Shanks R 2005 Composition, structure and thermal degradation of hemp cellulose after chemical treatments *Polymer Degradation and Stability* **89** 327–35
[7] Wang B, Panigrahi S, Tabil L and Crerar W 2007 Pre-treatment of Flax Fibers for use in Rotationally Molded Biocomposites *Journal of Reinforced Plastics and Composites* **26** 447–63
[8] Cai M, Takagi H, Nakagaito A N, Katoh M and Ueki T, Waterhouse G I and Li Y 2015 Influence of alkali treatment on internal microstructure and tensile properties of abaca fibers *Industrial Crops and Products* **65** 27–35

A comparison of the structural strength between fiberglass and jute fiber in the Acehnese Traditional Boat *Jalo Kayoh* using finite element method

Akram[1], Iskandar Hasanuddin[1], Nazaruddin[2], M. Syahril Anwar[1], Zulfan[2], and Norhafizan Ahmad[3]
[1]Department of Mechanical and Industrial Engineering, Faculty of Engineering University of Syiah Kuala, Banda Aceh, Indonesia.
[2]Department of Informatics, Faculty of Mathematics and Natural Sciences, University of Syiah Kuala, Banda Aceh, Indonesia.
[3]Department of Mechanical Engineering, Faculty Engineering, University of Malaya, Kuala Lumpur, Malaysia.

E-mail: akram@unsyiah.ac.id

Abstract. The Acehnese traditional boat, known as *Jalo Kayoh*, is a mean of transportation used by Acehnese fishermen. The main constituent of the boat is wood. However, due to the decline of high-quality wood supply and as a preventative measure of illegal logging, fiberglass and jute fiber are used instead of wood. This study compares the strength of the two materials using finite element method. The *Jalo Kayoh* model plan stands at 4m in length, 0.6 m in width, and 0.4 m in height. A 2500 N static load is applied to the surface, using a C3D10 quadratic tetrahedron 0.02 mesh. The result of the simulation to the fiberglass is a maximum displacement of 7.123 x 10^{-5}m, while the jute fiber has a maximum displacement of 2.255 x 10^{-4} m. The maximum stress stands at 1.612 x 10^{6} Pa for the fiberglass and 1.523 x 10^{6} Pa for the jute fiber. The maximum strain occurs at 1.654 x 10^{-5} for the fiberglass and 4.581 x 10^{-5} for the jute fiber. To conclude, fiber glass has more stress 1.05 % and less strain 2.76 % than jute fiber and both the materials can sustain the load given.

1. Introduction

The Acehnese traditional fishing boat, known as *Jalo Kayoh*, is a mean of transportation used by the fishing community to work and perform activities on the water, be it in a river, a lake, or the sea. *Jalo Kayoh* is traditionally built using wood as the main material. The quality of wood that degrades over time makes *Jalo Kayoh* susceptible to degradation and decay. This cannot be prevented as the fishing community are dependent on wood in the construction and maintenance of the *Jalo Kayoh*.

In this study, the model plan of the *Jalo Kayoh* has the specification of the length of 4m, the width of 0.6 m, and height of 0.4m. Its main function is to help fishers in their water activities such as short distance transportation, catching fish, crabs, shrimp, and pond rehabilitation. *Jalo Kayoh* needs to have a good hydrodynamic design and a strong structural design. With the decrease of high-quality wood supply and as a preventative step to combat illegal logging. The durability of wood used as the main component of the boat lasts between two to three years [1], it makes wood a less ideal material to use in the long term.

Nowadays, people are starting to switch to synthetic materials such as fiberglass, using resin as the adhesive and fiberglass as the filling. The quality of the product made from synthetic fibers is comparatively good that the production has grown locally and globally. The previous study shows that a fiberglass based composite (chopped strand mat and woven roving) using the adhesive 157 resin polyester as the base material in constructing the Acehnese traditional boat *Jalo Kayoh* has a very good physical and mechanical property [2].

Jute fiber (*Corchorus olitorius*) is fiber from a plant that can grow well in Indonesia. It can grow all year long with a big number of production. Jute fiber is used in the covering and packaging of harvested goods, commonly known as gunny sacks. Due to its elasticity, researchers are interested in developing jute fiber as a material in constructing the traditional Acehnese boat *Jalo Kayoh*.

2. Literature Review

2.1 Composite Material Jute Fiber
The application of composite material jute fiber is very efficient in enduring load and force. Because of that, the composite material jute fiber is very strong and firm when the load is applied in the same direction as the fiber, yet it is inadequate when the load is applied perpendicular to the fiber direction [3]. The modification of the jute fiber surface with alkali is very effective in increasing the adhesive bond between the fiber and the matrix in the composite polypropylene with a jute fiber strengthener. The application of the alkali can remove the wax and fat residue on the fiber, which makes it difficult to apply the resin. Based on the observation using SEM-micrograph, the jute fiber without alkali looks softer compared to the one treated with alkali, which looks tougher. [4]

2.2 Fiber Reinforced Plastic (FRP)
The laminate of Fiber Reinforced Plastic (FRP) composition has high stiffness. Moreover, it has strength, tolerance of fatigue damage, fatigue strength for weight ratio, low thermal expansion coefficient, low corrosion, high internal damping and low maintenance cost. [5]

Table 1. Mechanical Properties of Natural Fiber [6]

Fiber	Tensile Strength (MPa)	Young's modulus (GPa)	Elongation at break (%)	Density (g/cm^3)
Abaca	400	12	3-10	1.5
Bagasse	350	22	5.8	0.89
Bamboo	290	17	-	1.25
Banana	529-914	27-32	5.9	1.35
Coir	220	6	15-25	1.25
Cotton	400	12	3-10	1.51
Curaua	500-1150	11.8	3.7-4.3	1.4
Flax	800-1500	60-80	1.2-1.6	1.4
Hemp	550-900	70	1.6	1.48
Jute	410-780	26.5	1.9	1.48
Kenaf	930	53	1.6	-
Pineapple	413-1627	60-82	14.5	1.44
Ramie	500	44	2	1.5
Sisal	610-720	9-24	2-3	1.34
E-glass	2400	73	3	2.55

2.3 Finite Element Method (FEM) Analysis
Finite element method (FEM) is a method that is used to analyze a construction. This method is used in the construction of ships, on-shore, and off-shore buildings. The application of this method is not limited to the field of steel construction, but also fluidics. Structural analysis using finite element method allows the researcher to understand the spread of the stress on the construct that is being analyzed. The failure of construction can be shown by using this analysis. It also reveals the exact point of such failure. This will, in turn, allow planners to make modifications and fortifications on the

construction where the failure has been identified. The load applied to the *Jalo Kayoh* is 2500 N/m^2. The type of the load is body force.

Figure 1. Load display and the boundary condition of the boat's structure

2.4 Element and Structure Flexibility
An element's flexibility is the ability of an element or a structured system to follow the shape change of the element or the structure due to an external load. It can be formulated that the relationship between the external element internal force (F), the element's flexibility (d) and the displacement/deformation (Δ).

$$\Delta = d.\ F \tag{1}$$

Where,

$$d = \frac{1}{k} \tag{2}$$

Flexibility (d) is the opposite of rigidity (K) [7].

3. Result of Simulation

From the simulation using the material fiber reinforced plastic, the result is a maximum displacement of 7.123 x 10^{-5}m at the node 1.8922. The displacement contours can be seen in Figure 2 below, and also Table 2 shows the maximum displacement value.

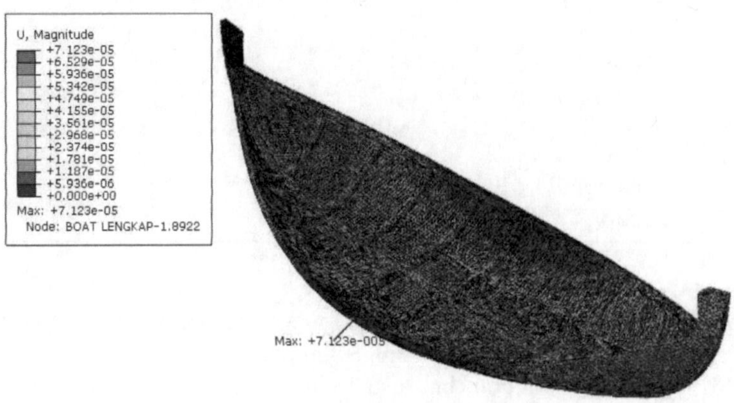

Figure 2. Displacement contours in the material fiber reinforced plastic

From the simulation using the material jute fiber, the result is a maximum displacement of 2.255 x 10^{-4} m at the node 1.8922. The displacement contours can be seen in figure 3 below.

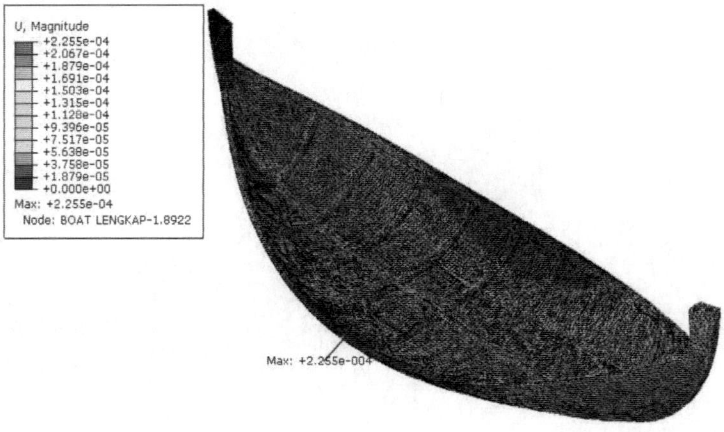

Figure 3. Displacement contours in the material jute fiber

Table 2. Maximum displacement value

Force (N)	Fiber Glass (m)	Material Jute (m)
0	0	0
2.49	7.20E-06	2.26E-01
4.98	1.42E-05	4.51E+00
8.71	2.49E-05	7.89E+00
14.32	4.09E-05	12.96 E-05
22.72	6.49E-05	20.57 E-05
24.90	7.12E-05	22.55 E-05

The Figure 4 below shows the comparison between the displacement that occurs in the fiberglass and the one that occurs in jute fiber, where it is known that natural fiber jute is stronger than the natural fiberglass.

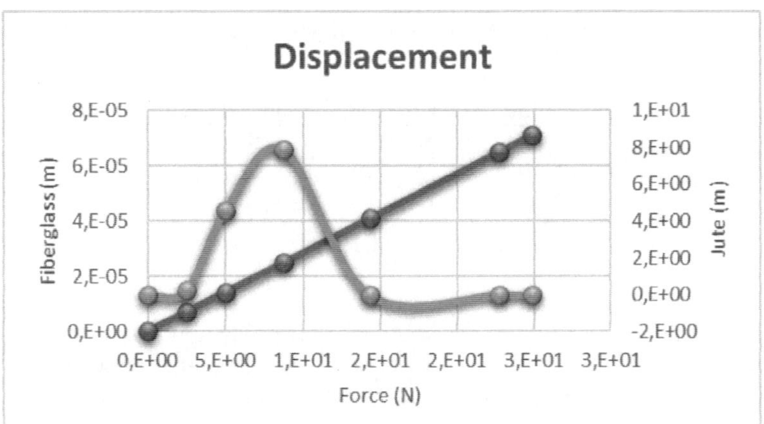

Figure 4. Comparing displacement between fiberglass and jute fiber

From the simulation using the material fiberglass and jute fiber, the result is the maximum stress of 1.523 x 10^6 Pa and a maximum strain 4.581 x 10^{-5}. The contours for each number are shown in figure 5 and figure 6 below.

Figure 5. Stress contours (*von Mises*) of the material fiber reinforced plastic and jute fiber

Figure 6. Strain contours of the material fiberglass and jute fiber

The table 3. below shows the maximum of the stress and the strain between the material fiberglass and jute fiber.

Table 3. Maximum of stress value *(von Mises)* and strain between fiberglass and jute fiber

Material	Stress	Strain
Fiberglass	1.612 x 10^6	1.654 x 10^{-5}
Jute fiber	1.523 X 10^6	4.581 x 10^{-5}

4. Conclusion
Some things that be concluded from the simulation are:
1. Fiber glass has more stress 1.05 % and less strain 2.76 % than jute fiber.
2. Both materials in *Jalo Kayoh* simulation can endure the load applied with 2500 N.

Acknowledgment
Thank you to Indonesian Ministry of Research, Technology and Higher Education for the funding of research with the No. 15/SP2H/LT/DPRM/IV/2017.

References

[1] Badan Statistik Provinsi Aceh 2015 *Aceh In Figures* BPS Provinsi Aceh [Province of Aceh] (in Indonesian)

[2] Akram and Udink Aulia 2015 Pembuatan perahu nelayan berbahan plywood diperkuat serat ijuk dan resin [Manufacture of Fishing Boats Made From Plywood In Reinforcing Ijuk Fiber and Resins] *Jurnal Pengabdian Kepada Masyarakat* [*Jurnal of Community Service*] (Banda Aceh : Syiah Kuala University) (in Indonesian)

[3] Budianto 2015 Analisis kekuatan struktur pada Kapal Wisata Sungai Kalimas [Analysis of Structural Strength on The Tourist Boat of Kalimas River] *Teknik Desain dan Manufaktur* [*Engineering Design and Manufacturing*] (Politeknik Perkapalan Negeri Surabaya, Indonesia) (in Indonesian)

[4] Liu X Y and Dai G C 2007 Surface modification and micromechanical properties of jute fiber mat reinforced polypropylene composites *eXPRESS Polymer Letters* 1 299-307

[5] Jauhari N, Mishra R and Thakur H 2015 Natural Fibre reinforced composite laminates – A review *International Conference on Materials Processing and Characterization* **2** 2868-2877

[6] Sanjay M R, Arpitha G R & B Yogesha 2015 Study on Mechanical Properties of Natural - Glass Reinforced Polymer Hybrid Composites : A Review International Conference on Materials Processing and Characterrization 2959 – 2967

[7] Suharjanto and Prasetyo 2016 *Analisis Struktur dengan Metode Matriks* [Structural Analysis with Matrix Method] 33 (in Indonesian)

IOP Conf. Series: Materials Science and Engineering **352** (2018) 012021 doi:10.1088/1757-899X/352/1/012021

Study on pipe deflection by using numerical method

Husaini, Amir Zaki Mubarak and Rizki Agustiar
Mechanical and Industrial Engineering Department, Engineering Faculty, Syiah Kuala University, Jl. Tgk. Syech Abdurrauf No. 7 Darussalam – Banda Aceh 23111, Indonesia

E-mail: husainiftm@unsyiah.ac.id

Abstract. Piping systems are widely used in a refinery or oil and gas industry. The piping system must be properly designed to avoid failure or leakage. Pipe stress analysis is conducted to analyze the loads and critical stress occurred, so that the failure of the pipe can be avoided. In this research, it is analyzed the deflection of a pipe by using Finite Element Method. The pipe is made of A358 / 304SS SCH10S Stainless Steel. It is 16 inches in size with the distance between supports is 10 meters. The fluid flown is Liquid Natural Gas (LNG) with the range of temperature of -120 ° C to -170 ° C, and a density of 461.1 kg / m 3. The flow of LNG causes deflection of the pipe. The pipe deflection must be within the permissible tolerable range. The objective is to analyze the deflection occurred in the piping system. Based on the calculation and simulation, the deflection is 4.4983 mm, which is below the maximum limit of deflection allowed, which is 20.3 mm.

1. Introduction

Piping systems are widely used in flowing fluids from one to another component. They are used in simple mechanical applications, such as draining system and agricultural drainage pumping system [1][2]. They are also used in industrial applications such as in refineries, oil and gas industries. For most industrial purposes, the piping system must be properly designed to avoid a crack, failure or leakage [3] [4][5][6][7] [8].

In designing a piping system, the behavior of the piping system due to loading should be understood while considering the design codes and standards. The behavior of piping systems includes displacement, acceleration, voltage and force. In the world of engineering, it is known as pipe stress analysis [3]. Pipe stress analysis is conducted to determine the critical stress and loads that occur in the pipeline so that the failure of the pipe can be avoided.

In Arun LNG plant in the city of Lhokseumawe in Aceh province, there are many piping systems from the upstream sector to the distribution sector. The piping system has a very long track and is supported by slipper type support. The fluid flown inside the pipe is Liquid Natural Gas (LNG), with the range of temperature is from -120 ° C to -170 ° C and the density is 461.1 kg / m^3. The flow of LNG becomes a burden for the pipe that causes the deflection. This research analyzes the effect of pipe deflection by using Finite Element Method. The deflection of the pipe must be within the permissible tolerable range.

IOP Conf. Series: Materials Science and Engineering **352** (2018) 012021 doi:10.1088/1757-899X/352/1/012021

2. Method

2.1. Material specification

The pipe analyzed is located in Unit 68 of the LNG piping system (Loading Unit) as can be seen in Figure 1. It is a ten meters pipe supported by slipper pedestals at both edges with a flange in between. The model of the analyzed pipe can be seen in Figure 2.

Figure 1. Analyzed piping system.

The material of the pipe is Stainless Steel A358 / 304SS SCH10S (the detail is shown in Table 1). In addition, the code of Flange material is A182-F304, CL 150, SCH10S, WN-RF and the code of bolt material is A320-B8 and A194-BT, (FOR 4"-150LB FLG JOINT). For the gasket, the material code is COMP-ASB RF gaskets 1.5mm LGIBTAZ T # 1000 Harrow COMP - NON-ASBESTOS / REINFORCED PIPE, ISO LB (T = 1, 5 - 2.5mm THK).

Table 1. Chemical composition Stainless Steel A358 / 304SS [9].

Grade	Chemical Composition %					Tensile Test		
	C	Si	Mn	P	S	Ultimate Tensile Strength (N/mm^2)	Yield Strength (N/mm^2)	Elongation
304	0.08	0.75	2.00	0.045	0.030	586	241	55

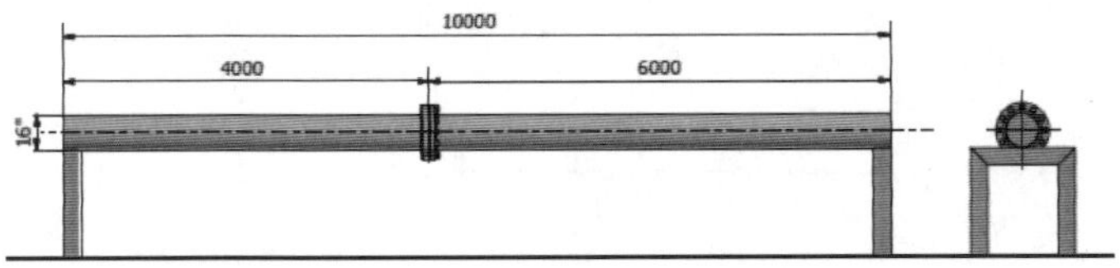

Figure 2. 2D model of the pipe.

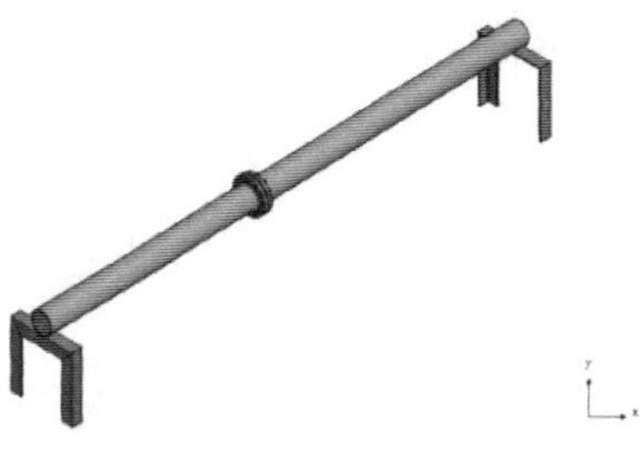

Figure 3. 3D model of the pipe.

2.2. Modeling

The pipe is modeled by using Autodesk Inventor. The 3D model is shown in Figure 3. The simulation is conducted by using ANSYS. The material properties inputted in the modeling process is shown in table 2.

IOP Conf. Series: Materials Science and Engineering **352** (2018) 012021 doi:10.1088/1757-899X/352/1/012021

Table 2. Material Properties [9].

Property	Value	Unit
Poisson's Ratio	0.26	N/A
Modulus Elasticity Tension	193	GPa
Mass Density	8.03	g/cm^3
Yield Strength	290	MPa
Ultimate Strength	621	MPa

3. Result and Discussion

3.1. Analytical Analysis

The deflection is calculated analytically based on ASTM (American Society of Testing Materials) pipe standards and then analyzed numerically. The diameter of the analyzed pipe is 16 inches, pipe density is 8.03 g/cm^3, the temperature inside the pipe is -162 ° C, and the pressure inside the pipe is 9.0 MPa.

Table 3. Analytical result

Property	Value
Weight of LNG	559,48 N
Weight of the Pipe	4948,889 N
Total Weight of Pipe dan LNG	5508,369 N
Moment of Inertia	$1.276 \times 10^{-4} m^4$
Weight per Unit	$550,83\ {}^{N}/_{m}$
Modulus of Elasticity	$193 \times 10^{9}\ {}^{N}/_{m^2}$

The deflection of the pipe is obtained initially by calculating the weight of LNG and the pipe. Afterward, the moment of inertia and the modulus of elasticity is calculated. The result is tabulated in Table 3. Based on the result, the deflection (y) is calculated:

$$
\begin{aligned}
y &= -\frac{5ql^4}{384\,EI} \\
&= -\frac{5\ .\ 559{,}8\ .\ (10)^4}{384\ .\ (193\,x\,10^9)\ .\ 1{,}275\ .\ 10^{-4}} \\
&= -\frac{5\ .\ 550{,}83\ .\ (10)^4}{9{,}44928\ \mathrm{x}\ 10^9} \\
&= \mathbf{-2{,}91\ mm} \\
y &\approx \mathbf{-3\ mm}
\end{aligned}
$$

Whereas:

y = Deflection
q = Weight per unit
l = Length of pipe
E = Modulus of elasticity
I = Moment of Inertia

Based on the book "Steel Pipe: A guide for Design and Installation, 4th Ed." By AWWA Staff, the maximum allowed the limit of flexible lined and coated = 5% of the outer diameter of the pipe [10]. The maximum amount of deflection allowed is

$= 5\% \, x \, \text{OD}$
$= 5\% \, x \, 406 \text{ mm}$
$= 20{,}3\text{mm}$

Whereas:
OD = Outer Diameter of Pipe

Compared with the deflection calculation result, which is 3 mm, the occurred deflection is below the allowed deflection, which is < 20.3 mm that the construction is safe.

3.2. Numerical Analysis

3.2.1. Mesh
The process of meshing is performed on the entire range of pipe components with automatic meshing. The mesh application is shown in Figure 4 and 5.

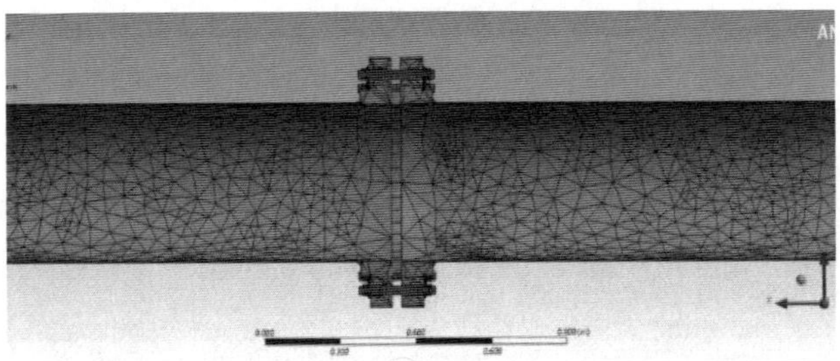

Figure 4. Meshing In the middle section of the pipe.

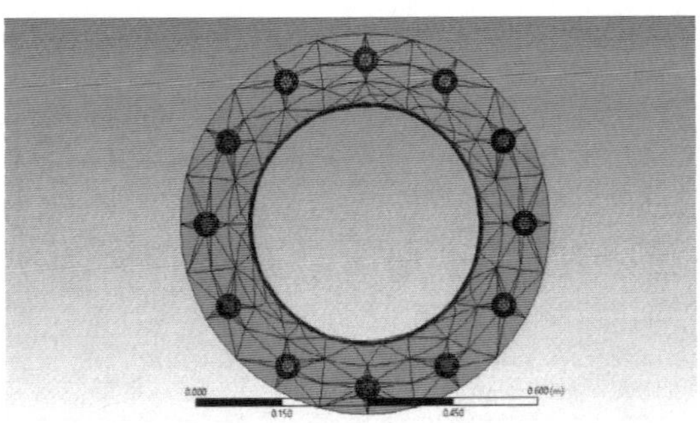

Figure 5. Meshing on the cross-section view of the pipeline.

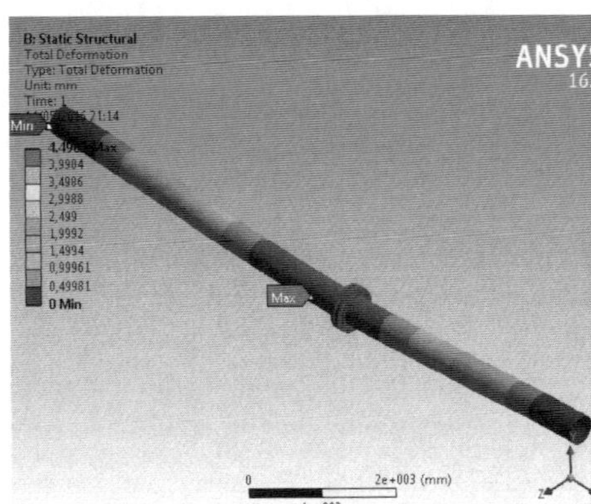

Figure 6. Deflection with maximum load on the pipe.

3.2.2 Deflection
The result of the deflection simulation for maximum load on the pipe can be seen in Figure 6. The image shows a color change generated due to voltage conditions caused by deflection in the pipe, it

IOP Conf. Series: Materials Science and Engineering **352** (2018) 012021 doi:10.1088/1757-899X/352/1/012021

appears that the maximum deflection occurs in the middle of the pipe, with the deflection of **4.4983 mm.** The deflection is within the allowed limits.

Figure 6 and 7 show the value of the largest deflection in the middle of the pipe with a distance of 500 cm from the edge. The deflection is **4.4983 mm.** The smallest visible deflection is at the end of the pipe, which is **0.49981 mm.**

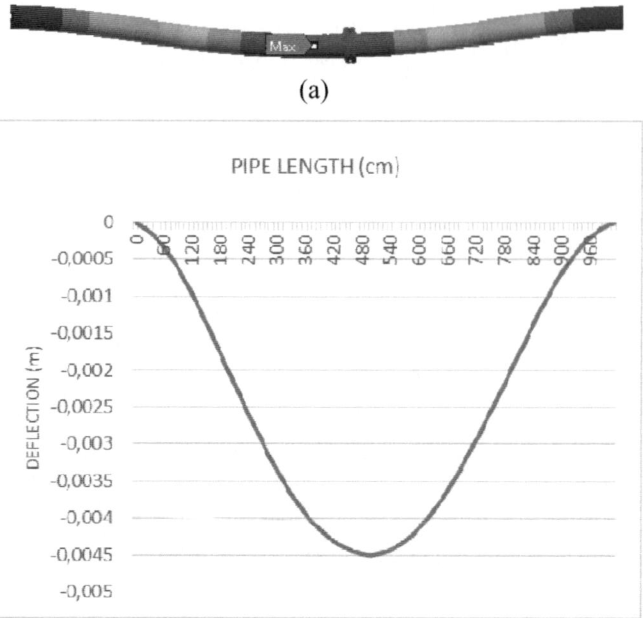

(a)

(b)

Figure 7. (a) Pipe deflection on the XY axis. (b) The deflection the pipe that occurs in every 10 cm on the XY axis.

3.2.3 Stress

The simulation results of maximum equivalence stress (Von Misses) that occur on the pipe can be seen in Figure 8. The maximum equivalence is 100.39 MPa. The maximum shear stress is 38.435 MPa. The shear stress result can be seen in Figure 9.

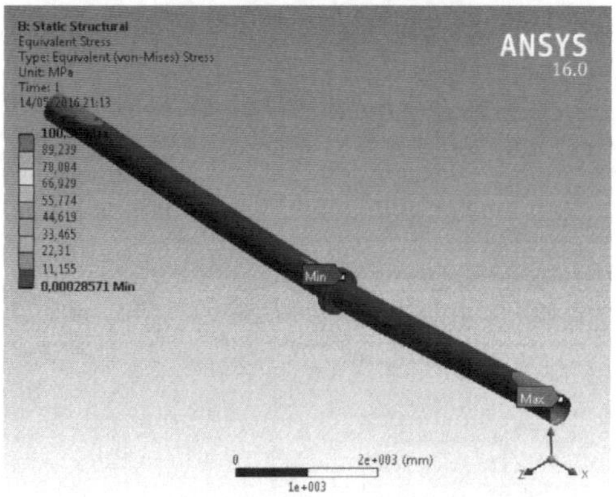

Figure 8. Equivalence stress (Von Misses) that occurs in the pipeline.

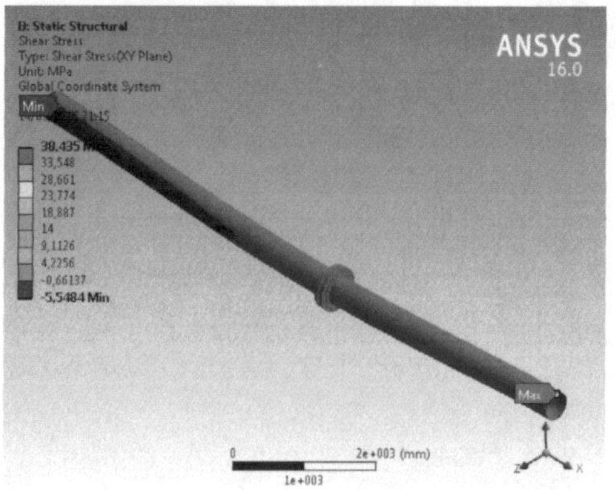

Figure 9. The shear stress result.

The simulation result *of normal stress* can be seen in Figure 10. The figure shows that the maximum normal stress is 79.176 MPa.

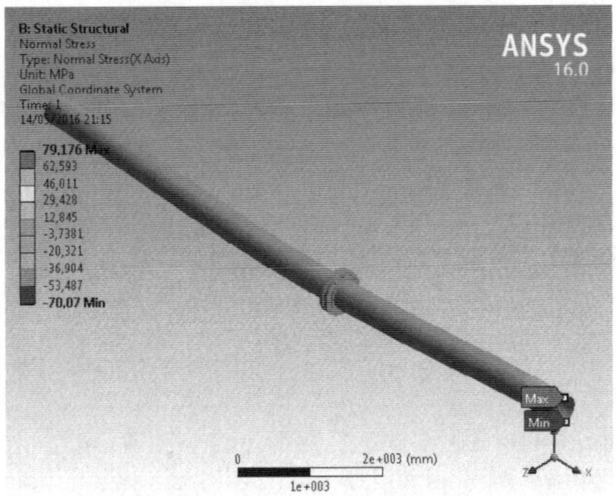

Figure 10. Normal stress result.

4. Conclusion

Based on the calculation and simulation, the occurred deflection is safe as it is below the maximum deflection limits, which is 20.3 mm.

The result of the pipe deflection numerical calculation, by the distance between the pedestals of 10 meters, shows the deflection of 4.4983 mm. The maximum equivalence values stress (Von Misses) is 100.39 MPa. The maximum shear stress is 38.435 MPa, and the maximum Normal Stress is 79.176 MPa.

References

[1]. Munarak A Z, Maulana M I, Syuhada A 2016 Design and testing of fish drier system utilizing geothermal energy resource in Ie Suum, Aceh Besar, *AIP Conf. Proc.* **1717** 050011-1 – 050011-7

[2]. Syuhada A, Mubarak A Z, Maulana I 2016 Study of hybrid power system potential to power agricultural water Pump in Mountain Area, *AIP Conf. Proc.* **1717** 050013-1 – 050013-8

[3]. Thesis UMY 2016 Sistem Perpipaan https://thesis.umy.ac.id/datapublik/t25134.pdf. (Accessed 26 January 2016) (in Indonesian)

[4]. Husaini, Mubarak A Z and Qadri N 2016 Analysis of deformation effect on the Support of Steel Pipe A106 by using Finite Element Method *Proceedings of SNTR III* 118-125

[5]. Husaini, Mirza, Ali M and Mahmud M N 2014 Analisis kegagalan daerah lasan pipa stainless steel sebagai media reboiler pabrik pupuk *Prosiding SNTTM XIII* 223-228 (in Indonesian)

[6]. Tsyss V G, Sergaeva M Yu 2016 Finite element analysis of discharge antivibrational pipe stress state of the Piping System Flexible Joint *Procedia Engineering* 251-257

[7]. G Mathan, N Siva Prasad 2012 Study of dynamic response of piping system with Gasketed Flanged Joints using Finite Element Analysis *International Journal of Pressure Vessels and Piping* 28-32

[8]. Husaini, and Kishimoto K 2001 Mixed mode fracture behavior of PC/ABS blends, Proceedings of SPIE - The International Society for Optical Engineering **4317** 111 - 116

[9]. AK Steel 2016 A354/304SS www.aksteel.com. (Accessed 27 January 2016)

[10]. AWWA Staff 2004 *Steel Pipe: A guide for Design and Installation, 4th Ed*

IOP Conf. Series: Materials Science and Engineering **352** (2018) 012022 doi:10.1088/1757-899X/352/1/012022

Ceramic tools insert assesment based on vickers indentation methodology

Husni, Muhammad Rizal, M. Aziz M and M. Wahyu
Department of Mechanical and, Industrial Engineering, Faculty of Engineering
Syiah Kuala University, Banda Aceh, Indonesia

Email: husniusmn@gmail.com

Abstract. In the interrupted cutting process, the risk of tool chipping or fracture is higher than continues cutting. Therefore, the selection of suitable ceramic tools for interrupted cutting application become an important issue to assure that the cutting process is running effectively. At present, the performance of ceramics tools is assessed by conducting some cutting tests, which is required time and cost consuming. In this study, the performance of ceramic tools evaluated using hardness tester machine. The technique, in general, has a certain advantage compare with the more conventional methods; the experimental is straightforward involving minimal specimen preparation and the amount of material needed is small. Three types of ceramic tools AS10, CC650 and K090 have been used, each tool was polished then Vickers indentation test were performed with the load were 0.2, 0.5, 1, 2.5, 5 and 10 kgf. The results revealed that among the load used in the tests, the indentation loads of 5 kgf always produce well cracks as compared with others. Among the cutting tool used in the tests, AS10 has produced the shortest crack length and follow by CC 670, and K090. It is indicated that the shortest crack length of AS10 reflected that the tool has a highest dynamic load resistance among others insert.

1. Introduction

In practice, the machining process is optimized based on factors, such as tool failure, cutting temperature, cutting force or power, chatter, surface finish and dimension errors left on the workpiece. The performance of the cutting tool, or the machinability, is normally measured on the basis of the life of tool edge, surface texture, and chip volume ratio as a function of tool geometry. In relation to ceramic tools, especially in the application in interrupted cutting, failure of the ceramic tool is occurred due to chipping or fracture. In interrupted turning, the tool undergoes a cyclic load and when the tool enters the work material, the cutting edge suffers a shock load or the impact force.

Ceramic cutting tools are very popular in engineering industries because of their better performance in high cutting speeds. This tool can be operated at high material removal rates. In comparison with usual cutting tool materials such as high-speed steel and cemented carbides, the most important aspects of ceramics are their higher wear resistance, higher hot hardness and higher chemical stability. Desperate limitation s of these tools are the ceramic properties of low tensile strength and poor toughness [1].

Vickers Indentation method has widely used in the evaluation of hardness of the hard material, when this method applied to the brittle material the impression of indentation along with crack length produced is used to measure the fracture toughness. The main advantage of the indentation technique is its simplicity, usefulness, and the fact that the fracture initiating cracks approximate natural defects in the brittle material [2]. The results obtained are typically almost equal to the direct test method but minimal specimen preparation, and less amount of material [3, 4].

2. Fracture Toughness

The fracture toughness is an important property required to evaluate the mechanical performance of structural materials. One method used to measure fracture toughness is using Vicker's hardness indentation which yields critical intensity factor (K_{IC}) value of Mode I. This method is principally useful for low fracture toughness of brittle material since it is simple, rapid and permit small sample size. Furthermore, fracture toughness also represents the capability of brittle material to resist the growth of an existing micro crack or flaw. Principally, the indentation technique needs to measure the length of cracks formed around the indentation of the load in excess of the critical load, required to induce cracks surrounding the indentation. The fracture toughness (K_{IC}) of a material is presented by the energy per unit area that is required to produce new crack surfaces and thereby propagate a crack through the material. This value is also known as the critical stress intensity factor represents the effective local stress at the crack tip. It should be emphasized that the fracture toughness value measured using this method is usually not as accurate as those from other more conventional tests. This method relies heavily on the use of indentation models. On top of that, there are a variety of expressions available in the literature and the prospect of selecting the most suitable one for a particular test condition can be varied. Furthermore, most of the models are based on ideal crack patterns that may or may not be present in the tested due to the difference of work materials and indenter to be used. This will affect the result of the indentation test, causing inconsistency between fracture toughness values obtained by the indentation method and those obtained from the conventional techniques.

Normally, fracture toughness (K_{IC}) carried out either using double cantilever beam, a single edge notched beam or chevron-notch test, these methods need complex testing procedure, the samples require a relatively large size of specimens and time-consuming. However, indentation methods require a small amount of specimen in the size of the cracks formed surrounding the indentation impression and at loads in excess of the critical load required to initiate the cracks [4].

When an indenter penetrates a brittle, three different zones can be found in the material. Just below the wedge, where a hydrostatic core produced due to the high compressive stress induced beneath the indenter. Outside the hydrostatic core, the large strain develops due to the pushing action of the core. In this area, tensile cracks may be initiated starting from pre-existing flaws in the material.

Indentation crack may be categorized into two groups according to the geometrical shape of crack just below the indentations surface. First is in the shape of the halfpenny crack system (also called median or radial- median crack system) and the other is in the shape of the Palmqvist crack system [2, 4-6]. To recognize between them is to polish the surface layers of the cracks produced in the indentation zone. The median cracks system, the cracks remain connected to the corner of the diagonal while the Palmqvist crack system, the cracks become detached as shown in figure 1. Lateral cracks are generated in the form of fracture of materials on the surface due to microscopic chipping occurred around an indentation. This fracture is ultimately caused by the propagation of cracks just below the plastic deformation zone and penetrates to the surface specimen during the unloading sequence. Lateral crack has been studied very rarely, as they are usually considered as a secondary indentation crack system that does not impart the strength and seems difficult to model.

From *Mikio Fukuhara et al.* [7], the value of fracture toughness K_{IC} were determined from the hardness indentation tests using hardness-testing machine with a Vickers' diamond pyramid indenter. The K_{IC} value was used to show the performance of cutting tool in machining of Cast Iron with high speed and feed. The K_{IC} value was calculated by the following equation for the median crack:

$$K_{IC} = \frac{0.203 \cdot H \cdot a^{1/2}}{(c/a)^{3/2}} \dots\dots\dots \quad (1)$$

Where H is the Vickers hardness indentation value, 2a is the length of diagonal impression and 2c is the total of indentation crack length. The crack is examined under a scanning electron microscope.

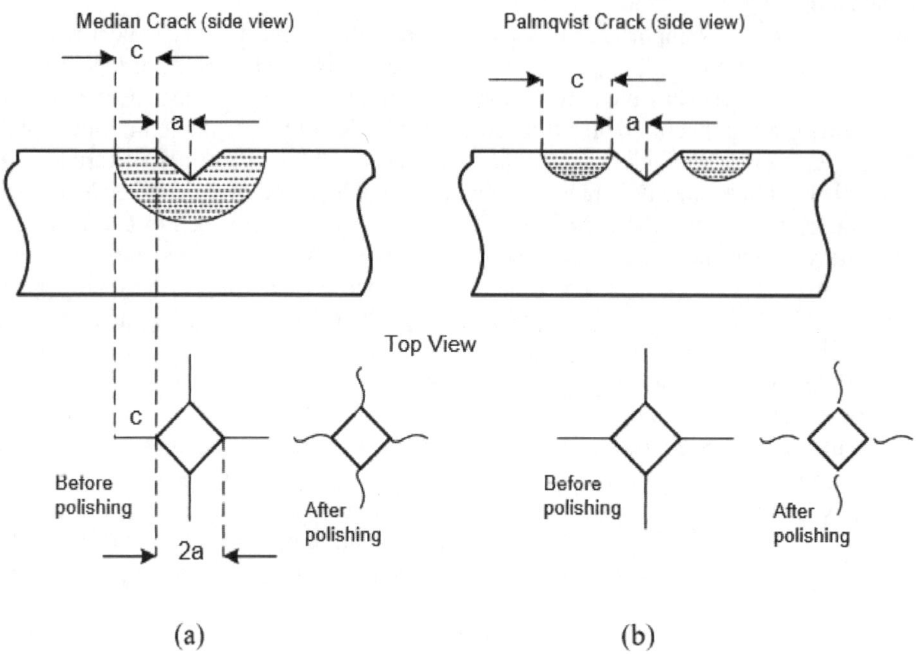

Figure. 1 (a) Median crack system around Vicker Indentor (b) Palmqvist Crack System around indentor

3. Experimental Procedure

In indentation tests, the square pyramidal diamond indenter has been used to evaluate the ceramic tool inserts. Four different loads were used during indentation tests; 0.2, 0.5, 1, 2.5, 5 and 10 Kgf. The cutting tool used were AS10 (S_3N_4) denoted by sample A, CC650 (70%Al_2O_3 +22.5% TiN+7.5%TiC) denoted by sample B, and K090 (70%Al_2O_3 +30% TiC) denoted by sample C as shown in figure 2. Each sample was polished with diamond paste started with 2.5, 1 and finished with 0.5 μm before being tested. The purpose of polishing the tool is to eliminate the residual stress zone on the surface [5]. All indentation were conducted with a constant indenter dwell time of 15s at room temperature. After indentation, the lengths of the cracks formed are immediately measured. The cracks length arising at the four corners of indentation impression were measured, and the total of four crack length of the indentation was divided by four. However, small size cracks around the indentation and cracks branched from the main corner cracks were ignored [4]. The value of fracture toughness K_{IC} was analyzed by using Eq.1.

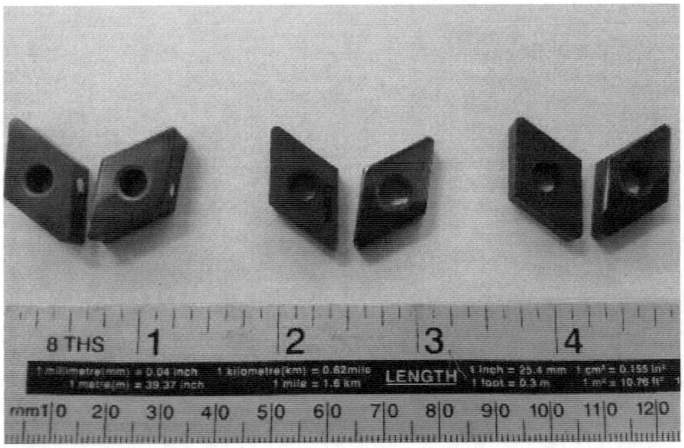

Figure 2. Types of cutting tools used

4. Results and Discussion

In indentation, the cracks are initiated at the corner of the impression. Nevertheless, only well behave cracks are used for analysis. Based on visual observation during indentation tests, the crack is formed only after unloading. The median crack systems were formed on all types of the tool. The lateral cracks are formed around the surface of the indentation impression figure 2, 3 and 4 and in certain cases, microscopic chipping occurs as seen in figure 5. Microchipping also occurs after unloading. During penetration of the indenter into the body, plastic deformation or destructive of the material occurs beneath the surface. However, the indenter brings about high compressive stresses in the contact region at the same time, inhibiting the separation of atoms. However, as soon as the indenter leaves the penetration point, the situation changes. The plastically deformed thin surface layer maintains tensile stresses under the surface. Since no compressive stress acts in this region now, these tensile stresses may generate one or more cracks despite their small magnitude. Owing to the elastic rebounding of the relieved material, even minute fragments may split off the surface.

4.1. Selection of Suitable Indentation Load

The results show that, the magnitude of indentation load is influenced by the hardness impression and crack formation significantly. Figure 3 (a), (b), (c), (d), (e) and (f) show the results of Vickers indentation of AS10 (Si_2N_3) ceramic with different indentation loads. From the figure, it is seen that the crack length and the hardness impression of Si_2N_3-based ceramic increases with increasing indentation load. The indentation impression induced by indentation loads of 5 kgf is produced better crack length compare to other loads. At lower indentation loads, measurement uncertainly encounter since the small hardness load having difficulties in producing surface crack on each corner of indentation impression. At the higher loads, the breakage and cracking of cutting tool surface make the measurement tending impossible. There is also influence the indentation size, wherein the hardness of the material decreases with increasing indentation load.

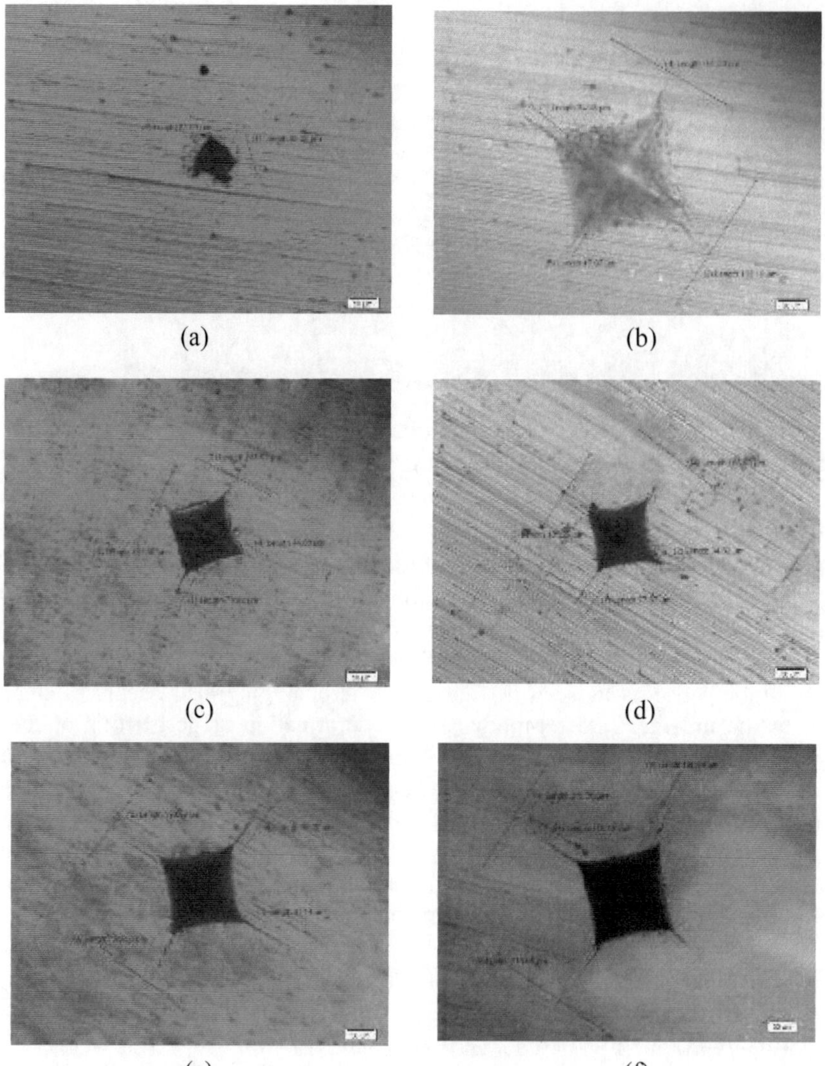

(a) (b)

(c) (d)

(e) (f)

Figure 3. Identation Impression for different load of AAS10 (a) 0.2 kgf, (b) 0.5 kgf, (c) 1.0 kgf, (d) 2.55 kkgf, (e) 5.0 kgf dan (f) 10 kgf

IOP Conf. Series: Materials Science and Engineering **352** (2018) 012022 doi:10.1088/1757-899X/352/1/012022

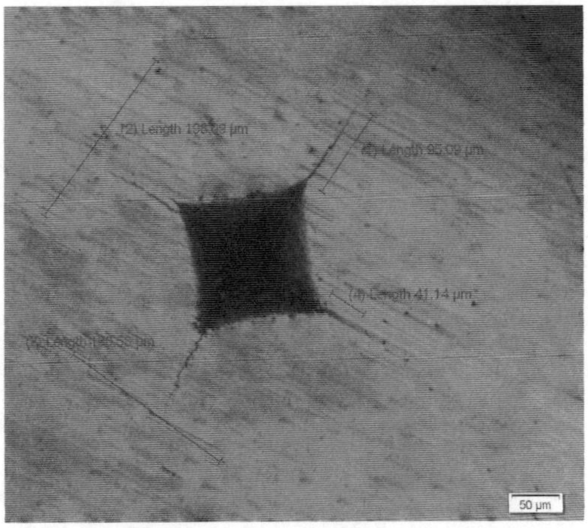

Figure 4. Indentation made at 5.0 kgf load on sample A showing four corner cracks

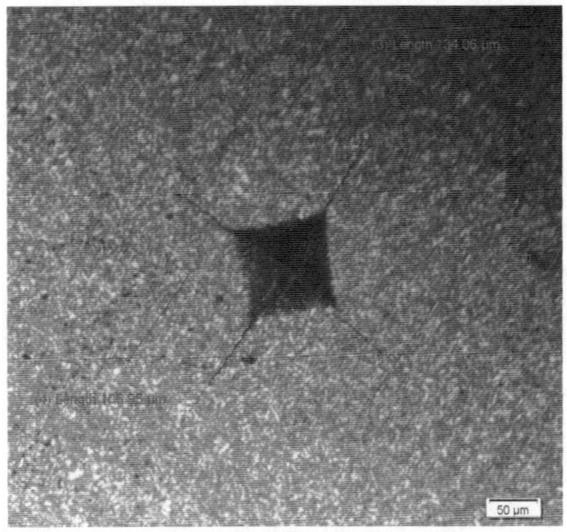

Figure 5. Indentation made at 5.0 kgf load on sample B showing four corner cracks

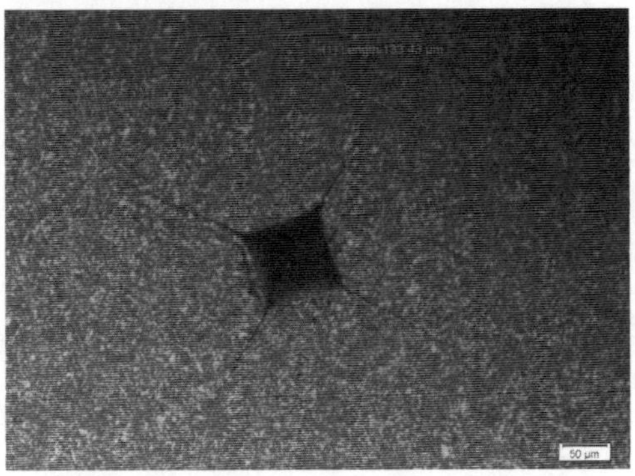

Figure 6. Indentation made at 5.0 kgf load on sample C showing four corner

Figure 7. Microchipping occurs after unloading on sample C made at 10 kgf.

In general, the indentation impression induced by indentation load of 5 kgf is always well developed as compared to other loads. Figure 4,5, and 6, also show the indentation impression of 5 kgf load for different cutting tool material, all the cutting tool develop well crack length. However, the indentation-induced by indentation load of 10 kgf chipping was found around the indentation impression.

4.2. Crack Length

The indentation study was concentrated on understanding a fracture mechanics interpretation of the crack length and load relation. The crack length increases as the load increases (figure 8). The crack length of sample A is shorter compared to the other samples. It is indicated that the cutting tool has higher fracture toughness but lower hardness. The higher of the crack length meaning that has lower the fracture toughness and the cutting tool has higher hardness and lower fracture toughness, the cutting tool with lower fracture toughness cannot be used for cutting process produced cyclic load such as interrupted cutting.

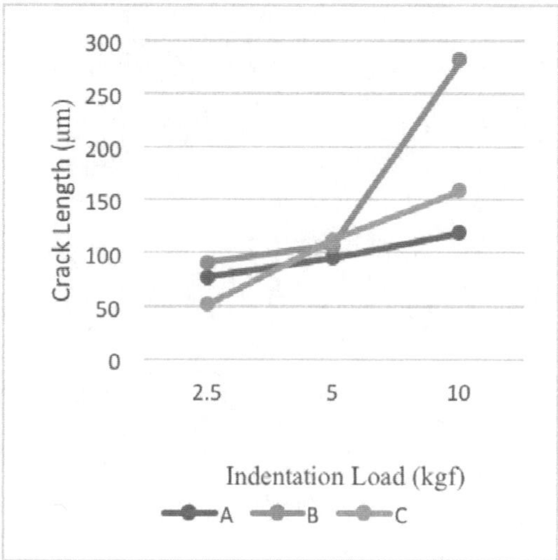

Figure 8. The relation between crack length and indentasion load

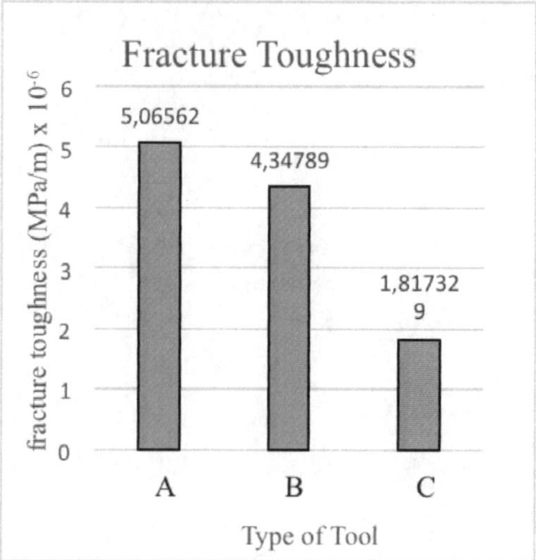

Figure 9. Fracture toughness of tool

4.3. Fracture Toughness (K_{IC})

To determine the K_{IC} of ceramic tool inserts the equation (1) developed by Nihara et al. (1984) was used. The indentation impression induced by indentation loads of 5 kgf is well developed as compared to other loads. Figure 9 shows that the tool AS10 has the highest fracture toughness and follow by CC 650 and K090. It is indicated that higher fracture toughness means the better performance of cutting tool and vise versa.

5. Conclusion

From the result, we can conclude that, Based on visual observation during indentation tests, it can be concluded that, the crack is formed only after unloading. The median crack systems were formed on all types of the tool. Lateral cracks on the surface are formed around the indentation and microscopic chipping usually occurs after unloading. The indentation impression induced by indentation loads of 5 kgf is produced better crack length compare to other loads. At lower indentation load, the problem altered relay on the load dependence of hardness tester and from measurement uncertainly owing to the small indentation impression size. At higher load, the cutting tool pronounces to chipping around indentation impression. The crack length of sample A is shorter compared to the other samples. It is indicated that the cutting tool has higher hardness and lower fracture toughness, the cutting tool with lower fracture toughness cannot be used for cutting process produced cyclic load such as interrupted cutting.

Acknowledgements

The authors would like to gratitude the financial support from the Ministry of Research, Technology and Higher Education, under the Applied Product financing scheme. This paper contains the findings obtained from the Project No: SP DIPA-042.03.1.401516/2017 8.

References

[1] Husni, Choudhury I A, Taha Z and Philips P K 2002 *Fundamental Study of Early Fracture of Ceramic Tools in Interrupted Cutting* Proceedings ETM 106-110

[2] Song Y K and Varin R A 1998 Indentation Microcracking and Toughness of Newly Discovered Ternary Intermetallic Phases in the Ni-Si-Mg System *Intermetallics* **6** 379-393

[3] Huang C Z and Wang J and Ai X 2000 Development of new ceramic cutting tools with alumina Coated Carbide Powder *International Journal Of Machine Tools & Manufacture* **40** 823-32
[4] Shetty D K, Wright I G, Mincer P N and Clauer H A 1985 Indentation fracture of WC-Co cermets *Journal of Materials Science* **20** 1873-82
[5] Liang K M, Orange G and Fantozzi G Evaluation by indentation of fracture toughness of ceramic materials *Journal of Materials Science* **25** 207-14
[6] Warren P D 1995 Determining the fracture toughness of brittle material by hertzian indentation *Journal of the European Ceramic Society* **15** 201-207
[7] Fukuhura M, Fukazawa K and Fukawa A 1985 Physical Properties And Cutting Performance of Silicon Nitride Ceramic *Wear* **102** 195-210

IOP Conf. Series: Materials Science and Engineering **352** (2018) 012023 doi:10.1088/1757-899X/352/1/012023

Permanent magnet DC motor control by using arduino and motor drive module BTS7960

SSyukriyadin, SSyahrizal, G Mansur, H P Ramadhan
Department of Electrical and Computer Engineering, Faculty of Engineering, Syiah Kuala University, Banda Aceh, 23111, Indonesia

E-mail: syahrizal.ee@unsyiah.ac.id

Abstract. This study proposes a control system for permanent magnet DC (PMDC) motor. PMDC drive control system has two critical parameters: control and monitoring. Control system includes rotation speed control and direction of rotation of motor using motor drive module BTS7960. The PWM signal has a fixed frequency of waves with varying duty cycles (between 0% and 100%), so the motor rotation can be regulated gradually using a potentiometer already programmed on the Arduino Uno board. The motor rotation direction setting uses the H-bridge circuit method using a 3-way switch to set the direction of forward-reverse rotation of the motor. The monitoring system includes measurements of rotational speed, current, and voltage. Motor rotation speed can be adjusted from the armature voltage settings through the duty cycle PWM setting so that the motor speed can be increased or decreased by the desired duty cycle. From the unload PMDC motor test results it has also been shown that the torque of the motor is relatively constant when there is a change in speed from low rpm to high rpm or vice versa.

1. Introduction

Permanent magnet DC motors (PMDC) have been widely used in industries as mechanical drivers. They are also applied in transport vehicles, biomedical equipmentand household appliances. These type of motors have several advantages,i.e.good speed regulation, simple starting system, low cost,small physical size, high efficiency and they do not need high current and no excitation at the start. The torque of these motor is relatively constant at any speed after reaching their nominal speed [1]. Besides these advantages, there are some disadvantages of these motors, i.e. the flux density generated in the air gap is so limited and such as having the limitation of field flux density generated in the air gap, and due to the armature reactions,the demagnetizing effects are generated in the motor. This condition causes the motor inoperative[2].

The speed of the PMDC can be controlled in the range below and above its nominal speed. The most common method to control the speed of the PMDCis by varying the input voltage to the motor. Studies have been done to investigate the speed control methods of these PMDC such as the AC-DC power supply control using single phase matrix converter (SPMC) [3], the speed control based on the uncertain disturbances [4], automatic sensorless speed control based on PMDC characteristics compensation [5], and low rotation speed setting performance was investigated to maintain the stability of the PMDC torque [6]. Arduino and LabView have been applied to control the speed of the PMDC [7]. It used a L298motor driver to control the speed of the PMDC. The input voltage to the motor is regulated by changing the duty cycle of the pulse width modulation (PWM) signal to the motor. The LINX package is used as an interface between LabView and Arduino. As several devices are involved in controlling the speed of the motor, these make the control system of this method complicated. The design can be simplified by eliminating the LabView in the design. Therefore, in this paper a simple method is proposed to control the speed of PMDC motor by using Arduino as a microcontroller and a BTS7960 module as a motor driver.

2. Methods of PMDC speed control

DC motor speed control can be achieved by adjusting the armature voltage or using electronic power components to acquire voltage drops on the armature. The relation of speed and voltage can be described from the equivalent circuit of PMDC motor,whereas the equivalent PMDC is similar to DC shunt motor with the exception that the field is given by a permanent magnet as shown in figure 1[1].

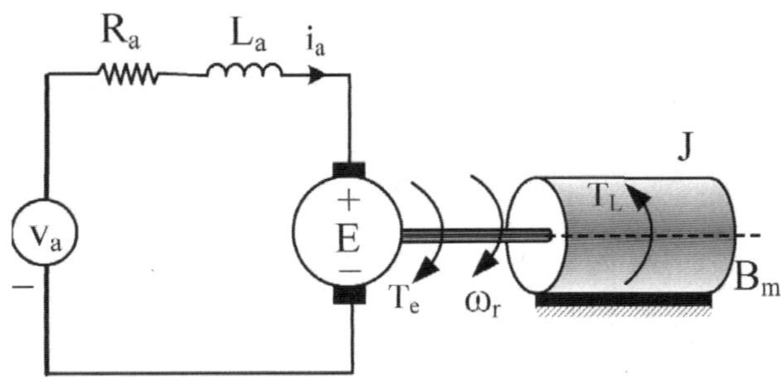

Figure 1. Equivalent circuit of PMDC motor.

As seen in figure 1 above, the dynamic equivalent equation of PMDC motor can be described as below [1]:

$$V_a = R_a . i_a + L_a \frac{di_a}{dt} + E \quad (1)$$

$$E = k_v \omega_m \quad (2)$$

$$T_e = k_T . i_a = J \frac{\omega_m}{dt} + B_m . \omega_r + T_L \quad (3)$$

$$T_L = K_n . {\omega_m}^2 \quad (4)$$

Equation (2) above has described the characteristics of speed versus PMDC motor voltages as shown in figure 2 [2].

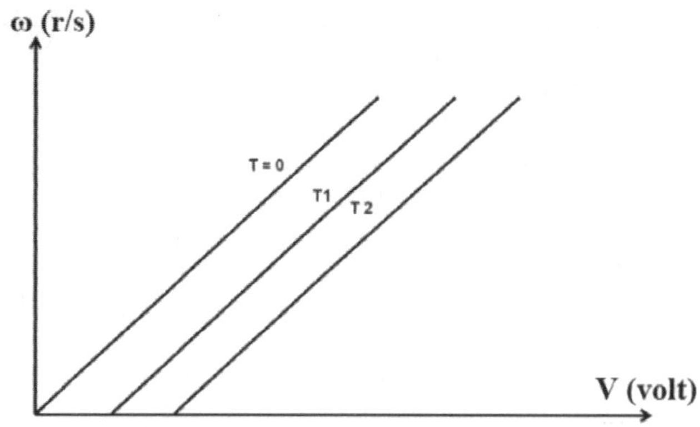

Figure 2. Speed vs voltage characteristics

It is also described in [2] the characteristic of torque versus speed of PMDC motor at varying armature voltages under steady-state conditions as shown in figure. 3.

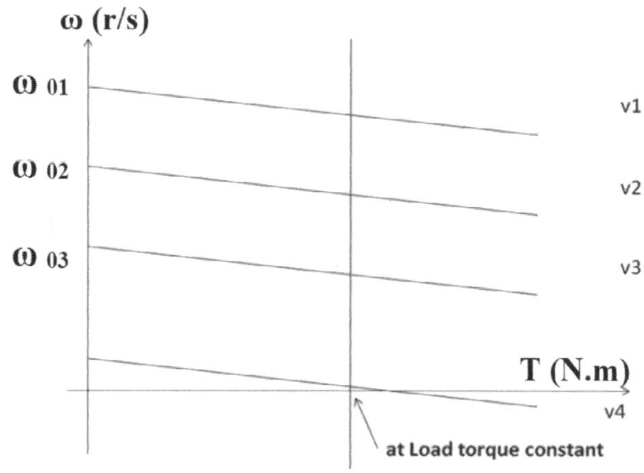

Figure 3. Torque vs speed characteristics.

Pulse Wave Modulation (PWM) is a method used to regulate motor rotation speed by adjusting the percentage of high pulse width to the period of a square signal in the form of the periodic voltage applied to the motor as a power source[2], [8], [9]. A recent article in [7] described the using of PWM technique in Arduino board to adjust PMDC motor rotation speed. The direction of PMDC motor rotation (forward - reverse) can be adjusted by varying RPWM (Right PWM) and LPWM (Left PWM) to change switching mode in H-bridge motor drive module.

3. Proposed System

The block diagram of PMDC control system proposed in this articleis shown in figure 4.The system using power supply 24 volt DC as a voltage source on PMDC motor type MY1016 (350 Watt). The motor itself has a speed rate of 2750 RPM and a torque rate of 1.2 N.m in current rate of 19.2 A.The potentiometer is used as a regulator to set a setpoint value in Arduino board in which the one who adjuststhe speed of PMDC motor.The 3-way switch is used as forward or reverseguidance of DC motor rotation, and also be as a function to switch off the PMDC motor. Arduino will then compared the setpoint by using the magnitude of the voltage obtained from the potentiometer, and the result value is then processed to produce the desired data. Furthermore, Arduino will instruct the motor drive module BTS7960 to rotate the motor to the right or left with desired motor rotation speed by the input of potentiometer. In the monitoring system, there are several sensors that will monitor the parameters of the motor system such as current, voltage, and a rotational speed of the motor.Sensors will then transmit Analog to Digital(ADC) signals to Arduino to display on the LCD.

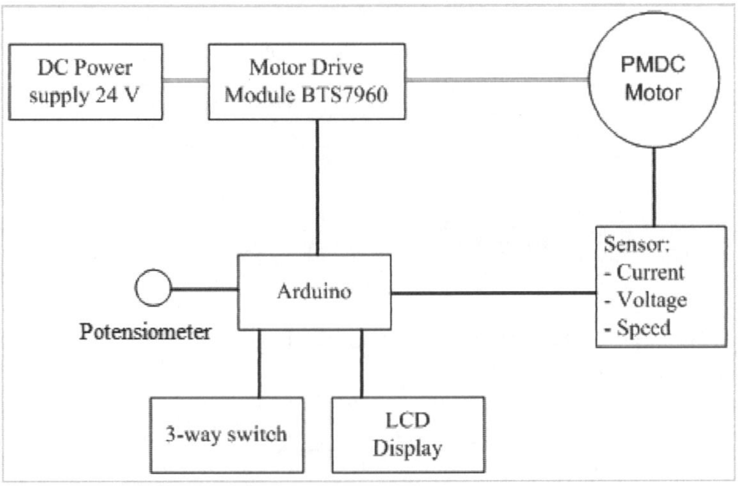

Figure 4. Proposed System.

4. Result and analysis

4.1 Direction of PMDC motor rotation

The result of PMDC motor rotation direction testing is shown in table 1. PMDC motor will rotate forward when it is given logic value 1 to RPWM and 0 to LPWM,and will rotate counterclockwise direction when given logic value otherwise (0 to RPWM and 1 to LPWM). The motor is turned off in any set logic value other than above mentioned (1 to RPWM and 1 to LPWM, 0 to RPWM and 0 to LPWM). Thus it can be concluded that the test control of DC motor rotation direction can run well on the system.

Table 1. Results of motor rotation PMDC direction control test.

Output Logic		Direction
RPWM	LPWM	
0	0	Off
1	0	Clockwise*(forward)*
0	1	Anti-Clockwise*(reverse)*
1	1	Off

4.2 Speed of PMDC motor

The speed test is performed by fluctuating the value of PWM duty cycle from 0–100%. The test is used to acknowledge the relationship between PWM value and average voltage associated with the speed of the PMDC motor.The motor rotation speed is set gradually from the minimum to maximum speed by adjusting the duty cycle value using a potentiometer as a determinant of the signal setpoint value. The results of PMDC motor speed testing is shown in figure 5.

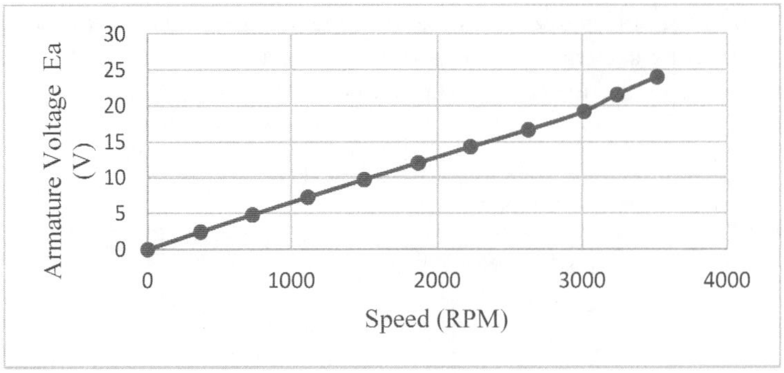

Figure 5. Speed versus voltage.

4.3 Unload torque PMDC motor

The results of unload test are in the formed of armature voltage, armature current, armature speed and torque of PMDC motor as shown in table 2.

Table 2. PMDC testing of the un-load motor.

No.	Armature VoltageEa (Volt)	Armatur CurrentIa(Ampere)	Speed (RPM)	Torque (Nm)
1.	0	0	0	0
2.	2.45	1.98	374	0.126286
3.	4.8	1.973	732	0.123609
4.	7.25	1.823	1114	0.113352
5.	9.69	1.761	1503	0.108472
6.	12.07	1.736	1869	0.107112
7.	14.3	1.734	2231	0.106188
8.	16.66	1.732	2632	0.104744
9.	19.2	1.724	3014	0.104927
10.	21.55	1.724	3243	0.10939
11.	24.05	1.724	3517	0.11824

As derived from table 2,the relation between torque and motor speed is shown in figure 6.

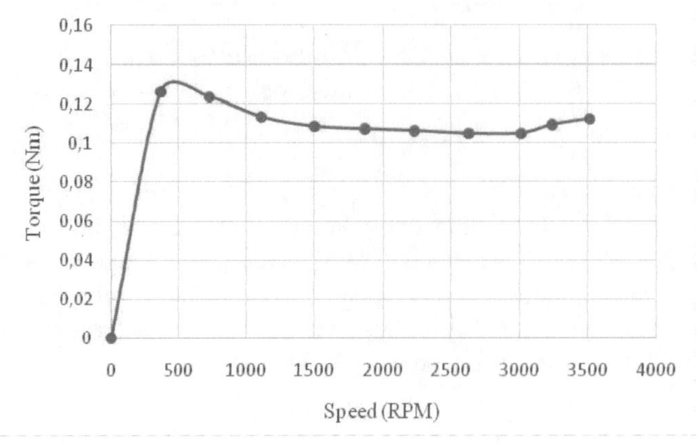

Figure 6. The relation between torque and speed of PMDC motor underunload conditions.

It is clearly shown that when the armature voltage E_a is varied, the current is relatively constant but motor speed changes dynamically following the change of armature voltage.As also shown in figure 6, it can be concluded that when a speed change occurs between 500 rpm to 3500 rpm the PMDC torque motor is relatively constant.

5. Conclusion

Based on the results obtained from several tests performed in this research, the direction of motor rotation can be performed using RPWM technique for the forward direction and LPWM for the reverse direction. Motor rotation speed can be adjusted from the armature voltage settings through the PWM duty cycle arrangement so that the motor speed can be increased or decreased by the desired duty cycle.The result of PMDC motor testing on un-load condition also shows that torque motor is relatively constant when the speed change from low rpm to high rpm or vice versa.

References

[1] Mahghani M and Afjei E 2017 Carrier-based with unipolar switching technique current control of PMDC motors in electric vehicle applications*Iranian Conference onElectrical Engineering (ICEE)* 1060

[2] Tarnini M Y 2014 Microcontroller based PMDC motor control for driving 0.5 KW scooter*Power Electronics and Motion Control Conference and Exposition (PEMC), 16th International*1204

[3] Idris A, Baharom R, Hamzah, N R and Hamzah M K 2013 Controlled AC-DC converter fed permanent magnet DC motor using single-phase matrix converter *Power Engineering and Optimization Conference (PEOCO), 2013 IEEE 7th International* 764

[4] Chu H, Tao W, Gao B, Liu Q and Chen H 2016 Speed control of the permanent-magnet DC motor subjected to uncertainty and disturbance*Control Conference (CCC), 35th Chinese* 4664

[5] Saurav K and Potluri R 2013 Sensorless speed control of a permanent magnet dc motor by compensating the plant nonlinearities*Industrial Electronics (ISIE) International Symposium* 1

[6] Chu H, GaoB, Gu W and Chen H *2017* Low-Speed Control for Permanent-Magnet DC Torque Motor Using Observer-Based Nonlinear Triple-Step Controller*IEEE Transactions on Industrial Electronics* **64** 3286

[7] Angalaeswari S, KumarA, Kumar D and Bhadoriya S 2016 Speed control of permanent magnet (PM) DC motor using Arduino and LabVIEW*Computational Intelligence and Computing Research (ICCIC), IEEE International Conference* 1

[8] Li Q, Huang H and Yin B 2008The study of PWM methods in permanent magnet brushless DC motor speed control system Electrical Machines and Systems*ICEMS International Conference* 3897

[9] Yurkevich V D and Stepanov N A 2014 PWM speed control of DC motor based on singular perturbation technique*Ultra Modern Telecommunications and Control Systems and Workshops (ICUMT), 6th International Congress on* 434

IOP Conf. Series: Materials Science and Engineering **352** (2018) 012024 doi:10.1088/1757-899X/352/1/012024

The integration of elastic wave properties and machine learning for the distribution of petrophysical properties in reservoir modeling

T C Ratnam[1], D P Ghosh[1] and B M Negash[2]
[1]Petroleum Geoscience Department, Universiti Teknologi PETRONAS, Malaysia
[2]Petroleum Engineering Department, Universiti Teknologi PETRONAS, Malaysia

E-mail: teresaclare.ratnam@gmail.com

Abstract. Conventional reservoir modeling employs variograms to predict the spatial distribution of petrophysical properties. This study aims to improve property distribution by incorporating elastic wave properties. In this study, elastic wave properties obtained from seismic inversion are used as input for an artificial neural network to predict neutron porosity in between well locations. The method employed in this study is supervised learning based on available well logs. This method converts every seismic trace into a pseudo-well log, hence reducing the uncertainty between well locations. By incorporating the seismic response, the reliance on geostatistical methods such as variograms for the distribution of petrophysical properties is reduced drastically. The results of the artificial neural network show good correlation with the neutron porosity log which gives confidence for spatial prediction in areas where well logs are not available.

1. Introduction

Reservoir modeling is an inherent part of the field development process. It integrates various disciplines such as petroleum geoscience, reservoir engineering and petrophysics. The aim of a reservoir model is to mimic the subsurface performance to the best of our knowledge. Conventional reservoir modeling employs various geostatistical methods to predict the spatial distribution of facies and reservoir properties in between wells. The model is correct at well location, however elsewhere the variograms may not depict the true geological subsurface. The uncertainty is then carried forward in the following procedure of reservoir simulation. This study aims to use elastic wave properties, aided by an artificial neural network, to derive a geologically constrained static reservoir model.

2. Literature Review

2.1. Reservoir Modeling

One of the important steps in the life of a field is reservoir modeling. It is a platform for the assimilation of available data and geology [1]. The conventional method of constructing a geological model combines various sources of data such as well logs, seismic data, rock physics and petrophysical data. Well log data is a good source of geological data which possesses the vertical resolution for reservoir simulation [2]. The issue of the conventional method arises due to the lack of data of rock and fluid properties in between the wells. Therefore, an array of geostatistical tools are incorporated to populate the grid cells.

Geostatistics is based on stochastic optimal linear estimation [3]. In static reservoir modeling, geostatistical methods are used to distribute properties such as lithology, permeability and porosity. Variogram analysis and kriging are examples of geostatistical tools [4]. Due to the limited number of wells in a reservoir, several realizations are made through stochastic simulation which relies on a vertical proportions curve established from well logs [5]. This means that the proportions of lithology which is seen at well locations are assumed to be representative of the rest of the field. Conventional geostatistical reservoir modeling methods are incapable of portraying reservoir heterogeneity that represents geological processes in temporal sequence [6].

2.2. Elastic Wave Properties

The main factors that play a role in determining the velocity of P-wave (V_p) and S-wave (V_s) are density, ρ, and structural strength. Apart from that, the effect of lithology also affects the velocities. Hence, seismic velocities obtained from sonic logs can be used to discriminate between sand and shale. The ratio of V_p/V_s is often used to distinguish between rock and sediment types as it is usually more sensitive than just V_P alone. Furthermore, changes in porosity and fluid saturation are also reflected in the seismic velocities.

2.3 Scaled Inverse Quality Factor

New seismic attributes called the scaled inverse quality factor of P-wave (SQp^{-1}) and scaled inverse quality factor of S-wave (SQ_s^{-1}) were developed by Hermana et al. [7] for the separation of lithology and pore fluids. The response of SQp^{-1} is similar to the gamma-ray log whereas SQ_s^{-1} is similar to the resistivity log. These attributes are formulated from the Quality Factor of P-wave and S-wave. Eq. 1 and Eq. 2 show the definition of SQp^{-1} and SQ_s^{-1} respectively.

$$SQ_p^{-1} = \frac{5}{6}\frac{1}{\rho}\frac{\left(\frac{M}{G}-2\right)^2}{\left(\frac{M}{G}-1\right)} \tag{1}$$

$$SQ_s^{-1} = \frac{10}{3}\frac{1}{\rho}\frac{\left(\frac{M}{G}\right)}{\left(\frac{3M}{G}-2\right)} \tag{2}$$

where M and G are the compressional and shear modulus respectively, and (M/G) can be approximated with $(V_p/V_s)^2$. Therefore, the variables in the equations are the P-wave velocity (V_p), S-wave velocity (V_s) and density (ρ) which are obtained from seismic inversion.

2.4. Artificial Neural Network (ANN)

An artificial neural network (ANN) is similar to the complex network of neurons in a brain whereby it is connected by multiple processors called nodes, each producing a sequence of real-valued activations [8]. The first neuron, which is the input neuron, is triggered by data collected from the environment. The consecutive neurons, on the other hand, are activated by weighted connections of preceding neurons. ANNs are applied in various situations including regression analysis, classification or pattern recognition, data processing and forecasting time series [9].

3. Methodology

The workflow of this study aims to predict the petrophysical properties between well locations by introducing elastic wave properties. These elastic wave properties were obtained through seismic inversion. The samples that were collected were constrained to the reservoir interval of the field. The supervised learning approach was used whereby the targets (or response) were neutron porosity and resistivity whereas the features (or predictors) were the elastic wave properties.

From the results of seismic inversion, the P-wave velocity (V_p), S-wave velocity (V_s) and density (ρ) were extracted. The elastic wave properties were then evaluated and processed to remove outliers and anomalies, if any. Here, the 3σ algorithm is employed to detect outliers [10]. The elastic wave properties were analyzed through feature expansion to improve the correlation with the targets. Among the expanded features is the scaled inverse quality factor of P-wave (SQp^{-1}) and scaled inverse quality factor of S-wave (SQ_s^{-1}).

Finally, the feature selection is shown in Eq. (3) below:

$$\text{Feature } (x) = [V_p, V_s, \rho, SQ_p^{-1}, SQ_s^{-1}, \rho*V_p, \rho*V_s, V_p/V_s] \tag{3}$$

4. Results & Discussion

The features obtained from seismic inversion are used as input for the neural network. The neutron porosity log is taken from a section of the wells to be used as training data in the neural network. The results of the training are shown in Figure 1. As observed, the correlation coefficient is 0.95711. Then, the neural net is tested using a section of wells that were not initially included in the training dataset. The result of the training is shown in Figure 2 where the correlation coefficient is 0.90435. This satisfactory correlation is applicable for the distribution of properties in a static reservoir model, thus reducing the need to rely solely on geostatistics between wells.

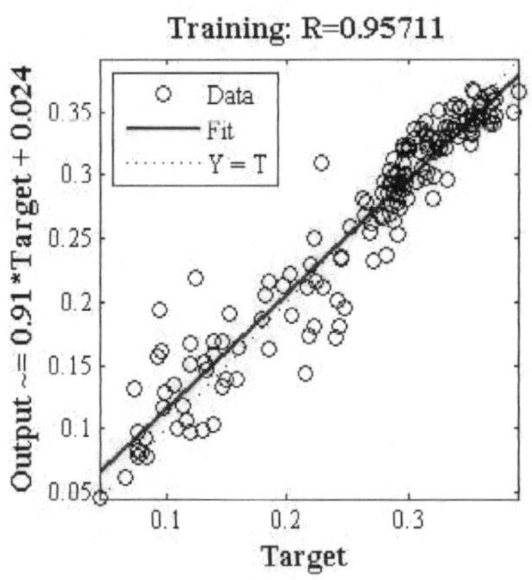

Figure 1. The result of the ANN training with the neutron porosity log as the target.

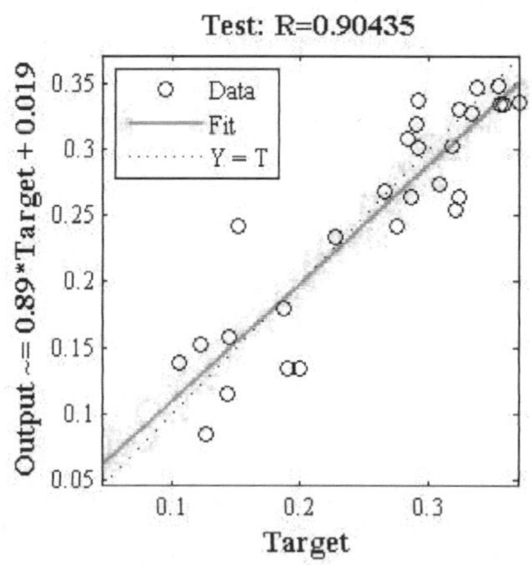

Figure 2. The results of testing the ANN using samples that were not included in the training dataset.

4.1. Comparison with SQ_p^{-1} Attribute

The SQ_p^{-1} attribute is used as a lithology identifier tool as demonstrated by Hermana. Figure 3 shows the cross-plot of neutron porosity vs. SQ_p^{-1}. The correlation coefficient is 0.5418. This is done with as a single attribute. The results may be improved when combined with multiple attributes.

4.2. Comparison with single and multiple attribute regression

The proposed method is compared to conventional methods that are available such as single attribute regression (Figure 4) and multiple attribute regression (Figure 5). The correlation coefficients are 0.6841 and 0.7862 respectively. The multiple attribute regression produces better results than the single attribute regression, however it is less accurate in comparison to the proposed method.

IOP Conf. Series: Materials Science and Engineering **352** (2018) 012024 doi:10.1088/1757-899X/352/1/012024

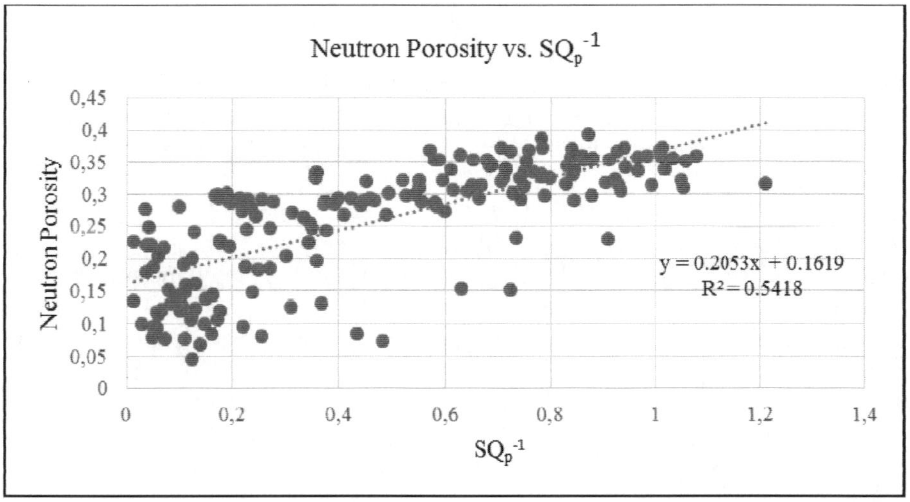

Figure 3. A cross-plot of neutron porosity versus SQp-1. The cross-plot shows the ability of the attribute to predict neutron porosity.

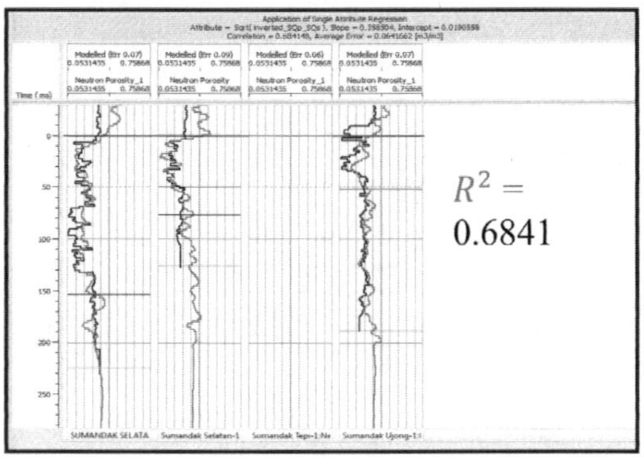

Figure 4. Application of single attribute regression on three wells.

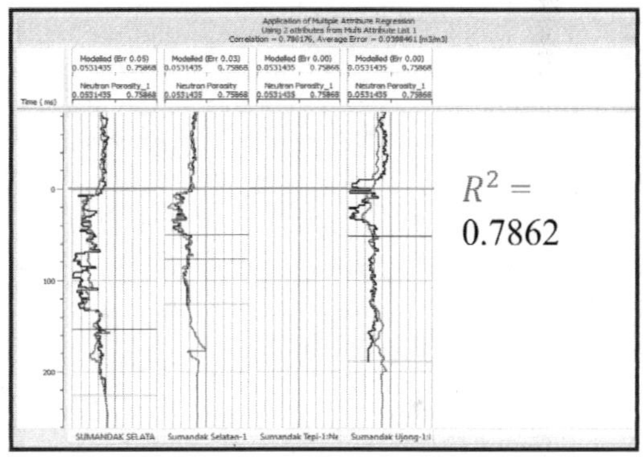

Figure 5. Application of multiple attribute regression on three wells.

5. Conclusion

The incorporation of elastic wave properties from the seismic response improves the distribution of reservoir properties in between well locations. The use of an artificial neural network can establish a strong correlation between the features and the desired output, which is neutron porosity. This approach provides a geological reference in between well locations by converting each seismic trace into a pseudo-well log. Hence, the information that is distributed between wells conforms to the geology and not merely a statistical approach.

Acknowledgment

We would like to thank Universiti Teknologi PETRONAS for funding this research under the Graduate Research Assistantship scheme and PETRONAS for providing the necessary data.

References

[1] Branets L V, Ghai S S, Lyons S L and Wu X-H 2009 Challenges and technologies in reservoir modeling *Commun. Comput. Phys.* **6** 1–23

[2] Behrens R A, MacLeod M K, Tran T T and Alimi A C 1998 Incorporating seismic attribute maps in 3D reservoir models *Society Pet. Eng.* **1** 122–6

[3] Hristopulos D T 2015 Stochastic Local Interaction (SLI) model: Bridging machine learning and geostatistics *Comput. Geosci.* **85** 26–37
[4] Bohling G 2005 Introduction to geostatistics and variogram analysis *Kansas Geol. Surv.* 1–20
[5] Sacchi Q, Borello E S, Weltje G J and Dalman R 2016 Increasing the predictive power of geostatistical reservoir models by integration of geological constraints from stratigraphic forward modeling. *Mar. Pet. Geol.* **69** 112–26
[6] Pyrcz M J, Sech R P, Covault J A, Willis B J, Sylvester Z, Sun T 2015 Stratigraphic rule-based reservoir modeling. *Bull. Can. Pet. Geo*l. **63** 287–303
[7] Hermana M, Ghosh D P, Sum C W 2016 Optimizing the lithology and pore fluid separation using attenuation attributes *Offshore Technol. Conf.*
[8] Schmidhuber J 2015 Deep learning in neural networks: An overview *Neural Networks* **61** 85–117
[9] Tealab A, Hefny H and Badr A 2017 Forecasting of nonlinear time series using artificial neural network *Futur. Comput. Informatics J.*
[10] Kadlec P, Gabrys B and Strandt S 2009 Data-driven soft sensors in the process industry *Comput. Chem. Eng.* **33** 795-814

IOP Conf. Series: Materials Science and Engineering **352** (2018) 012025 doi:10.1088/1757-899X/352/1/012025

Graphene nanoplatelets as high-performance filtration control material in water-based drilling fluids

Syahrir Ridha[1], Arif Ibrahim[1], Radzi Shahari[1], Syarizal Fonna[2]
[1] Petroleum Engineering Department,
Universiti Teknologi PETRONAS, Bandar Seri Iskandar, 32610,
Perak Darul Ridzuan, Malaysia.
[2] Mechanical Engineering Department, Universitas Syiah Kuala,
Banda Aceh, 23111, Indonesia.

E-mail: mohamad.arif_g03640@utp.edu.my

Abstract. The main objective of this work is to evaluate the effectiveness of graphene nanoplatelets (GNP) as filtration control materials in water based drilling fluids. Three (3) general samples of water based drilling fluids were prepared including basic potassium chloride (KCl) drilling fluids, nanosilica (NS) drilling fluids and GNP drilling fluids. Several concentrations of NS and GNP were dispersed in controlled formulations of water based drilling fluids. Standard API filtration tests were carried out for comparison purposes as well as High Temperature High Pressure (HTHP) filtration tests at 150 °F (~66 °C), 250 °F (~121 °C) and 350 °F (~177 °C) at a fixed 500 (~3.45MPa) psi to study the filtration trend as a function of temperature. Mud cake samples from several tests were selectively chosen and analyzed under Field Emission Scanning Electron Microscope (FESEM) for its morphology. Results from this work show that nanoparticle concentrations play a factor in filtration ability of colloid materials in water based drilling fluids when studied at elevated temperature. Low temperature filtration, however, shows only small differences in volume in all the drilling fluid samples. 0.1 ppb concentrations of GNP reduced the fluid loss of 350 °F by 4.6 mL as compared to the similar concentration of NS drilling fluids.

1. Introduction

Drilling fluids are a mixture of solids, liquids and chemicals with the liquid being the continuous phase. During conventional overbalance drilling, hydrostatic pressure applied by drilling fluid is higher than the formation pore pressure as a primary well control mechanism. Thus, there is a flow of liquid from the wellbore into the formation leaving solid particles suspended in the fluid initially, on the wall of the wellbore forming a layer of the impermeable zone known as filter cake [1]. At the early stage of filtration also known as spurt loss, both large and small particles will be deposited on the face of the wellbore; because the drag force is driving the particles to the cake surface is high. Afterwards, only smaller and smaller particles are deposited [2]. The amount of fluid loss which is directly proportional to the cake growth rate gradually decreases until the cake becomes very impermeable at which no particles small enough to be deposited on the smallest pores on the cake are available in the suspension. Thus, theoretically having a smaller size of particles in drilling fluid is desirable to form a more impermeable filter cake.

The applications of nanomaterials in the petroleum industry have always been thought since the emergence of nanotechnology with the invention of scanning tunneling microscope in 1981 [3] and the discovery of fullerene in 1985 [4]. The beginning of the research on nanomaterials in drilling fluids however is largely unknown in literature. The interest is mainly due to its size, these materials have access to the smallest pores and act as sealing agent in any lithology types [5]. Up to now, several laboratory applications of nanomaterials in drilling fluids have been reported in literature such as the applications in filtration control [6-11], mitigation of pipe sticking [10], torque and drag [9] and shale stabilization [11, 12].

Sensoy et al. [11] studied the potential of using nanoparticles to plug the pore throat of shales and reduce its permeability. The idea of this study come from the result of their previous works about the average pore throat sizes of shales in which they found it to be around 10 to 30 nanometers and is smaller than the conventional colloid materials in drilling fluids. Sensoy et al. [11] used 20nm silicon dioxide (NS) and found that with the use of silicon dioxide in water based drilling fluids, the permeability of shales was reduced by a factor of 5 to 50 [11]. In 2011, Javeri et al. [10] also used NS in the water based drilling fluid and tested for its filtration properties at 80°F and 100 psi differential pressure though the amount of the NS used was reported. The results show that with the addition of NS in the drilling fluid, the filter cake thickness was reduced by 34%. However, no filtrate volume was reported from this study [10]. In 2015, Taraghikhah et al. [12] and Wahid et al. [13], further the study of NS in drilling fluids. Though Taraghikhah et al. continue to test it on water based fluids with a slightly different modification for shale stabilization property, the latter tested the filtration properties of NS in the synthetic based drilling fluid. The results of both researches show that the conventional drilling fluids system are improved with the addition of NS.

Although discovered in 2004 by Andre K. Geim and Konstantin Novoselov [14], application or invention research on graphene have only started recently after its fundamental properties studies have been proven to exceed most materials in the world in every aspect. Graphene is the world's first two-dimensional (2D) crystalline material has a theoretical thickness of 0.34 nm [15, 16]. Graphene consists of a single layer of carbon atoms in the sp^2 hybridization state with each atom being covalently bonded to three other carbon atoms in a hexagonal lattice with a carbon to carbon distance of 0.142nm [14, 17].

Two reports from early 2016 by Aftab et al. [18] and Ismail et al. [19], presented the comparison in rheological properties, lubricity, and filtration properties of several nanomaterials at Low Temperature Low Pressure (LTLP) conditions. Aftab et al. used the same concentration on multiwalled carbon nanotubes (MWCNT), NS and GNP in water based drilling fluids while Ismail et al. studies the effect of different concentration of MWCNT, NS and Glass Beads (GB) in water based drilling fluids as well. While both kinds of literature show improvement in the performance of drilling fluids with nanomaterials, the highlight here is the performance of GNP which is significantly better than both MWCNT and NS in every aspect [18, 19]. Both studies however were carried out under LTLP conditions which would yield unrealistic scenarios to be extrapolated to field conditions.

In this study, experimental works have been conducted to study the effectiveness of GNP as filtration control materials under different temperatures, compared with NS and conventional KCl drilling fluid systems. Filter cake produced from filtration tests were analyzed under FESEM for its morphology.

2. Materials and Method

2.1. Materials

Conventional drilling fluid additives such as viscosifiers, alkalinity control and weighting materials were all procured from Scomi Oiltools Sdn. Bhd. Nanosilica (7 – 24 nm, purity >99.8 wt. %) was procured from PlasmaChem GmbH. Graphene Nanoplatelets-SigmaAldrich (diameter 40 – 70 nm, length 2 – 5 nm, purity > 95 wt. %) procured from Advance Altimas (M) Sdn. Bhd. was used as received. Surfactants used to functionalize GNP were procured from Avantis Laboratory Sdn. Bhd. and was used as received.

2.2. *Functionalization of GNP*

500 ppm surfactants were prepared in 500 mL of distilled water in an Erlenmeyer flask. A fixed amount of GNP was then dispersed in the solutions at a concentration of 0.1 mg / mL. The resultant solution was then ultrasonicated in an ultrasonic bath for 30 minutes to get surfactant-coated GNP.

2.3. *Water based drilling fluids formulations and preparations*

350 mL of water based drilling fluids were prepared by mixing distilled water, potassium chloride, xanthan gum, polyanionic cellulose, caustic soda and barite. Each formulation was designed to achieve 11.5 pounds per gallon (ppg). Table 1 shows the approximate drilling fluid formulation with mixing time and order of each additive. Each formulation may have small differences in the amount of distilled water and barite to achieve a consistent 11.5 ppg due to the difference in nanomaterials concentrations.

Table 1. Formulations of drilling fluids

Materials	Basic WBM	WBM + NS	WBM + GNP	Mixing time	Mixing order
Distilled water, (ml)	300	300	300	-	1
Potassium Chloride, (ppb)[a]	20.0	20.0	20.0	2	2
Xanthan gum, (ppb)	1.0	1.0	1.0	5	3
Polyanionic cellulose, (ppb)	3.0	3.0	3.0	5	4
Caustic soda, (ppb)	0.1	0.1	0.1	5	5
Barite, (ppb)	160	160	160	30	6
NS, (ppb)		0.01,0.05,0.1, 0.2,0.3		15	7
GNP, (ppb)			0.01,0.05,0.1, 0.2,0.3	15	7

[a] ppb – pounds per barrel

2.4. *Filtration Tests*

2.4.1. Standard API filtration test. Standard API filtration tests followed API RP 13B-1 Recommended Practice for Field Testing Water-Based Drilling Fluids. Tests were conducted using a OFITE standard LPLT API Filter Press and Whatman quantitative filter paper, hardened low-ash grade 50 (estimated pore size 2.7 μm). All filtration loss tests were conducted at 25 °C. In a standard run, 100 mL of drilling fluids were placed in the apparatus with 100 ± 1.0 psi differential pressure supplied across a filter paper. Filtrate volume was then measured every one minute for 30 minutes. The filter paper was then removed, lightly washed, and set out at room temperature for 24 hours to allow for the water to evaporate from the filter cake. After the filter cake has dried up, its thickness was determined using a Vernier caliper to the nearest millimeter.

2.4.2. HTHP filtration tests. HTHP filtration tests followed API RP 13B-1 Recommended Practice for Field Testing Water-Based Drilling Fluids. Tests were conducted using OFITE 4-unit HTHP Filter Press with Regulators and Temperature Controller with Whatman quantitative filter paper, hardened low-ash grade 50 (estimated pore size 2.7 μm). Three (3) different temperatures were studied; 150 °F (~66 °C),

250 °F (~121 °C) and 350 °F (~177 °C) for each formulation of drilling fluids to study the filtration trend as a function of temperature. In a standard run, 100 mL of drilling fluids were placed in the apparatus with 500 ± 1.0 psi differential pressure between top and minimum recommended back pressure supplied using nitrogen gas across a filter paper. Filtrate volume was then measured every one minute for 30 minutes. After the tests, the cell was allowed to cool a minimum of one hour before releasing the pressure inside the cell. The filter paper was then removed, lightly washed, and set out at room temperature for 24 hours to allow for the water to evaporate from the filter cake. After the filter cake has dried up, its thickness was determined using a Vernier caliper to the nearest millimeter.

2.5. Microstructure analysis

Several cakes were chosen and the top solids were removed for the plugged particles to be studied under Field Emission Scanning Electron Microscope (FESEM). The microstructures were observed using Zeiss Supra 55VP in Universiti Teknologi PETRONAS' Centralized Analytical Laboratory (CAL). After the samples have been dried, a small part of the filter cakes was cut and coated with a thin layer of gold prior to vacuum and observation at an accelerated voltage of 5.00 keV.

3. Results and Discussion

3.1. Fluid Loss

Table 2. Filtration test results of different drilling fluid formulation

Drilling Fluids	API (ml)	HPHT @ 150 °F (ml)	HPHT @ 250 °F (ml)	HPHT @ 350 °F (ml)
WBM	5.1	9.6	15.8	70.0
WBM + 0.01 ppb NS	5.0	9.0	14.8	70.0
WBM + 0.05 ppb NS	4.8	8.6	14.6	70.0
WBM + 0.1 ppb NS	4.7	8.5	14.5	26.6
WBM + 0.2 ppb NS	4.7	8.7	13.4	26.4
WBM + 0.3 ppb NS	4.6	7.6	13.0	22.4
WBM + 0.01 ppb GNP	5.1	8.8	14.6	70.0
WBM + 0.05 ppb GNP	4.8	8.5	12.5	70.0
WBM + 0.1 ppb GNP	4.6	8.4	12.2	22.0
WBM + 0.2 ppb GNP	4.5	7.6	12.0	21.4
WBM + 0.3 ppb GNP	4.4	6.9	11.2	21

Table 2 summarizes the total drilling fluid filtration tests results for (a) API, (b) HTHP, 150 °F, (c) HTHP, 250 °F and (d) 350 °F for each formulation. At 250 °F, similar trends were observed by Aftab et al. [18] for 0.1 ppb concentration of NS and GNP although using a different surfactant and surfactant concentrations.

Looking at the trend, generally fluid loss decreases with increasing amount of both nanomaterials even on the highest concentration used in this study. As can be seen fluid loss deteriorated with increasing temperatures. This deterioration becomes significant especially at HTHP conditions (350 °F). Although there was not much difference in fluid loss between 0.05 ppb and 0.1 ppb of both NS and GNP at 250 °F, the difference is significant at 350 °F suggesting the minimum concentrations of both nanomaterials needed for high temperature conditions is between 0.05 and 0.1 ppb.

The effect of the addition of nanomaterials on the decreasing trend of fluid loss can simply be explained by plugging the small pores on the filter cake created after larger particles have deposited on the surface of the filter paper during spurt loss making the filter cake less permeable after some time. Thus, decreasing the overall total fluid loss.

In addition, the stability of fluid loss at high temperature is also affected by the addition of nanomaterials. This could be explained by understanding colloidal dispersion. The properties of colloidal dispersions are closely related to its surface area to mass ratio. Large colloids are mainly affected by gravitational forces while other forces affect smaller colloids such as van der Waals forces and electrostatic forces. The enormous difference between its surface area and mass is the reason the properties of the surface including the shape is very important for the colloidal solution. The greater the degree of subdivision of a solid body, the greater will be its surface area per unit mass and therefore the greater the influence of the surface are on properties of colloidal dispersions [20]. At static high temperature filtration, nanomaterials may have help in the suspensions of other colloids materials preventing it from sagging thus reducing the overall fluid loss. However, to confirm this, high temperature dynamic filtration need to be carried out to study the cross flow of colloids in the fluid. Although factors such as thermal stability of the nanomaterials itself may also play a factor as seen by the difference in fluid loss volume between NS and GNP. This phenomenon was also observed by Aftab et al. [18].

Although the decreasing trend of fluid loss with increasing nanomaterials may seem beneficial, having excessive nanomaterials may impose detrimental effect; not only on its cost effectiveness but also on drilling induced formation damage. It was observed during filtration tests, the spurt loss fluid is different in color, suggesting that some nanomaterials which is much smaller than the filter cake pore size (~2.7 μm) may have invaded through the filter papers. It is believed at some point, increasing the concentrations of nanomaterials any further will no longer reduce the fluid loss suggesting its optimum concentrations. In this study, the optimum concentrations of nanomaterials have yet to be found.

3.2. Mud Cake Thickness and Sample Characterization

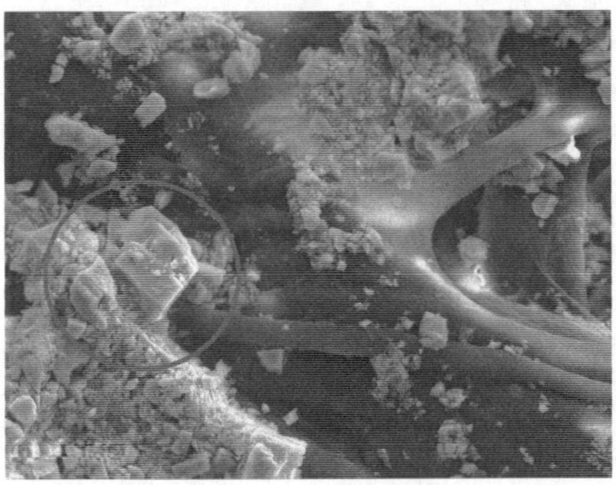

Figure 1. SEM image of NS filter paper after top solids have been removed

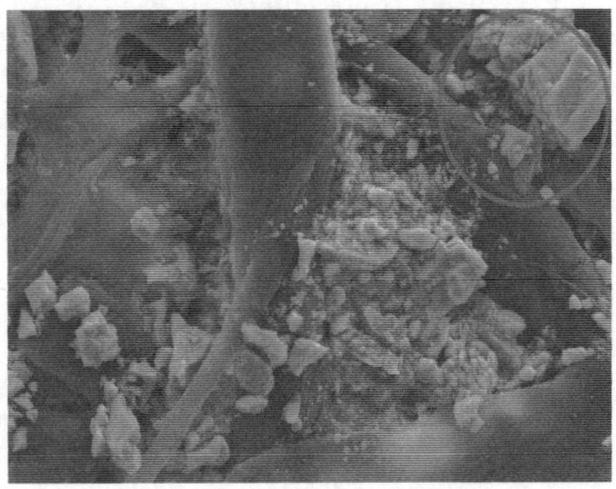

Figure 2. SEM image of GNP filter paper after top solids have been removed

For every test temperatures, the thickness of filter cake remains consistent among all drilling fluid formulations; < 2 mm for API, 4 mm for 150 °F, 6 mm for 250 °F and 9 mm for most 350 °F test temperatures. Figure 1 and Figure 2 shows the SEM image of NS and GNP filter paper after top solids have been removed to study the particles that have been plugged into the filter paper at 5,000 magnification size. From the raw observation and mapping, it was difficult to discern which particle is the nanomaterials itself especially GNP which is made of the same material as the filter paper; carbon. Thus, spectrum processing which studies the concentration of fundamental elements were employed at different pore throats. On average basic WBM filter paper shows 36.27 wt. % of carbon, 0.37 wt. % of silicon while NS filter paper shows only 31.07 wt. % of carbon due to the increase of 0.84 wt. % of silicon. In contrast, GNP filter paper shows a carbon content of 37.93 wt. % and only 0.43 wt. % of silicon. From this, it is evident that nanomaterials do in fact plug the small pore throats of the filter papers.

4. Conclusion

This work was devoted to evaluating the effectiveness of GNP as a filtration control material in water based drilling fluid with NS which has been extensively studied as a benchmark. The results from all sets of tests and concentrations indicate that GNP is an effective filtration control material in water based drilling fluids. This is especially evident at the higher temperature as compared to NS. By adding a small 0.3 ppb of GNP in a similar formulation of basic WBM, the 70 mL spurt loss was slowed down to 21 mL which is 1.4 mL less than similar concentrations of NS for 30 minutes period interval at 350 °F. Even though the minimum concentration of nanomaterials has been discovered for high temperature applications, the optimum concentration in term of cost effectiveness and formation damage due to its small sizes is still under investigation. Therefore, further research with the aim to study its cost effectiveness and formation damage effect should be considered to get a broader understanding of the effectiveness of GNP as a filtration control material.

Acknowledgements
The authors would like to express their gratitude to Universiti Teknologi PETRONAS and Universitas Syiah Kuala for providing financial assistance on this project under grant no 0153AB-M57.

References

[1] Azar J J and Samuel G R 2007 *Drilling Engineering* (Oklahoma: PenWell Corporation)
[2] Jiao D, Sharma M M 1994 Mechanism of cake buildup in crossflow filtration of colloidal suspensions *J. Colloid Interface Sci.* **162** 454-62
[3] Binnig G and Rohrer H 1983 Scanning tunneling microscopy *Surf. Sci.* **126** 236-44
[4] Kroto H W, Heath J R, O'Brien S C, Curl R F and Smalley R E 1985 C 60: buckminsterfullerene *Nat.* **318** 162-3
[5] Contreras O, Nygaard R and Alsaba M 2014 Application of In-house prepared nanoparticles as filtration control additive to reduce formation damage *SPE Int. Symp. Exhibition on Formation Damage Control (Louisiana, USA 26–28 February 2014)*
[6] Jung Y, Barry M, Lee J K, Tran P, Soong Y, Martello D and Chyu M 2011 Effect of nanoparticle additives on the rheological properties of clay-based fluids at high temperature and high pressure *AADE National Techical Conf. Exhibition (Houston, Texas 12-14 April 2011)*
[7] Kosynkin D V, Ceriotti G, Wilson K C, Lomeda J R, Scorsone J T, Patel A D, Friedheim J E and Tour J M 2011 Graphene oxide as a high-performance fluid-loss-control additive in water based drilling fluids *ACS App. Materials Interface* **4** 222-7
[8] Cai J, Chenevert M E, Sharma M M and Friedheim J 2012 Decreasing water invasion into atoka shale using nonmodified silica nanoparticles *SPE Drilling and Completion* **27** 103-12
[9] Hareland G, Wu A, Lei L, Husein M M and Zakaria M F 2012 Innovative nanoparticle drilling fluid and its benefits to horizontal or extended reach drilling *SPE Canadian Unconventional Resources Conf. (Alberta, Canada 30 Oct – 1 Nov 2012)*
[10] Javeri S M, Haindade Z W and Jere C B 2011 Mitigating loss circulation and differential sticking problems using silicon nanoparticles *SPE/IADC Middle East Drilling Technology Conf. Exhibition (Muscat, Oman 24-26 October 2011)*
[11] Sensoy T, Chenevert M E and Sharma M M 2009 Minimizing water invasion in shale using nanoparticles *SPE Annual Technical Conf. Exhibition (Louisiana, USA 4-7 October 2009)*
[12] Taraghikhah S, Mohammadi M K and Nowtaraki K 2015 Multifunctional nanoadditive in water based drilling fluid for improving shale stability *IPTC (Doha, Qatar 6-9 December 2015)*
[13] Wahid N, Aslam M and Hanafi N H 2015 Optimum nanosilica concentration in synthetic based

mud (SBM) for high temperature high pressure well *SPE/IATMI Asia Pacific Oil & Gas Conf. Exhibition (Bali, Indonesia 20-22 October 2015)*

[14] Gogotsi Y and Presser V 2013 *Carbon nanomaterials* (Florida: CRC Press)

[15] Geim A K and Novoselov K S 2007 The rise of graphene *Nat. Materials* **6** 183-91

[16] Neto A C, Guinea F, Peres N M, Novoselov K S and Geim A K 2008 The electronic properties of graphene *Phys. Rev.* **81** 109

[17] Tanaka K and Iijima S 2014 *Carbon nanotubes and graphene* (Japan: Elsevier)

[18] Aftab A, Ismail A and Ibupoto Z 2016 Enhancing the rheological properties and shale inhibition behavior of water-based mud using nanosilica, multi-walled carbon nanotube, and graphene nanoplatelet *Egyptian J. Petroleum* **26** 291-9

[19] Ismail A, Aftab A, Ibupoto Z and Zolkifile N 2016 The novel approach for the enhancement of rheological properties of water-based drilling fluids by using multi-walled carbon nanotube, nanosilica and glass beads *J. Petroleum Sci. Engineering* **139** 264-75

[20] Pashley R and Karaman M 2005 *App Colloid Surface Chem.* (Australia: John Wiley & Sons)

IOP Conf. Series: Materials Science and Engineering **352** (2018) 012026 doi:10.1088/1757-899X/352/1/012026

Response spectra analysis of the modal summation technique verified by observed seismometer and accelerometer waveform data of the M6.5 Pidie Jaya Earthquake

Irwandi[1,3], Ibnu Rusydy[2,3], Umar Muksin[1,3], Ariska Rudyanto[4], Daryono[4]
[1]Physics Department, Syiah Kuala University, Banda Aceh, Indonesia
[2]Geology Department, Syiah Kuala University, Banda Aceh, Indonesia
[3]Tsunami and Disaster Mitigation Research Center, Syiah Kuala University, Banda Aceh, Indonesia
[4]Indonesian Agency for Meteorology, Climatology and Geophysics

E-mail: irwandi@unsyiah.ac.id

Abstract. Wave vibration confined in the boundary will produce stationary wave solution in discrete states called modes. There are many physics applications related to modal solutions such as air column resonance, string vibration, and emission spectrum of the atomic Hydrogen. Naturally, energy is distributed in several modes so that the complete calculation is obtained from the sum of the whole modes called modal summation. The modal summation technique was applied to simulate the surface wave propagation above crustal structure of the earth. The method is computational because it uses 1D structural model which is not necessary to calculate the overall wave propagation. The simulation results of the magnitude 6.5 Pidie Jaya earthquake show the response spectral of the Summation Technique has a good correlation to the observed seismometer and accelerometer waveform data, especially at the KCSI (Kotacane) station. On the other hand, at the LASI (Langsa) station shows the modal simulation result of response is relatively lower than observation. The lower value of the reaction spectral estimation is obtained because the station is located in the thick sedimentary basin causing the amplification effect. This is the limitation of modal summation technique, and therefore it should be combined with different finite simulation on the 2D local structural model of the basin.

1. Introduction

Most of the physics application is to find the solutions by implementing differential equations. The initial value and boundary condition will lead to unique solutions with specific patterns. For instance, when a solution of the wave equation in boundary value condition is confined to a finite region of spaces of the boundary, some pattern will happen. Space that is filled up by vibrating patterns is called standing wave. There are many physics applications related to modal solutions such as air column resonance, string vibration, and emission spectrum of the atomic Hydrogen. Confining the wave could quantize the frequency of the vibration into several modes. The quantized frequency is distributed into spectral of energy. Therefore, if we know the spectral of energy we can find the complete solution by summing the modes of the wave. Thus, the similar method is used to find seismic wave propagation on the surface of earth structure known as the surface wave.

IOP Conf. Series: Materials Science and Engineering **352** (2018) 012026 doi:10.1088/1757-899X/352/1/012026

In the framework of the linear theory of elastic media, the balance of the forces such as inertia, body forces, and surface forces acting on a cubic element within the continuum is considered. Therefore Newton's laws of motion could be expressed as follows,

$$\begin{aligned}\rho\frac{\partial^2 u_x}{\partial t^2} &= \rho X + \frac{\partial\sigma_{xx}}{\partial x} + \frac{\partial\sigma_{yx}}{\partial y} + \frac{\partial\sigma_{zx}}{\partial z}\\ \rho\frac{\partial^2 u_y}{\partial t^2} &= \rho Y + \frac{\partial\sigma_{xy}}{\partial x} + \frac{\partial\sigma_{yy}}{\partial y} + \frac{\partial\sigma_{zy}}{\partial z}\\ \rho\frac{\partial^2 u_z}{\partial t^2} &= \rho Z + \frac{\partial\sigma_{xz}}{\partial x} + \frac{\partial\sigma_{yz}}{\partial y} + \frac{\partial\sigma_{zz}}{\partial z}\end{aligned} \tag{1}$$

Where a Cartesian coordinate system (x, y, z) is adopted. sij(**x**,t) (i=x, y, z; j=x, y, z) indicates the second-order stress tensor, ρ is the density of the material, and X, Y, Z are the components of body forces for a unit mass.

In general, the relation between stress and deformation can be complex because it is influenced by several parameters including pressure, temperature, and the amount of the variability of the stress. Nevertheless, by assuming the deformations and the stresses in short duration (the conditions mostly satisfy the problems in the ground motion estimation) are small, we can assume that solid behaves linearly, and the constitutive relation is linking to the stresses and the deformation could be based on the Hooke's laws of the stress tensor. After applying some symmetrical properties and assuming the locally isotropic stress, we obtain,

$$\sigma_{ij} = \lambda e_{kk}\delta_{ij} + 2\mu e_{ij} \tag{2}$$

where λ is the quantities and μ are called *Lamè parameters*.

Let us consider a halfspace system where the vertical z-axis is positive downward from the free surface. The vertical stresses are null which is defined by the plane z=0 (Figure 1).

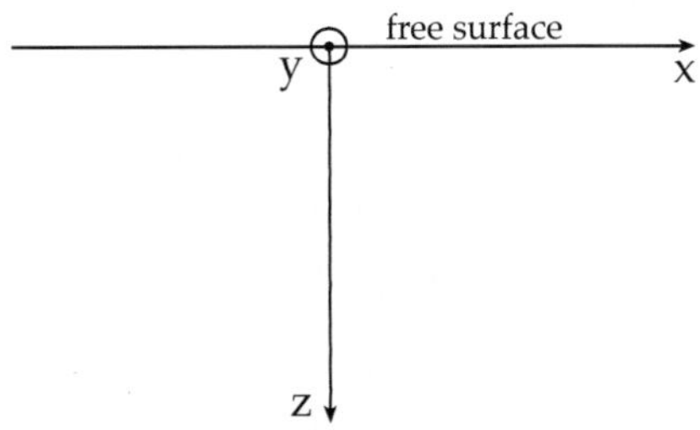

Figure 1. Adopted reference for a vertically heterogeneous halfspace system

By assuming that l, m, and r are continuous functions of z and the body wave velocities are $\alpha = \sqrt{\frac{\lambda+2\mu}{\rho}}$ and $\beta = \sqrt{\frac{\mu}{\rho}}$, the values of α_H and β_H are the largest when z > H remain constant for greater depths. Here, the elastic parameters depend only on the vertical coordinate [1-2].

2. Method of Modal Summation Technique

By neglecting the body forces, we can consider solutions of Equation 1 to form harmonic wave plane propagating along the positive x-axis:

$$u(x,t) = F(z)e^{i(\omega t-kx)} \quad (3)$$

The seismic wave source is introduced in the medium representing the fault which is supposed to be planar, as a discontinuity in the displacement and shear stresses fields, concerning the fault plane. On the contrary, the normal stresses are supposed to be continuous across the fault plane. Maruyama [3] and Burridge and Knopoff [4] demonstrated the representation theorem the rigorous equivalence of the effects between a faulted medium with a discontinuity in the displacements and shear stress fields and an unfaulted medium where the proper body forces are applied.

Following the procedure proposed by Kausel and Schwab [5], we assume that the periods and the wavelengths, which we are interested in, are large compared to the rise time and the dimensions of the source. Therefore, the source function, describing the discontinuity of the displacement across the fault, can be approximated by a step function in time and a point source in space. Furthermore, if the normal stress is continuous across the fault, then based on the representation theorem the equivalent body force in an unfaulted medium is double-couple with a total null moment. With this assumption, the eigenvalues and eigenfunctions of the problem are already determined. We can write the expression for the displacement with varying time, i.e., the synthetic seismogram, for the three components of motion. The asymptotic expression of the Fourier transforms of the displacement **U** = (Ux, Uy, Uz), at a distance r from the source, can be written as $U = \sum_{m=1}^{\infty} {}^{m}U$, where m is the mode index and:

$$\begin{aligned} {}^{m}U_x(r,z,\omega) &= \frac{e^{\frac{-3}{4}i\pi}}{\sqrt{2\pi}}\left[\chi_R(h_s,\varphi)S(\omega)\frac{\sqrt{k_R}e^{-ik_R r-\omega r C_{2R}}}{\sqrt{r}}\frac{\epsilon_0 u_x(z,\omega)}{2c_R v_{gR} I_{1R}}\right]_m \quad {}^{m}U_y(r,z,\omega) \\ &= \frac{e^{\frac{-3}{4}i\pi}}{\sqrt{2\pi}}\left[\chi_L(h_s,\varphi)S(\omega)\frac{\sqrt{k_L}e^{-ik_L r-\omega r C_{2L}}}{\sqrt{r}}\frac{u_x(z,\omega)}{2c_L v_{gL} I_{1L}}\right]_m \quad {}^{m}U_z(r,z,\omega) \\ &= {}^{m}U_x(r,z,\omega)\frac{e^{\frac{-i\pi}{2}}}{\epsilon_0} \end{aligned} \quad (4)$$

The suffixes R and L refer to the quantities associated with Rayleigh and Love modes, respectively. In Equation 4, S(ω) =|S(ω)|exp[i·arg(S(ω))] is the FT of the source time function while χ(h_s,φ) represents the azimuthal dependence of the excitation factor expressed by [6]:

$$\chi_R(h_s,\varphi) = d_0 + i\chi_L(h_s,\varphi) = i \quad (5)$$

with:

$$\begin{aligned} d_0 &= \frac{1}{2}B(h_s)sin\lambda sin2\delta d_{1R} & d_{1L} &= G(h_s)cos\lambda sin\delta d_{2L} \\ &= -C(h_s)sin\lambda cos2\delta d_{2R} & &= -G(h_s)sin\lambda cos2\delta d_{3L} \\ &= -C(h_s)cos\lambda cos\delta d_{3R} & &= \frac{1}{2}V(h_s)sin\lambda sin2\delta d_{4L} \\ &= A(h_s)cos\lambda sin\delta d_{4R} & &= V(h_s)cos\lambda sin\delta \\ &= \frac{-1}{2}A(h_s)sin\lambda sin2\delta \end{aligned} \quad (6)$$

Where φ is the angle between the strike of the fault and the strike- receiver angle that is measured anticlockwise, hs is the focal depth, δ is the dip angle and λ is the rake angle (see Figure 2). The function of hs that appear in Equation 6 depends on the values assumed by the eigenfunctions at the hypocentre. The details in proving the above equation was done by Panza et al. [2]. The seismogram is computed by summing the time series radiated by the single point-sources with the appropriate time-shifts that are defined by the rupture process. The resultant time series shows great influence on the directivity and the distribution of energy released in time may have influenced on the synthesized ground motion [2, 7].

At each site, the horizontal components (P–SV radial and SH transverse) synthetic seismograms are first computed for a seismic moment of 10^{-7} N m and then scaled to the magnitude of the earthquake using the moment–magnitude relation from Kanamori [8]. The finiteness of the source is accounted by scaling the spectrum using the spectral scaling law proposed by Gusev [9] as reported by Aki [10]. For the period between 1 and 2 s, the Gusev spectral fall-off produces higher spectral values than the ω^{-2} spectral fall-off, and thus guarantees a conservative hazard computation [11].

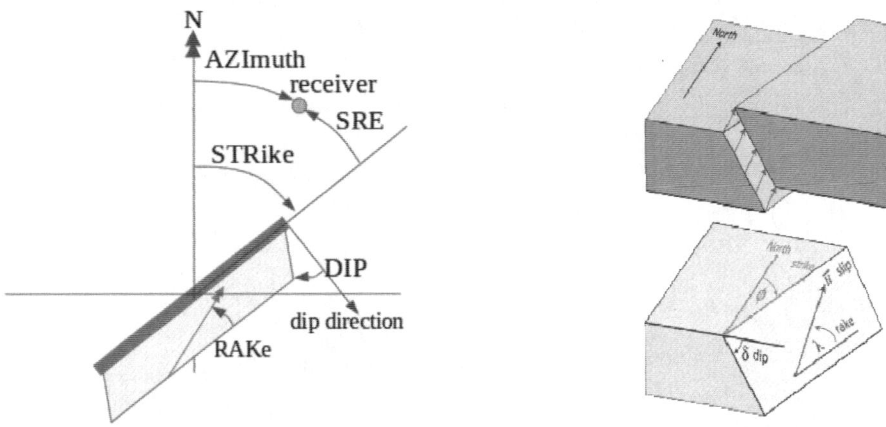

Figure 2. Angle conventions used for the seismic source system.

3. Theoretical Comparison and Observation of the Response Spectral

Parameters for a seismic source system are origin time, source location, strike, dip, rake, and magnitude. The value is taken from the Pidie Jaya earthquake data with origin time 2016-12-06 22:03:33 UTC or 2016-12-07 05:03:33 local time, at 5.283°N, 96.168°E 13.5 km depth, focal mechanism parameters: Strike=243°, dip=81°, rake=33°, and magnitude = 6.5 Mw. The source information of the Pidie Jaya earthquake is obtained from USGS [12]. The fault that generated the Pidie Jaya earthquake seems to be the branch of the Northern part of the Great Sumatran Fault [13-14]. Figure 3 shows the seismic source location, the focal mechanism on the map, and the available station of BMKG.

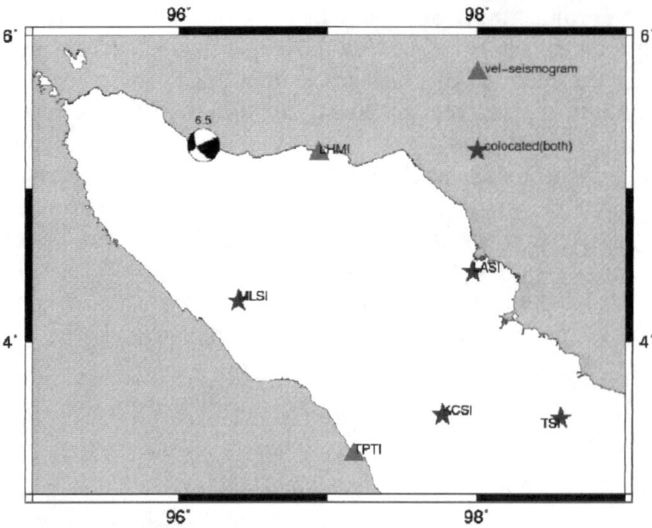

Figure 3. The source Location of the M6.5 Pidie Jaya and the seismic stations closed to the epicenter (blue stars).

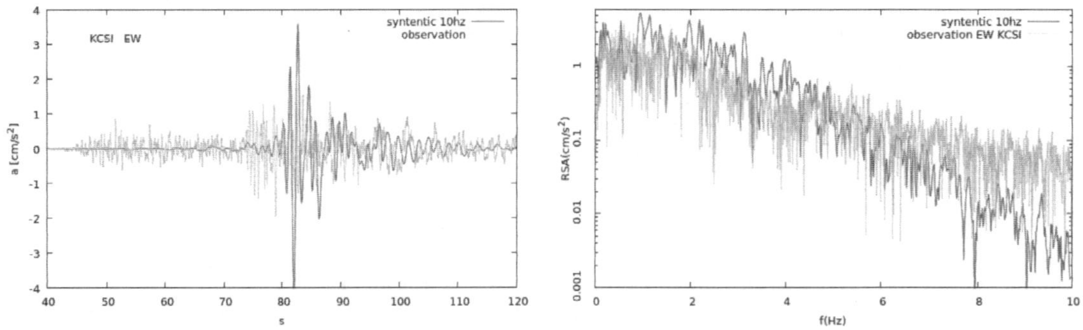

(a) (b)

Figure 4. (a) The synthetic waveform and observed accelerogram from the KCSI station recorded during the Pidie Jaya earthquake (b) related response spectral of the waveform.

The waveform solutions expressed by equation 4 depend on the 1D geological structure$u_x(z,\omega)$ that form dispersion curve as a function of depth. The accelerogram waveform solution for the KCSI station from the Pidie Jaya earthquake is shown in Figure 4a. Because the waveform depends on the geological structure, the spectrum in the frequency domain is also called response spectra as shown in Figure 4b.

We investigate the response spectra of the observed and computed signals for accelerogram and seismogram from the two stations KCSI and LASI as shown in Figure 5 and 6. The Response Spectral of Acceleration (RSA) and the Response Spectra of Velocity (RSV) analyses are useful for practical engineering purposes.

Figures 5a, 5c, and 5e show the comparison of the accelerogram spectral response for the KCSI of EW, NS, and Z components, respectively. The low-frequency accelerograms have relatively good suitability between synthetic and observation compared to those of high frequency. The seismogram waveforms which could be interpreted as velocities have good suitability not only at low frequency but also for low high frequency as shown in figure 5b, 5d, and 5f. The coda and noise waves which naturally work at high-frequency acceleration contribute the difference between the synthetic and the observed.

Figure 6 shows the comparison between accelerogram (figure 6a, 6c, and 6e) and seismogram (figure 6b, 6d, and 6f) for the LASI station. For the LASI station, the observed waveform responds spectra is higher than the synthetics seismogram. The location of the LASI station is located in Langsa Aceh at 4.457°N and 97.970°E which is near the coastline on top of the thick sedimentary layer.

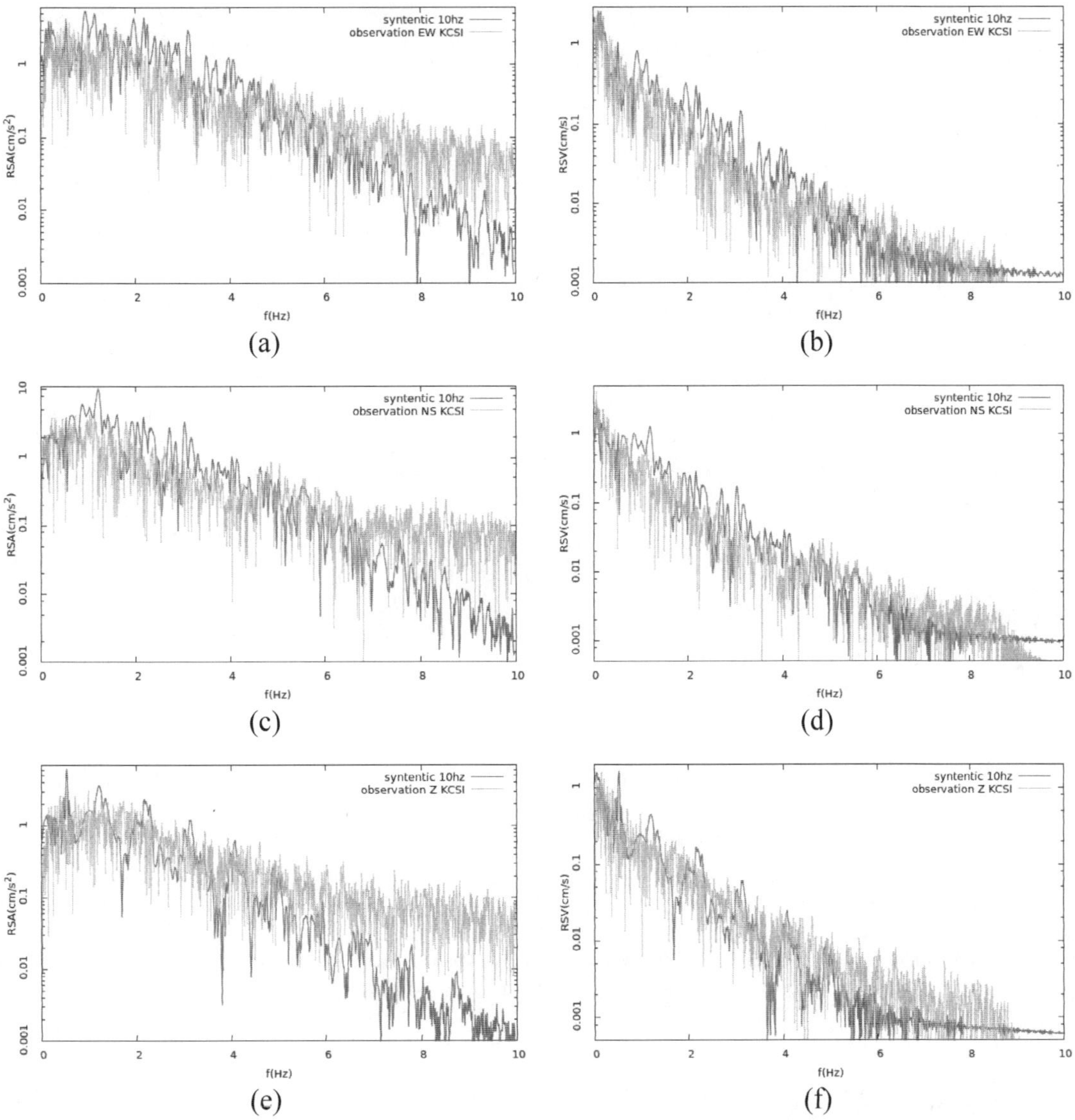

Figure 5. The synthetic and the observed accelerogram (left) and seismogram (right) respond spectral of the KCSI station which the epicenter distance of 263 km and located at 3.522°N and 97.772°E in the Kutacane City. Geologically, the station is located on the thin sediment of the Paleozoic metamorphic bed rock of Alas formation.

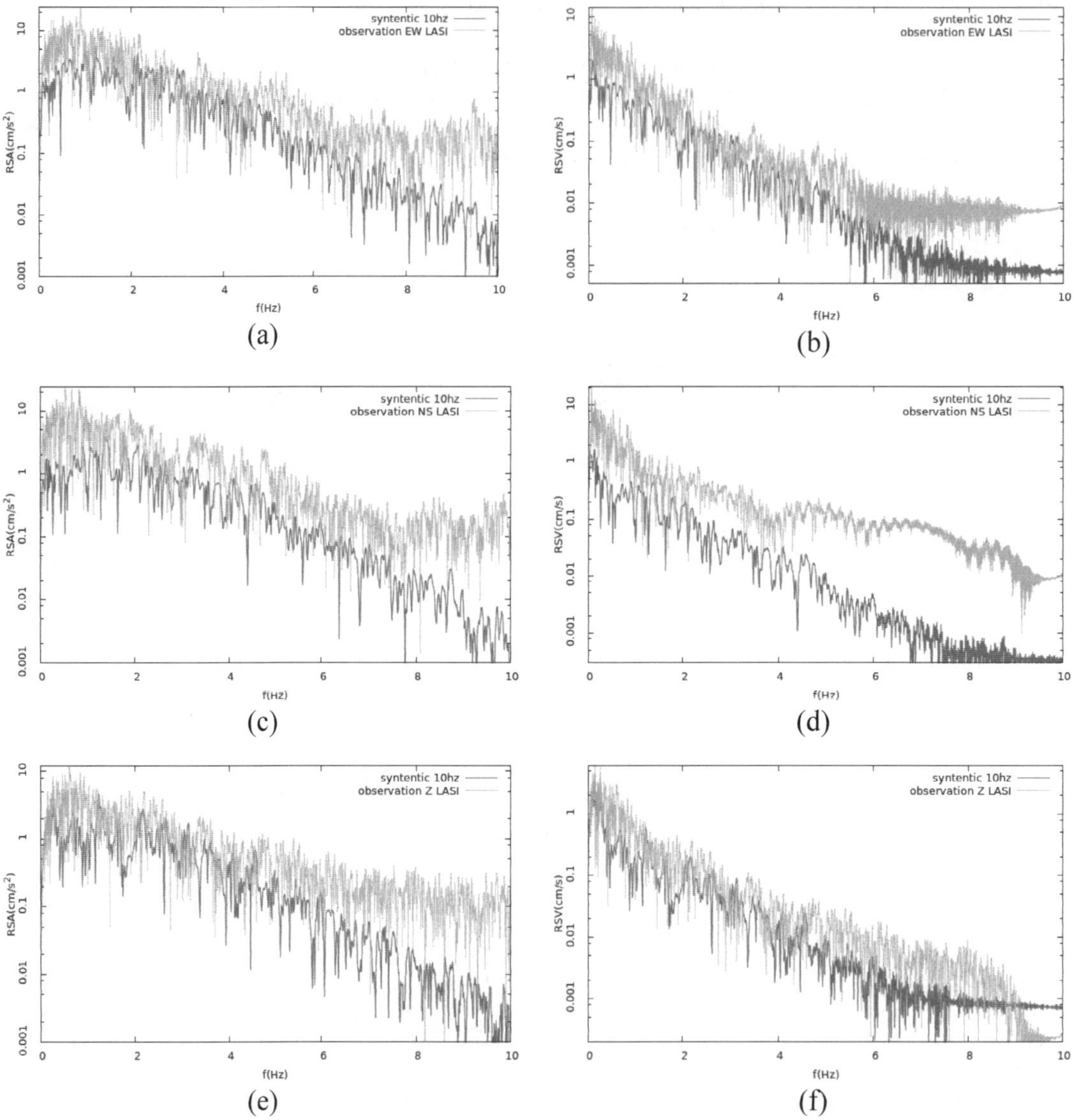

Figure 6. The synthetic and the observed accelerogram (left) and seismogram (right) response spectra of the LASI station with epicenter distance of 219.78 km at 4.457°N and 97.970°E in the Langsa City. Geologically, the station is located on top of the thick sedimentary basin of younger alluvium formation.

4. Conclusions and Suggestion

The respond spectral of seismogram and accelerogram has successfully produced by modal summation technique for thin sedimentary layer located on bedrock as shown at the KCSI station during the Pidie Jaya earthquake. However, the modal summation technique has limitation to produce appropriate respond spectra for thick sedimentary basin because the 1D approach cannot handle the amplification effect of the 2D sedimentary basin. To reach the appropriate respond spectra for civil engineering purposes, the 1D model of summation technique must be hybridized with a 2D finite different method to handle the amplification effects. Research on the synthetic seismogram needs to be compared to several other methods to increase the precision of the ground respond spectral calculation for the geotechnical purposes.

Acknowledgments
I wish to acknowledge the assistance from the Department of Geoscience, Trieste University, and International Centre for Theoretical Physics ICTP. I would like to thank colleagues from the Physics Department, the Geophysics Engineering Department, TDMRC and BMKG for encouragement.

References

[1] Irwandi 2017 Advantages of realistic model based on computational method: NDSHA versus Standard PSHA *IOP Conf. Series: Earth and Environmental Science* **56** (doi:10.1088/1755-1315/56/1/012007)
[2] Panza G F, Romanelli F and Vaccari F 2001 Seismic wave propagation in laterally heterogeneus anelastic media: theory and applications to seismic zonation *Adv. Geophys.* **43** 1–95
[3] Maruyama T 1963 On the Force Equivalents of Dynamical elastic dislocations with reference to the earthquake *Mechanism. Bull. Earthquake Res. Inst. Tokyo Univ.* **41** 467-86
[4] Burridge R and Knopoff L 1964 Body force equivalents for seismic dislocations *Bulletin of the Seismological Society of America* **54** 1875-88
[5] Kausel E and Schwab F 1973 Contributions to love-wave transformation theory: Earth-flattening transformation for love waves from a point source in a sphere *Bulletin of the Seismological Society of America* **63** 983-93
[6] Ben-Menahem A and Harkrider D G 1964 Radiation patterns of seismic surface waves from buried dipolar point sources in a flat stratified earth *Journal of Geophysical Research* **69** 2605-20
[7] Hassan H M, Romanelli F, Panza G F, ElGabry M N and Magrin A 2017 Update and sensitivity analysis of the neo-deterministic seismic hazard assessment for Egypt *Engineering Geology* 218 77–89
[8] Kanamori H 1977 The Energy Release in Great Earthquakes *Journal of Geophysical Research* **82** 2981-7
[9] Gusev A A 1983 Descriptive statistical model of earthquake source radiation and its application to an estimation of short-period strong motion *Geophysical Journal International* **74** 787-808
[10] Aki K 1987 Strong Motion Seismology *Strong Ground Motion Seismology* (Springer Netherlands) 3-39
[11] Vaccari F 1995 LP-Displacement Hazard Evaluation in Italy *Proc. 24th General Assembly of the European Seismological Commission* 1489-98
[12] USGS (United Stated Geological Survey) 2016 Earthquake Hazard program M 6.5-14km WNW of Reuleuet, Indonesia https://earthquake.usgs.gov/earthquakes/eventpage/us10007ghm#moment-tensor
[13] Tabei T, Kimata F, Ito T, Gunawan E, Tsutsumi H, Ohta Y, Yamashina T, Soeda Y, Ismail N, Nurdin I, Sugiyanto D and Meilano I 2015 Geodetic and Geomorphic Evaluations of Earthquake Generation Potential of the Northern Sumatran Fault, Indonesia *GENAH* 21-28
[14] Irwandi, Marwan, Muksin, and Fashbir 2017 Applications of the VLF-EM method for rapid Sumatran fault identification in Leuser national park, Aceh AIP Conference Proceedings **1861** 030050 (doi: 10.1063/1.4990937)

The implementation of unit price of work standard SNI 7394: 2008 for the construction of reinforced concrete beam

Tripoli, Mubarak, Nurisra, Mahmuddin
University of Syiah Kuala, Jl. Syekh Abdurrauf 10, Banda Aceh, Indonesia

E-mail: tripoli@unsyiah.ac.id

Abstract. This paper discusses the implementation of Indonesian National Standard (SNI) 7394: 2008 on procedures for calculating the unit price of concrete work for the construction of building and housing. The standard provides some reinforced concrete constructions unit price (UP) analysis by specified the total number of reinforcing uses. Related to reinforced concrete beam work (Analysis No. 6.31), the reinforcement requirement is stated at 200 kg/m^3 of concrete. Once the implementation considers various earthquake zoning, the question will arise about the extent to which the standard is feasible to apply. Therefore, this research aimed to analyze the possibility of UP standard implementation by certain earthquake zonation. This research is focused on the construction of reinforced concrete beam for buildings with function as educational, residential and office buildings. The data used are sourced from 21 buildings in two zones in Aceh Province, covering Zone 10 and Zone 15 based on earthquake map of SNI 1726: 2012. The analysis results indicate that the UP standard for reinforced concrete beam cannot be applied to all zoning. The UP standard is only possible on buildings constructed in Zone 10 or zonation with seismic spectral response *0.6g to 0.7g* or lower.

1. Introduction

Cost estimation plays an important role in the implementation of construction projects. The activities are covering the process of estimates or judgments, based on certain facts and assumptions, of the final cost of a project, product, or process [1]. The estimation is the process of preparing one of the most comprehensive documents in a project, which affects the overall project team and project related activities. The inaccuracy of the estimation will bring disadvantage to the parties involved. The cost estimation process usually begins with analyzing the unit price (UP) of work. Some standards are set by the government by the scope of work and type of construction [2]. About reinforced concrete work, the unit price could be analyzed by referring to SNI 7394: 2008 on the procedure for calculating the unit price of concrete work for the construction of buildings and houses [3].

The building is a construction formed by several main components, namely structural components, non-structural components, and utility components [2]. Structural components are components that have the largest cost portion of a building. These components include construction of foundations, columns, beams, floor plates, and roof structures. This construction can be made using concrete, steel, wood, or composite materials such as reinforced concrete. Associated with the beam, this construction is one of the structural components designed to hold and move the transverse load across space to the supporting element [4]. A beam is a part of a structure used as a floor slab holder and a column top binder. Reinforced concrete composite material commonly used for beam construction. This construction formed by two main components, namely concrete and steel for reinforcement.

Some factors determine the planning of the reinforced concrete beam structure dimensions. One important factor is related to earthquake loading. The planning standard used is SNI 1726: 2012 on earthquake resistance planning procedures for buildings and non-buildings [5]. This SNI describes the potential difference of earthquake risk by the earthquake-zoning map. This map shows the position of the entire zones of Indonesia based on different seismic spectral response value. The zoning indicated by difference colors which informed upper and lower values of spectral response limits in gravity (*g*). Spectral value is used as a reference to design the earthquake load in the planning of a building in Indonesia.

Some studies related to the cost estimation and unit price of works in buildings in Aceh Province have been done. The research develops estimation model at the conceptual stage (pre-construction design) [6, 7, 8]. Associated with earthquake risk factors, research has been undertaken for modelling rehabilitation costs [9] and reinforcement requirement analysis for reinforce concrete slab [10]. However, there is no special research discussing the issues related to the application of UP for the construction of reinforced concrete beam in some earthquake zoning. The existence of this study will provide answers to the extent to which SNI 7394: 2008 can be applied to the zoning and function of certain buildings.

2. Methods

2.1. Location and object of research

Survey on this research was conducted on building projects built in Aceh Province. The research object's review is focused on building with simple and non-simple classification as defined in Ministry of Public Work Regulation No. 45/PRT/M/2007 [2]. The object of the building to be studied can be classified according to the function and location based on the earthquake zonation map of the year SNI 1726: 2012. The criteria location and object of research as follows:

- The object locations of the research used based on SNI 1726: 2012 earthquake zonation maps, namely Zone 10 (the region with the spectral response of *0.6g to 0.7g*) and Zone 15 (the region with the spectral response of *1.2g to 1.5g*), as shown in Figure 1.
- The project used the minimum of two-story buildings.
- The functions of the buildings studied are for offices (OFC), residence (RES), and education (EDU).

2.2. Data collection

The required data are as follows:

- Document of the contract of building project constructed from 2012 until 2015, containing quantity and bill of quantity data, and drawings (site plan, building plan, and building view).
- Indonesia earthquake zonation map SNI 1726: 2012.

2.3. Data processing and analysis

The process of data processing and analysis through the stages as follows:

- The data grouping is based on earthquake zoning division, covering Zone 10 and Zone 15. Data of each earthquake zone are reclassified based on office function (OFC), residence (RES), and education (EDU).
- The requirement analysis of reinforcement material is done by calculating the need of reinforcement of the beam done based on the dimension of the cross-section, the type of reinforcement (deformed/ribbed bar and plain bar), the reinforcement function (main bars and binders) and the reinforcement diameter used.
- The ratio of reinforcement used per m^3 of concrete by comparing the total weight of the beam reinforcement volume with the volume of concrete in each building.
- Analyze deviation from the UP of SNI standard.

Statistical analysis is used to determine the mean value ($\bar{x}$) and standard deviation (σ) of the results of the analysis of the ratios of each building (x) for the total number of buildings (n), with the following equation:

$$\bar{x} = \frac{\sum x}{n} \tag{1}$$

$$\sigma = \sqrt{\frac{\sum (x-\bar{x})^2}{n-1}} \tag{2}$$

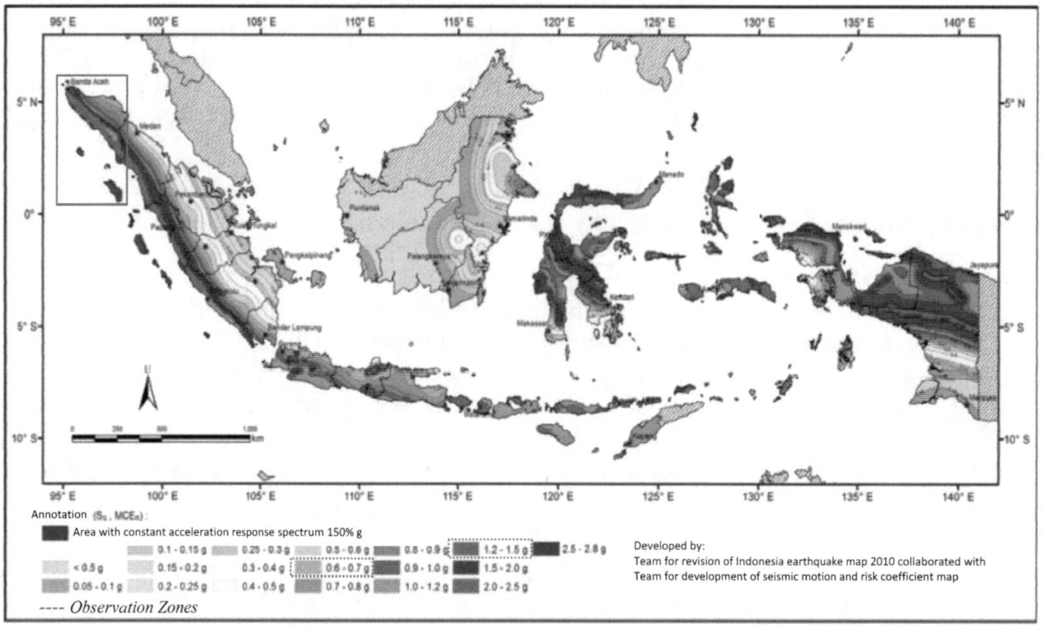

Figure 1. Indonesia Earthquake Zoning Map of SNI 1726: 2012

3. Results and discussions

3.1. Description of the buildings

The building that is used for data is 21 buildings consist of 2 and 3-storey buildings, which are distributed at 28.6% in Zone 10, and 71.4% in Zone 15. Based on building function, the study object consists of 28.6% for each of office buildings and education buildings, and the remaining 42.8% are residential buildings. The analysis of beam constructions was only focused on floor beam (primary beams and secondary beams) and excludes tie beam and roof beam. The full description is shown in Table 1.

3.2. The actual ratios of reinforcement use per 1 m^3 of reinforced concrete beam

The actual beam reinforcement ratio is analyzed based on the ratio of reinforcement use (in kg units) to the use of concrete (in units of m^3). The average reinforcement ratios for Zone 10 were 194.41 kg/m^3 increased by 6.4% to 206.92 kg/m^3 for Zone 15. This increase was consistent with the increase of the load due to the earthquake corresponding to the spectral response in the two zones. The calculation results of beam reinforcement ratios can be seen in Table 2.

Table 1. Description of the buildings

Zone	Building function			Number of building
	Office	Education	Residence	
10	2	2	2	6 (28.6%)
15	4	4	7	15 (71.4%)
Total	6 (28.6%)	6 (28.6%)	9 (42.8%)	21 (100.0%)

Table 2. The actual reinforcement ratio of reinforced concrete beam

No	Zone	Name of building	Building function	Concrete qty (m^3)	Reinforce. qty (kg)	Reinforce. ratio (kg/m^3)	Average (kg/m^3)	Min (kg/m^3)	Max (kg/m^3)	Deviation standard (kg/m^3)
1	10	OFC-1	Office	21.73	4,330.08	199.27	194.41	184.28	212.67	10.52
2		OFC-2	Office	7.50	1,594.99	212.67				
3		EDU-1	Education	47.33	8,722.59	184.28				
4		RES-1	Residence	55.29	10,320.03	186.65				
5		EDU-2	Education	14.05	2,739.38	194.97				
6		RES-2	Residence	18.67	3,520.98	188.62				
7	15	EDU-3	Education	20.70	3,754.48	181.37	206.92	181.37	227.71	15.67
8		EDU-4	Education	144.01	26,335.16	182.87				
9		RES-3	Residence	15.25	3,460.51	226.92				
10		EDU-5	Education	14.23	2,847.02	200.11				
11		RES-4	Residence	14.80	3,296.86	222.76				
12		RES-5	Residence	13.08	2,944.23	225.06				
13		OFC-3	Office	21.00	4,137.02	197.00				
14		RES-6	Residence	9.16	1,781.82	194.52				
15		EDU-6	Education	16.61	3,391.64	204.19				
16		RES-7	Residence	29.81	5,862.25	196.65				
17		OFC-4	Office	65.10	14,823.99	227.71				
18		RES-8	Residence	21.24	4,394.36	206.87				
19		RES-9	Residence	20.63	4,135.19	200.49				
20		OFC-5	Office	10.18	2,282.78	224.35				
21		OFC-6	Office	19.60	4,173.20	212.92				

3.3. Comparison of the actual ratios and SNI

SNI 7394: 2008 provides that for reinforced concrete beam work used reinforcement of 200 kg/m^3 of concrete. The results of the analysis at Zone 10 show an average deviation of 5.59 kg or 2.80% below the value of the needs of the SNI reinforcement. This condition explains that this standard can still be applied to this zone. However, when viewed from the standard deviation values that arise, the use of this standard still needs to consider the possibility of deviation. The indication of such deviation is indicated by one building data with a positive deviation value greater than the standard deviation value. In Zone 15, the general deviation value indicates that the actual reinforcement ratio is greater than the value of the SNI bone needs ratio. The average deviation was 6.92 kg/m^3 with a standard deviation of 15.67 kg/m^3. This condition indicates that the UP application on SNI for this zone is not feasible because it has the potential to cause cost estimation with the amount below the actual requirement. The complete results of the ratio of actual reinforcement to the UP standard are shown in Table 3.

Table 3. The actual reinforcement comparing to SNI

No	Zone	Name of building	Building function	Reinforcement ratio (kg/m^3)		Deviation	
				Actual	SNI	kg/m^3	%
1	10	OFC-1	Office	199.27	200.00	-0.73	-0.37
2		OFC-2	Office	212.67	200.00	12.67	6.33
3		EDU-1	Education	184.28	200.00	-15.72	-7.86
4		RES-1	Residence	186.65	200.00	-13.35	-6.67
5		EDU-2	Education	194.97	200.00	-5.03	-2.51
6		RES-2	Residence	188.62	200.00	-11.39	-5.69
					Average	-5.59	-2.80
				Deviation Standard		10.52	5.26
7	15	EDU-3	Education	181.37	200.00	-18.63	-9.31
8		EDU-4	Education	182.87	200.00	-17.13	-8.57
9		RES-3	Residence	226.92	200.00	26.92	13.46
10		EDU-5	Education	200.11	200.00	0.11	0.05
11		RES-4	Residence	222.76	200.00	22.76	11.38
12		RES-5	Residence	225.06	200.00	25.06	12.53
13		OFC-3	Office	197.00	200.00	-3.00	-1.50
14		RES-6	Residence	194.52	200.00	-5.48	-2.74
15		EDU-6	Education	204.19	200.00	4.19	2.10
16		RES-7	Residence	196.65	200.00	-3.35	-1.67
17		OFC-4	Office	227.71	200.00	27.71	13.86
18		RES-8	Residence	206.87	200.00	6.87	3.43
19		RES-9	Residence	200.49	200.00	0.49	0.25
20		OFC-5	Office	224.35	200.00	24.35	12.18
21		OFC-6	Office	212.92	200.00	12.92	6.46
					Average	6.92	3.46
				Deviation Standard		15.67	7.83

3.4. Comparison of Reinforcement Ratio Based on Building Function

The reinforcement ratio analysis needs to be examined more specifically related to the three building functions in each zone. Potential standard UP applications are possible for buildings with residential functions (Zone 10) and educational functions (Zones 10 and Zone 15). Concerning the function of office buildings throughout the zone, the mean deviation indicates an actual ratio value greater than the UP standard. Similarly, it is seen in the residence function in Zone 15. The positive deviation value explains that in this functions and zones, the use of UP cannot meet the needs of actual reinforcement that should be equipped. In other words, the required of reinforcement materials are still below the actual needs. When associated with the estimated cost value, the estimated amount generated by UP usage is potentially insufficient for the completion of construction. An illustration of the comparison of the reinforcement ratio based on the building function is shown in Table 4.

Table 4. The deviation of reinforcement based on building function

Building Function	No of Building			Average deviation of reinforcement per zone			
	Zone 10	Zone 15	Total	Zone 10		Zone 15	
				kg/m^3	%	kg/m^3	%
Office	2	4	6	5.97	2.98	15.50	7.75
Residence	2	7	9	-12.37	-6.18	10.47	5.23
Education	2	4	6	-10.37	-5.19	-7.87	-3.93

4. Conclusions

Implementation of standard unit price (UP) of work from SNI 7394: 2008 for the construction of reinforced concrete beams can be concluded by considering the following two conditions:

- Implementation of the UP standard based on zoning is only possible for building buildings built in Zone 10 or areas with spectral responses below *0.6g*. The reinforcing material index on the UP standard of 200 kg/m^3 of concrete for beam construction is still adequate to meet actual reinforcement usage in this zone.
- Implementation of UP standards based on building functions is possible on certain functions only. However, the application associated with the building function should not necessarily be used. Still, zonation-related implementation as a whole need to consider the indication of standard deviation values result.

The results of the analysis on Zone 10 are based on relatively limited data, so that the accuracy of the research results is still possible to be improved by the addition of more relevant data. The implementation of standards in Zones 11 and Zone 12 has the potential to be done. This condition is seen from the results of the analysis at Zone 10 that indicates the margin ratio in the range of 183.89 - 200 kg/m3.

Acknowledgement

This research was granted for funding support from *Hibah Penelitian Produk Terapan* No. 55/UN11.2/PP/SP3/2017 of Fiscal Year 2017, Directorate General of Research and Development, Ministry of Research, Technology and Higher Education. The authors express high appreciation to all parties who have contributed to this research, especially to Ichwan and other research teams that involved from data collection to completion of the research.

References

[1] Aaron A L 1997 The Estimating Process *the Engineer's Cost Handbook: Tools for Managing Project Costs* ed R E Westney (New York: Marcel Decker) Chapter 1 p 1–40
[2] *Ministry of Public Work Regulation No. 45/PRT/M/2007 on the Technical Guidelines of State Building Development*
[3] *SNI 7394:2008 on Procedures for calculating the unit price of concrete work for the construction of buildings and housing*
[4] Ching F D K 2014 *Building construction illustrated* Fifth ed. (New Jersey: John Wiley & Sons)
[5] *SNI 1726: 2012 on Procedures of earthquake resistant planning for buildings and non-buildings*
[6] Mubarak 2010 Studi Harga Satuan Bangunan Gedung di Provinsi Nanggroe Aceh Darussalam *Jurnal Teknik Sipil* **1** p 43-52 (in Indonesian)
[7] Mubarak and Tripoli 2011 Penggunaan variabel harga material untuk memprediksi harga satuan *Prosiding Seminar API 2011* (in Indonesian)
[8] Fachrurrazi, Husin S, Tripoli and Mubarak 2017 Neural network for the standard unit price of the building area *Proc. Engineering* 171 p 282 – 293
[9] Mahmuddin and Mubarak 2014 Housing Rehabilitation Budget Estimate due to Earthquake Disaster in Aceh Province using the Multiple Linear Regression Models *Proc. 9th AIWEST-DR 2014* p 52 – 59
[10] Tripoli, Nurisra, and Mubarak 2017 Analisis Kebutuhan Tulangan Pelat Lantai Beton Bertulang Pada Konstruksi Bangunan Gedung *Prosiding Konferensi Nasional Teknik Sipil 11* p 1 – 9 (in Indonesian)

IOP Conf. Series: Materials Science and Engineering **352** (2018) 012028 doi:10.1088/1757-899X/352/1/012028

Fresh and mechanical properties of self-compacting concrete with coarse aggregate replacement using Waste of Oil Palm Shell

Hakas Prayuda, Fadillawaty Saleh, Taufiq Ilham Maulana, Fanny Monika
Department of Civil Engineering, Faculty of Engineering, Universitas Muhammadiyah Yogyakarta, Lingkar Selatan Street, Bantul, Yogyakarta, Indonesia 55183

e-mail: hakas.prayuda@ft.umy.ac.id ; taufiq.im@ft.umy.ac.id

Abstract. Self-compacting Concrete (SCC) is a real innovation that can solidify itself without the help of tools to ease field practice. In its implementation, SCC can use alternative materials to reduce waste, such as Oil Palm Shell (OPS). In this research, OPS used as a replacement of crushed stone as the main coarse aggregate. The concrete mixture used consists of cement, sand, crushed stone, OPS as a variation of aggregate substitutes, palm oil fuel ash, and superplasticizer. OPS used were variated with 0%, 5%, 10%, 25% and 50% of crushed stone aggregate weight with age up to 28 days. Tests were conducted on fresh and mechanical properties. From the results, it is known that replacement of aggregate using OPS meets fresh properties criteria and although the compressive strength of OPS concrete mixture is lower than normal SCC, OPS still can be an alternative in making SCC and reducing palm oil industrial waste.

1. Introduction

Self-compacting concrete (SCC) can be defined as concrete that can compact itself without the need to be compacted by tools such as the vibrator or manual manpower. The use of SCC has many advantages, for example, SCC can reach areas that are difficult to be reached as in the corner of cast and SCC can pass through the concrete with a complicated and tight rebar spaces. Many research has been done on SCC, one of them concerning to process of standardization in SCC testing and the composition of concrete mixtures in SCC [1].

In its development, SCC concrete does not only use the general concrete material, but also added various kinds of additional materials, both organic and non-organic. This addition has a variety of purposes, such as reducing waste and increasing strength of the concrete. Additional or partial replacements of fine aggregate or coarse aggregate that have been done previously are using recycled concrete aggregates [2-3], iron slag [4], coal fly ash and rice husk ash [5], and also from palm waste which is palm oil fuel ash [6-7] and oil palm shell [8-10].

In Indonesia, there are much waste produced and not utilized properly, one of which is oil palm shell (OPL) waste. This country is one of many countries that produce a lot of palm oil [11] which is commonly produced on Sumatra Island and Kalimantan Island. Waste from the palm oil industry is oil palm shell (OPL) and palm oil fuel ash (POFA). In this paper, it will be discussed using OPL waste as a substitute for an aggregate of SCC to reduce the use of gravel as coarse aggregate and to utilize the available waste. The study was conducted at Universitas Muhammadiyah Yogyakarta to find out new and mechanical properties of SCC by replacing partially coarse aggregate using OPL and using POFA added materials.

2. Materials and Test Method

2.1. Palm Oil Fuel Ash (POFA) and Oil Palm Shell (OPS)

The POFA and OPS materials are obtained from the waste produced by one of the palm oil industries in Riau Province, Indonesia as depicted in figure 1 and figure 2. The POFA is firstly dried in an oven for 24 h with 100 ^{0}C and the material that is used passes the No. 200 filter size (particle size < 75 μm). Oxide composition test is also conducted, and the result is served in table 1 which has a great amount of silica (SiO_2) that good for the concrete mixture. The OPS material used has the size of 2 mm - 15 mm. Test of particular material was not conducted but referred to previous research [12] and compared with stone aggregate that is used in this research as written on table 2.

Figure 1. Picture of Palm Oil Fuel Ash (POFA)

Figure 2. Picture of Oil Palm Shell (OPS)

Table 1. Chemical composition in Palm Oil Fuel Ash (POFA)

Material	Oxide composition (%)										
	Al_2O_3	CaO	Fe_2O_3	MgO	Na_2O	K_2O	MnO	SiO_2	P_2O_5	TiO_2	LOI
POFA	8,87	3,24	1,06	1,42	0,57	3,22	0,03	52,63	1,86	0,31	27,7

Table 2. Characteristic of OPS [12] and coarse aggregate used

Parameter	Value	
	OPS [12]	coarse aggregate
Specific gravity (SSD)	1,17	1,62
Water absorption (24 h), %	23,32	6
Aggregate abrasion (with Los Angeles), %	4,8	9,9
Bulk density (compacted), kg/m^3	592	1480
Fineness modulus	6,24	-
Aggregate impact value (AIV), %	7,86	3,7

2.2. Concrete materials

SCC materials used consist of cement, fine aggregate (sand), coarse aggregate (gravel), water and superplasticizer. The cement used is Portland-pozolan cement (PPC) that meets ASTM C-150 [13] while fine aggregate and coarse aggregates are passing with ASTM C-33[14] requirements. Gradation values of fine aggregate and coarse aggregates grains can be found in figure 3 and figure 4. Superplasticizer is also added to the SCC mixture with viscocrete-10 from SIKA.

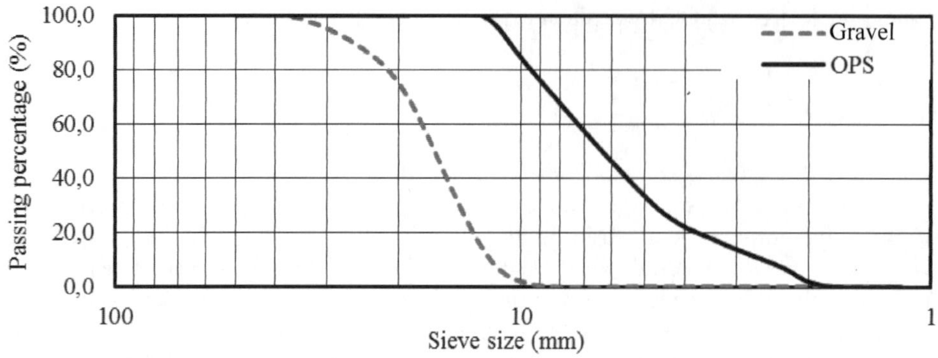

Figure 3. Grain size distribution of crushed stone aggregate/gravel and OPS

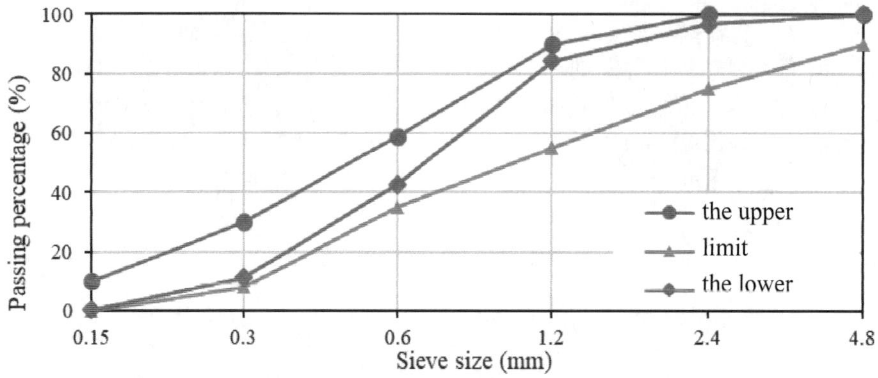

Figure 4. Grain size distribution of fine aggregate

2.3. Detail of variation samples and concrete mix design

In this research, Oil Palm Shell (OPS) is used as a replacement of gravel with variations of 0%, 5%, 10%, 25%, and 50% with curing time up to 28 days to find out the compressive strength value (f'_c) produced. In addition to Palm Oil Shell (OPS), the Palm Oil Fuel Ash (POFA) is added with a fixed value for all variations of 27.7% of cement and also superplasticizer viscocrete-10 added with 1.14% of cement. For the mix design, it follows the proportion of the European Federation of National Trade Associations (EFNARC) [15] as given in Table 3.

Table 3. Mixed design of SCC for every various replacement aggregate

Material	Aggregate Replacement by Oil Palm Shell				
	0%	5%	10%	25%	50%
Fine aggregate (kg)	5,18	5,18	5,18	5,18	5,18
Cement (kg)	2,58	2,58	2,58	2,58	2,58
Coarse aggregate (kg)	2,98	2,83	2,68	2,23	1,49
OPS (kg)	0	0,15	0,3	0,75	1,49
Superplasticizer (kg)	0,0375	0,0375	0,0375	0,0375	0,0375
POFA (kg)	0,716	0,716	0,716	0,716	0,716
Water (kg)	1,65	1,65	1,65	1,65	1,65
Number of samples	5	10	10	10	10

2.4. Testing method on fresh and mechanical properties of SCC

Several tests which are V-Funnel Test, L-Box Test, and J-Ring Test were performed to seek the fresh properties of SCC. Those following methods are conducted based on EFNARC [15] with some demand of V-Funnel Test value between 6-12 seconds, L-Box Test value (which is H2 / H1)> 0.8 and J-Ring Test value> 50 cm. The test equipment is presented in figure 5. For mechanical properties testing, the compressive strength of the concrete and the classification of collapse type refers to ASTM C-39 [16].

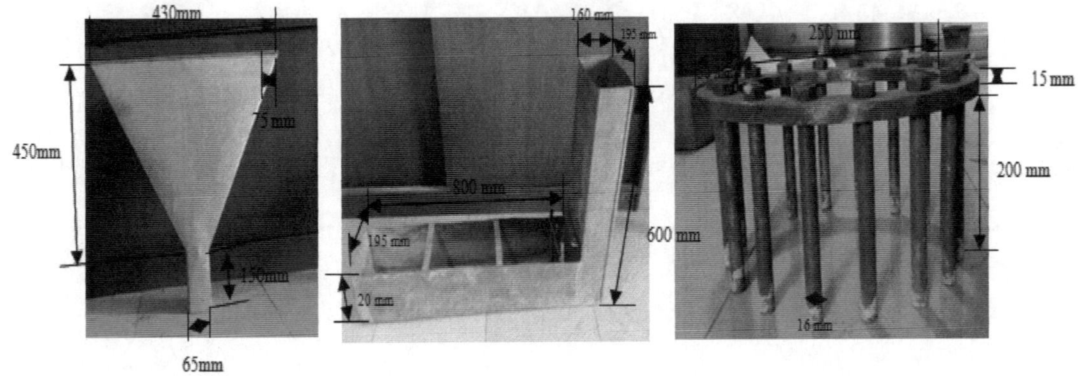

(a) (b) (c)

Figure 5. Tools on fresh properties test (a) V-Funnel, (b) L-Box, dan (c) J-Ring

3. Result and discussion

3.1. Properties of fresh concrete

From the test result, it was obtained for all samples with all the percentage of coarse aggregate replacement variation with OPS complied with V-Funnel Test, L-Box Test, and J-Ring Test according to EFNARC 2002 requirements. The results of fresh concrete properties test are presented in table 4.

Table 4. Result of fresh concrete properties test

OPS Percentage	V-Funnel (seconds)	L-Box H_2/H_1	J-Ring (cm)
0 %	7	0,8	51
5 %	5	0,8	51
10 %	6	0,875	51
25 %	8	0,9	51
50%	10	1	51

3.2. Compressive strength of samples

The summary of the compressive strength result is presented in table 5. From comparison graph between curing periods and compressive strength shown in figure 6a, it can be seen that all of compressive strength variations increase along with increasing age logarithmically, whereas in the comparison graph between compressive strength and variations of coarse aggregate replacement with OPS presented in figure 6b, it is known that at the early age of 1-3 days, there are several compressive strength values outperforms the normal SCC which are 5% and 10% OPL variation. However, for the final compressive strength value at 28 days, all variations have lower value compared to normal SCC. For all types of failures that occur, it can be classified as columnar vertical cracking through both ends that are classified as type 3 on ASTM C-39. Illustration of failure of the specimen can be seen in figure 7.

Table 5. Value of compressive strength for every age and variation of replacement aggregate by OPL

Code name	Age (days)	Compressive strength of SCC with OPL (MPa)								
		0%	5%	Mean	10%	Mean	25%	Mean	50%	Mean
1A	1	6,1	8,41	8,79	9,35	9,87	6,69	6,33	5,46	4,7
1B	1		9,16		10,39		5,96		3,93	
3A	3	12,82	13,54	13,43	14,47	13,37	12,53	11,6	8,76	8,68
3B	3		13,31		12,26		10,67		8,59	
7A	7	15,8	15,28	15,91	17,02	15,72	16,02	13,97	10,69	11,33
7B	7		16,53		14,42		11,92		11,96	
14A	14	20,37	13,87	15,88	19,67	18,55	16,87	14,91	11,54	11,48
14B	14		17,89		17,43		12,95		11,42	
28A	28	25,06	22,25	24,22	19,4	20,04	16,87	19,24	11,03	11,63
28B	28		26,19		20,68		21,6		12,13	

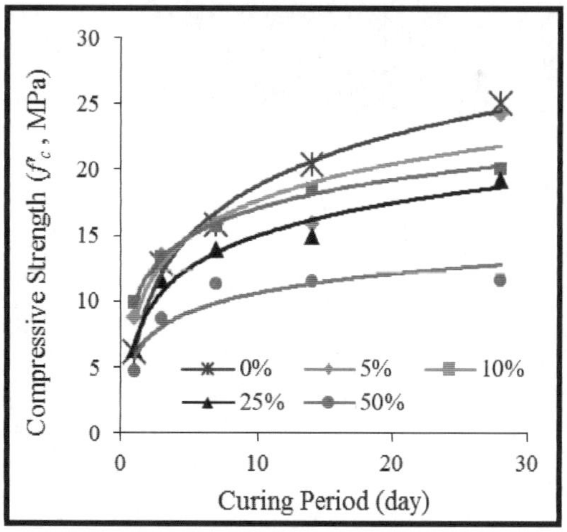

(a)

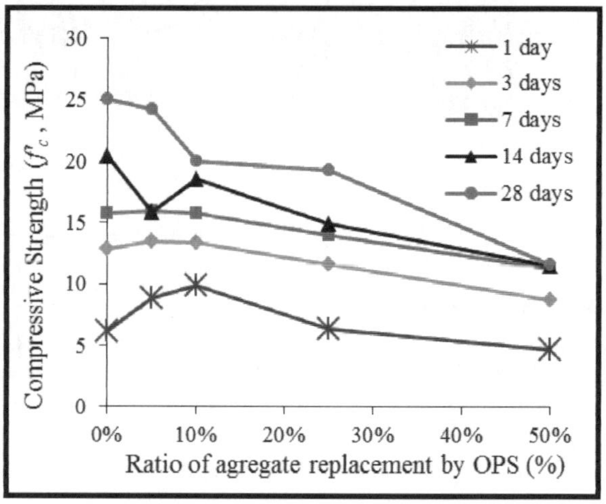

(b)

Figure 6. Compressive strength of samples based on (a) curing period and (b) various of aggregate replacement

(a) (b) (c) (d) (e)

Figure 7. Failure result of samples with aggregate replacement (a) 0%, (b) 5%, (c) 10%, (d) 25%, (e) 50%

4. Conclusion

To sum up, the replacement of rough aggregates using OCL as well as the addition of POFA meet on the properties of fresh properties. The use of palm oil waste that is OCL and POFA on SCC can be an alternative in reducing mining products usage (crushed stone aggregate). However, the normal SCC compressive strength at 28 days is still higher than the compressive strength resulting from rough aggregate replacement using OPC.

Acknowledgement
Thank you to Jezi Firnanda that helped to research material construction laboratory and to Mr. Sumadi as a laboratory assistant in Material Construction Laboratory, Department of Civil Engineering, Faculty of Engineering, Universitas Muhammadiyah Yogyakarta.

References
[1] Okamura H, Ouchi M 2003 Self-Compacting Concrete *J. Adv. Concr. Technol.* **1** 5-15
[2] Tuyan M, Mardani-Aghabaglou A, Ramyar K 2014 Freeze–thaw resistance, mechanical and transport properties of self-consolidating concrete incorporating coarse recycled concrete aggregate *Mater. Des.* **53** 983-991
[3] Omrane M, Kenai S, Kadri E, Aït-Mokhtar A 2017 Performance and Durability of Self Compacting Concrete using Recycled Concrete Aggregates and Natural Pozzolan *J Clean. Prod.* **165** 415-430
[4] Singh G, Siddique R 2016 Effect of iron slag as partial replacement of fine aggregates on the durability characteristics of self-compacting concrete *Constr. Build. Mater.* **128** 88-95
[5] Sua-iam G, Makul N 2014 Utilization of high volumes of unprocessed lignite-coal fly ash and rice husk ash in self-consolidating concrete *J. Clean. Prod.* **78** 184-194
[6] Soraj M 2013 Effect of Palm Oil Fuel Ash (POFA) on Strenght Properties of Concrete *Int. J. Sci. Res.* **3** 1-7
[7] Mohammadhosseini H, Awal A, Ehsan A 2015 Influence of palm oil fuel ash on fresh and mechanical properties of self-compacting concrete *Sadhana* **40** 1989-1999
[8] Khanje E, Salim M, Mirza J, Hussin M, Rafieizonooz M 2016 Properties of sustainable lightweight pervious concrete containing oil palm kernel shell as coarse aggregate Const *Constr. Build. Mater.* **126** 1054-1065
[9] Syafigh P, Jumaat M, Mahmud H 2010 Oil Palm Shell as a lightweight aggregate for producing hight strenght lightweight concrete *Constr. Build. Mater.* **25** 1848 - 1853
[10] Mannan M, Ganapathy C 2002 Engineering properties of concrete with oil palm shell as coarse aggregate *Constr. Build. Mater.* **16** 29-34
[11] Basiron Y 2007 Palm oil production through sustainable plantations *Eur. J. Lipid Sci. Technol.* **109** 289-295
[12] Mannan M, Ganapathy C 2001 Long-term strengths of concrete with oil palm shell as coarse aggregate *Cem. Concr.* **31** 1319-1321
[13] ASTM C150 2017. *Standard Specification for Portland Cement* (West Conshohocken : ASTM International)
[14] ASTM C33 2016. *Standard Specification for Concrete Aggregates*, (West Conshohocken : ASTM International)
[15] EFNARC 2002 *Specifications and guidelines for self-consolidating concrete* (Hampshire UK: European Federation of National Trade Associations Representing Producers and Applicators of Specialist Building Products)
[16] ASTM C39 2017 *Standard Test Method for Compressive Strength of Cylindrical Concrete Specimens* (West Conshohocken : ASTM International)

IOP Conf. Series: Materials Science and Engineering **352** (2018) 012029 doi:10.1088/1757-899X/352/1/012029

Stress and deformation analysis of tapered cantilever castellated beam using numerical method

Taufiq Ilham Maulana, Bagus Soebandono, Beta Satria Jagad, Hakas Prayuda
Department of Civil Engineering, Faculty of Engineering, Universitas Muhammadiyah Yogyakarta, Lingkar Selatan Street, Bantul, Yogyakarta, Indonesia, 55183

Email: taufiq.im@ft.umy.ac.id; bagus_soe@yahoo.com; hakas.prayuda@ft.umy.ac.id

Abstract. The castellated beam is often used in buildings because of its lighter weight compared with a normal steel beam. There are many types of an opening in the castellated beam, one of which is hexagonal openings. This paper will discuss the analysis of stress and deformation on castellated beam with a variation of openings diameter, space between holes, and angle of hexagonal openings. Furthermore, stress distribution on specimen will be seen under static loading. This study used IWF section 150x75x5x7 with 4 variations of the span with one fixed support, and yield strength is 400 MPa. Linear finite element analysis is used with 10-node tetrahedron solid element, by observing von Misses stress. The software used in this study are freeware, which is LISAFEA 8.0 for analyzing and FreeCAD for drawing. The result shows that value of stress and deformation for each sample is quite volatile, but it can be concluded that stress distribution around the opening is larger than in web and flange.

1. Introduction

The beam is a structural element that supports shear force and bending moment more dominant than axial load. Many materials are utilized as a beam, one of which is steel materials. Steel has many advantages of usage, for example, tensile and compressive strength steel is higher than concrete, and its strength is more uniform than concrete because it is fabricated in the factory. But on some occasion, steel is still considered expensive rather than concrete. To reduce its cost, there are a lot of innovations, one of them is to create steel beam as a castellated beam. The castellated beam is a normal steel beam modified by cutting into the half with specific pattern and united by welding so that its height increases significantly with holes in its web [1]. Experimental and numerical tests were done in previous research such as looking for elastic-plastic behavior [2], optimizing by size and distance of hole [3], and improving analysis of castellated beam using finite element analysis[4], but all of those studies used simply-supported so that implementation of prismatic steel beam is still relevant.

There are many kinds of supports, one of which is fixed support in cantilever beam structure. Cantilever beam internal forces are concentrated in the supported area so that practicing prismatic section steel beam in the field would cost much. Some studies already conducted about the usage of the structural member in cantilever support for example steel-concrete composite[5], high steel reinforced concrete[6], and checked by the crack with modeling [7], but so far the further research of castellated beam as cantilever structure has not been conducted yet.

Furthermore, non-uniform section steels are usually used for reducing the cost of construction and increasing strength purposes. This innovation already tried to be applied for many applications, for example in dome-like structures [8]. Also, there is already research about numerical modeling based on an experimental test for bucking behavior of tapered steel section [9], but the application in castellated beam have been done before.

This paper discusses stress and deformation that happened in castellated cantilever beam with hexagonal openings which is used in cantilever beam and also applying non-prismatic section as tapered steel member to optimize steel beam usage. It also shows stress distribution surround body of the structure. Numerical test was conducted using linear finite element method.

2. Samples and Analysis Method

2.1. Samples

Steel material that used in this research refers to ANSI/AISC 360-10 [10] with yield strength 400 MPa, Modulus of Young 200 GPa and Poisson's ratio 0.3. All of the samples have its final height 255 mm at fixed support. The number of samples is 72 with 4 variations which are span length, the angle of openings, openings space and the hole diameter of openings. The detailed of variations are written in table 1.

Table 1. Detailed variations of samples that are used in this study

Span Length (m)	Angle (^{0})	Openings space (mm)	Hole diameter (mm)
2	55	60	50
2.5	60	80	75
3		100	100
3.5			

Samples are drawn in 2 dimensions using AutoCAD program to get the final length nearest its span length requirement. For example, in figure 1, it is shown how to cut the steel section from prismatic with hexagonal pattern and reunited to be a non-prismatic section. The cutting of steel section is done diagonally and only takes 120 mm from the total height of 150 mm (IWF 150x75x5x7) because the other 30 mm were not cut to keep the flange part and corner part as original. Each sample is given a code name, for example, D100-S60-JA100 which means this sample has openings diameter 100 mm, angle of opening 60^0 and space between holes 100 mm.

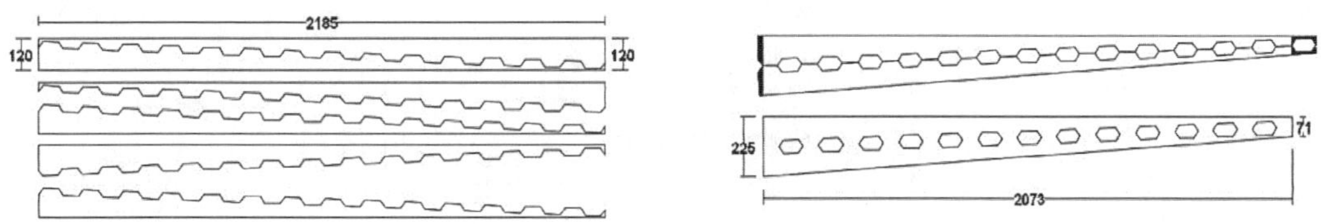

Figure 1. Creating non-prismatic castellated beam section sample from regular steel beam

2.2. Analysis method

In this study, an analysis is conducted using finite element method with 10-node tetrahedron solid element. Before all samples are analyzed, those were modelled as 3D solid steel beam in FreeCAD as shown in figure 2 and after that solid samples are imported to finite element analysis program which is LisaFEA and in that program, several actions were done such as defining material, determining number of element that is used by determining size element, running the program, and resuming the result of stress and deformation that happened. The process of modeling samples in LisaFEA is depicted in figure 3. The stress of samples is calculated using Von-misses criterion [11] as stated in Equation 1. After all those steps are done, the number of elements used for each length is determined by calculating convergence method to seek an optimum number of the element. This method is comparing between the number of element and displacement under the same load and chooses the

IOP Conf. Series: Materials Science and Engineering **352** (2018) 012029 doi:10.1088/1757-899X/352/1/012029

smallest number of element that is resulting from the best displacement which approaches real value with small error (< 5%).

$$(\sigma_1-\sigma_2)^2+(\sigma_2-\sigma_3)^2+(\sigma_1-\sigma_3)^2=2\sigma_y^2 \quad (1)$$

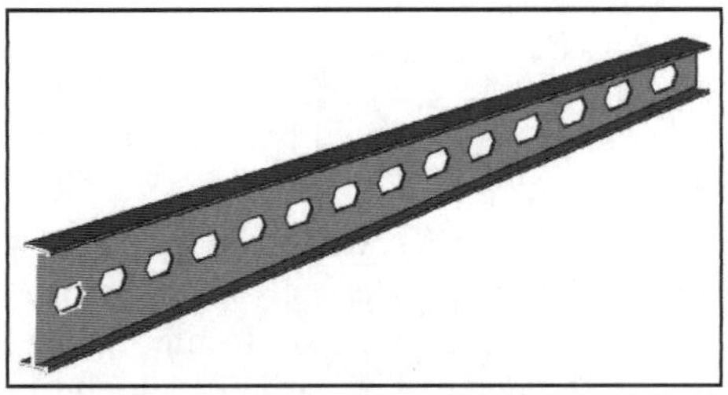

Figure 2. Solid 3-dimensional drawing in FreeCad

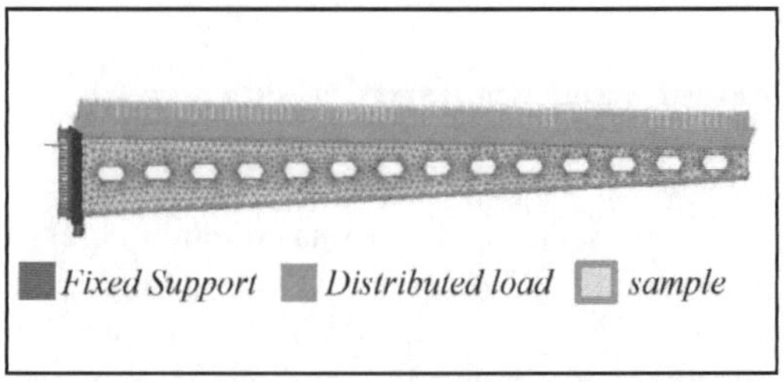

Figure 3. Analyzing samples in LisaFEA

3. Result and Discussion

Convergence tests are done for each length variation. For example, in figure 4, the convergence test is done for length variation 2 meters, and the number of an element is chosen 10,000 because the displacement of samples does not change much after modeled with more elements. The load is given to samples based on the result of stress which approaches yield stress for every length variation. Those are resumed in table 2. Stress and deformation result from the test are fluctuating for each sample, which is shown in figure 5. From figure 6, it can be seen that stress is concentrated at corner of hexagonal openings nearby support which is the weakest and vulnerable area.

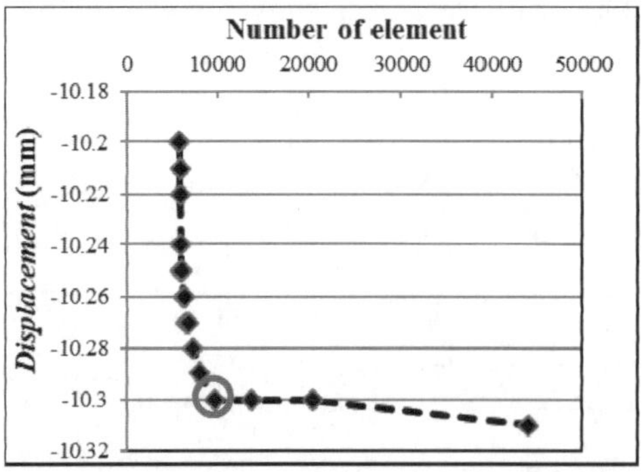

Figure 4. Convergence test result and choosing number of element

Table 2. Resume of load given in every sample in the same span length

Span length (m)	Code name	Optimum load for analyzing (ton)	Stress reached (MPa)
2	D100-S60-JA100	2.5	385.4
2.5	D100-S60-JA100	2.3	385.2
3	D100-S60-JA100	2.1	378.1
3.5	D100-S55-JA100	1.9	398.3

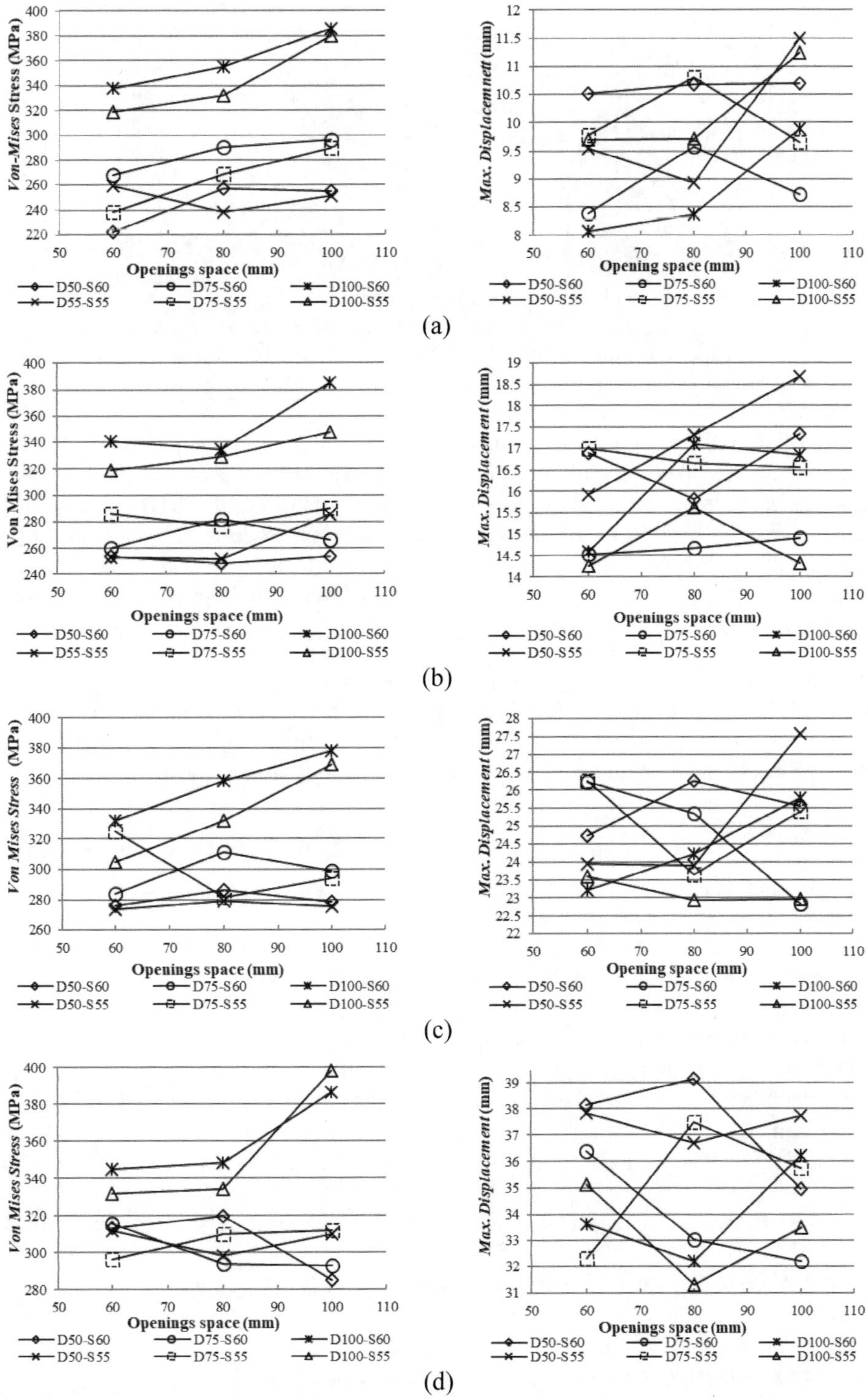

Figure 5. Stress and displacement resulted from the test with span length of (a) 2 m; (b) 2.5 m (c) 3 m; (d) 3.5 m

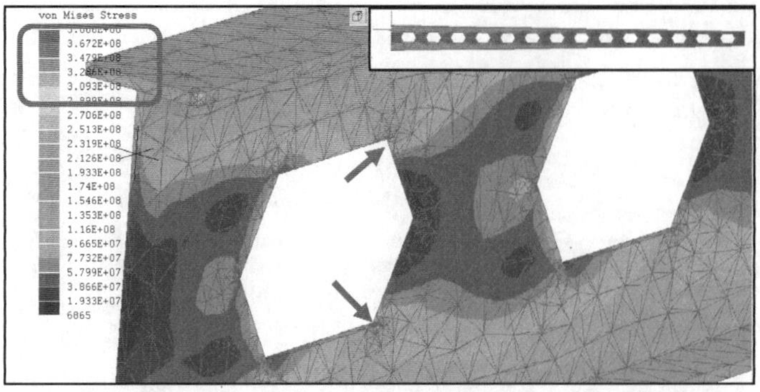

Figure 6. Stress distribution in body of castellated beam

4. Conclusion

To sum up, the value of stress and distribution for each sample are fluctuating. Those samples do not have any pattern because several factors namely different result openings amount in every sample, the different distance between support and the first hole for each sample and different height of the beam near free support resulted from the cutting method. Further study should be conducted to predict the value of stress and distribution based on the result. Nevertheless, it can be known that stress distribution in every sample is concentrated to the corner of hexagonal openings which is the nearest from the fixed support.

Acknowledgement

Thanks to LP3M UMY that gave us funding research in 2016-2017 with No. contract 1095/SP2-PL/LP3M-UMY/IV/2017 and thanks to all of the staff at Department of Civil Engineering, Faculty of Engineering, Universitas Muhammadiyah Yogyakarta for the help in this research.

References

[1] Ajim S, Priyanka D R, Nikita S R, Punam T L, Komal S S and Kalyani P B 2017 Structural analysis and design of different types of castellated beam *International Research Journal of Engineering and Technology* **4** 793-795

[2] Jun L and Jiawei X 2011 Elastic-plastic analysis of the circle-castellated steel beam using Finite element method *Advanced Materials Research* **422** 854-857

[3] Budi L, Sukamta and Partono W 2016 Optimization analysis of size and distance of hexagonal hole in castellated steel beams *Procedia Engineering* **171** 1092-1099

[4] Zhu D, Cao X, Bai Y and Zhang Q 2012 A discussion about the improvement of finite element analysis units of Castellated beam *Advanced Materials Research* **446** 648-651

[5] Gu X, Chen T, Li H, Zhang W and Wang H 2012 Behavior of steel-concrete composite cantilever box beams under Negative Moment *International Journal of Steel Structures* **12**(4) 509-521

[6] Kewei D and Aiyu Y 2012 Stress analysis of tall steel reinforced concrete cantilever beam *Applied Mechanics and Materials* **151** 286-290

[7] Altunıs¸ık A, Okur F, Kahya V 2017 Structural identification of a cantilever beam with multiple cracks: Modelling and Validation International Journal of Mechanical Sciences **130** 74-89

[8] Liu S W, Bai R, Chan S L, Liu Y P 2016 Second-Order Direct analysis of domelike structure consisting of tapered members with I-Section *J. Struct. Eng. ASCE* **142** 1-11

[9] Tankova T, Martins J P, Silva L S D, Marques L 2017 Numerical model for the buckling behaviour of tapered steel members based on experimental tests. *EUROSTEEL 2017* **1** 1106-1115

[10] ANSI/AISC 360:2010 *Specification for Structural Steel Buildings* (Chicago(IL) : American Institute of Steel Construction)

[11] Srinath L 2009 *Advanced Mechanics of Solid* (New Delhi : Tata McGraw-Hill Publishing)

IOP Conf. Series: Materials Science and Engineering **352** (2018) 012030 doi:10.1088/1757-899X/352/1/012030

A review of BIM (Building Information Modeling) implementation in Indonesia construction industry

Abdi Suryadinata Telaga
Lecturer, *Informatics Management*, *Politeknik Manufaktur Astra*,
Jl Gaya Motor Raya No. 8 Sunter II, Jakarta, 14330, Indonesia

E-mail: abdi.telaga@polman.astra.ac.id

Abstract. Construction projects in Indonesia have been growing rapidly in the last three years. Therefore, construction management is very important to ensure completion of construction projects are within schedule and budget. Utilization of building information modeling (BIM) can increase the efficiency of a construction project. However, the implementation of BIM in Indonesia is still not known. This paper is intended to review the implementation of BIM in Indonesia through literature analysis. To find BIM articles in Indonesia, Firstly, searching was limited to English articles published in reputed journals or conferences. However the results were limited, then the search was expanded to the article using Indonesian languages that published in journal and conference. Based on the number of articles, the results showed that BIM research in Indonesia is still in a dearth. Furthermore, BIM study cases were conducted in a limited location and within a small population. Nevertheless, the literature shared the conclusion that BIM can increase project efficiency, but the implementation was hindered by high initial investment cost, inadequate human resources, small demand, and technology resistant. The research contributes to providing a current reported level of BIM implementation in Indonesia. In the future research to study of BIM implementation comprehensively in Indonesia is eminent.

1. Introduction

Construction projects in Indonesia have been growing rapidly since 2014. Particularly, the current government has set priority to develop Indonesia infrastructure. Consequently, investment in a construction project has been estimated to grow 5.2% per year [1]. Project management is one of the critical factors to assure a project can be finished on schedule. In Indonesia of the construction projects, 38% were late, 47% were on time, and 15% were faster than schedule [2]. The project owner is usually more focus on budget overrun than time delay because direct effect of a delayed project is less obvious. However, in an infrastructure project consequence of a delayed project is not only directly affecting the project cost, but the area of a construction site is also severely affected. In a major city, traffic congestion increase in the affected area eventually leads to huge economic loss. During the development of LRT project and elevated toll road in existing Cikampek toll road, the traveling time from Jakarta to Cikampek take 2 hours longer than usual. Consequently, the logistic cost increases significantly because of doubling traveling time. Therefore, adopting technology to make a project on schedule is eminent.

Building information modeling (BIM) has been widely known for having many benefits for the construction industry. Improvement of project cost control and conflict reduction are among the benefits associated with BIM [3]. Therefore, BIM has been widely adopted in many countries to increase the productivity of construction projects. However, level of BIM utilization in construction industry varies among countries. While developed countries are among the early implementer of BIM, BIM implementation in developing countries still poses many challenges. Construction companies in Indonesia have used computer-aided design (CAD) to draw 2-D design and construction work. The 2-D drawing becomes the source of conflict during the construction phase as misinterpretation among designer and contractor frequently occurred. To solve the conflict, designer and contractor have to clarify and change the drawing. The process requires many meetings before all the misinterpretation were solved. If the problem cannot be managed, it could lead to delay and cost overrun. As the BIM has potential to reduce such integration problem, BIM implementation in Indonesia is promising. However, the high cost of BIM software, lack of skill of employee, and unclear benefits of BIM to the company are among the BIM implementation challenges in developing countries [4]. Therefore, studying about BIM research in Indonesia can provide information about the level of BIM implementation and challenges posed in Indonesia construction industry.

2. Methodology

Firstly, searching is focused on obtaining BIM related article from reputed peer-review academic publishers. Query using the keyword "BIM Indonesia" from ScienceDirect only resulted from one article. As the number of articles was very low, the searching had to be expanded using Google scholar with the similar keyword. Then the searching only produced another two articles that specifically address BIM implementation in Indonesia. As the literature about BIM implementation in Indonesia are very limited and hardly found. The search had to be expanded again to include peer review the Indonesian language articles using the keyword "implementasi BIM" and "penerapan BIM". The search only returned four relevant articles in the Indonesian language. BIM articles from final project, thesis, and dissertation were omitted because the articles were not accessible and not open to the public. The collected articles then were analyzed using descriptive statistic.

3. Results and Discussion

The collected articles are presented in figure 1. Only seven articles associated with BIM were published until 2017. The first article about BIM in Indonesia was published in 2013 explaining BIM implementation experience in several construction projects in Indonesia [5]. Indonesia companies just started to use BIM in 2012. It shows that Indonesia is lack behind developed countries which have been using BIM since 2000. Only two articles have published between 2013 and 2015 can be interpreted that BIM has not been widely known Indonesia. However, five articles have been published in the last two years indicating that there is growing interest in BIM research in Indonesia.

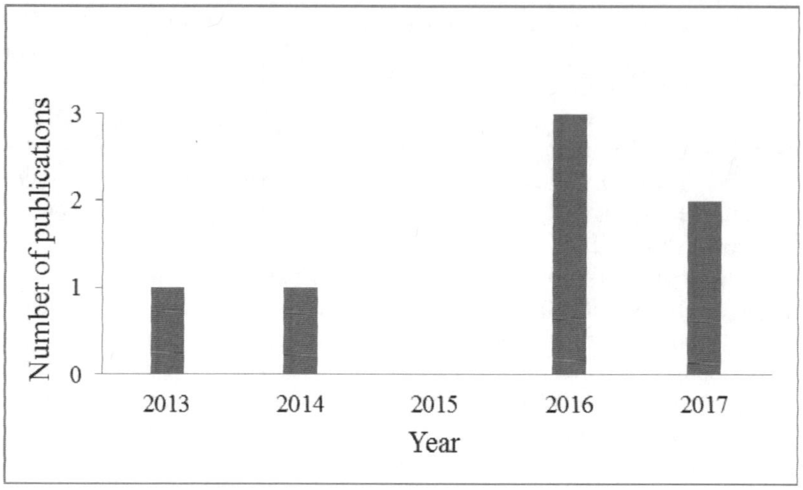

Figure 1. Number of publications per year

Some articles published based on publication type is presented in figure 2. There are not many articles have been published in peer review international journal indicate that BIM research is not yet mature in Indonesia. Indonesian scholar only published three international articles since 2013. Therefore, Indonesian scholars must be encouraged to publish English articles to attract international readers.

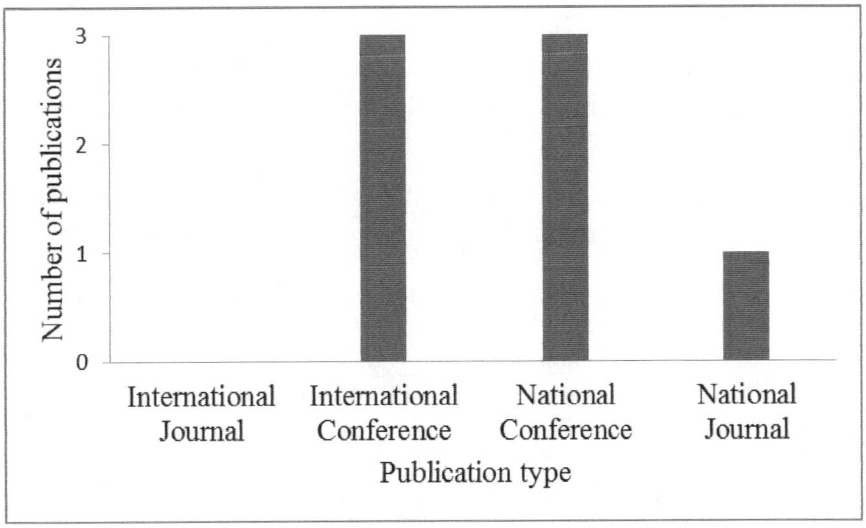

Figure 2. Number of articles based on publication type

BIM research in Indonesia is summarized in table 1. Each article is categorized by BIM implementation framework dimension suggested by Jung and Jo [6]. The dimensions are technology, perspective, and construction management. Strong focus on technology is shown in 71.43% articles. The number can be understood that people are more interested to see the real benefit when using new technology in construction projects. Moreover, the studies are focusing on the benefit of BIM and the implementation challenges. The studies confirmed that BIM could improve construction project in Indonesia [5] in construction planning [7, 8], construction cost due to time efficiency and human resources efficiency [8–10], and project documentation [9]. However, successful implementation of BIM requires technological support [10]. But the implementation was hindered by companies' internal factor such as lack of BIM skill worker and technology resistance, and external factor like low demand from clients [9]. Companies also need to have adequate capital to cover initial high licensing cost [9] and capable of handling incompatibility issue among BIM software [10].

While the academicians have more understanding than practitioners about BIM as integration software, practitioners only use BIM to obtain project information [11]. The situation was driven by low demand from clients. Therefore, the companies do not need to utilize all capabilities of BIM. Eventually, the companies do not require high BIM skilled users as BIM usages are for 3-D modeling and visualization [12].

Based on the companies' utilization of BIM, the maturity level in Indonesia companies are still in level 0 and 1. [13]. The maturity level 0 is a pre-BIM stage, in this stage companies uses 2-D documentation to explain 3-D object. Furthermore, detail documents explaining the object such as specification, quantities, and cost estimation are separated from the documentation [14]. Indonesian companies still use 2-D documentation in their construction project. Maturity level 1 is object-based modeling stage, in this stage companies use BIM object for visualization. Currently, Indonesian companies only utilize BIM for modeling and visualization [12], and information [11].

Table 1. Summary of BIM research in Indonesia

Theme	Finding	Dimension	Year
Design and construction of prefabricated building [5].	BIM has benefit to streamline construction projects in Indonesia	Technology	2013
Application of BIM for pre-construction [7].	BIM provides information about pre-construction planning, particularly construction logistic planning.	Technology	2014
Awareness and implementation of BIM among academicians and practitioners [12].	1. High awareness of BIM (70%) but the low implementation (38%). 2. BIM's main uses are 3-D modeling and visualization.	Perspective	2016
Perception of academicians and practitioners of BIM technology implementation in architecture [11].	Practitioners perceive BIM as informative software while academicians perceive BIM as integration software.	Perspective	2016
Comparison between Project using BIM and project using a conventional method in term of time efficiency, cost, and human resources [8].	Savings due to using BIM compared to conventional: 1. Project planning time by 50%. 2. Human resources by 26.66%. 3. Personnel cost by 52.25%.	Technology	2016
Exploration of technological support for BIM implementation, benefits of BIM, challenges of BIM in construction project [10].	1. The need of technological support for BIM is eminent. 2. BIM benefit is to reduce cost 3. Incompatibility between different BIM software.	Technology	2017
Challenges of BIM implementation in small-medium enterprise architecture [9].	1. Main benefits of BIM are time efficiency, better communication and coordination, and improved project documentation. 2. Main challenges are lack of skilled BIM user, low demand from a client, high investment cost, and resistance to technological change.	Technology	2017

IOP Conf. Series: Materials Science and Engineering **352** (2018) 012030 doi:10.1088/1757-899X/352/1/012030

4. Conclusion

The first documented use of BIM in Indonesia construction industry was in 2012 [5]. Since then the progress of BIM is still slow. Low international journal articles show that BIM research is not yet mature in Indonesia. The perspective studies also show that BIM implementation in Indonesia is still in its infancy. Although stakeholders in Indonesia construction industry is aware of the benefit of BIM to their business, the penetration is still low. Several companies have promoted to elevate BIM usage in industry. However, the implementation is hampered by skill worker and technology resistance, and external factor like low demand from clients, high investment cost, and BIM software incompatibility. Particularly in small and medium companies (SMC) when the project is simple, the investment cost outweighs the benefit of BIM. Therefore, integrating BIM course into university curriculum could increase the availability of BIM skill worker. While licensing cost is free for educational purpose, SMC could ask for a special price through an agreement between association and Software Company to reduce investment cost.

In conclusion, the research contributes to providing a current reported level of BIM implementation in Indonesia. Therefore, in the future to provide a broad view of BIM implementation, research to study of BIM implementation in various construction sectors is eminent. Furthermore, to overcome the future shortage of BIM skill among construction worker, study about BIM in Indonesia higher education is a necessity.

References

[1] Indonesia's Construction & Building Materials Sector On the Up & UpGlobal Business Guide Indonesia 2017 http://www.gbgindonesia.com/en/property/article/2016/indonesia_s_construction_andamp_building_materials_sector_on_the_up_andamp_up_11586.php (2016)

[2] Trigunarsyah B 2004 Constructability Practices among Construction Contractors in Indonesia *J. Constr. Eng. Manag.* **130**

[3] Ghaffarianhoseini A, Tookey J, Ghaffarianhoseini A, Naismith N, Azhar S, Efimova O and Raahemifar K 2017 Building Information Modelling (BIM) uptake: Clear benefits, understanding its implementation, risks and challenges *Renew Sustain Energy Rev.* **75** 1046–1053

[4] Bui N, Merschbrock C and Munkvold B E 2016 A review of building information modelling for construction in developing countries. *Procedia Eng.* **164** 487–494

[5] Sidjabat H R, Tambun R L and Tambunan R 2013 Application of Building Information Modelling on Design and Construction of Prefabricated Building *6th Civil Engineering Conference in Asia Region* (Jakarta Indonesia: Asian Civil Engineering Coordinating Council)

[6] Jung Y and Joo M 2011 Building information modeling (BIM) framework for practical implementation *Automation in Constr.* **20** 126–133

[7] Rayendra and Soemardi B W 2014 Studi Aplikasi Teknologi Building Information Modeling untuk Pra-Konstruksi *Simposium Nasional RAPI XIII - 2014 FT UMS* http://hdl.handle.net/11617/5528 (in Indonesian)

[8] Permatasari C A B, Adhi R P, Hidayat A and Nugroho H 2016 Perbandingan Efisiensi Waktu, Biaya, dan Sumber Daya Manusia antara Metode Building Information Modelling (BIM) dan Konvensional (Studi Kasus: Perencanaan Gedung 20 Lantai) *J. KARYA Tek SIPIL* **5** 220–229 (in Indonesian)

[9] Wilis F A, Larasati D and Suhendri 2016 The challenges of implementing building information modeling in small-medium enterprises architecture firms in Indonesia *Int J Archit. Environ. Eng.* **4** 828

[10] Chandra H P, Nugraha P and Putra E S 2017 Building Information Modeling in the architecture-engineering construction project in Surabaya *Procedia Eng.* **171** 348–353

[11] Irwanuddin I, Indraprastha A and Kusuma H E 2016 Persepsi praktisi dan akademisi terhadap

penerapan teknologi BIM di arsitektur *Prosiding Temu Ilmiah IPLBI 2016* (Malang: Program Studi Arsitektur Fakultas Teknik Sipil dan Perencanaan Institut Teknologi Nasional (ITN) Malang) 16–172 (in Indonesian)

[12] Hanifah Y 2016 Awareness dan pemanfaatan BIM : Studi Eksplorasi *Temu Ilmiah IPLBI* (Malang: Program Studi Arsitektur Fakultas Teknik Sipil dan Perencanaan Institut Teknologi Nasional (ITN) Malang) 49–54 (in Indonesian)

[13] Khosrowshahi F and Arayici Y 2012 Roadmap for implementation of BIM in the UK construction industry *Eng. Constr. Archit. Manag.* **19** 610–635

[14] Succar B 2009 Building information modeling framework: A research and delivery foundation for industry stakeholders *Autom Constr.* **18** 357–375

IOP Conf. Series: Materials Science and Engineering **352** (2018) 012031 doi:10.1088/1757-899X/352/1/012031

Analysis of pedestrian walking characteristics at vertical facilities in underground train station

Nur Sabahiah Abdul Sukor[1,2], Farah Hanin Mohd Nain[1], Irwina Ismail[1] and Mohd Ashraf Mohamad Ismail[1,2]

[1] School of Civil Engineering, Engineering Campus, Universiti Sains Malaysia, 14300 Nibong Tebal, Pulau Pinang, Malaysia

[2] Senior Lecturer, School of Civil Engineering, Universiti Sains Malaysia

E-mail: cesabahiah@usm.my

Abstract. Kuala Lumpur is the national capital of Malaysia as well as the center of attraction for people looking for jobs. Thus, in the future people will tend to use public transport including the Light Rail Transit (LRT) more often than driving their vehicle. This is especially so due to critical traffic congestion on the road especially during peak hours. The increasing population in Kuala Lumpur could lead the LRT stations to become more crowded form day to day due to excessive demand. Due to this problem, the facilities provided will not be enough and suitable to cater the space demand. Unfortunately, the limited study was done regarding the pedestrian spaces and facilities at underground LRT in Malaysia. Thus, this paper aims to establish the fundamental understanding of pedestrian behavior at vertical facilities in underground train station.

1. Introduction

The study of pedestrian movement is important in order to facilitate the pedestrian with comfort and safe environment. The previous study in Malaysia figured out that the mean walking speed for the pedestrian at signalized and non-signalised crosswalk is 1.39 m/s and 1.31 m/s respectively [1]. Meanwhile, the average free flow speed and the estimated jam density for the pedestrians on the sidewalk in Dhaka, Bangladesh are 1.2m/sec and 3.32peds/m^2 respectively [2]. However, less study reported on the pedestrian movement and their characteristics in underground spaces in Kuala Lumpur. Also, usually during the peak hour, there is a huge volume of the pedestrian using underground spaces and the facilities provided might be limited in the future if the demand for the pedestrian spaces increases.

Therefore, to provide comfort and safest facilities, the pedestrian walking facilities supposed to be designed in such a way that they can be used conveniently and be easily assembled, built and installed with no confusion involved in the process of their usage [3]. The analysis of pedestrian travel time, free flow speed, speed, flow, density and the capacity of facilities are required to provide the best and pleasant facilities for the pedestrian. One of the methods to determine all of these elements is through a video camera survey to gather the detail of walking characteristics of the pedestrian [4,5]. This survey can record all of the movement of the pedestrian in the facilities including the stairways and escalators. In addition, to understand the pedestrian behavior and their characteristics at the underground train station, the analysis on pedestrian travel time is important. Travel time refers to the amount of time required to travel from one point to another point [6]. The estimated travel time on different walking facilities such as stairway and escalator will be helpful in designing the size of the facility required that suitable for the demand and the walking characteristics of the pedestrian.

2. Methods

A video survey was held on 27th to 29th January 2015 in KLCC LRT station during the peak hours and non-peak hours to investigate the pedestrian walking speed-density-flow. The aim of this survey is to collect pedestrian travel time, flows, density on the stairway and escalators. The physical dimensions of the vertical facilities such as length and width of the stairway and escalator were also determined. A survey plan for data collection at KLCC LRT station is illustrated in Figure 1. The figure shows the location of the video cameras that used to collect the data. All of the video cameras were synchronized to determine the exact entry and exit time of the chosen pedestrian.

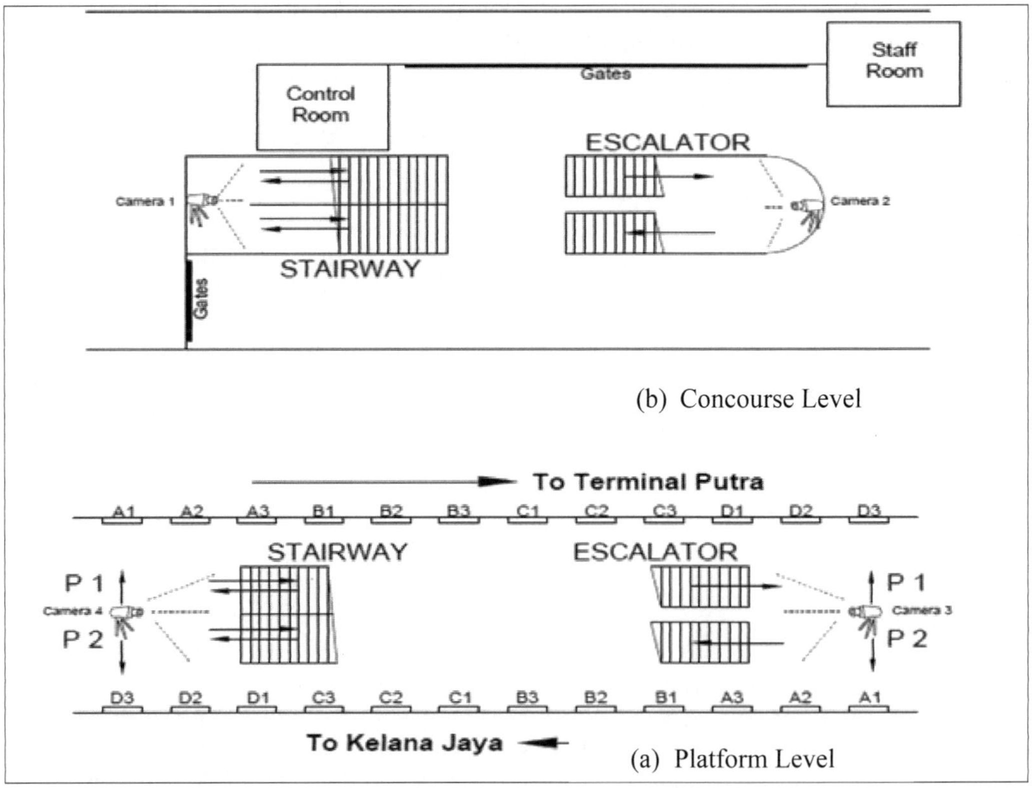

Figure 1. Location of video cameras for data collection at KLCC LRT Station.

Then, data analysis was carried out to extract the required data from the recording by using media player. This method has its advantage as it can be repeated and the data can be reviewed. Moreover, pedestrian manual counting was also done during the video camera surveys for calibration purpose.

Three parameters of pedestrian behavior such as flow, speed, and density, were extracted manually by playing the video recordings on a computer screen. The two hours video recording was divided into 15-minutes intervals. Due to the limitation of images in the video camera recording, 3-s intervals and 23-s for the stairway and escalator facilities respectively was adopted because most of the people could finish walking the distance of the observed length. In this study, pedestrian flow was determined by counting the numbers of the pedestrian passing a line of sight across the width of the facilities. For stairway, the number counted is the total number of pedestrian for both ascending and descending directions and then divided by 3-s and the width of a facility to obtain the flow rate. For escalator, the number counted is the total number of pedestrian for both ascending and descending directions and then divided by 23-s and the width of a facility. The flow rate was expressed as ped/s/m.

To obtained the pedestrian density, the number of pedestrians occupying the observed area for 3-s was counted and then divided by the observed area. The density was expressed as ped/m². Speed can be obtained by taking the travel time of pedestrian who traversed the observed length. The average travel time of the three random pedestrians who walk following the crowd is taken. The speed was computed as the observed length of facilities divided by the travel time and expressed as m/s. Moreover, a random pedestrian was chosen during the non-peak hour to obtain their time travel along the facility. The data of the pedestrian travel time was obtained using manual counting during the video camera surveys.

3. Results and Discussion

Figure 2 shows the total number of pedestrian per hour at the KLCC LRT Station for the three days of the survey. Observation time for pedestrian counting from the video camera recording based on the peak hour obtained from the graph below. The physical characteristics of the facilities are shown in table 1.

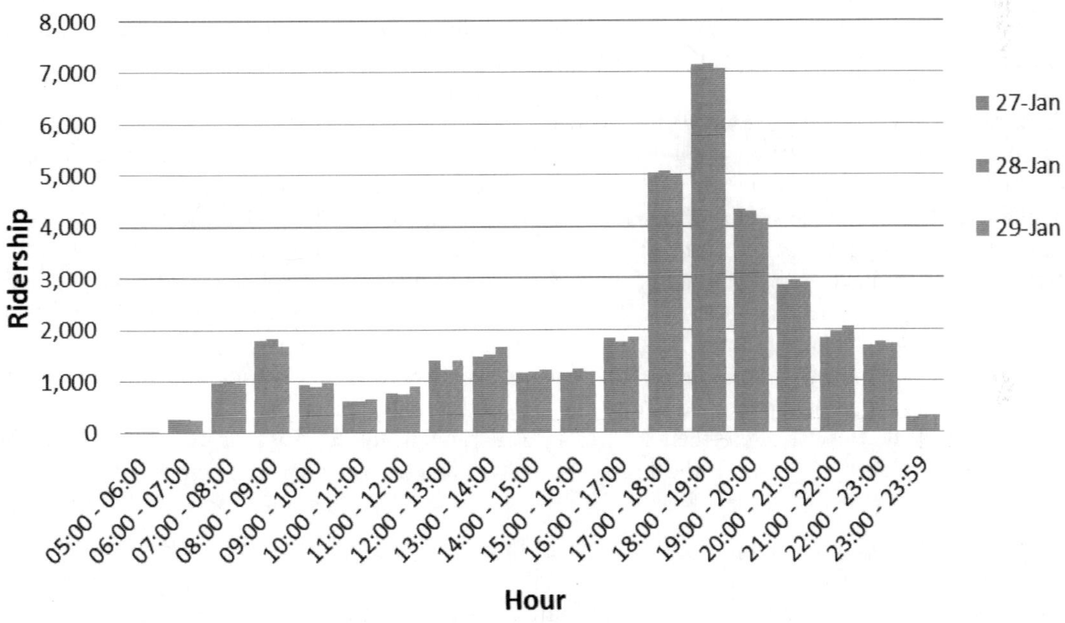

Figure 2. KLCC hourly station ridership.

Table 1. Observation times and physical characteristics of vertical facilities.

Walking Facility	Observation Time	Physical Characteristics				
		Width, mm	Length Observed, mm	Grade, °	Tread width, mm	Step Riser Height, mm
Stairway	7.30 a.m. – 9.30 a.m.	3810	12962	/	299	162
Escalator	12.00 p.m. – 2.00 p.m. 5.30 p.m. – 7.30 p.m.	1150	13900	30	/	/

IOP Conf. Series: Materials Science and Engineering **352** (2018) 012031 doi:10.1088/1757-899X/352/1/012031

3.1. Pedestrian free flow speed

From the data obtained, the free flow speed in descending direction is higher than ascending direction for stairway and escalator (Table 2). Ascending direction refers to flow from platform level to the concourse level, whereas descending direction refers to flow from concourse level to platform level. This phenomenon occurred because pedestrian needs more effort to travel in ascending direction compare to descending direction. At stairway, pedestrian walks more freely with their desired speed when traveling from concourse level to platform level. They tend to increase their walking speed at the facilities when the train has arrived at the station.

Table 2. Pedestrian free flow speed at stairway and escalator.

Free Flow Walking Speed (m/s)			
Stairway		Escalator	
Ascending	Descending	Ascending	Descending
0.806	0.965	0.755	0.786

A similar situation occurred at the escalator where the pedestrian-only increase their walking speed when the train arrived. Due to less flow of pedestrian at the escalator, pedestrian tends to control their walking speed, and they were walking at the escalator, and their speed was affected by the escalator velocity which is 0.5 m/s. It was shown in Table 2 that the pedestrian walked at the escalator as the average walking speed were 0.755m/s and 0.786 for ascending and descending direction respectively.

3.2. Speed-Flow-Density Relationship at Stairway

The scatter plots of pedestrian speed-density at the stairway for the different time of peak hours, including (a) morning peak, (b) noon peak and (c) evening peak hour can be seen in Figure 3. It shows that the speed is decreased as the increase of the density in the facility. Most of the pedestrian at Figure 3(a) walks at speed below 1 m/s differ with the pedestrian at Figure 3(c) because they walk in a different direction. Pedestrian at Figure 3(a) walks in ascending direction (platform level to concourse level) whereas Figure 3(c) walks at descending direction (concourse level to platform level). Thus, pedestrian in Figure 3(a) need more effort to walk upward compare to Figure 3(c).

Apart from that, pedestrian in Figure 3(b) tends to walk more slowly compare to Figure 3(a) and Figure 3(c). This phenomenon may be due to the time pressure that cause the pedestrian in Figure 3(a) and Figure 3(c) to walk at high speed such as in a rush going to the office or meeting whereas the lower speed at Figure 3(b) due to the less important walk purpose such as going out for lunch or shopping.

Figure 4 indicates the relationship of pedestrian flow-density at the stairway for the different time of peak hours including morning, noon and evening peak hour. The scatter plots of pedestrian flow-density for stairways are plotted as shown in Figure 4 below for three different peak hours, including (a) morning peak, (b) noon peak and (c) evening peak. Before the pedestrian density increases to a critical value, the variation of flow between Figure 4(a) and 4(b) is almost the same. At the same density, Figure 4(c) gives less value of flow due to the increased conflicts among pedestrian especially during the arriving of the train. There is less conflict occurred in Figure 4(c) because of most of the pedestrian walk in ascending direction and only a few in descending direction.

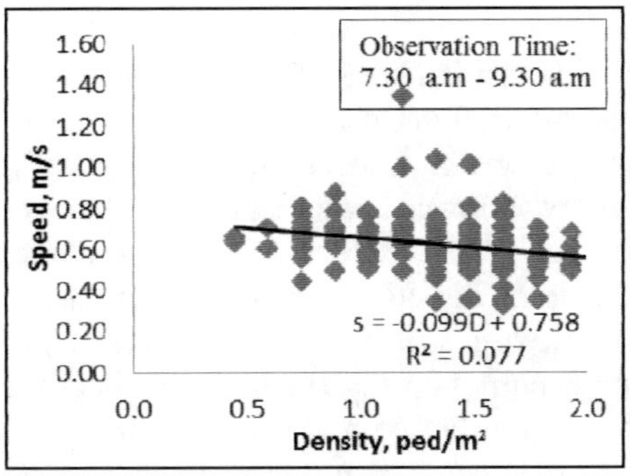

(a) Morning Peak

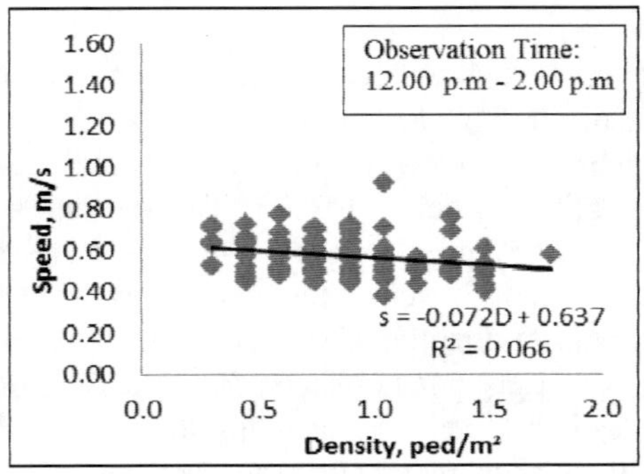

(b) Noon Peak

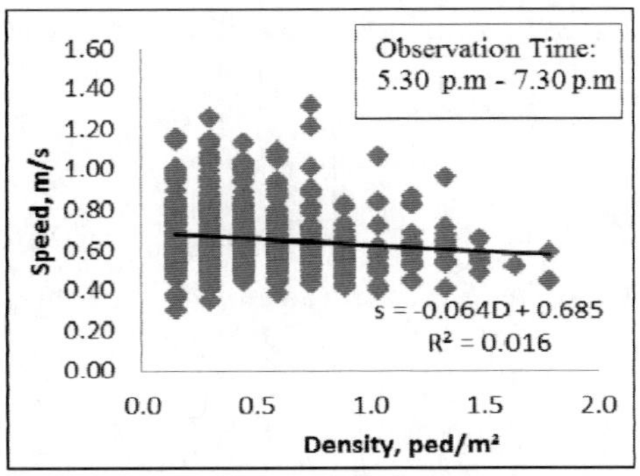

(c) Evening Peak

Figure 3. Scatter plots of pedestrian speed-density at the stairway.

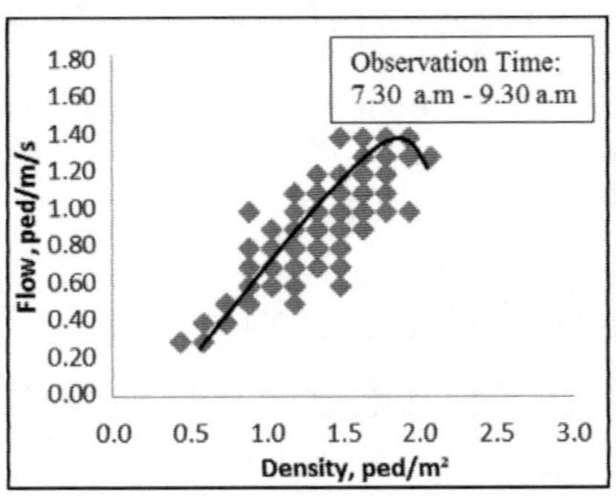

(a) Morning Peak

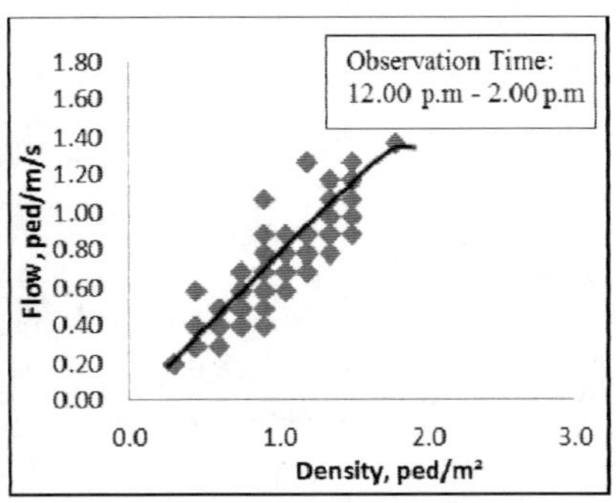

(b) Noon Peak

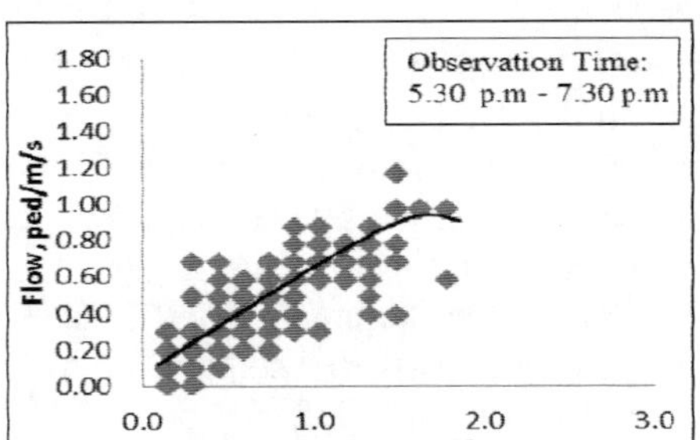

(c) Evening Peak

Figure 4. Scatter plots of pedestrian flow-density at the stairway.

IOP Conf. Series: Materials Science and Engineering **352** (2018) 012031 doi:10.1088/1757-899X/352/1/012031

Moreover, as it was shown in Figure 5, the graph of pedestrian speed-flow relationship at stairway is illustrated. The highest pedestrian flow on morning and evening peak is 1.4 ped/m/s, whereas, on morning peak, the flow was below 1.2 peds/m/s. Pedestrian walking speed that obtained in Figure 5(a) and Figure 5(b) was below 1.0 m/s. The walking speed was lower because of most of the pedestrian travel in an ascending direction. When the pedestrian arrived at the station, the conflict will occur because the walking speed of the major flow from ascending direction was affected by the pedestrian from the opposite direction. At evening peak, most of the pedestrian was travel at the range speed of 0.5 – 1.7 m/s. High speed obtained due to less conflict occurred during evening peak because of most of the pedestrian travel in descending direction with less conflict with the pedestrian in the opposite direction. This phenomenon occurred because, through the observation, high demands in descending direction due to the most of the pedestrian were the workers that used the train after office hour.

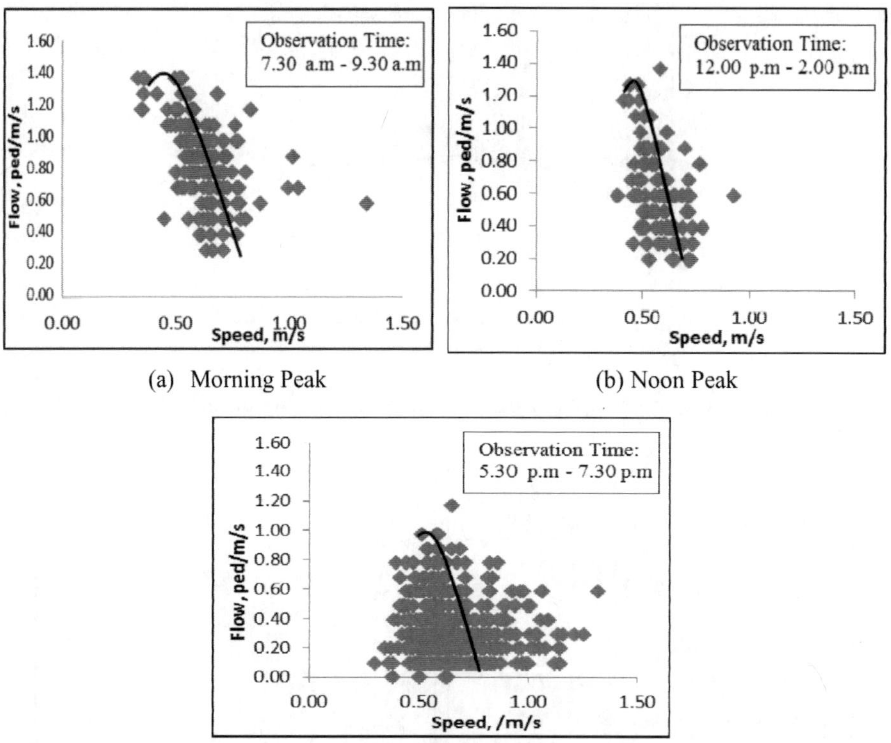

(a) Morning Peak (b) Noon Peak

(c) Evening Peak

Figure 5. Scatter plots of pedestrian speed-flow at the stairway.

3.3. Speed-Flow-Density Relationship at Escalator

The scatter plots of speed-flow relationship in escalator from three different peak hours are displayed in Figure 6. Since the escalator on morning peak only operating in ascending direction, it can be compared that pedestrian who has chosen the right side escalator will travel at range speed of 0.5 – 1.7 m/s which higher than left side of an escalator.

In addition, based on the observation, the preferable side of the escalator will be affected by the position of the arriving train. The pedestrian will choose the nearer escalator and based on the graph below, the travel speed of the pedestrian was not affected by the flow of the facility. This happens because based on the observation, the flow on the escalator will be split into two: pedestrian who following the escalator velocity will stay on the left side and pedestrian can travel with their desire speed will travel along the right side. To get the variation of the scatters plot for the escalator, the observation for pedestrian speed based on the pedestrian travel time from the right side.

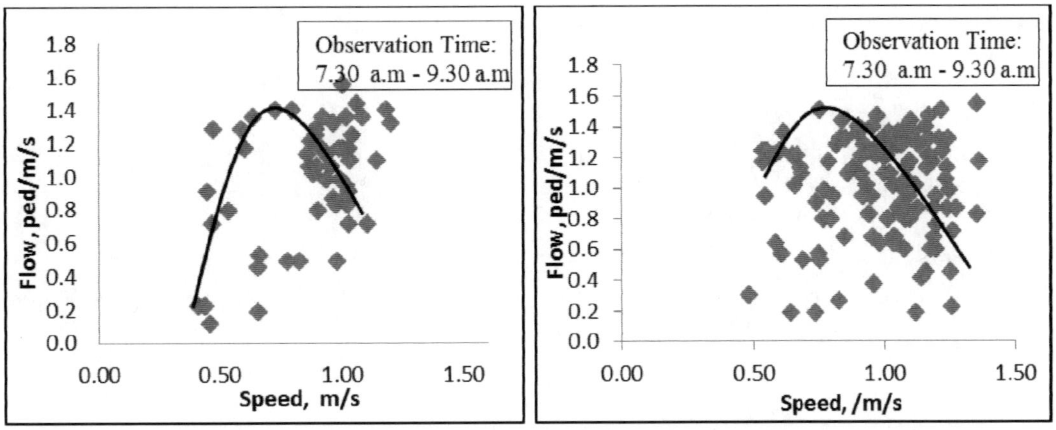

(a) Left Side (b) Right Side

Figure 6. Scatter plots of pedestrian speed-flow for ascending escalator.

In Figure 7 the scatter plots for speed-flow density in escalator for both directions during noon peak and evening peak respectively are presented. During noon peak, the pedestrian flow in Figure 7(a) was 0.8 m/s which is lower compared to pedestrian flow in Figure 7(b). This is because lower demand observed during noon peak, whereas the high flow observed in Figure 7(a) due to the pedestrian flow in ascending direction that arrives at the station. The highest value of noon peak and evening peak are 0.7 ped/m/s and 1.4 ped/m/s respectively. High flow obtained in Figure 8(b) compared to the flow in Figure 8(a) because of the high demand occurred in the evening peak. The high flow was observed during evening peak due to the end time of office hour. Most of the people are heading back home. In addition, from the observation during the video camera survey, it is found that pedestrian in descending direction will increase their walking speed during the arriving of the train.

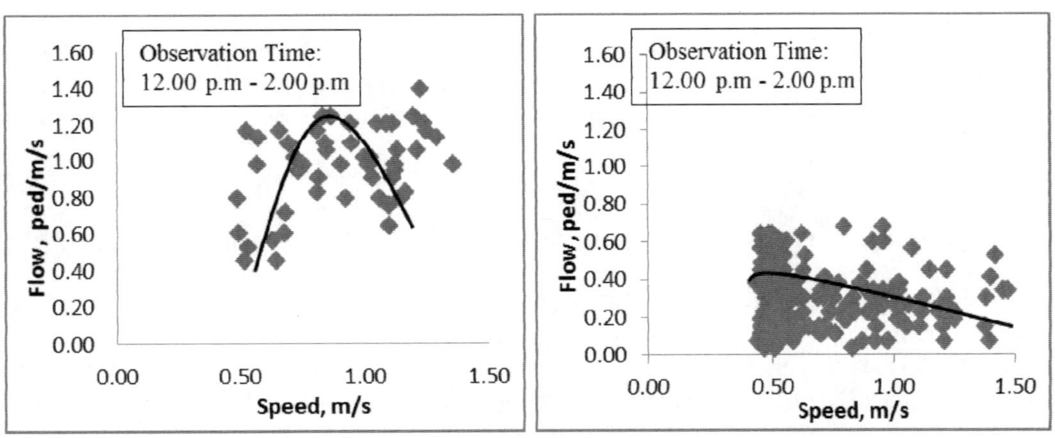

(a) Ascending Direction (b) Descending Direction

Figure 7. Scatter plots of pedestrian speed-flow for noon peak escalator.

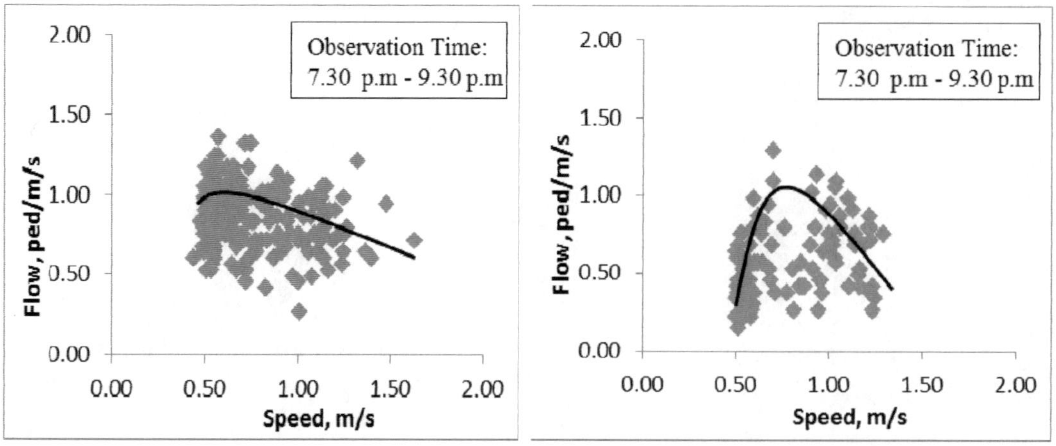

(a) Ascending Direction (b) Descending Direction

Figure 8. Scatter plots of pedestrian speed-flow for evening peak escalator.

4. Conclusion

Pedestrian walking characteristics in KLCC Station were different between morning, noon and evening peak. During morning and evening peak, pedestrian tends to walk at high speed compared to noon peak. This is because they were in a rush hour for the important purpose such as going to the office. This is different with the noon peak as the pedestrian more relax and travel with their desire speed with fewer obstacles from another pedestrian. In stairway, pedestrian in minor flow will have to decrease they walking speed or to stop at one point when high flow is coming from the opposite direction. In addition, as the pedestrian density increase, the movement of the faster pedestrian is likely to be impeded by the slower pedestrian. This phenomenon occurred for both facilities. Though the scope of this study is limited, the survey result can still be a reference for the future study on pedestrian flow, especially in Malaysia. The result from this study can be useful for evacuation planning in underground spaces. Furthermore, it is believed that the study on pedestrian walking facilities will be beneficial for future planning of design facilities.

Acknowledgement

This work was funded by Fundamental Research Grant Scheme (FRGS) 203/PAWAM/6071279 with the collaboration between Universiti Sains Malaysia (304/PAWAM/6050336/A119) and PRASARANA Malaysia Berhad. The preparation of this research paper was funded by CRA AUNSEED-Net CR1701.

References

[1] Goh B H, Subramaniam K., Wai Y T and Mohamed A A 2012 Pedestrian crossing speed: the case of Malaysia *International Journal for Traffic and Transport Engineering* **2** 323-332

[2] Rahman K, Ghani . A, Kamil A A, Mustafa A. and Chowdhury M A K 2013 Modelling pedestrian travel time and the design of facilities: a queuing approach *PLoS ONE* **8**

[3] Guo Z 2009 Does the pedestrian environment affect the utility of walking? A case of path choice in downtown Boston. *Transportation Research Part D: Transport and Environment* **14** 343-352

[4] Ye J, Chen X, Yang C and Wu J 2008 Walking behavior and pedestrian flow characteristics for different types of walking facilities. *Transportation Research Record: Journal of the Transportation Research Board* **2048** 43-51

IOP Conf. Series: Materials Science and Engineering **352** (2018) 012031 doi:10.1088/1757-899X/352/1/012031

[5] Do T, Haghani M, and Sarvi, M 2016 Group and single pedestrian behavior in crowd dynamics *Transportation Research Record: Journal of the Transportation Research Board* **2540** 13-19
[6] Garber N J and Hoel L A 2010 *Traffic and Highway Engineering. Fourth Edition* (Canada: Cengage Learning)

Method for the visualization of landform by mapping using low altitude UAV application

N. Sharan Kumar[1], Mohd Ashraf Mohamad Ismail[1], Nur Sabahiah Abdul Sukor[1], William Cheang[2]
[1]School of Civil Engineering, Engineering Campus, Universiti Sains Malaysia, Penang, Malaysia
[2]Regional Technical Manager, Plaxis Asia (Plaxis bv), Plaxis AsiaPac Pte.Ltd

E-mail: ceashraf@usm.my

Abstract. Unmanned Aerial Vehicle (UAV) and Digital Photogrammetry are evolving drastically in mapping technology. The significance and necessity for digital landform mapping are developing with years. In this study, a mapping workflow is applied to obtain two different input data sets which are the orthophoto and DSM. A fine flying technology is used to capture Low Altitude Aerial Photography (LAAP). Low altitude UAV (Drone) with the fixed advanced camera was utilized for imagery while computerized photogrammetry handling using Photo Scan was applied for cartographic information accumulation. The data processing through photogrammetry and orthomosaic processes is the main applications. High imagery quality is essential for the effectiveness and nature of normal mapping output such as 3D model, Digital Elevation Model (DEM), Digital Surface Model (DSM) and Ortho Images. The exactitude of Ground Control Points (GCP), flight altitude and the resolution of the camera are essential for good quality DEM and Orthophoto.

1. Introduction

After a decade, the instruments for data acquisition in geodesy have been rapidly improved. The instruments become cheaper, smaller, more accurate and can gather a huge number of data within a very short time interval. These devices are light, mobile, completely automated, providing access to even most unavailable study area. It is simple to be used so that, in addition to the mapping, they are also used for various purposes of concept and other kinds of designs application. [1]

UAVs as remote sensing platforms have given many research groups the chance to obtain information at adequately minimal cost to legitimize the utilization of remote sensing in any case. These platforms may, therefore, become the catalyst for many new clients and uses of remote sensing, and will turn out to be so significantly when airspace regulations have been adapted to acknowledge them as a regular airship. UAV Photogrammetry provides information used for image stitching. Autopilot system guarantees planned flight path, camera triggered auto-control to take a picture at every waypoint. Geodetic coordinates of each image projection center are available. UAV Photogrammetry is a very promising technology that needs to be better investigated A UAV is the prominent part of an entire framework which is very important to fly machine around. Physically there is no pilot on board to exhibit the aircraft, it flies autonomously with the aid of applications designed to control the aircraft. [2] [3][4][5]

IOP Conf. Series: Materials Science and Engineering **352** (2018) 012032 doi:10.1088/1757-899X/352/1/012032

Miniaturized sensor, enhance battery innovation, and the accessibility of (UAV) have provided a new chance in engineering applications. The UAV gives a stage to short proximity and low height aerial photography. Landform hazard assessment, remake, and evaluation of degradation process are gradually being studied through digital surface modeling of landscapes. Nevertheless, the panorama of any area will reliably be extra abounding compared to any PC portrayal. The main reason for modelers isn't photorealistic depiction, yet rather convincing ones that recount story required for correspondence with customers, partners, and anybody occupied with the landscape arrangement procedure. [6][7][8]

The photogrammetric UAV technique permits to study unsafe range with dependability, accuracy, and safety. The post-elaboration methodology must be carried out, to bring up complex geometric surfaces attributes identified with geological and hydrological information. LAAP provides 3 ways of information; forward information, documentation, and the preliminary investigation. The conventional photogrammetric documentation technique delivers good quality Geographical information with high spatial resolution, especially on the utilization of the aerial platform of flight elevation and the degree of territories. The conventional photogrammetric documentation technique provides very quality and high spatial resolution Geographical Information, particularly considering the degree of the territories examined and flight elevations of the aerial platform utilized. [9][10]

2. UAV and technical specification

The DJI Phantom 4 Pro quadcopter UAV was used for the image acquisition as shown in figure 1. The drone automation weight around 1.388 kg makes it a powerful, low cost, low weight UAV class. The cruising speed of UAV is on average 45mph with 28 minutes flying period in low wind condition.
It is viable to cover a 5km radius from take-off point. Flight altitude can be differently and must follow the national aviation regulation which is 120 m from ground level, obviously for the need of image resolution, the altitude should be adjusted to achieve the preferable Ground Sampling Distance (GSD). Imagery has been taken using high-resolution 1-inch 20-megapixel CMOS sensor. It has integrated GPS facilitates every image with geodetic coordinates with maximum frame size of 5472 × 3078 pixels.

Figure 1. DJI Phantom 4 Pro quadcopter.

3. Image acquisition

The normal workflow accepted for image acquisition has been used by many practitioners. Following are the steps for image acquisition as shown in figure 2.

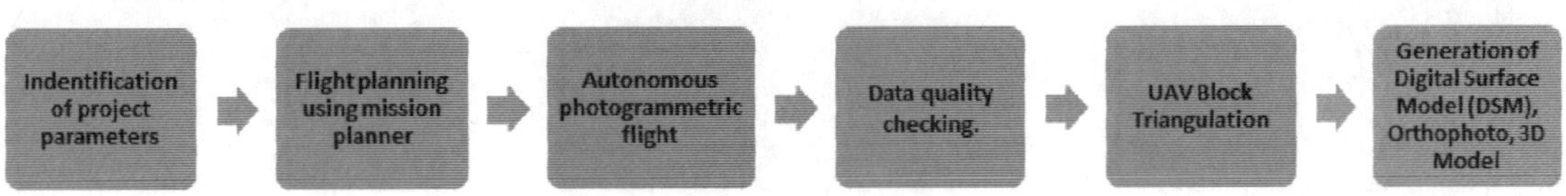

Figure 2. Workflow of image acquisition.

The UAV mounted camera receives capturing pictures during autonomous flight. The smartphone application, Map for Pilot (MAP) or Drone Deploy mission planner and autopilot board are used during autonomous flight. The Google Earth feature is used to imply waypoint and flight. This feature activates self-sufficient pictures and autonomous take-off and landing. Mission Planer highlights waypoint passage by point-click and alters the coveted flight parameter before transfer the mission, so that the UAV can take off to perform the mission. Vital component of nature of the information checking is image resolution and clarity. The captured pictures are examined right after the consummation, and if needed, a flight is rehash. All the data collections and examinations are done in real time.

Experimental flight using DJI Phantom 4 Pro plan to conduct a short flight and gather imagery. Arranging for the flight relied upon great climate, still wind and optimal altitude. UAV flight was carried out over the airspace domain of Universiti Sains Malaysia Nibong Tebal (USM) and Universiti Malaysia Perlis (UNIMAP), the northern region of Malaysia. About 5 ha and 7 ha of flight territory in USM and 15 ha in UNIMAP. Landform with same geomorphologic was specially picked as the focus was on the use of photogrammetric means for image capture in USM. Meanwhile, in UNIMAP the image captured at the landform with differential geomorphic features. The flight strip was produced as square polygons with obtained imageries. It was conducted about 60m above ground, which enables to capture GSD of the 3.5cm image.

4. Photogrammetry Image processing

The photogrammetric technique includes many processes for example orientation, aerial triangulation, bundle adjustment, build dense cloud, build mesh, build texture. The camera parameters, focal length, principal points coordinate, and the pixel sizes are essential for interior orientation. The detailing of these camera parameters is outlined right before the handling stage. The points between images are exchangeable through image connection algorithm at relative orientation. The Tie points are used to adjust images in alike conditions.

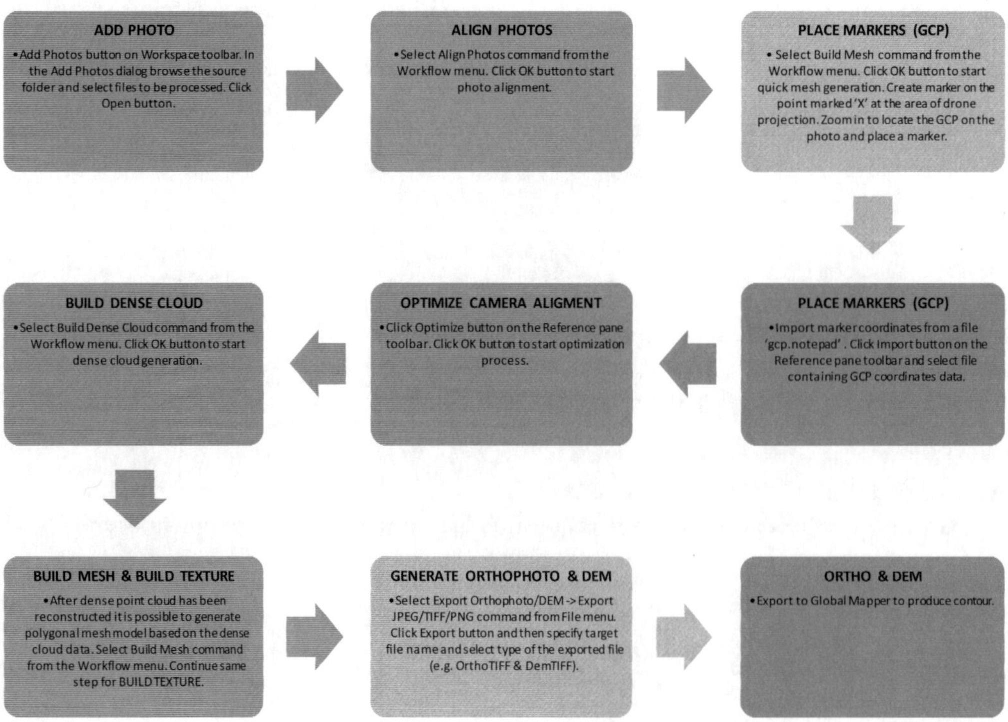

Figure 3. Flowchart of Image Processing Using Agisoft Phothoscan.

IOP Conf. Series: Materials Science and Engineering **352** (2018) 012032 doi:10.1088/1757-899X/352/1/012032

The outputs were extended to local coordinate system by Ground control points (GCP) prior to image processing. Whiles Real Time Kinematic Global Positioning System (RTK-GPS) collects the GCP, and it is essential to adjust the height elevation value because drone mapping has high error in z elevation compared to the x and y coordinates. Landform with different morphology requires the use of GCP with at least more than 5 readings. The outcome of this study is being discussed in digital orthophoto and digital slope model. These both input can be exported into other softwares such as Global Mapper for further analysis. The figure 3 below shows the flowchart of image processing using Agisoft Phothoscan software to extract the orthophoto and digital surface model.

5. Results and discussion

The study is presented in digital orthophoto and digital surface model. The tie points and GCP are used to generate Digital surface model prior to aerial triangulation process. Independent orthoimages are used to generate digital orthophoto in photogrammetry process. figure 4, 5, and 6 shows the example of DSM and digital orthophoto after photogrammetric processing. Based on these outputs, the study area contour also can be created by underlying the digital orthophoto and DSM in Global Mapper Software.

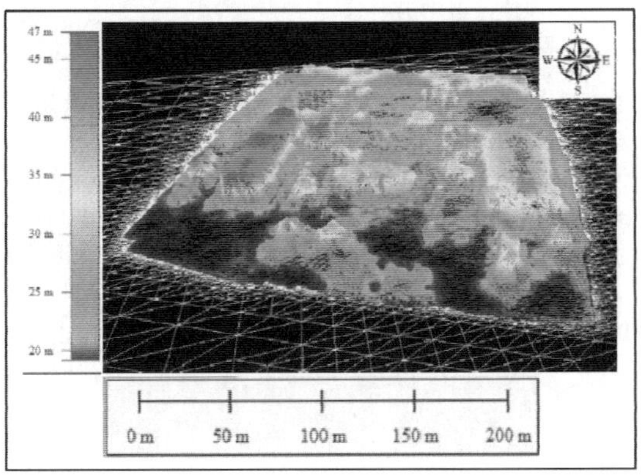

(a) (b)

Figure 4. (a) Digital Surface Model (b) Digital orthophoto in USM Site A.

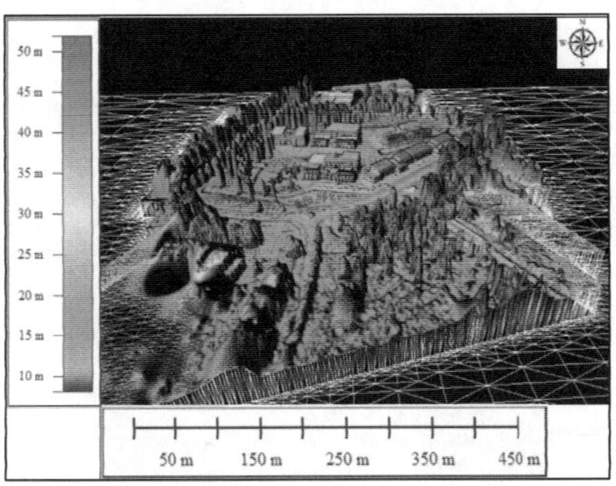

(a) (b)

Figure 5. (a) Digital Surface Model (b) Digital orthophoto in USM Site B.

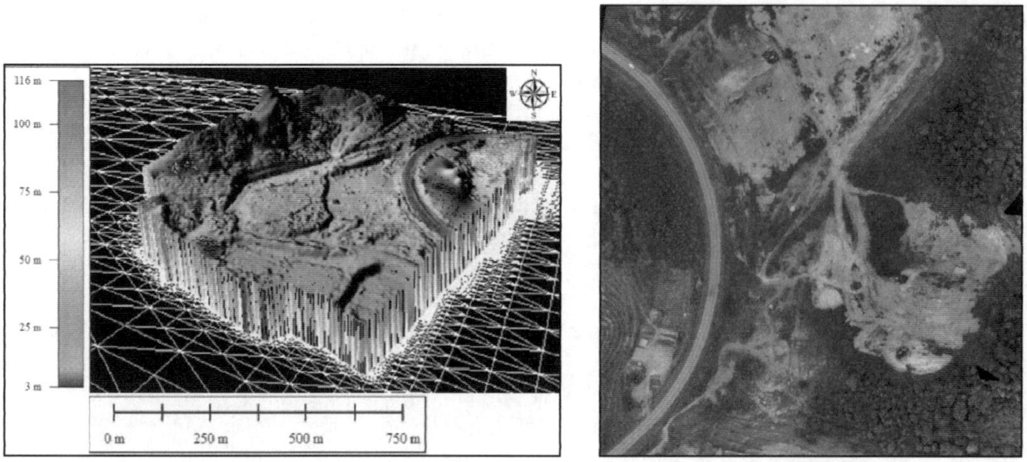

(a) (b)

Figure 6. (a) Digital Surface Model (b) Digital orthophoto in UNIMAP.

The landform can be categorized based on the visualization of the 3D model and digital orthophoto. A better understanding of the overall landscaping of the area can be achieved through this method. This technique could be a preliminary step for remodeling or restoration of the landform with a complete understanding of the overall study territory. The remodeling and restoration/improvement scheme can be designed by considering evolved local topography, enhancement of the economic and ecological value. The application of digital photogrammetric workstation is partially functional. Therefore, cartographic data collection and the UAV image processing needing spatial software. DEM analysis detaches from the generated produce, shows appositeness of DSM adjusting precision mapping tool.

6. Conclusion

Low altitude UAV capable to produce data to further understanding and analyzing the landform by covering the whole study area. Modern devices such as GPS satellites can be very helpful for photogrammetry process and advanced visualization techniques, these help to fulfilling the landscape documentation and visualization. Geographic framework, finished 3D digital reconstruction and rendering technique constitute a prevalent modern and coordinated method for depicting a landscape and also manage with its protection and progression.

Acknowledgments.

The author would like to express their appreciation to Universiti Sains Malaysia Research University(RU) Grant (1001/PAWAM/814192) for the financial support to carry out this research.

References

[1] Vasić D, Ninkov T, Bulatović V, Sušić Z and Marković M 2014 Terrain mapping by applying unmanned aerial vehicle and lidar system for the purpose of Designing in Serbia *Ingeo 2014, 6th International Conference on Engineering Surveying* 217–222

[2] Everaerts J 2008 The use of Unmanned Aerial Vehicles (Uavs) for remote sensing and mapping 1187–1192

[3] Berteška T and Ruzgienė B 2017 Photogrammetric mapping based on UAV imagery **6991**

[4] Tahar K N 2012 Aerial terrain mapping using unmanned aerial vehicle approach *Int. Arch. Photogramm. Remote Sens. Spat. Inf. Sci. - ISPRS Arch.* **39** 493–498

[5] Tahar K N 2012 A new approach on slope data acquisition using Unmanned Aerial Vehicle **13** 780–785

[6] Harwin S and Lucieer A 2012 Assessing the accuracy of georeferenced point clouds produced via multi-view stereopsis from Unmanned Aerial Vehicle (UAV) imagery **4** 1573–1599

[7] Fadda S, Fiori M and Matzuzzi C 2017 Developing rehabilitation design for the abandoned mine excavations in Central Sardinia, Italy **930**
[8] Eisank C 2012 Object-based landform mapping at multiple scales from digital elevation models (DEMs) and aerial photographs
[9] Cara S and Matzuzzi C 2016 Assessment of landscape by photogrammetry proximity Uav survey technique : A case study of an abandoned mine site in the Furtei
[10] Jose Angel Martinez D P, Cerrillo-Cuenca E and Salas-Tovar E 2017 Low altitude aerial photography applications for digital surface models creation in archaeology low altitude aerial photography applications for digital surface models creation in archaeology

IOP Conf. Series: Materials Science and Engineering **352** (2018) 012033 doi:10.1088/1757-899X/352/1/012033

Geohazard reconnaissance mapping for potential rock boulder fall using low altitude UAV photogrammetry

N. Sharan Kumar[1], Mohd Ashraf Mohamad Ismail[1], Nur Sabahiah Abdul Sukor[1], William Cheang[2]
[1]School of Civil Engineering, Engineering Campus, Universiti Sains Malaysia, Penang, Malaysia
[2] Regional Technical Manager, Plaxis Asia (Plaxis bv), Plaxis AsiaPac Pte.Ltd

E-mail: ceashraf@usm.my

Abstract. This paper discusses potential applications of unmanned aerial vehicles (UAVs) for evaluation of risk immediately with photos and 3-dimensional digital element. Aerial photography using UAV ready to give a powerful technique for potential rock boulder fall recognition. High-resolution outputs from this method give the chance to evaluate the site for potential rock boulder falls spatially. The utilization of UAV to capture the aerial photos is a quick, reliable, and cost-effective technique contrasted with terrestrial laser scanning method. Reconnaissance of potential rock boulder susceptible to fall is very crucial during the geotechnical investigation. This process is essential in the view of the rock fall hazards nearby site before the beginning of any preliminary work. Photogrammetric applications have empowered the automated way to deal with identification of rock boulder susceptible to fall by recognizing the location, size, and position. A developing examination of the utilization of digital photogrammetry gives numerous many benefits for civil engineering application. These advancements have made important contributions to our capabilities to create the geohazard map on potential rock boulder fall.

1. Introduction

Recently, the utilization of unmanned UAVs in the civilian and commercial term are ending up progressively normal, as well as for the use of concerning anthropic and natural disasters. The use of UAVs can be considered as a quick and low-cost effort to reach the zone which is frequently troublesome to investigate in safe conditions. Without a doubt, UAVs can be utilized to acquire information on the unstable region, by taking incalculable of photos and recordings from several points and distinctive edges of view [1] [2][3].

UAV's these days an essential wellspring of information for inspection, reconnaissance, mapping, and 3D modeling demonstrating issues. New applications in the short range proximity territory are presented, being the UAVs a low-cost option to the traditional manned aerial photogrammetry. UAVs, equipped for producing the photogrammetric data acquisition with mounted digital cameras. It can take off in manual, semi-automated and autonomous modes. With the photogrammetry output such as 3D results like DSM/DTM, contour lines, textured 3D models, vector data, and so on can be created, in a sensible automated manner [4][5][6].

UAVs is proven development in the field of mapping and research. Reconnaissance and monitoring should be possible with straightforward, cheaper, and precise than using different platform, for example satellites and light aircraft. The UAV capabilities are awesome and have a great deal of points of interest for the civilian purpose for monitoring, precision agribusiness, and urban development. The high-quality images can give the extraordinary outcome for different applications which can do visual translation in view of the pictures. This image is reasonable for mapping reason since it is cover little region and can deliver the updated map, for example update the Google Earth images with overlay the layer with UAV image and the user can recognize the changes on that territory [7] [8][9].

As of late, the flight of UAVs works with high degrees of autonomy by the global position system and onboard digital camera and computer. Capabilities of UAVs output to produce high accuracy topographic map by means of the image of quad-rotors UAV and ground control points (GCPs). UAV basically comes to quick riddles; little includes the high precision topographic mapping tasks. As engineers and planners, we have to see whether UAV-based photogrammetry would be sufficiently precise to map high accuracy topographic to replace current GPS and total station [2][10].

2. Site description

The study area is located at Bandar Baharu, Kedah. The study area is the new site for the development of motocross track. Fig 1 (a) and (b) shows the condition of the study area with rock boulder and its position on the small hills.

(a) (b)

Figure 1. Overview of site with rock boulders location on (a) front and (b) side views

3. Methodology

The study took place on 17 July 2017, utilizing DJI Phantom 4 Pro quadcopter. Its objective was to distinguish the potential rock boulder fall at the examination range. The territory that was measured is shown in figure 2 with all the ground control point marked. The flight plan was plotted using DJI Ground Pro Station mission planner application in which flying altitude, zone and flight direction was loaded. The flight took off 75 meters from the ground and total of 89 photos were gathered for the desired GSD of 3.5 cm.

Figure 2. The site with all the location of marked GCP.

Measurements of ground control point were taken utilizing RTK method with receivers Trimble R6 and 5800 GNSS. Ground control points reading was taken using maximum number of satellites and taking into consideration all the ground obstacles in the encompassing region. In the survey, 150cm x 150cm plastic marking sheets were utilized. The drone able took off autonomously in autopilot mode due to the on-load up the control system. The information obtained was processed with the aid of photogrammetry software Agisoft Photoscan, which is proficient in creating a geo-referenced 2D orthomosaic and 3D point-cloud from the imagery obtain during the flights. The software has three stages as shown in (Figure 3)

a) Initial processing, where it is conceivable to choose a region of orthomosaic with cloud points and include a ground control points, accordingly performing areal-triangulation.
b) Point Cloud Densification is to create a 3D point cloud.
c) DSM and orthomosaic generation to create the model of the ground and orthomosaic.

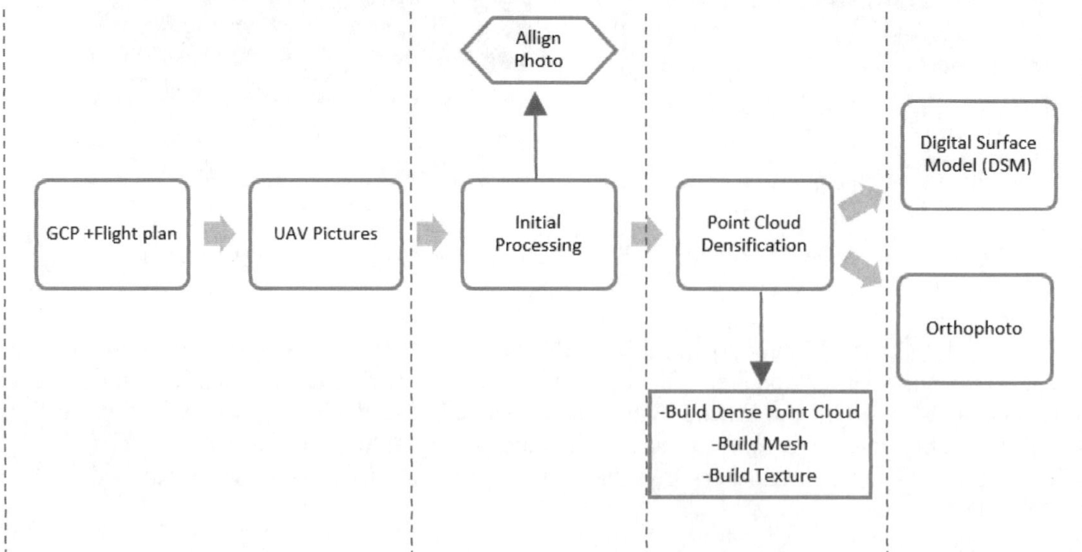

Figure 3. Example of the Agisoft Photoscan workflow for UAV photogrammetry process.

4.Results and discussion

The result of this research is Digital Surface Model (DSM), digital orthophoto, digital orthophoto as geohazard map. The Agisoft Photoscan software was used to process the images. For analysis processing, Global Mapper was used to create the 3D model. Analysis can be performed based on the final product and software capability.

4.1. Orthophoto

The Agisoft Photoscan software was also used to produce the orthoimages. For this process, the fundamental prerequisites are the digital image, the interior and exterior orientation parameters of each image and the DSM of the study area. The orthoimages were produced utilizing image resolution equal to one GSD which 3 cm/pixel. Figure 4 indicates one of the generated orthoimages.

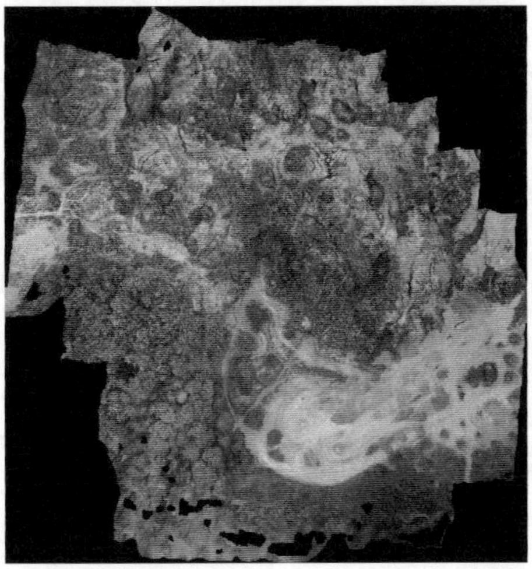

Figure 4. Generated orthoimage for the whole region.

4.2. Digital Surface Model

The triangulated Digital Surface Model (DSM) of the block region was obtained by photogrammetric technique utilizing digital correlation of pixels in the Agisoft Photoscan software. The method to choose only ground points to create the DSM was done based on the stereoscopic view and manual methodology. The number ground points chose to produce the DSM was almost to 44 380 and around 1 point for each square meter. The DSM region has the minimal elevation near 55 m and 103 m for the most elevated esteem. Based on height values of the 5 surveyed control points, the precision for the DSM was approximately to 0.9 m. Figure 5 shows the 3D representation of DSM region.

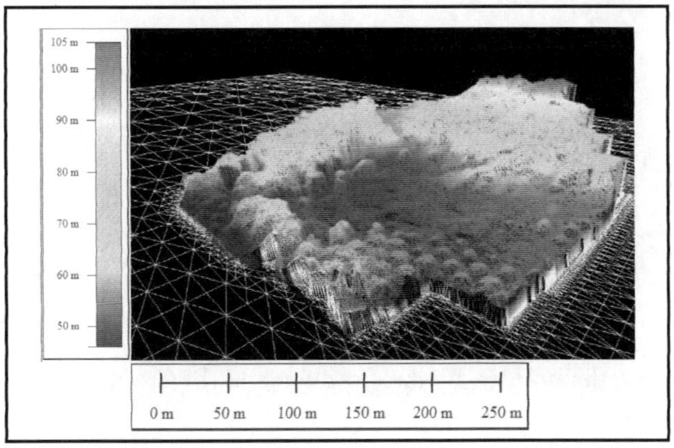

Figure 5. Show the 3D representation of the generated DTM

4.3. Geohazard Map

The digital orthophoto was used to mark the potential rock boulder which may fall based on three categories. Their categories are overhanging, positioned on loosely ground and size more than 3m rock boulder. The boulder was classified based on these 3 categories with the aid of Digital Elevation model and 3D model. Figure 6 below is the geohazard map of potential rock boulder fall.

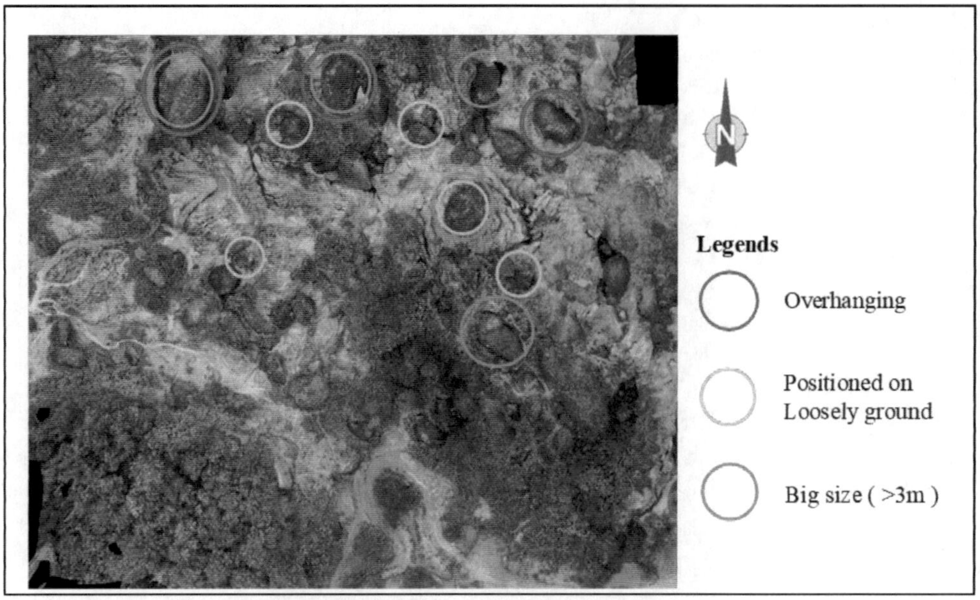

Figure 6. Geohazard map on potential rock boulder fall.

4.4. Risk and hazard of rock boulder fall

For this section, the analysis was carried out to identify the possible zone where the rock fall may occur. Few lines were drawn in the global mapper software to identify the possible direction where it may topple based on the elevation line. Figure 7 shows the example of rock boulder chosen to carry out the analysis to identify the section elevation and rockfall zone. Based on the elevation, the slope angle of the drawn section can be identified, and the critical zone is identified through the possibility of rock fall through slope angle (table 1).

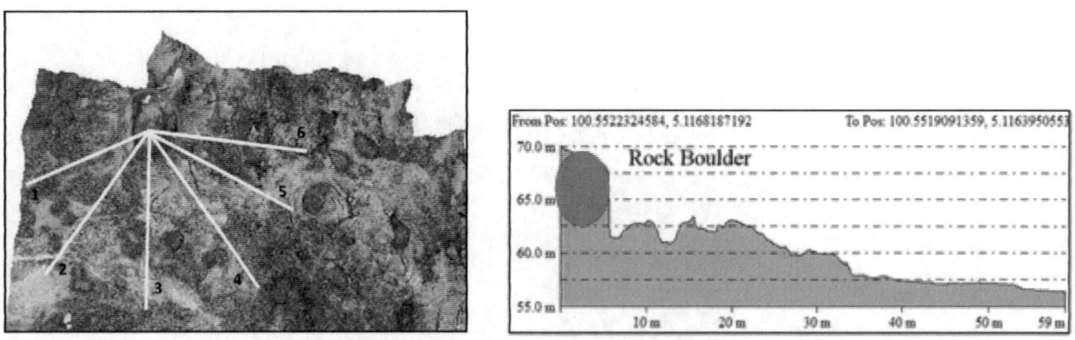

(a) (b)

Figure 7. (a) Marked section to identify the elevation and (b) Cross section of Line 1 elevation.

Table 1. Shows the angle of the slope and the possible hazard of rock fall.

Line	Angle of rock slope (◦)	Possibility of hazard through rock fall
1	30	Yes, because the angle of the slope is steeper and the distance for it to roll is longer.
2	40	Yes, because the angle of the slope is steeper and the distance for it to roll is longer.
3	20	Yes, because the angle of the slope is steeper.
4	10	No, because the distance for it to roll is shorter.
5	5	No, because the distance for it to roll is shorter.
6	5	No, because the distance for it to roll is shorter.

5.Conclusion

In this study, we have demonstrated the outcome obtained by utilizing UAVs to survey areas on potential rock boulders fall as one of the preliminary studies. The results were used to produce geohazard map on the rock boulder before the start of any work at the proposed development of motocross track. The products outlined have been considered to help the decision making to manage the potential risk of rock boulder fall. The presented study evidenced that both qualitative and quantitative evaluations of potential rock boulder fall can be conducted with the aid of UAVs, where the survey results are accessible in a fast and straightforward manner.

Acknowledgments

The author would like to express their appreciation to Universiti Sains Malaysia Research University(RU) Grant (1001/PAWAM/814192) for the financial support to carry out this research.

References

[1] Giordan D, Manconi A, Facello A, Baldo M, dell'Anese F, Allasia P and Dutto F 2015 Brief communication : The use of an unmanned aerial vehicle in a rockfall emergency scenario *Nat. Hazards Earth Syst. Sci* **15** 163–169

[2] Chi Y, Lee Y F and Tsai S 2016 Study on high accuracy topographic mapping via UAV-based images *IOP Conf. Ser.. Earth Environ. Sci.* **44** 032006

[3] Eid B M, Chebil J, Albatsh F and Faris W F 2013 Challenges of integrating unmanned aerial vehicles in civil application *IOP Conf. Ser.. Master. Sci. Eng.* **53** 012092

[4] Remondino F, Barazzetti L, Nex F, Scaioni M and Sarazzi D 2011 UAV photogrammetry for mapping and 3d modeling – current status and future perspectives *ISPRS - Int. Arch. Photogramm. Remote Sens. Spat. Inf. Sci.* **XXXVIII-1** 25–31

[5] Berteška T and Ruzgienė B 2013 Photogrammetric mapping based on UAV imagery *Geodesy and Cartography* **39** 158-163

[6] Everaerts J 2008 The Use Of Unmanned Aerial Vehicles (Uavs) for remote sensing and mapping *In The International Archives of the Photogrammetry, Remote Sensing and Spatial Information Sciences* **XXXVII** 1187–1192

[7] Samad A M, Kamarulzaman N, Hamdani M A, Mastor T A and Hashim K A 2013 The potential of Unmanned Aerial Vehicle (UAV) for civilian and mapping application *In Proceedings - 2013 IEEE 3rd International Conference on System Engineering and Technology, ICSET* 313-318

[8] Liba N and Berg-Jürgens J 2015 Accuracy of orthomosaic generated by different methods in example of UAV platform MUST Q *IOP Conf. Ser.. Master. Sci. Eng.* **96** 12-41

[9] Restas A 2016 Drone applications for supporting disaster management drone applications for supporting disaster management *World Journal of Engineering and Technology* **3** 316-321

[10] Graca N, Mitishita E and Goncalves J 2014 Photogrammetric Mapping Using Unmanned Aerial Vehicle *In The International Archives of the Photogrammetry, Remote Sensing and Spatial Information Sciences* **XL-1** 129–133

IOP Conf. Series: Materials Science and Engineering **352** (2018) 012034 doi:10.1088/1757-899X/352/1/012034

The main factor affecting the competitiveness of Contractor Company

Nurisra, Nurul Malahayati, and Mahmuddin
University of Syiah Kuala, Jl. Syekh Abdurrauf 10, Banda Aceh, Indonesia

E-mail: nurisra@yahoo.com

Abstract. Contractor companies must have the competitive advantage to compete in maintaining the survival of the company. Problems arise because quite a lot of advantages can be used and these advantages must be used appropriately to produce competitiveness for the company to continue to compete and to win the competition. This study aims to determine the main factors affecting the competitiveness of medium-class contractors in Banda Aceh. Data collection was obtained through questionnaires distributed to 31 middle-class contractors in Banda Aceh. Data processing and analysis is done by using descriptive analysis. Based on the result of descriptive analysis, it can be concluded that the most important competitiveness factor with a mean score value 4.52 is the relationship, and the factor that has the highest mean score value is the relationship with the government of 4.97, while the result of the ranking analysis is obtained 25 factor that is critical to the competitiveness of medium-class contractors in Banda Aceh.

1. Introduction

In Indonesia, the competitiveness of construction service companies is still relatively weak. Currently, national contractors are faced with a globalization era marked by the enactment of Asian Free Trade Area (AFTA) in 2015. With the enactment of the AFTA, foreign contractors can freely compete for projects in the construction market in Indonesia. Whereas in the era of free trade, competitiveness is one measure of the company's successful performance. Based on the above phenomenon, should every Indonesian construction service company create the competitive advantage to maintain its survival in the global competition of construction service business world?.

Some of the problems that contractors face already summarized, both internal and external. Internal problems faced are essentially due to weak competitiveness and external problems faced by the challenge the free market due to globalization, monetary crisis, regional autonomy, and difficulties of financial support from the government. Understanding the competitiveness of organizations and the development of competitive strategies has become an important issue in the construction business, which has led to an increase in the number of studies related to the problem. The construction industry is dynamic and has special characteristics for each region and will have different impacts in the construction project competition. Similarly, local contractors with different backgrounds and experiences will have different power or competitiveness between different regions. It is important for contractors to realize the competitive advantage and can develop competitive strategies.

Competitiveness is the ability of the company to compete with its competitors. Therefore, companies must have the competitive strategy and competitive advantage that focus on dynamic processes. Porter [1] states that competitive advantage is key to achieving superior corporate, industrial and economic performance.

Indonesian construction service companies must be ready to compete and have competitive sources of resources that include physical resources, financial resources, organizational process structures and systems, and human resources (HR). There are many competitiveness factors of contractor companies that can be used to enhance a company's competitiveness such as project management, organizational structure, competitive strategy, relationships, supply, marketing, corporate image, technical and technological capabilities, as well as financial capabilities [2-3].

Many factors can affect the competitiveness of the company, therefore need to know what factors are dominant, so that helps contractors to maintain the company. The objective of the study was to determine the main factors affecting the competitiveness of middle-class contractors in Banda Aceh.

2. Methods

Data collection was done by distributing questionnaires addressed to contractor companies. The questionnaire was designed into two main sections, the first questionnaire containing questions about the characteristics of the respondents and also explaining the identity of the company in general. The second part questionnaire contains statements relating to factors affecting the competitiveness of contractors.

2.1 Location and object of research

The object of this study is the competitiveness of middle-class construction companies. The survey sites were conducted at Middle-Class construction companies that handle construction projects located in the city of Banda Aceh.

2.2 Data collection

The required data are as follows:

- Primary data is data obtained directly from the respondent or object under study, or something to do with the research. In this study, the primary data in question is the data obtained from the questionnaires.
- Secondary data is supporting data that was collected from LPJK Aceh. The secondary data in this study includes a list of names of middle-class construction companies located in the city of Banda Aceh.

2.3 Data processing and analysis

Data processing and analysis through the stages as follows:

- Conducting recapitulation and tabulation of the questionnaire survey results. Assessment of answers using the Likert scale where scale 1 (very unimportant), scale 2 (not important), scale 3 (neutral), scale 4 (important), and scale 5 (very important).
- Validity and reliability test are performed to determine the accuracy and accuracy of a measuring instrument (questionnaire) in performing the measuring function. Testing the validity of the instrument using a test of the validity of the construction where the t count should be greater or equal to the table is 1.701. The result of validity test of each factor is bigger than t table = 1,99. The reliability testing is done by using Cronbach Alpha method where Cronbach Alpha coefficient must be greater than 0.6. Reliability test results for the whole factor obtained Cronbach Alpha value = 0.925> 0.6. Thus, it can be concluded that the questionnaire in this study is valid and reliable so it can function as a measuring tool in this study.
- Data analysis is done by calculating the mean score value of each factor. The Mean Score Value (MS) is calculated by the following equation:

$$MS = \frac{\Sigma(f \; x \; s)}{n} \quad (1 \leq MS \leq 5) \qquad (1)$$

where :

f = number of answers for each factor
s = score value for each respondent's answer (1-5)
n = number of respondents

3. Results and discussions

3.1. Characteristics of respondents

Respondents selected in this study were directors or deputy directors of middle-class construction companies with 31 contractors regarding company experience, a number of some construction projects ever worked on the last five years and the average contract value of construction projects conducted each year. The full description is shown in Table 1.

Table 1. Characteristics of respondents

Company experience	**Frequency**	**Percentage (%)**
< 5 years	0	0.00
5-10 years	2	6.45
10-15 years	11	35.48
15-20 years	18	58.06
> 20 years	0	0.00
Number of construction projects worked over the last five years		
1-3	4	12.90
4-6	11	35.48
7-10	16	51.61
> 10	0	0.00
The average contract value of projects implemented per year		
Rp 100 - 500 million	2	6.45
Rp 500 million - Rp 1 billion	7	22.58
Rp 1 - 5 billion	15	48.39
Rp 5 – 10 billion	7	22.58

3.2 Descriptive analysis

Descriptive analysis is used to find the total and mean score of each competitiveness factor and presented in Table 2.

Table 2. Mean Score

Code	**Factors**	**Frequency**					**Total score**	**Mean Score**
	Project Management	1	2	3	4	5		
P_1	Site management	0	0	6	13	12	130	4.19
P_2	Cost management	0	0	0	13	18	142	4.58
P_3	Quality management	0	0	1	11	19	142	4.58
P_4	Time management	0	0	0	10	21	145	4.68
P_5	Contract management	0	0	4	20	7	127	4.10
P_6	Ability to resolve problems or disputes	0	0	12	15	4	116	3.74
P_7	Risk management	0	0	3	18	10	131	4.23
P_8	Supply chain and logistics management	0	0	7	15	9	126	4.06
P_9	Effectiveness in the management of work safety	0	0	10	14	7	121	3.90
P_{10}	Knowledge of local construction laws	0	0	13	14	4	115	3.71

P_{11}	Number of contracts successfully committed	0	0	8	9	14	130	4.19
							Mean	4.18
	Organizational structure							
P_{12}	Communication and coordination between functional areas within the company	0	0	11	15	5	118	3.81
P_{13}	Tasks and functions are described for each field	0	0	11	17	3	116	3.74
P_{14}	Motivation and job satisfaction	0	0	6	19	6	124	4.00
P_{15}	Retain core employees	0	0	4	20	7	127	4.10
							Mean	3.91
	Competitiveness Strategy							
P_{16}	There is a clear competitive strategy	0	0	7	15	9	126	4.06
P_{17}	Implement the strategy that has been made/planned	0	0	2	23	6	128	4.13
P_{18}	Understand in detail the strategy made/planned	0	0	6	17	8	126	4.06
							Mean	4.09
	Relationship							
P_{19}	Relationship with owner	0	0	0	7	24	148	4.77
P_{20}	Relationship with subcontractor or supplier	0	0	8	8	15	131	4.23
P_{21}	Relationship with government	0	0	0	1	30	154	4.97
P_{22}	Relations with the public	0	0	6	14	11	129	4.16
P_{23}	Relationship with consultants	0	0	1	14	16	139	4.48
							Mean	4.52
	Bidding							
P_{24}	Bidding strategy	0	0	2	13	16	138	4.45
P_{25}	Bidding experience	0	0	1	7	23	146	4.71
P_{26}	Bidding resources	0	0	6	11	14	132	4.26
							Mean	4.47
	Marketing							
P_{27}	The ability to collect and process information from new projects	0	0	9	18	4	119	3.84
P_{28}	Availability of product information and labor costs, materials, equipment, and other resources	0	0	6	21	4	122	3.94
P_{29}	Ability to forecast changes in construction market conditions	0	0	10	15	6	120	3.87
							Mean	3.88
	Corporate Image							
P_{30}	The credibility of the company	0	0	7	23	1	118	3.81
P_{31}	Company grade	0	0	9	18	4	119	3.84
P_{32}	Project performance (quality, safety, environment)	0	0	7	16	8	125	4.03

P_{33}	The level of bank confidence in the company	0	0	9	18	4	119	3.84
P_{34}	Professional qualification of project manager	0	0	4	20	7	127	4.10
							Mean	3.92
	Technical Capabilities							
P_{35}	Capacity of construction equipment and workers	0	0	3	12	16	137	4.42
P_{36}	Ability of technical and professional employees	0	0	6	16	9	127	4.10
P_{37}	Ability of technological innovation	0	0	13	12	6	117	3.77
							Mean	4.10
	Financial capability							
P_{38}	Payment to subcontractors/suppliers on time	0	0	18	11	2	108	3.48
P_{39}	The level of financial confidence is indicated by the bank's recommendations	0	0	16	15	0	108	3.48
P_{40}	Financial resources (capital, cash, equity, and retained earnings)	0	0	11	16	4	117	3.77
P_{41}	Financial stability in the face of crisis	0	0	14	16	1	111	3.58
							Mean	3.58

Based on the results on Table 2 obtained the highest mean score in the group of competitiveness factors of relationship (4.52), meaning that respondents strongly agree with the main competitiveness factor. Then followed by bidding capability 4.47, project management 4.19, technical and technological ability 4.10, a competitiveness strategy 4.09, corporate image 3.92, organizational structure 3.91, marketing 3.88, and financial ability 3.58.

3.3 Factors affecting the competitiveness of contractors

Table 3 shows the 25 highest ranking of contractor competitiveness factors. The relationship with government (P21) is ranked first, so it is important for contractors to maintain good relationships with partners, especially governments, owners, and consultants. In Banda Aceh alone the construction market is still very large from government projects as well as projects with large budgets widely available in government, to reduce risks and enhance the competitiveness of firms relationships must be maintained to increase the likelihood of winning project tenders.

Time management (P4), cost management (P2), and quality management (P3) become a very important factor in terms of management in improving the competitiveness of enterprises, by balancing between time, cost, and quality, the company can improve the quality of project performance to improve competitiveness internally and increase trust externally so that the company continues to gain the trust of the project owner to continue to get sustainable construction project work.

Another very important factor is the experience of offering (P19) and bidding strategies (P24), each contractor company must be able to win project bids to secure employment while maintaining the survival of the company, therefore experience in the bidding of the construction projects, bidding strategies are needed to compete for job projects. The company itself has its strategy so that this factor is considered important in the construction industry.

Table 3. Ranking of contractor competitiveness factors

	Competitiveness factors	**Mean Rank**
P_{21}	Relationship with government	4.97
P_{19}	Relationship with client	4.77
P_{25}	Bidding experience	4.71
P_{4}	Time management	4.68
P_{2}	Cost management	4.58
P_{3}	Quality management	4.58
P_{23}	Relationship with consultant	4.48
P_{24}	Bidding strategy	4.45
P_{35}	Capacity of construction equipment and workers	4.42
P_{26}	Bidding resources	4.26
P_{7}	Risk management	4.23
P_{20}	Relationship with subcontractor and supplier	4.23
P_{1}	Site management	4.19
P_{11}	Number of contracts successfully committed	4.19
P_{22}	Relationship with public	4.16
P_{17}	Implement the strategy that has been made/planned	4.13
P_{5}	Contract management	4.10
P_{15}	Retain core employees	4.10
P_{34}	Professional qualification of project manager	4.10
P_{36}	Ability of technical and professional employees	4.10
P_{8}	Logistics management and supply chain	4.06
P_{16}	Specific competitive strategy	4.06
P_{18}	Understand in detail the strategy made/planned	4.06
P_{32}	Project performance (quality, safety, environment)	4.03
P_{14}	Motivation and job satisfaction	4.00

4. Conclusions

This study presented the survey carried out among Banda Aceh construction companies. The critical factors leading to construction company competitiveness have been investigated through interviews among top-level managers and owners of the companies. Based on the descriptive analysis, it is found that the most influential competitiveness factor is cooperation relationship, while for the most chosen indicator is the relationship with the government. Based on the ranking analysis there are 25 competitiveness factors that can improve the competitiveness of a company.

Future researchers involving managers/owners of construction contractors can provide a comparison of the results of studies conducted in other cities in Indonesia

Acknowledgement

The authors express high appreciation to all parties who have contributed to this research, especially to Kamal Rachmanda and other research teams that involved from data collection to completion of the research.

References

[1] Porter M E 1980 Competitive strategy: Techniques for analyzing industries and competitors *New York: Free Press*

[2] Lu W Liyin S and Michael C H Y 2008 Critical success factors for competitiveness of contractors: China study *J. Const. Eng. Manage.* **134** 972-982

[3] Tan Y, Shen L Y, Michael C H Y and Ann A C L 2007 Contractor Key Competitiveness Indicators (KCIs): Hong Kong study *Surveying and built environment* **18** 33-46

IOP Conf. Series: Materials Science and Engineering **352** (2018) 012035 doi:10.1088/1757-899X/352/1/012035

Geotechnical approach for occupational safety risk analysis of critical slope in open pit mining as implication for earthquake hazard

Munirwansyah[1], Masyhur Irsyam[2], Reza P Munirwan [3], Halida Yunita[3] and M Zulfan Usrina[4]

[1] Professor and Lecturer of Civil Engineering, Syiah Kuala University. Banda Aceh.

[2] Professor and Lecturer of Faculty of Civil and Environmental Engineering, Institut Teknologi Bandung

[3] Lecturer of Civil Engineering, Syiah Kuala University, Banda Aceh

[4] Magister student of Civil Engineering-Management Infrastructure, Syiah Kuala University, Banda Aceh, Indonesia

E-mail: nir_geotechnical@unsyiah.ac.id

Abstract. Occupational safety and health (OSH) is a planned effort to prevent accidents and diseases caused by work. In conducting mining activities often occur work accidents caused by unsafe field conditions. In open mine area, there is often a slump due to unstable slopes, which can disrupt the activities and productivity of mining companies. Based on research on stability of open pit slopes conducted by Febrianti [8], the Meureubo coal mine located in Aceh Barat district, on the slope of mine was indicated unsafe slope conditions, it will be continued research on OSH for landslide which is to understand the stability of the excavation slope and the shape of the slope collapse. Plaxis software was used for this research. After analyzing the slope stability and the effect of landslide on OSH with Job Safety Analysis (JSA) method, to identify the hazard to work safety, risk management analysis will be conducted to classified hazard level and its handling technique. This research aim is to know the level of risk of work accident at the company and its prevention effort. The result of risk analysis research is very high-risk value that is > 350 then the activity must be stopped until the risk can be reduced to reach the risk value limit < 20 which is allowed or accepted.

1. Introduction

This study located in a local coal mine in Sumatera -Aceh, Meurebo District, Sumber Batu Village, of West Aceh. Coal mining activities in this area use open pit system (Open Pit). Open pit mining has entered complex mining environment in terms of high and step, deep, surface and underground combined mining which led the geotechnical researcher to face several challenges [1]. The failure of site investigation has led to catastrophic disaster serious problem for the safety of public [2].

Ramli, [3] noted that occupational safety and health (OSH) issues in the mining industry, as well as construction projects lately, continue to grow in line with the development of technology. Such progress has resulted in the emergence of intense competition between industry and construction service providers, often associated with issues of OSH that can be used as a means to enter national and international markets. The situation has changed the society's view of the importance of seriously implementing OSH management in applying the construction project.

Therefore, the authors examine the effect of landslides on the safety of the work by using Job Safety Analysis (JSA) method, to identify the hazard of work safety which will be analyzed for risk management so that it can be grouped as hazard level and its handling technique. The purpose of this research is to know the level of risk of work accident in the mining area of excavation of coal mining caused by the slope of unsafe slopes and prevention effort [3].

The results of this study will provide information on the level of occupational accidents, especially on the slopes of coal mines Meureubo Aceh Barat District of Aceh Province, as well as the proper implementation of OSH and Mitigation in case of Work Accidents on the slopes of a coal mine. Following figures are the condition of Meureubo coal mining in the west of Aceh Barat district in figure 1(a) and 1(b).

(a) (b)

Figure 1. Unsafe Vertical slope (a) Critical Condition of the Coal Mining in Meureubo, (b) Unsafe Equipment (Bulldozer and truck) mobilization. The condition of coal harvesting place in Meurubo. [13]

2. Slope Stability Analysis

According to Imran [4], slope stability analysis is sufficiently made with boundary equilibrium approach to obtain the value of safety factor against the smallest collisions. Various ways are offered for this analysis, including ordinary slices, bishop, janbu, morgen stren-prize, and sarma. At least two ways must be used in this stability and safety factor used in planning is derived from the average value of the results of all the means employed [4]. A numerical model of regional scale two-dimensional finite element methods for analysis slope stability was also conducted by Faruque [5]. The result shows that there is a high prospect to cause water flow into the tension zone that eventually creates slope instability. A deep drainage system as suggested by Bomont also being considered [6]. The drained residual shear strength is also applicable for slope analysis with preexisting shear surface [7].

Febrianty, D. [8], this slope stability analysis is calculated by plaxis 8.2 software with the assumption that the analysis is the same as the Limit Equilibrium Method (LEM) method as shown in figure. 2, by dividing the period of land bounded by the landslide into vertical slices deep in plaxis made in the shape of the triangular element as shown in Figure 3. The occurrence of the landslide can be assumed to be shaped circles of shadow (colored) and total deformation mesh in the form of the arrow (arrow direction movement) as shown in Figure 4 and Figure 5. This theory of equilibrium balance is based on the following assumptions;

(1) Land that behaves like a material, Mohr-Coulomb,
(2) The safety factor of the components of cohesion and the inner-angle component is as large,
(3) The safety factor for all slices of the soil mass is the same.

Safety factor is obtained by summing forces on each slice, as well as the momentary summation of the center point of the collapsed field [8].

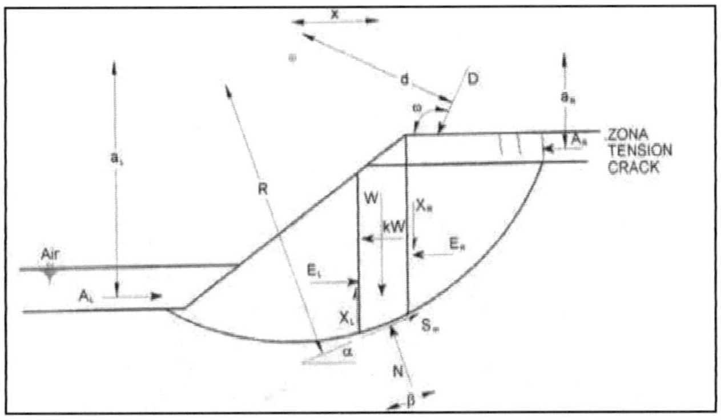

Figure 2. Vertical Circle Slices for Slope Analysis [8]

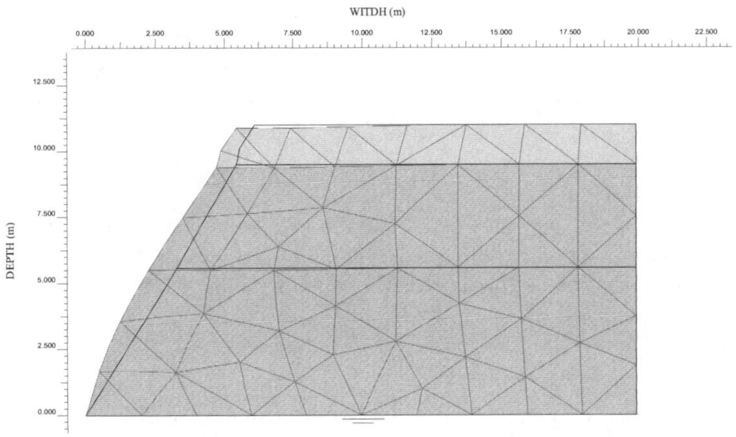

Figure 3. The triangular shape for the landslide field, Deformed Mesh Modeling of Plaxis 8.2 [8]

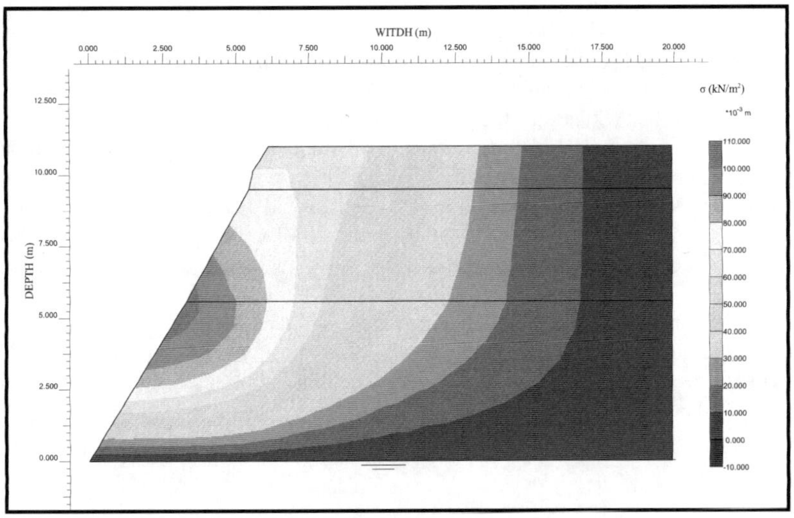

Figure 4. Slope failure identification of a shadow circle [8]

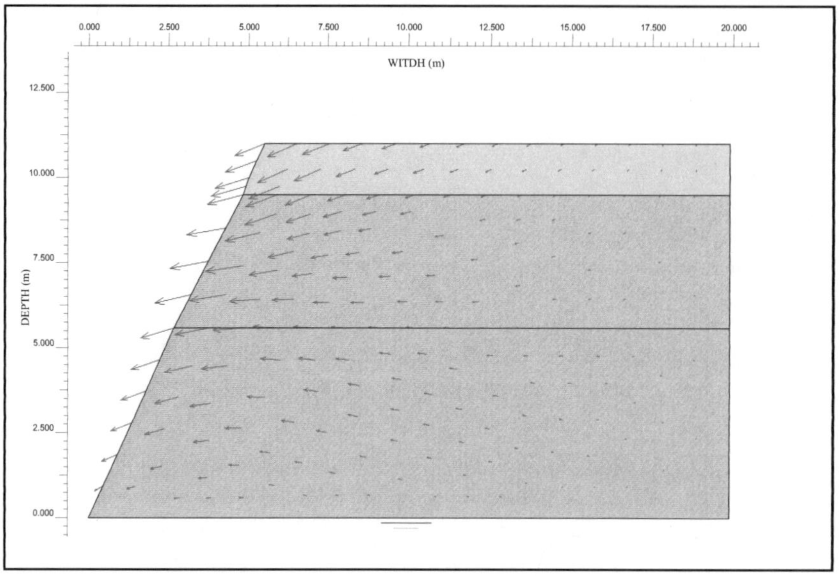

Figure 5. Direction of movement in the form of arrow displacement failure [8]

3. Job Safety Analysis (JSA)

Said [9], Job Safety Analysis or also known as Job Hazard Analysis is an effort to study/ analyze and recording each sequence of work steps of a job, followed by identification of potential hazards in it then solved by determining the best effort To reduce or eliminate/control the dangers to the work being analyzed [9].

4. Risk Analysis

According to the Australian Standard [10], the risk is the possibility or probability of occurrence of something that can lead to an impact on a goal. The risk is measured based on the possibility of a case or consequences that can be caused. While an understanding of Occupational Health and Safety Australian Standard (OHSAS) 18001 risk as cited from Ramli [3], is a combination of the possible happening of dangerous event or exposure with the severity of injury or health problem caused by the event or exposure [10]. The effect of drill vibration from mining site has also given an impact to the surface as [11] which lead to accelerate the landslide possibility surrounding the pit site. This circumstance is also part of risk analysis.

According to Australian Standard [12], risk analysis is a systematic activity using existing information to determine the extent and consequence of frequency of an event. This analysis should consider the range of potential consequences and how risks can occur. The purpose of the risk analysis is to distinguish between small risk and large risk and provide data to assist evaluation and risk management. The semi-quantitative analysis yields more detailed priorities than qualitative analysis, since risk is divided into categories. This method is similar to qualitative analysis, the difference lies in the description or description of the parameters that exist in the semi-quantitative analysis is expressed with a certain value or score. According to Australian standard 4360: 1999, the semi-quantitative analysis considers the possibility of combining two elements, namely probability (likelihood) and exposure (exposure) as frequency. There is a strong relationship between the frequency of exposure and the probability of occurrence of risk [12].

IOP Conf. Series: Materials Science and Engineering **352** (2018) 012035 doi:10.1088/1757-899X/352/1/012035

Table 1. Risk Level Semi Quantitative Analysis Method (Risk Management) [12]

Risk Level	Category	Action
> 350	Very High	3.1cm
180 – 350	Priority I	1.8 cm
70 - 180	Substansial	2.8 cm
20 – 70	Priority 3	2.8 cm
< 20	Acceptable	0 cm

5. Analysis qualitative and quantitative

Analysis of data obtained from the Plaxis program is completed qualitatively and quantitatively to obtain results in order to be in further analysis in submitting suggestions. Plaxis program analysis has three stages as below.

Table 2. Material Set in Plaxis

	Parameter	Name	Yellowish Light Clay 1-4 m	Darkys Black Soil 4-7 m	Sandy Silt >7 m	Unit
Natural Soil Undisturbed Sample	Material Model	Model	MC	MC	MC	-
	Type of Behaviour	Type	Undrained	Undrained	Undrained	-
	Dry Soil Weight	dry	16.0	18.2	18.52	kN/m3
	Wet Soil Weight	wet	12.5	12.5	20.53	kN/m3
	Horizontal	K_H	8.64×10^{-5}	0.0864	2.80 E-7	m/day
	Permeability	K_P	8.64×10^{-5}	0.0864	2.80 E-7	m/day
	Vertical Permeability	E	2000	4500	980.7	kN/m2
	Young's Modulus		0.3	0.4	0.2	-
	Poisson's Ratio	C	0.5	2.1	17.75	kN/m2
	Cohesion		6.2	8.8	1.45	0
	Friction Angel		0	0	0	0
	Dilatancy Angel					

5.1 Data input

The stages in data input, as follow;

(a) Entry of geometry of the slopes reviewed,
(b) Selection of analysis type and element type.
(c) Inclusion of material properties and soil coating, which corresponds to the Table. 2,
(d) Preparation of mesh as a whole,
(e) Run from the initial stresses of the analysis, and
(f) Run the calculation phase.

5.2 Data Output Stage

The data output stage is the stage of the results obtained from the calculation stage with the Plaxis program. The results obtained at the data output stage are:

1. Calculate the safety factors obtained from the relationship graph ΣMsf and displacement; and
2. Deformation of the slope shown from total displacement.

Analysis the stability of slope by means of trial and error against slopes by using Plaxis program which requires appropriate slope angle modeling. This modeling is completed to obtain slope stability from the actual condition in the field so that the result obtained from the slope is safe from the sliding.

IOP Conf. Series: Materials Science and Engineering **352** (2018) 012035 doi:10.1088/1757-899X/352/1/012035

Table 3. Modeling of Slope Stability

Modeling	Varian θ			Varian C			Hasil/ Output
	θ_{nt}+ 10%	θ_{nt}	θ_{nt}– 10%	Cnt + 10%	C_{nt}	C_{nt} – 10%	
df_n + 10%	+	OSH	OSH	OSH	OSH	OSH	
df_n	+	+	OSH	+	+	+	
df_n – 10%	+	+	OSH	+	+	+	

Inform: (+) Good Results, (OSH) Out Put Less Good

6. OSH Issues Analysis

Analysis of the data obtained from the Plaxis program is completed qualitatively and quantitatively to obtain results in order to be in further analysis in submitting suggestions.

7. Stages of Effect of Landslide on Occupational Safety

Analyze the impact of a landslide on work safety using JSA method. The steps in preparing the JSA (Job Safety Analysis) include:

Determining Job Type

1. Jobs that have a history of occupational accidents are the most severe or often a top priority for analyzed safety. Factors to consider in determining the work to be analyzed are as follows;
   (a) Occupational accident rate,
   (b) The level of accidents that caused the defect,
   (c) Potential severity of work accidents,
   (d) New work,
   (e) Jobs that have a history of near misses.
2. Detailing the sequence/work steps from start to work until completion of work.
3. Identify the hazards and potential of work accidents on each work sequence performed.
4. Determine the control measures against the dangers of each work sequence performed.

8. Stages of Analyzing Hazards with Risk Analysis Methods

There are 3 ways in the risk assessment: Qualitative, Semiquantitative and Quantitative. These methods are in principle almost equal to qualitative analysis, the difference in this method description of the existing parameters expressed with a certain value/score.

(1) Risk analysis,
(2) Risk handling,
(3) Risk received.

9. Results and Discussion

9.1 Field observation results

Based on field observations and soil sampling on the slopes of excavated tested in the laboratory, it is found that the surface soil which is the slope of the excavation can be categorized as vulnerable to stability, the soil is quite hollow, less dense composed of yellowish light clay with cohesion values and low shear angle values respectively, c = 0.5 kN/m^2, = 6.2°, σ = 0.3 kN/m^2 and darkys black soil with cohesion values and the shear angle values are also low respectively c = 2.1 kN/m and = 8,8°, σ = 0.4 kN/m^2.

9.2 Results of Slope Stability Analysis

Based on the calculation of slope stability using Plaxis program on cross-section of excavation that has the height and slope of the slope that has been determined, the results presented by the Plaxis program is safe and meets the standards of the predetermined slope safety factor. Which is shown on natural conditions, while the conditions under which the modeling shows the condition of safe and unsafe conditions.

9.3 Results of Plaxis Analysis

Based on the calculation of slope stability using the Plaxis program, the results displayed by the Plaxis program are partly safe, and some do not meet the criteria of the safety factor. This can be seen in the table below.

Table 4. Plaxis calculation results

Modeling	**Safety Factor Varian θ**			**Safety Factor Varian C**		
	θ_{nt} +10%	θ_{nt}	θ_{nt} – 10%	C_{nt} +10%	C_{nt}	C_{nt} – 10%
αf_n + 10%	1,18	0,94	0,78	0,97	0,94	0,92
αf_n	1,42	1,03	0,82	1,04	1,03	1,02
αf_n – 10%	1,56	1,05	0,87	1,12	1,05	1,06

In table 4 shows the lower the degree of slope (α) then the Safety Factor (SF) is getting bigger. Based on the results of the stability slip analysis with Plaxis program on the natural conditions (αfn and θnt) indicates stable slope conditions because of SF (1.03) > 1, as shown in Figure 6 and Figure 7.

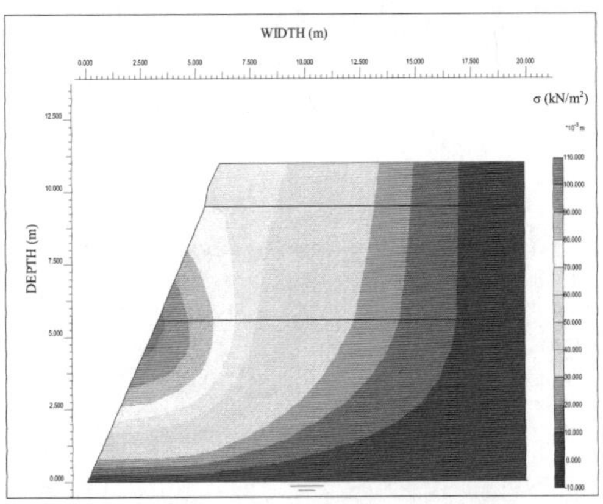

Figure 6. Plaxis Total Displacement Results Modeling of Cohesion Variants (αf_n) dan (θ_{nt})

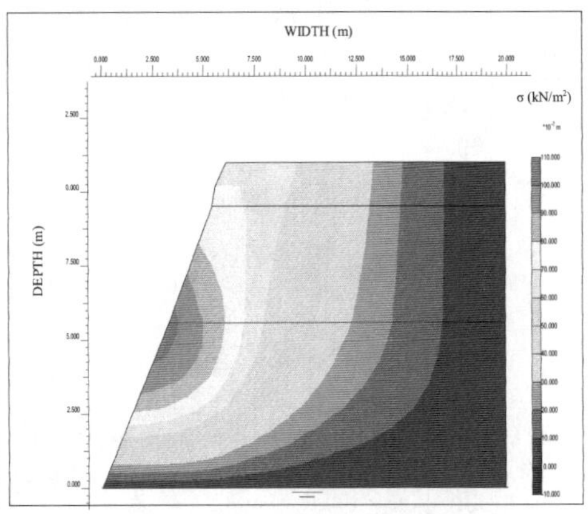

Figure 7. Plaxis Total Displacement Results Modeling of Cohesion Variants (αfn) and (C_{nt})

The geometry of slope stability analysis with Plaxis program under natural conditions (αfn and cnt) shows stable slope conditions because SF (1.03) > 1 is shown in figure 8.

In table 3. it can be seen that the highest security factor number reaches 1.56 at the time of modeling of shear variant which gains 10% (± 0 + 10%) and slope angle 54 degrees (αfn-10%). Plaxis results show safe conditions as shown in figure 9.

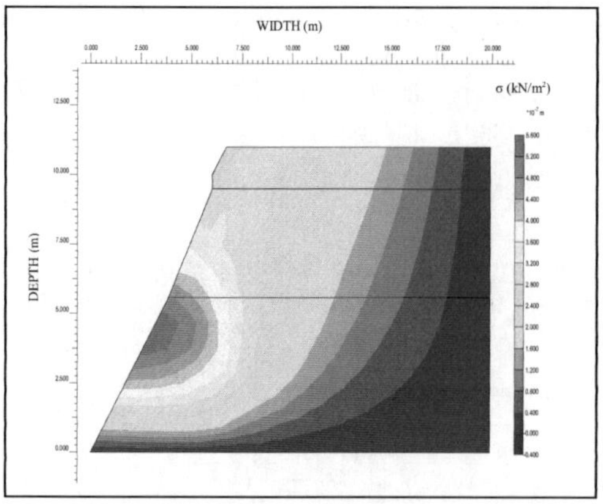

Figure 8. Results Plaxis Total Displacement Sliding Angle Variant (αf_n – 10%). (θ) and (θ_{nt} +10%).

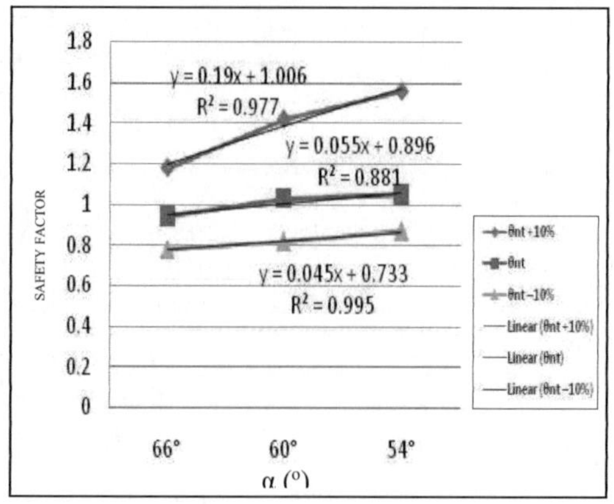

Figure 9. Plaxis Safety Factor Charts for Modeling of Angular Variants.

Table 3 shows that the relationship of variant of shear angle and cohesion to the value of security factor which can be seen in graphic form in figure 10.

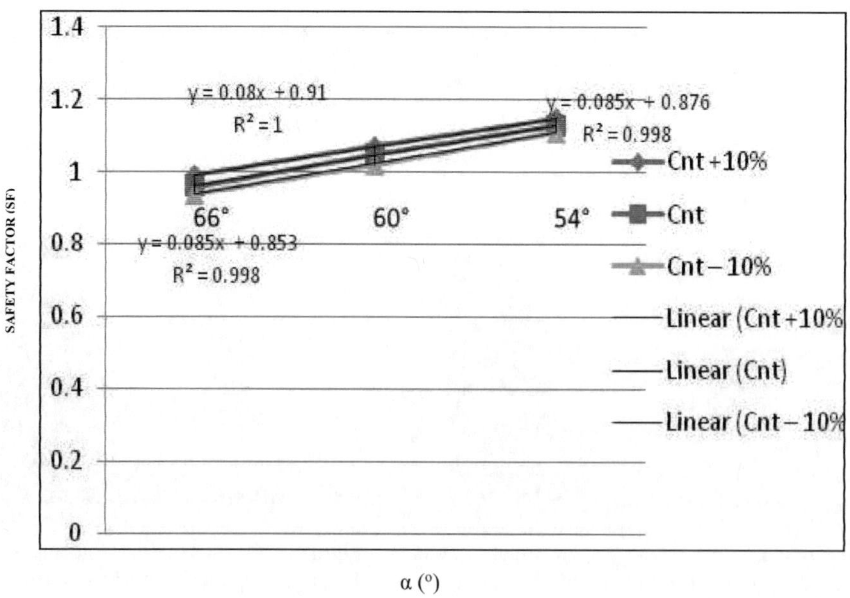

Figure: 10 Plaxis Safety Factor Chart For Modeling of Cohesion Variants (C kN/m^2)

Figure 10 shows the variant of the cohesion variant (c). The graph line depicts the lower angle of the slope, the higher the security factor. The safety factor values obtained in the cohesion variant are very close together, so that the graph line is visible on the image above as a thick line.

IOP Conf. Series: Materials Science and Engineering **352** (2018) 012035 doi:10.1088/1757-899X/352/1/012035

9.4. Analysis of K3 with Job Safety Analysis Method (JSA)

Analysis of K3 using JSA method in mine detected by landslide, the result obtained that there are some landslide point which have an effect on the safety can inhibit the process of taking coal even fatal. Based on JSA analysis result there is work sequence done where in the case of safety assessment on open mine slope is only analyzed on two items of work that is, Coal getting coal mining and Coal hauling coal hauling. Hazard potential is based on the calculation of slope stability using plaxis program on the quarry where some slopes do not meet established safety factor criteria.

9.5 Risk Analysis

Results of workplace risk analysis at Coal getting and Coal hauling stage. When clearing the soil mixture from the coal surface, the risk of a landslide pumped machine has a value of 3 acceptable risks, its countermeasures of the intensity that poses a reduced risk seminal. Then at the stage of coal scavenging and coal gathering there is a risk value of 750 and 450 so that enter in the category of very high-risk level, which means the activity is stopped until the risk can be reduced to reach the allowable or acceptable limits. The result of the risk analysis is shown in table 5.

Table 5. Risk Analysis Results

No	Job details	Risk	Risk Value	Level of risk	Action
	Coal Getting				
1	Cleaning impurities from coal surfaces.	The water pump machine is buried in landslides and coal mixed with soil.	3	*Acceptable*	The intensity that poses a risk is reduced to a minimum
2	Loose (hard coating)	Backhoe or hammer buried by landslide	750	*Very high*	Activity is stopped until the risk can be reduced to the limits allowed or accepted.
3	Collection / accumulation at the mine site for loading onto conveyance means	Backhoe or bulldozer slipped and rolled over	450	*Very high*	Activity is stopped until the risk can be reduced to the limits allowed or accepted.
	Coal Hauling				
1	Loading of coal to the conveyance & preparation of road access.	Dump trucks buried by landslides and dozer	450	*Very high*	Activity is stopped until the risk can be reduced to the limits allowed or accepted.
2	Transportation of coal from the mine site (pit) to the processing unit	Roads buried by landslides, cars overturned	50	*Priority 3*	It needs to be watched and watched on an ongoing basis.

Kramer [14], based on site visit survey there are no any hazard anticipation actions during the cut and hauling process. As a result, risk and level risk value from Table 5 cannot be reduced [14].

10. Conclusion

Based on the results of risk analysis on the open pit slopes using JSA method and calculate the stability of the slope using plaxis 8.2 software it can be explained some conclusions and suggestions as follow,

(1) Based on field observations and soil sampling on the slopes of excavated tested in the laboratory, it is found that the surface soil which is the slope of the excavation can be categorized as vulnerable to stability, the soil is quite hollow, less dense composed of yellowish light clay with cohesion values and low shear angle values respectively; $c = 0.5$ kN/m^2, $= 6.2^o$, $\sigma = 0.3$ kN/m^2 and darkys black soil with cohesion values and the shear angle values are also low respectively $c = 2.1$ kN/m^2 and $= 8{,}8^o$, $\sigma = 0.4$ kN/m^2.

(2) Analysis of slope stability at the excavation of coal mining quarry at the mine at Meurebo sub-district, West Aceh Regency, resulted in FS > 1 condition which means that the condition indicates the condition of the slopes is safe and there are also results showing FS < 1 which means the slopes are not safe.

(3) From unsafe conditions, the safety risk risk analysis using JSA method and risk analysis method, from the analysis of the coal extraction stage Coal getting concluded the existence of the acceptable small risk and there is a very high risk to safety, when cleaning the soil mixture Which is derived from the coal surface, the risk of pumping machine buried by landslide has a risk value of 3 acceptable, then at the stage of coal combustion and coal collection there is a risk value of 750 and 450 so that enter the category of very high-risk level, which means the activity is stopped until the risk Can be reduced to the limits allowed or accepted.

(4) At the Coal hauling coal hauling stage there is a risk value of 450 very high in coal loading activities to the transportation equipment & the preparation of road access, and in the transportation activities of coal from the mine site to the processing unit there is a risk value of 50 (priority) monitored and cared for on an ongoing basis.

Suggestion

The suggestions that the author exposes for the perfection of safety studies on the open slopes include the following:

- In the analysis of slope stability that has a very high-risk value it is necessary to do technical risk mitigation such as the installation of site pile, soil reinforcement, making the slopes more sloping, so that landslides that cause risks do not occur.

Acknowledgment

The authors would like to thank their respective universities, Geotechnics and Soil Mechanics Laboratory staff of Syiah Kuala University for their support in all possible ways. And also would like to extend a special thanks to the sponsored by Indonesian Ministry of Research and Technology and Syiah Kuala University Research Center for Professor Grant Scheme Year 2017 No: 1442/ UN11/SP/PNBP/2017 date 18 Mei 2017.

References

[1] Yang Z, Gao Q, Li M and Zhang G 2014 Stability analysis and design of open-pit mine slope in China: a review *Electronic Journal of Geotechnical Engineering* **19**

[2] Moh Z C 2004 Site Investigation and Geotechnical Failures *The International Conference on Structural and Foundation Failures, Singapore* 58-71

[3] Ramli 2010 *Practical Guide of of K3 OHS Risk Management* (Jakarta: Dian Agung)

[4] Imran I 2014 *Application Guide of Building Construction and Tsunami*

[5] Islam Md Rafiqul and Faruque M O 2012 Numerical Modeling of Slope Stability Consideration of an Open-Pit Coalmine in the Pulbari Coal Basin, NW Bangladesh *Electronic Journal of Geotechnical Engineering* **17** 3717-3729

IOP Conf. Series: Materials Science and Engineering **352** (2018) 012035 doi:10.1088/1757-899X/352/1/012035

[6] Bomont S and Mvirk O 2008 Back Experience from Two Cases of Stabilization of Coastal Landslides by Innovative Deep Drainage Systems *Proc. Of the 11th Baltic Sea Geotechnical Conference* **1** 19-26

[7] Stark T and Hussain M 2010 Drained Residual Strength for Landslides *GeoFlorida* 3217-3226

[8] Febrianti D 2015 *Slope Stability Analysis of Open-Pit Mining from Geotechnical Aspect and Work Safety* Thesis (Magister Infrastructure Engineering of Civil Engineering, Syiah Kuala University, Banda Aceh)

[9] Said A A 2012 *Technical Guide Analysis of Job Safety Analysis (JSA) in Identified of Dangerous of Work Area* (Bachelor Thesis, Syarif Hidayatullah Islamic University, Jakarta)

[10] Australian Standard/New Zealand Standard 4360:2004. 2004 *Risk management guidelines* (Sidney)

[11] Driad-Lebeau L, Lokmane N, Semblat J F, and Bonnet G 2009 Local amplification of deep mining induced vibrations Part 1: Experimental Evidence for Site Effects in a Coal Basin *Soil Dynamic and Earthquake Engineering* **29** 39-50

[12] Australian Standard/New Zealand Standard 4360:1999 *Risk management guidelines* (Sidney)

[13] Zulfan U M 2017 *Work Safety Analysis in Open-Pit by Using Plaxis Software* (Thesis Magister of Civil Engineering, Syiah Kuala University, Banda Aceh)

[14] Kramer S L 1996 *Geotechnical Earthquake Engineering* (USA: Prentice Hall)

IOP Conf. Series: Materials Science and Engineering **352** (2018) 012036 doi:10.1088/1757-899X/352/1/012036

Identifying inaccuracy of MS Project using system analysis

Fachrurrazi[1], Saiful Husin[1], Nurul Malahayati[1] and Irzaidi[2]
[1] Department of Civil Engineering, Syiah Kuala University, Jl. Teuku Nyak Arief, Darussalam, Kota Banda Aceh, Aceh 23111, Indonesia
[2] Department of Architecture Engineering, Syiah Kuala University, Jl. Teuku Nyak Arief, Darussalam, Kota Banda Aceh, Aceh 23111, Indonesia

E-mail: fachrurrazi@unsyiah.ac.id

Abstract. The problem encountered in project owner's financial accounting report is the difference in total project costs of MS Project to the Indonesian Standard (Standard Indonesia Standard / Cost Estimating Standard Book of Indonesia). It is one of the MS Project problems concerning to its cost accuracy, so cost data cannot be used in an integrated way for all project components. This study focuses on finding the causes of inaccuracy of the MS Projects. The aim of this study, which is operationally, are: (i) identifying cost analysis procedures for both current methods (SNI) and MS Project; (ii) identifying cost bias in each element of the cost analysis procedure; and (iii) analysing the cost differences (cost bias) in each element to identify what the cause of inaccuracies in MS Project toward SNI is. The method in this study is comparing for both the system analysis of MS Project and SNI. The results are: (i) MS Project system in Work of Resources element has limitation for two decimal digits only, have led to its inaccuracy. Where the Work of Resources (referred to as effort) in MS Project represents multiplication between the Quantities of Activities and Requirements of resources in SNI; (ii) MS Project and SNI have differences in the costing methods (the cost estimation methods), in which the SNI uses the Quantity-Based Costing (QBC), meanwhile MS Project uses the Time-Based Costing (TBC). Based on this research, we recommend to the contractors who use SNI should make an adjustment for Work of Resources in MS Project (with correction index) so that it can be used in an integrated way to the project owner's financial accounting system. Further research will conduct for improvement the MS Project as an integrated tool toward all part of the project participant.

1. Introduction

Contractors, as implementers in the project, should be supported with software applications to improve project implementation more effectively. All various management software, whether they have been available at the market for structured decision making, such as Winest [1], MS Project [2], Solver [3], Primavera [4], or even they have been developed by researchers for unstructured decisions for special purposes, such as Neural Network [5, 6], Fuzzy Logic [7], Expert Systems [8], demand a high degree of accuracy. This inaccurate problem of software must be solved immediately so that the use of the software will maximally support the performance and productivity of construction project in Aceh as maximum as possible.

One of the problems encountered, regarding the accuracy of using the MS Project software, is the difference in total project costs. This difference is seen in the inequalities between the nominal of the project cost plan and performance cost of the project (Realization). The project owner, as a responsible party for paying the contractor's work, does not tolerate this difference, especially in the project owner's financial accounting system. Data accuracy increases confidence in any decision-making decision [9]. Learning and knowing how to operate MS Project, as a tool to improve work productivity in the context of project management is not enough. The contractors should also consider whether the software would provide sufficient accuracy of results of an analytical tool for decision making in a project. The inaccuracy of software output can lead to the contractor's decision-making process to be hampered and even improper. This administrative problem, in its turn, will cause a delay in project completion. This is a case happened in Aceh for those who use MS Project as a tool for project management. This condition illustrates that on the one hand, the use of project management software will improve project management productivity and, on the other hand, its inaccuracies will lead to project administration matters.

The main purpose of this study is to find the cause of the inaccuracy of cost analysis in MS Project against standard cost analysis in Indonesia (referred to as SNI / Standard Nasional Indonesia). Several studies have been conducted such as [10, 11, 12, 13]. The studies are oriented towards the implementation of the MS Project in various object reviews. This research focuses on the accuracy analysis of MS Project. The method of System analysis will be used in this research. This method can also be applied to analyse inaccuracy in other software. This method will immediately find the cause in detail of the components that build the total cost [14]. This study will describe how we can find the cause of inaccuracy of MS Project and in what system elements of the cost estimating process it occurs. To get the correct answer, the MS Project cost and it analysis procedure will be compared to the current one as a manual cost analysis (using Indonesian National Standard). Operationally, the aims of this study are (i) identifying cost analysis procedures for both current methods (SNI) and MS Project; (ii) identifying cost bias in each element of the cost analysis procedure; And (iii) analysing the cost differences (cost bias) in each element to identify what is the cause of inaccuracies in MS Project toward SNI is.

This research has been able to identify the causes of differences in total project costs between MS Project and SNI. This difference is caused by the decimal limitation on the element of the Work of resources ($W_{i,j}$) in MS Project. The limitation on the number of decimals in this element (Work of resources/$W_{i,j}$) has been truncated to be two decimal digits. These truncation has caused a bias in the total cost using MS Project against SNI. As a consequence is the more resources used in a project, the more accumulating cost bias occurred. The result of identification of the cost bias problem in MS Project caused by truncation into two decimal digits in Work of Resources ($W_{i,j}$), encourages the contractor to adjust the cost (correction index) so that the MS Project will be used in an integrated way in the project owner's financial accounting system. It is a future research for MS Project improvement as an integrated tool for all of the participant project. In addition, there is a fundamental difference for the cost estimation procedure in MS Project against SNI, i.e., MS Project does not use the components of the Requirement of Resources ($R_{i,j}$) and Quantity of project activities (Q_i). Both components have blended in the Work of Resources ($W_{i,j}$). Improving the utilization and accuracy of MS Projects can be done by implementing the Requirement of Resources ($R_{i,j}$) and Quantity of project activities (Q_i) in the MS Project system.

2. Methods

This study was conducted with the intention to identify and to review the accuracy of the MS Project to support the performance in managing projects in Aceh. For this purpose, The framework has been built in this research, as shown in Figure 1. The stages and method used in the study are described operatively from the objectives of this study.

Method to identify the costing procedure for both current methods (SNI) and MS Project uses the system analysis approach, as shown in Figure 2. System analysis is to understand the complete composition of a system or procedures in order to be able identifying and evaluating various problems or obstacles that arise in that system [15]. System analysis could also be used to combine two or more systems into a structured system as well or evaluating problems that have been identified before. Describing the system analysis uses the data flow approach and showing how data moves through the system is, it will have four advantages:

1. It could further understanding of the interrelatedness of systems and subsystems
2. It could communicate the system knowledge to users through data flow diagrams
3. It could be to determine if the necessary data and processes have been defined.

Method to identify the bias in specific elements of the costing procedure is done using a unit cost case of the project activity, where each element of the unit cost of the activity in MS Project will be compared with the cost elements of the SNI. The elements and its connections in the cost system must be indicating their relationship to the cost theory of the body of knowledge, as detail description in the project cost structure [16]. This bias identification is started from the bottom element of the bottom-up procedure for the activity cost, as shown in figure 2. The mathematical equations of each cost element will be used to analyse for both SNI and MS Project. The abbreviation of the mathematical model is as shown in Table 1.

Table 1. Abbreviations

Notation	Description	Notation	Description
i	Number of Activities	OV_i	Overhead of Unit Price
j	Number of Element of each Activity	FC_i	Fixed Cost of each Activity
Q_i	Quantity of Activities	UP_i	Unit Price of each Activity
$S_{i,j}$	Standard Rate of Resources	Qi	Quantity of each Activity
$R_{i,j}$	Requirement of Resources	Cp	Cost of Project
$W_{i,j}$	Work of Resources of each Activity	CA_i	Cost of each Activity
DC_i	Direct Cost	Cp	Project Cost

The case used to identify the MS Project inaccuracy is the unit cost (unit price) of foundation work of a building as an object focus. It makes easier to detect cost differences [17]. The difference will be seen more clearly when uses the real cost data of the project activity in the costing procedure. A case of project activity, the foundation work activity, will be used to describe the bias of MS Project.

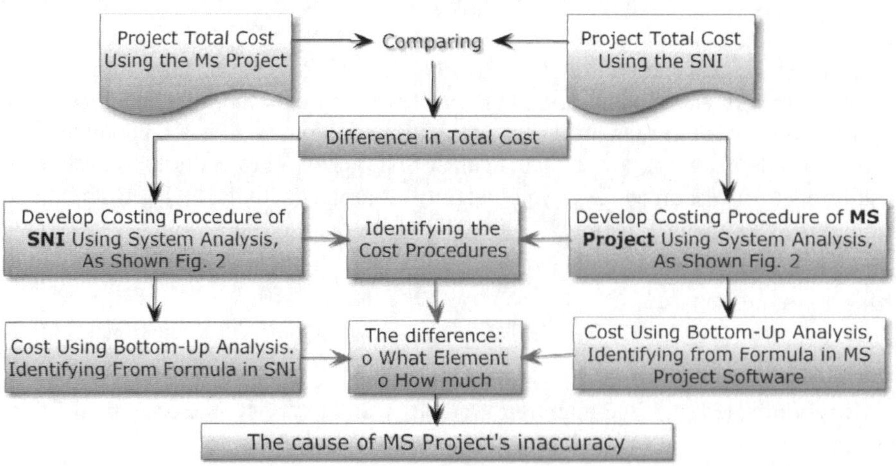

Figure 1. Framework for identifying the project cost accuracy on MS Project toward SNI

3. Result

3.1 Identifying the costing procedure

At this stage, we will identify the costing procedure for both of MS Project and SNI as the current method of project cost analysis. The procedure can be described as a system analysis of inputs, processes, and outputs. System analysis is the decomposition of a complete information system into its parts to identify and evaluate the problems, opportunities, constraints and expected needs so that the proposed improvements could be achieved, as shown in figure 2.

Based on figure 2, the MS Project cost system built on input, process, and output is based on Activity Based Cost (ABC) [18]. Process input is performed at the lowest level of WBS (Work Breakdown Structure) of the project. ABC method is a process to generate project cost from data input done at the activity level. The MS Project input data is on the Work of Resource element($W_{i,j}$), Standard Rate ($S_{i,j}$), both of which are used to generate Direct Cost and Fixed Cost (FC_i). While direct cost is sum product of $W_{i,j}$ and $S_{i,j}$. Direct Cost (DC_i) and Fixed Cost (FC_i) are to generate Activity Cost (CA_i). The Project Cost (Cp) can be generated by cumulative of all of the activity cost. This cost planning analysis method is referred to as bottom-up analysis.

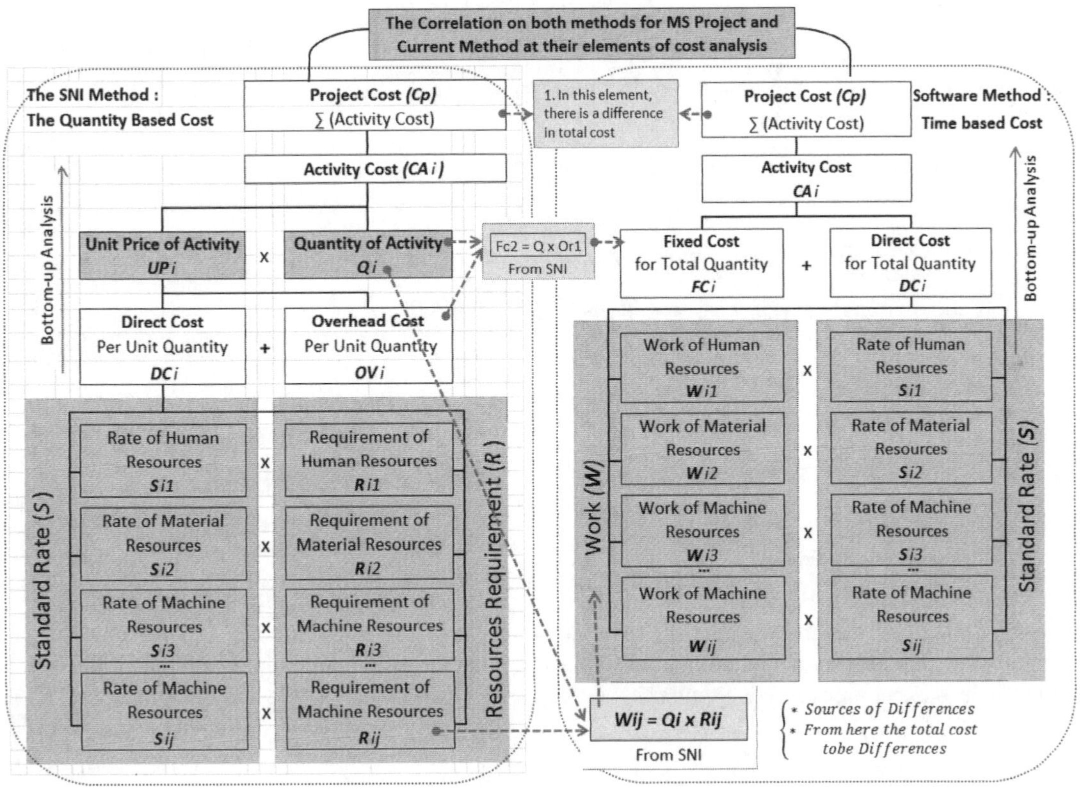

Figure 2. System Analysis for SNI and MS Project

MS Project input data is on the element Requirement of Resource ($R_{i,j}$), Standard Rate ($S_{i,j}$) both of which are used to generate Direct Unit Cost (DC_i) and Overhead Unit Cost (OV_i). Direct Unit Cost (DC_i) and Overhead Unit Cost (OV_i) are to generate Activity Unit Cost or Unit Price of each Activity (UP_i). Cost of each Activity (CA_i) is a multiplication of the Unit Price (UP_i) and it Quantity (Q_i). The result of Project Cost (Cp) is cumulative cost of each project activity ($\sum CA_i$) is cumulative of each activity. This cost planning analysis method is referred to as bottom-up analysis, as shown in figure 2.

3.2 Identifying cost bias in the specific elements of MS Project

Identifying cost bias in the specific element of MS Project can be done by comparing mathematical models of cost analysis to MS Project and SNI, as shown in Formula (15).

3.2.1 Costing Procedure of MS Project

The mathematical model for analysing the total project cost (Cp), which is based on bottom-up analysis, can be explained as follows:

For activity-i in project

$$Cp = \sum_{i=1}^{n} CA_i \quad (1)$$

$$CA_i = (DC_i + FC_i) \quad (2)$$

$$DC_i = \sum_{j=1}^{m} (W_{i,j} x S_{i,j}) \quad (3)$$

$$CA_i = \sum_{j=1}^{m} (W_{i,j} x S_{i,j}) + FC_i \quad (4)$$

For all activity in project

$$Cp = \sum_{i=1}^{n} \sum_{j=1}^{m} (W_{i,j} x S_{i,j}) + \sum_{i=1}^{n} FC_i \quad (5)$$

3.2.2 Costing Procedure of SNI

The mathematical model for analysing the total project cost (Cp), which is based on bottom-up analysis, can be explained as follows:

For activity-i

$$Cp = \sum_{i=1}^{n} CA_i \quad (6)$$

$$CA_i = Q_i x Up_i \quad (7)$$

$$For\ Up_i = \sum_{j=1}^{m} (R_{ij} x S_{ij}) + OV_i \quad (8)$$

$$CA_i = Q_i x (\sum_{j=1}^{m} (R_{ij} x S_{ij}) + OV_i) \quad (9)$$

$$CA_i = Q_i x \sum_{j=1}^{m} (R_{ij} x S_{ij}) + (Q_i x\ OV_i) \quad (10)$$

$$CA_i = \sum_{j=1}^{m} (Q_i x R_{ij} x S_{ij}) + (Q_i x\ OV_i) \quad (11)$$

$$Cp = \sum_{i=1}^{n} CA_i \quad (12)$$

For all activity in the project are:

$$Cp = \sum_{i=1}^{n} \sum_{j=1}^{m} (Q_i x R_{ij} x S_{ij}) + (Q_i x\ OV_i) \quad (13)$$

Based on the fact that there is a cost equivalent of the project activity on MS Project and SNI, the formula (4) of MS Project can be compared with formula (11) of SNI. The input element of the SNI is Requirement of Resources ($R_{i,j}$), Standard Rate ($S_{i,j}$), Overhead Cost (OV_i), and Quantity (Q_i).While the input of MS Project is Work of Resource ($W_{i,j}$), Standard Rate ($S_{i,j}$) and Fixed Cost (FC_i). In general, the similarity of both models for a single activity will be accurate if the Activity cost of the current method should be equal to an activity cost of MS Project.

$$(CA_i)\ of\ Current\ Methods = (CA_i)\ of\ MS\ Project \quad (14)$$

$$\sum_{j=1}^{m} (Q_i x R_{ij} x S_{ij}) + (Q_i x\ OV_i) = \sum_{j=1}^{m} (W_{i,j} x S_{i,j}) + FC_i \quad (15)$$

Based on the comparison of both methods it is seen that there are two components of cost procedure that are not analysed directly by MS Project, as shown in figure 2. Based formula (15), the element of Work of Resources ($W_{i,j}$) and Fixed Cost (FC_i) of MS Project defined as:

$$W_{i,j} = \sum_{j=1}^{m} (Q_i x R_{ij}) \quad (16)$$

This formula is also applied to Fixed Cost (FC_i), as shown in figure 2. In mathematically, as follows:

$$Fc_i = Q_i * OV_i \tag{17}$$

3.3 Identifying inaccuracy of MS Project

The data presented in Table 2, Table 3, and Table 4 are the results of cost estimating by using the unit price analysis for the foundation work using SNI. Table 2 illustrates resource assignment and Table 3 illustrates the calculation the unit price analysis at 1 M^3 of Foundation Activity. Table 4 illustrates bill of quantity of activity that is multiplying by Unit price and its quantities.

Table 2. Resource sheet as standard rate of resource

No	Resources name	Unit	Price per unit (in IDR)
1	Worker	Man days	60,000
2	Bricklayer	Man days	85,000
3	Head of Workers	Man days	110,000
4	Foreman	Man days	75,000
5	Mounting stone	M^3	165,000
6	Portland Cement	Kg	1,200
7	Sand	M^3	160

Table 3. Unit price analysis of project activities using the SNI standard

Unit Price Analysis of the Foundation Work						
Installing 1 M3 of *the Foundation Work*, with the mortar of 1 Portland Cement: 4 sand (Code: SNI DT:91:0007:2007-6.2)						
Components of the Activity	**Requirements**	**Unit**		**Standard Rate**		**Total**
I Materials						
Mounting stone	1.200	M3	x	165,000	=	198,000.00
Portland Cement	202.000	Kg	x	1,200	=	242,400.00
Sand	0.485	M3	x	160	=	77.60
		Subtotal of material			=	440,477.60
II Human Resources						
Worker	1.500	Man/days	x	60,000	=	90,000.00
Bricklayer	0.750	Man/days	x	85,000	=	63,750.00
Head of Workers	0.075	Man/days	x	110,000	=	8,250.00
Foreman	0.075	Man/days	x	75,000	=	5,625.00
		Subtotal of resources			=	167,625.00
Grand Total of unit price of Foundation work (1) + (2)					=	608,102.60
Overhead (15%)					=	91,215.39
Grand Total include overhead					=	699,317.99

Table 4. Estimate Sheet of BOQ (Bill of Quantity) of the Project in the SNI standard system

WBS	Activity	Quantity	Unit	Standard	Unit price	Total
1	The Foundation Work	1	M3	SNI (2013) A.3.2.1.1	699,317.99	699,317.99

The comparative analysis of the nominal cost of the activity between SNI and MS Project is as shown in Table 5. Table 5 shows the difference between total cost of the activity of 925.80 or 0.13% (925.80/699,317.99), which is the difference between total cost of MS Project (1,626,900) and SNI (1,627,340). It shows the value of MS Project bigger than SNI.

Table 5. Cost Comparison for the Foundation work (1 m^3) using MS Project and SNI

Resources	Unit	Requirement	Standard Rate	Unit Price of SNI	MS Project		Differences between MS Project and SNI
	Q	R	S		Work W	Total Ct_2	
(1)	(2)	(3)	(4)	(5)=(3)x(4)	(6)=(2)x(3)	(7)=(4)x(6)	(8)=(7)-(5)
Foundation Work	1 M^3						
Materials							
Mounting stone	M^3	1.200	165,000	198,000.00	1.20	198,000.00	-
Portland Cement	Kg	202.000	1,200	242,400.00	202.00	242,400.00	-
Sand	M^3	0.485	160	77.60	0.49	78.40	0.80
Human Resources							
Worker	Man/days	1.500	60,000	90,000.00	1.50	90,000.00	-
Bricklayer	Man/days	0.750	85,000	63,750.00	0.75	63,750.00	-
Head of Workers	Man/days	0.075	110,000	8,250.00	0.08	8,800.00	550.00
Foreman	Man/days	0.075	75,000	5,625.00	0.08	6,000.00	375.00
			Total	608,102.60		609,028.40	
			Fixed cost (overhead 15%)	91,215.39		91,215.39	
			Unit Price *(Up)* using SNI	699,317.99	Using MSP	700,243.79	
Total Foundation work with quantity 1 m3				(700,243.79 - 699,317.99):			925.80

Based on this identification, the cause of the differences in total cost is on the truncation decimal of the work of resource ($W_{i,j}$) component of the resources of the Foreman of MS Project. The value should be 0.075 but rounded up to 0.08. The rounding causes the difference in cost to 0.027%.

$$W_{i,j} = Q_i * R_{i,j} = 1 * 0.075 = 0.075 \sim 0.08 \quad (18)$$

While on other elements such as Mountain stone, Portland cement, sand, worker, and Bricklayer is no difference in results, because decimal digits on the element do not exceed two digits. Based on these conditions we have concluded that in MS Project the existence of rounding in a decimal digit of work of resources. Rounding in decimal digits of work of resources may occur in other activities of the project, so the cost may be greater or less than that value, depending on rounding up or down, resulting in the +/- condition in every activity. The cumulative can vary from one project to another project, which is caused by the difference of errors in each project activity and the number of activities of each project varies [19].

4. Discussion

4.1 Comparing the MS Project and SNI

The concept of cost plan for both SNI and MS Project is the same, i.e., using activity-based cost (ABC) concepts that using bottom-up analysis. Principles of cost analysis in MS Project are consistent and applicable in SNI platform, but there are some differences from the data input method. MS Project does not use the cost estimating unit price (unit cost) method, which is built from direct costs and overhead costs. The current method uses the cost estimating unit price (unit cost) method, which is constructed from direct cost units and overhead unit costs. This condition describes MS Project system can only be applied partially to SNI based system. The main difference lies in the use of unit price (unit cost) of SNI and it is not in MS Project. This will make difficult for MS Project that regard to the project implementation in the unit price contract type, it is in line with [20]. The unit price contract type uses the quantity as a basis for contract payment. The unit price will be used as the basis for determining the unit price of quantity performed. Therefore, quantity can be regarded as a basic estimate for project cost estimating. This is a cost plan analysis at an early stage and should be the basis during the implementation phase, as a benchmark of cost control and schedule. For MS Project implementation to be fully implemented, development of MS Project must also be used to analyse the unit price. This requires integration of Quantity of Activity (Q_i) and Requirement of Resource ($R_{i,j}$)

on MS Project system. Furthermore, it will also improve the accuracy of Work of Resource ($W_{i,j}$) calculation on MS Project, as formula (16).

Based on the comparison of both methods it is seen that there are two components/elements of the software that are not analysed directly by MS Project, but analysed separately manner (Beyond the MS Project system). These two components are Quantity of Activity (Q_i) and Requirement of Resource ($R_{i,j}$). This is one of the difficulties in implementing of MS Project. The formula that is used to calculate the Work of Resources ($W_{i,j}$) is the formula (16). The data of Quantity (Q_i) of the project are very important for measuring the progress of construction projects, especially for engineers. The engineer prefers to see the progress of a project that based on the Quantity (Q_i). They are compared to measure the amount of the actual cost. Besides that, both baseline and actual quantity are required to the owner as a validation for the accuracy of the designer consultant jobs. For the contractor, quantity is also needed on determining payment job order for groups of workers, especially for subcontract, as a basis to determine payment.

The Requirement of resources ($R_{i,j}$) is also used as a basis for the baseline of productivity. The Productivity [21] is used as a performance for the workers on the project. This productivity performance is the most important data for the contractor in the next projects. In addition, the contractor will use the requirement as an index to determine the productivity level of the resources plan to be compared with the resource performance. For the further, it will be used to measure the project performance, whether it was about what if analysis for the schedule and cost. Under these conditions, MS Project needs to be developed further, with respect to the Requirement of Resources ($R_{i,j}$). Based on reviews these discussions, in the future research of MS Project needs more developed for both item of Quantity of Activity (Q_i) and the Requirement of resources ($R_{i,j}$), especially for use on the projects in Indonesia that using the SNI platform.

4.2. Root cause the difference for both of MS Project and Current Method

The difference in total cost of the project in case one item of work is estimated at 925.80 (in IDR) or as much as 0.13% (925.80 /699,317.99), as shown in Table 5. This difference is due to the decimal truncation in resources assignment of sand material, head of workers and foreman item, as shown in Table 5, and formula (18). The rounding is 0.23 where it should be 0.225. The nominal value of this difference may vary from an activity to other activities, depending on the number of items and items that have been made. The maximum value that can be achieved from each project activity is ± 0.005 (± 0.5%). This value represents the midpoint value for truncation, the roundup and rounds down. This occurs because of the input data on the work of resource ($W_{i,j}$) are limited only the two decimal places. Under these conditions, the MS Project needs to be developed any further, with respect to the option of decimal places [22]. It is not restricted to a maximum of two digits only. Identification of the root causes the differences of the total cost has been made by comparing of the cost calculation method (cost analysis method) is done by comparing the real cost of single activity of MS Project to SNI. The limitation in a decimal digit of work of resources ($W_{i,j}$) of MS Project is two decimal digit maximum. It is a deficiency of MS Project and it must be resolved to increase the accuracy of the software.

5. Conclusion

Implementing the MS Project in Indonesia, especially for contractors in construction work, have some obstacles. This research has been able to identify the obstacle, namely: The two-digit decimal limitation on elements of Work of Resources in MS Project is the root cause of inaccuracy in the cost analysis of the MS Project. This condition causes MS Project cannot be used in an integrated manner in the project system, especially on the project owner's financial accounting system, which requires no difference in the total cost of the project cost plan and its payment realization. For optimal accuracy performance in Indonesia practice, it needs to do further improvement in the next research.

The contractor's job system in Indonesia using the Quantity-Based Costing (QBC) system, where Quantity of Activity (Q_i), and Requirements of Resources ($R_{i,j}$) is a major component of the analysis of unit price. The contrary, MS Project uses Time-Based Costing (TBC), where Work of Resources ($W_{i,j}$) becomes the main component. This study has identified that there is a relationship between Work of Resources ($W_{i,j}$) in MS Project is same as by multiplying both Quantities of Activity (Q_i), and Requirements of Resource ($R_{i,j}$) in SNI. For MS Project to adapt the SNI system, the MS Project needs to include both of Quantity (Q_i), and Requirements of Resource ($R_{i,j}$) into the operating system cost calculation in MS Project. The development will be carried out in the follow-up study.

System analysis with bottom-up technique has been able to identify the inaccuracy of MS Project to SNI. Thus, the method is also applied to identify on other software in the same cases. In the analysis of the system is needed an understanding of the interrelatedness of systems and subsystems, data flow diagrams. Thus, this research method can also be used to combine two or more systems into a structured system as well, by first having evaluated the problems that have been identified before.

Acknowledgment

The authors would like to thank and appreciate to parties involving in this research. Special thanks to Prof. Dr. Ir. Samsul Rizal, M.Eng., Rector of Syiah Kuala University and Dr. Ir. Taufiq Saidi, M.Eng., Dean of Engineering Faculty-Unsyiah for their contribution through “Hibah Laboratorium” Program based on letter number 2348/UN11/TU/2017; Construction Engineering Management (CEM) Laboratory - Civil Engineering of Syiah Kuala University.

References

[1] Niknam M and Karshenas S 2015 Integrating distributed sources of information for construction cost estimating using Semantic Web and Semantic Web Service technologies *Autom Constr.* **57** 222–38

[2] Fritz W 1992 Using the project management tool Microsoft Project for planning and controlling software development projects *Annu. Review in Autom Program* **16** 35–40

[3] Fachrurrazi F, Away Y, and Husin S 2014 The weights detection of Multi-criteria by using Solver *International Journal of Electrical and Computer Engineering* **7** 858-68

[4] Lester E I A 2014 Primavera P6 *Proj Manag Plan Control* 485–502

[5] Fachrurrazi, Husin S, Tripoli, and Mubarak 2017 Neural Network for the standard unit price of the building area *Procedia Engineering* **171** 282–93

[6] Fachrurrazi, Husin S, Munirwansyah and Husaini 2017 The subcontractor selection practice using ANN-Multilayer *Int. J. Technol.* **8** 761

[7] Aruna L and Aramudhan M 2016 Framework for ranking service providers of federated cloud architecture using Fuzzy sets. *Int. J. Technol.* **7** 643–53

[8] Hugeng H and Kurniawan R 2016 Development of the “healthcor” system as a cardiac disorders symptoms detector using an expert system based on Arduino Uno *Int. J. Technol.* **7** 78–87

[9] Keasey K and Watson R 1989 Consensus and accuracy in accounting studies of decision-making: A note on a new measure of consensus *Accounting, Organ Soc.* **14** 337–45

[10] Hebert J E and Deckro R F. 2011 Combining contemporary and traditional project management tools to resolve a project scheduling problem *Computers and Operations Research.* **38** 21–32

[11] Maroto C and Tormos P 1994 Project Management: An Evaluation of Software Quality *Int Trans Oper Res.* **1** 209

[12] Chen S M, Griffis F H, Chen P H, and Chang L M 2012 Simulation and analytical techniques for construction resource planning and scheduling *Autom Constr.* **21** 99–113

[13] Withall S 2007 Software requirement patterns (best practices) *International Journal of Project Management* **16** 384

[14] Al-Omari Z, Hamzeh A, Hamed SA, Sandouk A and Aldahim G 2015 A mathematical model for minimizing Add-On operational cost in electrical power systems using the design of

experiments approach. *International Journal of Electrical and Computer Engineering (IJECE)* 948–56

[15] Gibson J E, Scherer W T, Gibson W F 2007 *How to Do Systems Analysis* Hoboken NJ (USA: John Wiley & Sons, Inc.) 365

[16] Rose K H 2013 A guide to the project management body of knowledge (PMBOK® Guide)—Fifth Edition *Project management journal* **44**

[17] Al-Jibouri S H 2003 Monitoring systems and their effectiveness for project cost control in construction *Int. J. Proj. Manag.* **21** 145–54

[18] Foster G and Swenson DW 1997 Measuring the success of activity-based cost management and its determinants *Journal of management accounting research* **9** 109

[19] Fachrurrazi 2018 The End User Requirement for Project Management Software Accuracy *International Journal of Electrical and Computer Engineering (IJECE)* **8**.

[20] Llewellyn K N 1931 What price contract?. An essay in perspective *The Yale Law Journal* **40** 704-51

[21] Kaming P F, Olomolaiye P O, Holt G D and Harris F C 1997 Factors influencing craftsmen's productivity in Indonesia *Int. J. Proj. Manag.* **15** 21–30

[22] Resnick L B, Nesher P, Leonard F, Magone M, Omanson S, Peled I 1989 Conceptual bases of arithmetic errors: The case of decimal fractions *J. Res. Math. Educ.* **20** 8–27

IOP Conf. Series: Materials Science and Engineering **352** (2018) 012037 doi:10.1088/1757-899X/352/1/012037

Compression member response of steel angle on truss structure with variation of single and double sections

Arief Panjaitan, Purwandy Hasibuan
Civil Engineering Department, Faculty of Engineering, Syiah Kuala University, Jalan Tengku Syech Abdurrauf No. 7, Darussalam, Banda Aceh, Indonesia

E-mail: purwandy.hsb@unsyiah.ac.id

Abstract. Implementation of an axial compression load on the steel angle can be found at the various structure such as truss system on telecommunication tower. For telecommunication tower, steel angle section can be suggested as an alternative solution due to its assembling easiness as well as its strength. But, antennas and microwaves installation that keep increases every time on this structure demand reinforcement on each leg of the tower structure. One solution suggested is reinforcement with increasing areas section capacity, where tower leg consisted of single angle section will be reinforced to be double angle section. Regarding this case, this research discussed the behavior of two types of steel angle section: single angle of L.30.30.3 and double angles of 2L.30.30.3. These two sections were designed identically in length (103 cm) and tested by axial compression load. At the first step, compression member together with tension member was formed to be a truss system, where compression and tension member were met at a joint plate. Schematic loading was implemented by giving tension loading on the joint plate until failure of specimens. Experimental work findings showed that implementing double angle sections (103 cm) significantly increased compression capacity of steel angle section up to 118 %.

1. Introduction

Steel angles are members that commonly used in various structures especially truss systems such as truss system on bridge structures and telecommunication tower. For telecommunication tower, the needs of highness of the structure are needed to ensure telephone, radio, and television signal can be distributed well. Implementation of steel angle on telecommunication tower can be suggested as an alternative solution because of its assembling easiness as well as its strength. The quantity of antennas and microwave that keep increases during the time causes loading accommodated by tower feet (legs) also grows up. The common reinforcement method is by increasing area section capacity where tower feet consisted of single angle section will be reinforced to be double angle sections.

This research emphasized to discuss the behavior of two types of steel angle section: single angle section of L.30.30.3 and double angle sections of 2L.30.30.3. These two sections were designed identically in length (103 cm) and tested by axial compression load. Based on the experimental work, it can be observed member responses started from elastic condition until their failure.

Assembling process of truss system was started with forming tensile component on diagonal member and compression component on horizontal member. Both of these members were joined at the gusset plate as a plate that was implemented as loading location. Loading step was started by giving tensile load gradually through loading plate until horizontal member (compression component) achieved its failure. Observation record was carried out due to tensile force on loading plate and buckling deformation on horizontal member.

They were automatically collected and notified through experimental equipment such as tension load cell, data logger, and transducers. Based on these data, the response of two specimens can be specifically described and discussed in detail.

2. Specimen Preparation

2.1 Experimental Work Setting Up

This research implemented two types of steel member (diagonal and horizontal member), and they were formed to be a truss system. Both of these members were fastened with bolts on their supporting plates where these supporting plates were connected with end plates through the welded connection. End plates then were fastened to loading frame with the bolted connection. The tensile force that was worked on the gravitational direction and applied on the joint plate (joint plate of the horizontal and diagonal member) caused the diagonal member to be tensile component and horizontal member to be compression component. This experimental work was emphasized on horizontal member that accommodated compression force.

Load recording was carried out through *tension load cell and data logger*. Meanwhile, deformation recording was conducted through transducers placed on flange and web of steel angle either for single angle and double angle

2.2 Material Characteristics Determination

Determination of material characteristics is essential to be carried out because it will give information due to the compression member. For obtaining steel characteristics, coupon test was conducted in steel angle and give the result as yield stress (F_y) by 224 Mpa and ultimate stress (F_u) by 324 Mpa.

2.3 Transducers Placement Position

There were 4 transducers placed on different position. Transducer 1 was placed on the end plate of horizontal member and transducer 2 was placed on the end plate of the diagonal member. Placement of these transducers was aimed to observe possibility of buckling on end plate due to tensile load implementation on the joint plate. Transducer 3 was located on steel angle web. Meanwhile, transducer 4 was located on steel angle flange. Both of these transducers were located to measure compression member deformation. Regarding the slenderness of web and flange of steel angle, the failure modes predicted are local buckling and lateral- torsional-buckling.

2.4 Experimental Work Activities

Experimental work was carried out with giving tensile force at gusset plate (joint plate of the horizontal and diagonal member). This loading schematic was carried out by fastening iron ring on the threaded cylindrical bar so it will produce a tensile force on tension load cell that was continued to sling rope. This sling rope was clamped so it can distribute tensile force to the gusset plate. This mechanism caused tensile force worked on the diagonal member, and compression force worked on horizontal member.

The observation was emphasized on compression component (horizontal member). Tensile member had to be in elastic condition so it can ensure the failure would occur at compression member at first.

Data recording was carried out through data logger. These data were tensile force and deformation and they have collected automatically during the fastening of iron ring. This recording gave specimens responses started from elastic, yield, until failure condition.

2.5 Observation Planning

Observation planning during experimental work was emphasized on :

a. Increasing pattern of compression load due to deformation for double steel angles for both of single angle and double angles
b. Increasing mode of deformation in web and flange of steel angle for both of single angle and double angles.

c. Failure mode (buckling) for both of specimens.

3. Literature Study

One research with experimental work approach that discussed eccentricity effects on the equal-leg single angle was carried out by Liu (2007). Experimental work was conducted to single angle of L51.51.6,4 with variation of length members (900, 1200, and 1500 mm) and eccentricity distance in x and y-direction (e_x dan e_y). Experimental work was conducted by Hydraulic Universal Testing Machine to each specimen with giving compression force gradually (with increment of 0,5 kN). The test results showed that the greater eccentricity implemented on the specimen, it would give significant degradation in the compression strength of specimen.

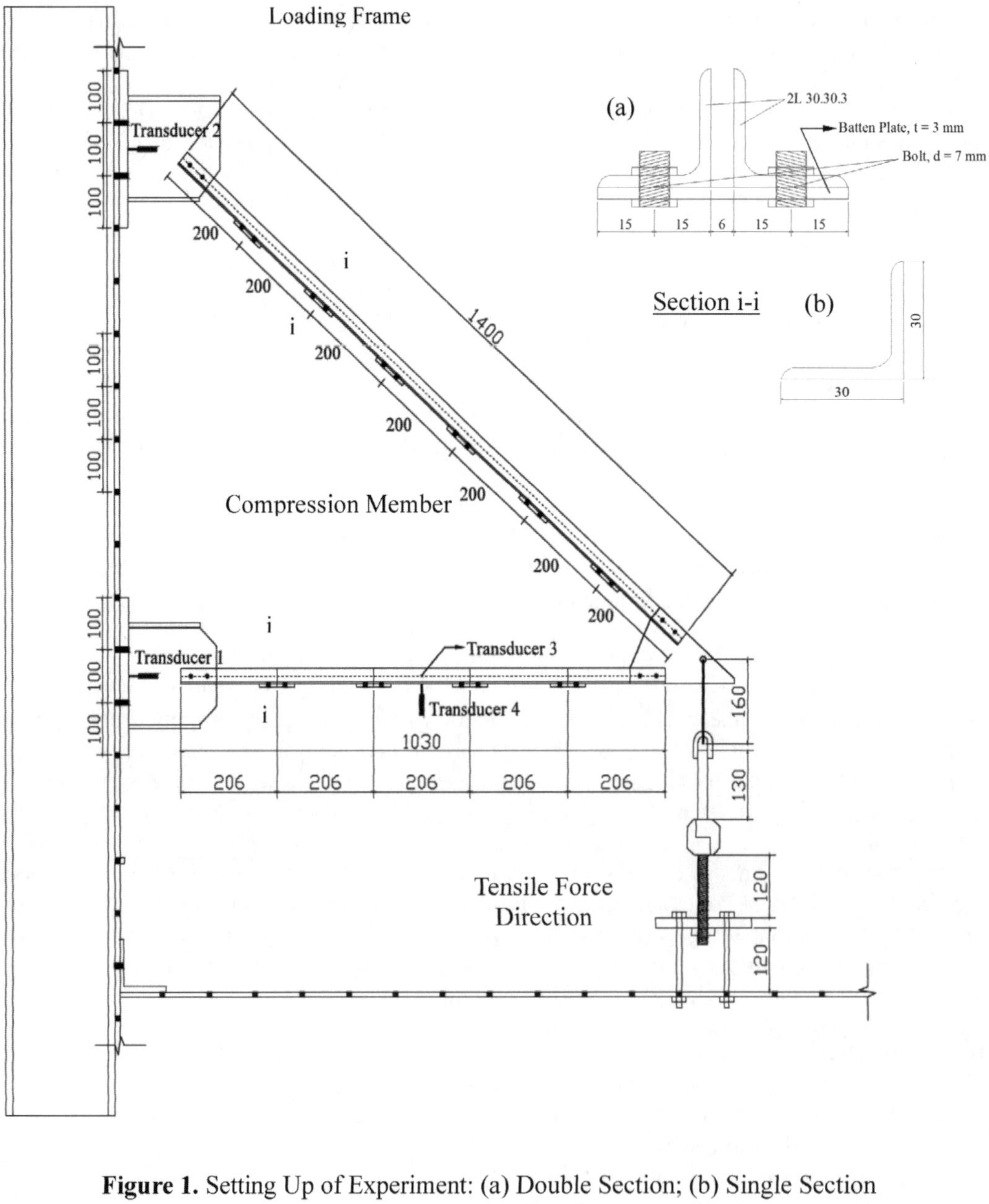

Figure 1. Setting Up of Experiment: (a) Double Section; (b) Single Section

Both of SNI 03-1729-2002 & SNI-03-1729-2015 asserted that a structural component subjected to ultimate concentric compression force has to fulfil $P_u \leq \phi_n P_n$ and Slenderness ratio of compression member must be less than 200 ($\lambda < 200$)

Compression strength of structural member (N_n) calculated based on SNI-03-1729-2002 for double steel sections in major and minor axis can be stated as :

$$\phi.N_{nx} = \phi.\frac{A_g.F_y}{\omega_x}, \ \phi.N_{ny} = \phi.\frac{A_g.F_y}{\omega_y} \quad (1)$$

Nominal strength of compression, P_n, calculated with code of SNI-03-1729-2015 for both single and double sections are :

$$\phi.P_{nx} = \phi.A_g.F_{crx}, \ \phi.P_{ny} = \phi.A_g.F_{cry} \quad (2)$$

Where : ϕ = reduction factor of compression member (ϕ = 0,85); P_n = Nominal strength of compression (N); Ag = Area section (mm^2); F_{crx} = Critical stress in major axis (MPa); F_{cry} = Critical stress in minor axis (MPa).

4. Test Results and Discussion

4.1 Tensile Force – Deformation Curve Of Transducer

Observation of this relationship curve aimed to gain information due to buckling on angle web of the compression member.

Deformation that occurred on web angle can be obtained through transducer 3. Experimental work showed that ultimate web deformation (buckling) of single angle (δ_{u1}= 41,155 mm) was higher than ultimate web deformation of double angles (δ_{u2}= 15,437 mm). This finding gave information that increasing section capacity to be double angles was effective to decline lateral buckling due to the slenderness of area section. Deformation of the flange can be measured through transducer 4 that was located on angle flange. Meanwhile, tensile force can be gained through tension load cell, and it would be transferred to the data logger.

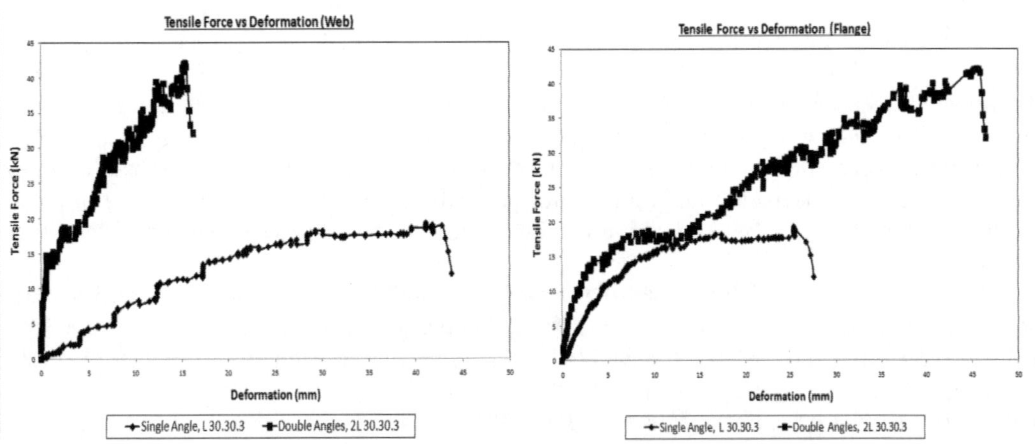

(a) (b)

Figure 2. Tensile force – deformation curve, (a) transducer 3, (b) transducer 4

Experimental work showed that higher ultimate deformation on flange was achieved by double angles (δ_{u2}= 45,49 mm) if it was compared to flange deformation of single angle (δ_{u2} = 25,465 mm). This was possible because double angles that were increased its section capacity would decrease its buckling possibility. So it performed optimally, and it would have higher buckling value on loading axis.

IOP Conf. Series: Materials Science and Engineering **352** (2018) 012037 doi:10.1088/1757-899X/352/1/012037

4.2 Comparison of Ultimate Member Forces and Ultimate Tensile Forces Due To Experimental Work Results, SNI 03-1729-2002, and SNI 03-1729-2015

Comparison of ultimate member forces and ultimate tensile force for single compression members and double compression members showed that double compression member performed ultimate member force (S_{u2} = 50172,83 N) and ultimate tensile force (P_{u2} = 42100 N). Both of these values were higher than single compression member values with ultimate member force (S_{u1} = 23060,43 N) and ultimate tensile force (P_{u1} = 19350 N). There were about 118 % in difference for ultimate member force and ultimate tensile force between double compression members and single compression member.

Besides, the analysis was also conducted with the assumption that compression member will only accommodate concentric force. Both of these analysis methods were done by code of SNI 03-1729-2002 and SNI 03-1729-2015.

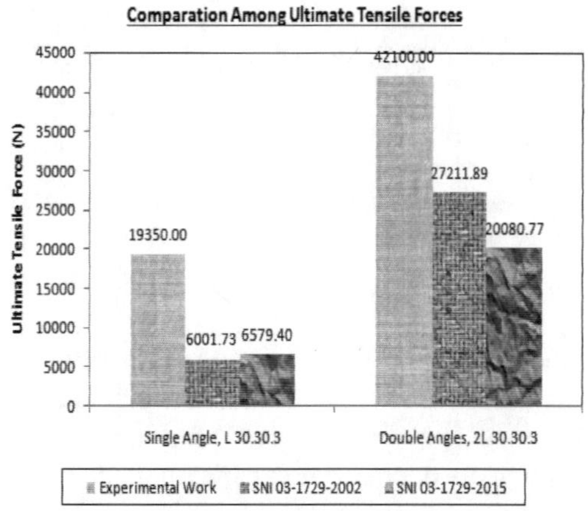

Figure 3. Comparation among ultimate tensile forces

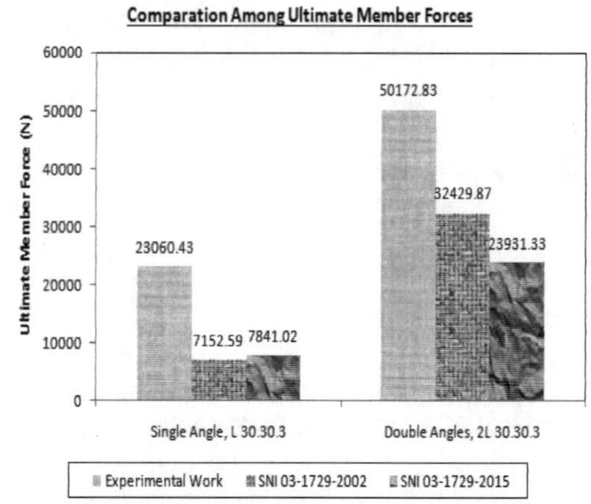

Figure 4. Comparation among ultimate member forces

Comparison of ultimate member force and ultimate tensile force of experimental work to the analysis result of SNI showed that either for ultimate member forces or ultimate tensile forces, experimental work results were higher than ultimate member force and ultimate tensile force of SNI analysis results. This was possible because SNI analysis commonly used for design needs referred to critical stress (F_{cr}) and based on coupon test on this research, critical stress value is smaller than yield stress. So it can be understood that SNI conducted analysis on the elastic range. Experimental work that was carried out over yield condition can be ensured to give higher compression load than SNI analysis.

Comparison between analysis results of SNI 03-1729-2002 and SNI 03-1729-2015 showed that there was increasing strength in compression from single angle to double angles. They were 353 % for SNI 03-1729-2002 and 205 % for SNI 03-1729-2015. These facts showed that the aspect of reinforcement in steel angles with increasing area section of the member, SNI 03-1729-2015 tended to show more conservative (safer) results than SNI 03-1729-2002 results.

4.3 Failure Modes of Compression Member

There were differences of failure modes of single angle and double angles. The failure mode of single angle was lateral- torsional-buckling on the mid-span of the member. It is possible because single steel angle has slender web and flange, so they did not have a good capacity to accommodate high compression load. The failure mode of double angles was local buckling to loading direction on the flange that occurred on the mid-span. This failure mode showed that improving section capacity to be double sections was effective to prevent lateral-torsional-buckling failure.

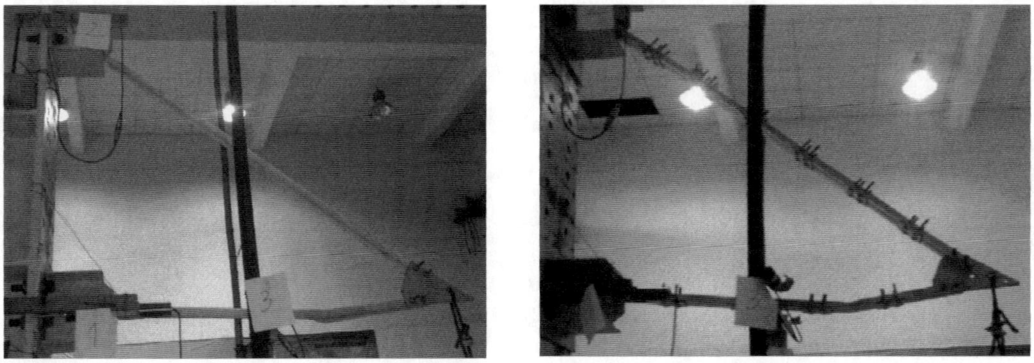

(a) (b)

Figure 5. Comparation of Failure Modes : (a) Failure mode of single angle ; (b) Failure mode of double angles

5. Conclusion and Suggestion

5.1 Conclusions

Based on the observation, it can be concluded that higher compression strength (Pu max = 42100 N) and higher deformation on the flange (δ_u flange = 45,49 mm) were achieved by double angles due to its area section capacity and its capability in accommodating compression load. Ultimate member force and ultimate tensile force obtained from experimental work results were higher than SNI analysis results due to SNI criteria refer to the elastic condition. Failure modes of each specimen showed that increasing area section capacity to be double sections was effective to prevent lateral-torsional-buckling failure.

5.2 Suggestions

For better understanding due to the behavior of compression member, it is suggested to conduct continued experimental work of single and double section in the form of the single element. Necessary to carry out experimental work of double sections with implementing diagonal plate (lacing) as substitution of batten plate.

References

[1] AISC 360-05 2005 *Specification for Structural Steel Building* (USA: AISC Chicago)
[2] Hasibuan P 2013 Identifikasi Identifikasi Gaya Batang Tekan Baja Profil Siku for Berbagai Macam Tumpuan melalui Metode Vibrasi *Laporan Penelitian Tesis Universitas Gadjah Mada Yogyakarta* (In Indonesia)
[3] Liu Y and Hui L 2008 Experimental Study of Beam-Column Behaviour of Steel Single Angle *Journal of Constructional Steel Research.* **64** 505-514.
[4] Sakla, S S. S 2001 Tables for the design strength of eccentrically-loaded single angle struts *Engineering Journal* (USA: AISC Chicago)
[5] SNI 03-1729.1-2015 Spesifikasi untuk gedung baja struktural *Badan Standarisasi Nasional* (in Indonesia)
[6] SNI 03-1729.1-2002 Spesifikasi untuk gedung baja struktural *Badan Standarisasi Nasional* (in Indonesia)

IOP Conf. Series: Materials Science and Engineering **352** (2018) 012038 doi:10.1088/1757-899X/352/1/012038

Compression member response of double steel angles on truss structure with member length variation

Purwandy Hasibuan[1], Arief Panjaitan[1], Muhammad Haiqal[2]
[1]Civil Engineering Department, Faculty of Engineering, Syiah Kuala University, Jalan Tengku Syech Abdurrauf No. 7, Darussalam, Banda Aceh, Indonesia
[2]Architecture Department, Faculty of Engineering, Syiah Kuala University, Jalan Tengku Syech Abdurrauf No. 7, Darussalam, Banda Aceh, Indonesia

E-mail: purwandy.hsb@unsyiah.ac.id

Abstract: One type of structures that implements steel angles as its members is truss system of telecommunication tower. For this structure, reinforcements on tower legs are also needed when antennas and microwaves installation placed on the peak of tower increases in quantity. One type of reinforcement methods commonly used is by increasing areas section capacity, where tower leg consisted of single angle section will be reinforced to be double angle sections. Regarding this case, this research discussed behavior two types of double angle steel section 2L 30.30.3 that were designed identically in area section but vary in length: 103 cm and 83 cm. At the first step, compression member together with tension member was formed to be a truss system, where compression and tension member were met at the joint plate. Schematic loading was implemented by giving tension loading on the joint plate, and this loading was terminated when each specimen reached its failure. Research findings showed that implementing shorter double angle (83 cm) sections, increased compression strength of steel angle section up to 13 %. Significant deformation occurring only on the flange for both of specimens indicated that implementing double angle is effective to prevent lateral-torsional buckling.

1. Introduction

Implementation of steel angles as members of truss system has been acknowledged until recent years. One type of structures built up with steel angles as its main components is telecommunication tower.

On telecommunications tower, some members form its structural systems such as tower feet (legs), diagonal members (braces), redundants, and horizontal members. The leg is tower element that has a function as the last receiver of load distribution from other members. Furthermore, the leg is a place where antennas and microwave as a spreader of radio waves will be placed. The leg has similar characteristics with a column where the leg is a compression member like column and legs placed on the bottom will accommodate the largest compression load.

During the time goes away, antennas and microwaves installation placed on the peak of the tower will increase in quantity, so reinforcement solution is absolutely needed on legs. One type of reinforcement methods commonly used is by increasing area section capacity, where tower leg consisted of single angle section will be reinforced to be double angle sections. In addition, redundant members installed to divide the effective length of tower legs demands optimal effective length of tower legs so it can ensure strength and economic value of telecommunication tower structure.

Regarding this case, this research discussed behavior two types of double angle steel sections 2L 30.30.3

that were designed identically in area section but vary in length: 103 cm and 83 cm. Based on the experimental work, it can be observed member responses started from elastic condition until their failure due to a compression load.

Table 1. Specimen Variations

Numb	Specimen	Member length (cm)
1	2L 30.30.3	103
2	2L 30.30.3	83

Truss system was formed with a tensile component on diagonal member and compression component on horizontal member. Both of these members were joined at the gusset plate as a plate that was implemented as loading placement. Loading step was started by giving tensile load gradually through loading plate until horizontal member (compression component) achieved its failure. Observation record was carried out due to tensile force on loading plate and buckling deformation on horizontal member. They were automatically collected and notified through experimental equipment such as tension load cell, data logger, and transducers and based on these data, the response of two specimens can be specifically described and discussed.

2. Specimens Preparation

2.1 Experimental Work Setting Up

In determining response description of the compression member of double steel angle in truss system, this experimental work implemented two types of steel member (diagonal and horizontal member), and they were formed to be a truss system. When forming truss system, both of these members were fastened with bolts on their supporting plates where these supporting plates were connected with end plates through the welded connection. End plates then were fastened to loading frame with the bolted connection. The tensile force that was worked on the gravitational direction and applied on the joint plate (joint plate of the horizontal and diagonal member) caused the diagonal member to be a tensile component and horizontal member to be compression component. This experimental work was focused on a horizontal member that accommodated compression force.

For achieving the comparative result, both of specimens were planned similar on their batten plate distances: 206 mm for 2L 30.30.3 (103 cm) and 207.5 mm for 2L 30.30.3 (83 cm). Load recording was carried out through tension load cell and data logger. Meanwhile, deformation recording was done through transducers placed on flange and web of double steel angles.

2.2 Material Characteristics Determination

In obtaining steel angle characteristics that will be observed, coupon test was conducted in steel angle. Through this test, steel characteristics (yield stress and ultimate stress) were obtained, and they can be used as a description of steel angle mechanical properties.

Table 2. Steel angle material characteristics

Component	Stress (MPa)	
	Yield	Ultimate
Steel Angle	224	324

2.3 Experimental Work Activities

Experimental work was started with giving tensile force at gusset plate (joint plate of the horizontal and diagonal member). This loading schematic was carried out by fastening iron ring on the threaded cylindrical bar so it would produce a tensile force on tension load cell that was continued to sling rope. This sling rope was clamped so it can distribute tensile force to the gusset plate. This mechanism caused tensile force worked on the diagonal member, and compression force worked on horizontal member.

The observation was emphasized on compression component (horizontal member). Tensile member had to be in an elastic condition so it can ensure the failure would occur at compression member at the first and the experimental work data can be optimally obtained.

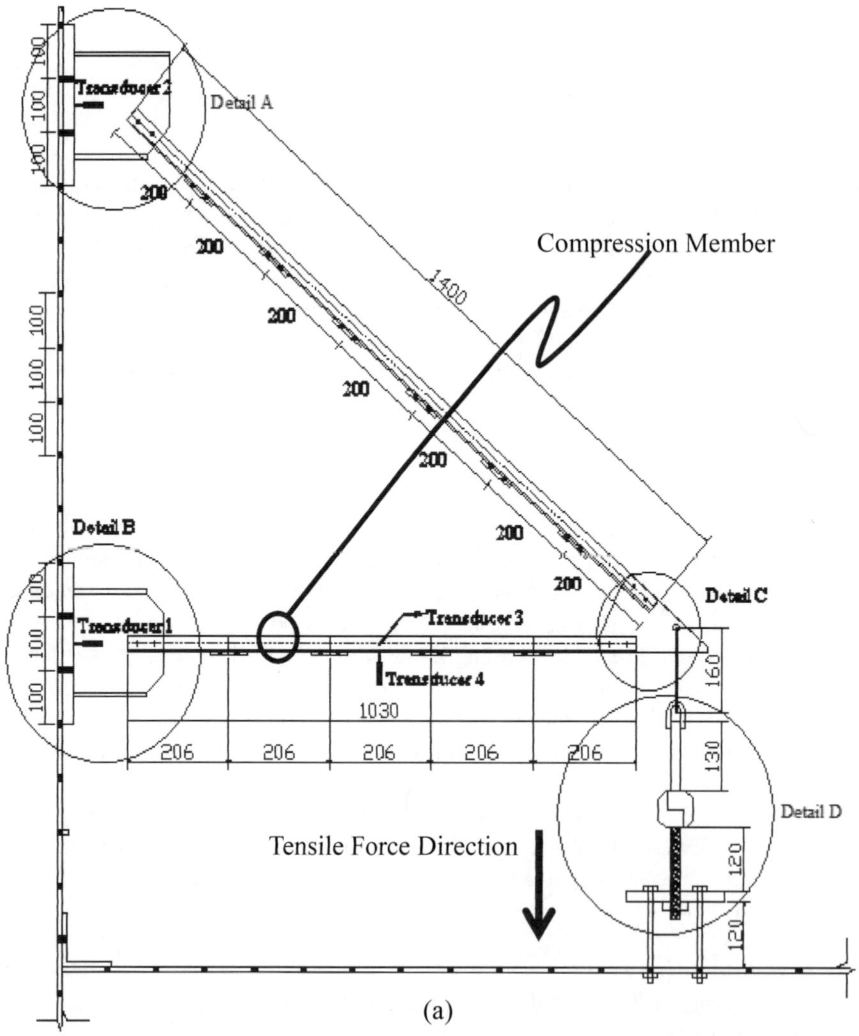

(a)

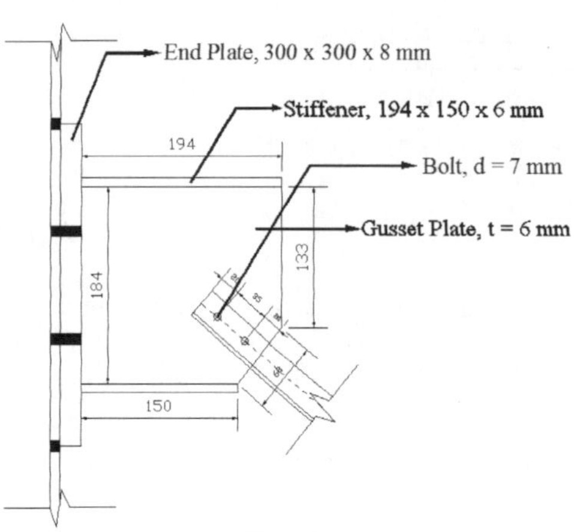

(b)

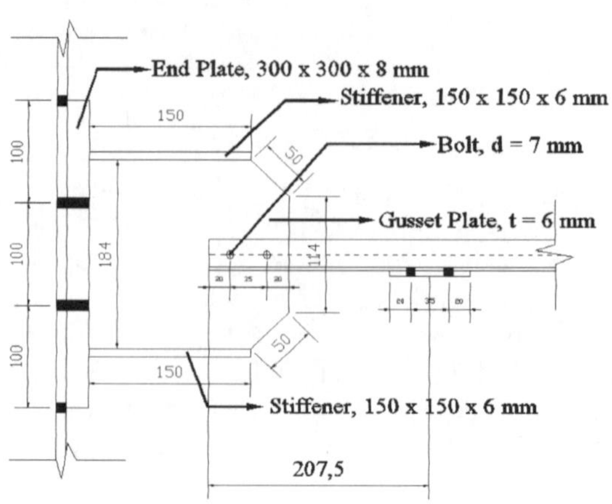

(c)

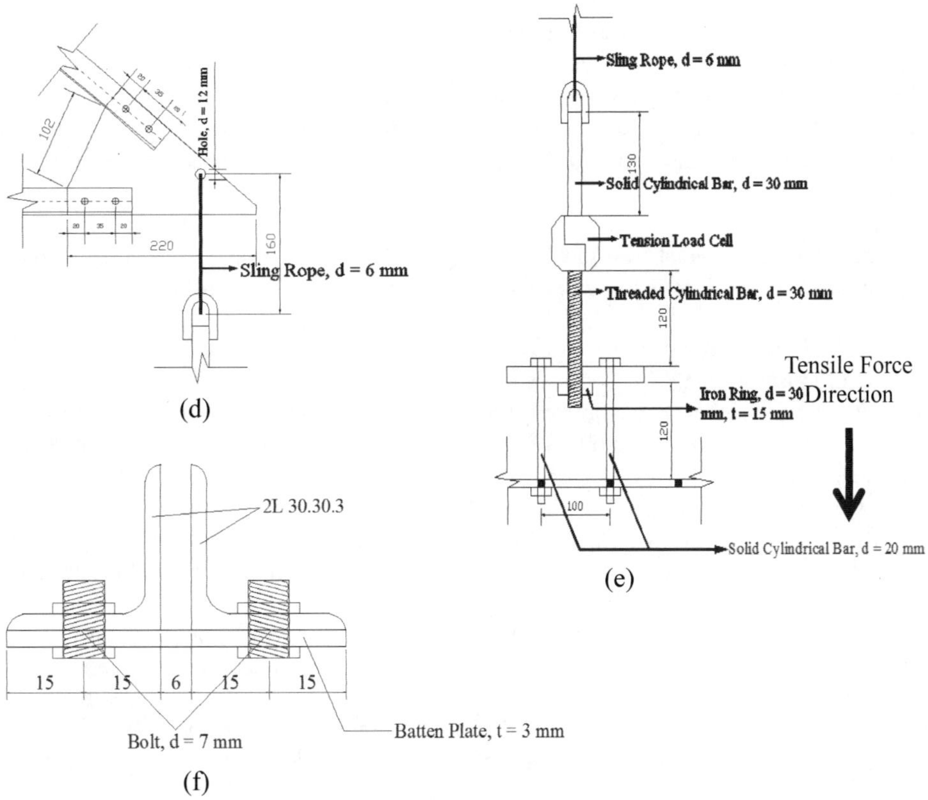

Figure 1. (a) Placement of transducers in truss system ; (b) Detail A : Upper gusset plate ; (c) Detail B : Bottom gusset plate ; (d) Detail C : Joint gusset plate ; (e) Detail D : Loading schematic of compression member ; (f) Specimen detail of 2L 30.30.3 (103 cm) and 2L 30.30.3 (83 cm)

3. Test Results and Discussion

3.1 Tensile Force – Deformation Curve Of Transducer 3 and Transducer 4

Deformation occurring on web and flange of angle can be obtained through transducer 3 and 4.

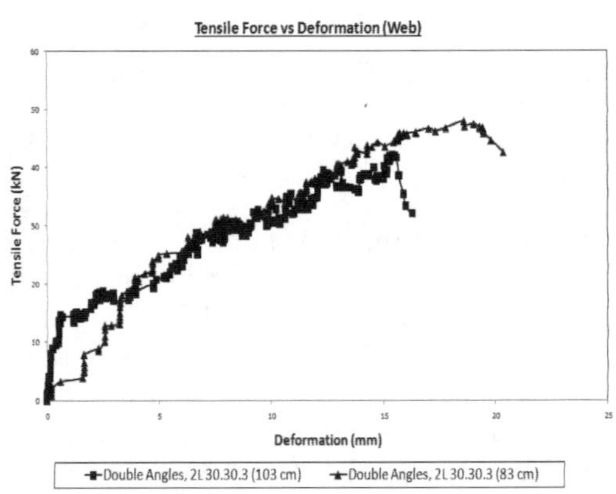

Figure 2. Tensile force – deformation curve of transducer 3

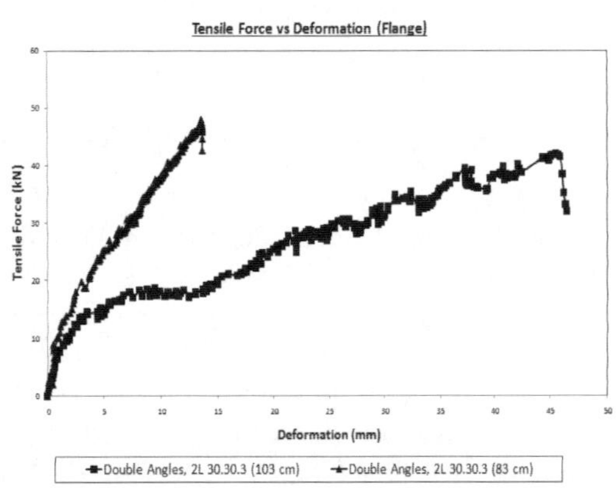

Figure 3. Tensile force – deformation curve of transducer 4

Experimental work showed that ultimate web deformation (buckling) of double angles (83 cm) was higher than ultimate web deformation of double angles (103 cm) with ultimate web deformation of double angles for

83 cm (δ_{u2}) was 18.61 mm and ultimate web deformation of double angles for 103 cm (δ_{u1}) was 15.437 mm. This was possible due to the failure mode of double angles (83 cm) that showed lateral buckling on a joint plate, so when buckling occurred on joint plate, member geometry will tend to follow the direction of plate buckling. Meanwhile, observation on angle flange showed that ultimate flange deformation (buckling) of double angles for 103 cm (δ_{u1}) was 45,49 mm. This value was higher than ultimate flange deformation (buckling) of double angles for 83 cm (δ_{u2}). It was 13,535 mm. This was possible due to the performance of double angles (103 cm) that was more optimal than double angles (83 cm), so it tended to give higher deformation (buckling) in compression member. It was different from double angles (83 cm). When premature buckling happened on joint plate, compression member was unable to increase its strength due to a compression force.

3.2 Comparison of Ultimate Member Forces and Ultimate Tensile Forces Due To Experimental Work Results, SNI 03-1729-2002, and SNI 03-1729-2015

Comparison of ultimate member forces and ultimate tensile force for double compression members (103 cm) and double compression members (83 cm) showed that compression members (83 cm) performed ultimate member force (S_{u2} = 57144,58 N) and ultimate tensile force (P_{u2} = 47950 N). Both of these values were higher than double compression member (103 cm) values with ultimate member force (S_{u1} = 50172,83N) and ultimate tensile force (P_{u1} = 42100 N). There were about 14 % in difference for ultimate member force and ultimate tensile force between double compression members (103 cm) and double compression members (83 cm).

Furthermore, the analysis was also conducted with the assumption that compression member will only accommodate concentric force. Both of these analysis methods were done by code of SNI 03-1729-2002 and SNI 03-1729-2015.Comparations of ultimate member forces and ultimate tensile force for double compression members (103 cm) and double compression members (83 cm) based on experimental work results, SNI 03-1729-2002, and SNI 03-1729-2015 were :

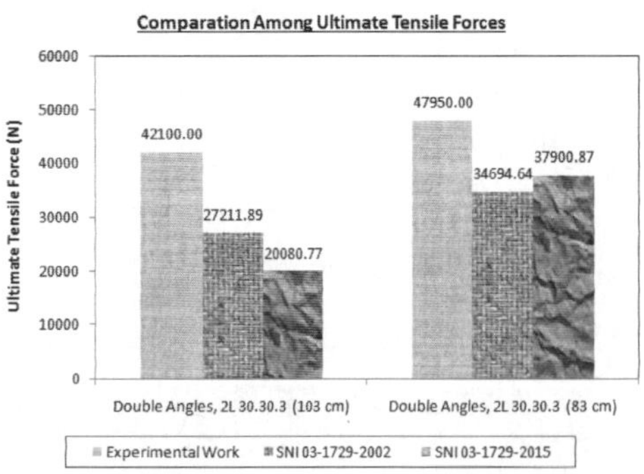

Figure 4. Comparation among ultimate tensile forces

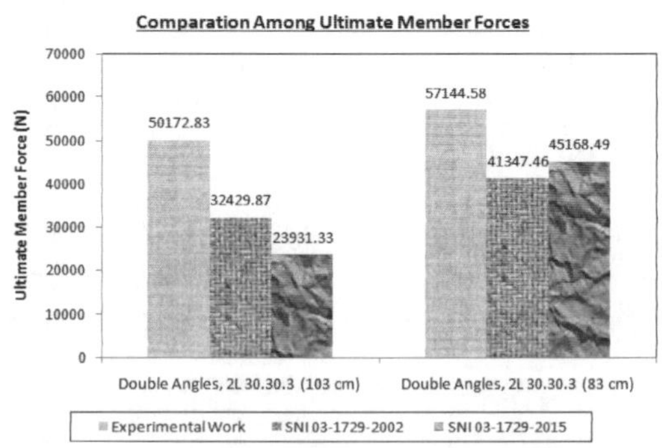

Figure 5. Comparation among ultimate member forces

Based on two bar charts above, it can be shown that either for ultimate member forces or ultimate tensile forces, experimental work results were higher than ultimate member force and ultimate tensile force of SNI analysis results. This is possible because SNI analysis commonly used for design needs referred to critical stress (F_{cr}) and based on coupon test on this research, critical stress value is smaller than yield stress. So it can be understood that SNI conducted analysis on the elastic range. Experimental work that was carried out over yield condition can be ensured to give higher compression load than SNI analysis. Comparison between analysis results of SNI 03-1729-2002 and SNI 03-1729-2015 showed that there was improving strength in compression from double angles (103 cm) to double angles (83 cm). Based on SNI 03-1729-2002, this improvement was 27 %. Meanwhile, based on 03-1729-2015, this improvement was 89 %. These facts showed that an aspect of reinforcement in double angles with shortening the effective length of the member, SNI 03-1729-2002 tended to show more conservative (safer) results than SNI 03-1729-2015 results.

IOP Conf. Series: Materials Science and Engineering **352** (2018) 012038 doi:10.1088/1757-899X/352/1/012038

3.3 Failure Modes of Compression Member

Failure modes for both of specimens were different each other. The failure mode of double angles (103 cm) was local buckling to loading direction on the flange that occurred on the mid-span. This failure mode showed that improving section capacity to be double sections was effective to prevent lateral-torsional buckling failure where this failure commonly occurs on steel section due to the slenderness of area section. The failure mode of double angles (83 cm) was lateral buckling on the joint plate due to lack of capacity of the joint plate in accommodating tensile force. This failure caused double angles (83 cm) was unable to perform optimally. It can be shown through its deformation either in flange or web that did not significantly occur when the failure happened on the joint plate.

(a) (b)

Figure 6. Comparation of Failure Modes : (a) Failure mode of double angles (103 cm) ; (b) Failure mode of double angles (83 cm)

4. Conclusion

Based on the observation, it can be concluded that higher compression strength (Pu max = 47950 N) was achieved by double angles (83 cm) due to the smaller slenderness of double angles (83 cm), meanwhile higher deformation on flange (δ_u flange = 45,49 mm) was reached by double angles (103 cm) due to better performance in accommodating compression force. Besides, it can be found that ultimate member force and ultimate tensile force obtained from experimental work results were higher than SNI analysis results due to SNI criteria refers to the elastic condition. According to the analysis, it was found that comparison between analysis results of SNI 03-1729-2002 and SNI 03-1729-2015 showed that SNI 03-1729-2002 was more conservative (safer) in the aspect of reinforcing double angles with shortening the effective length of member than SNI 03-1729-2015.

References

[1] AISC 360-05 2005 *Specification for Structural Steel Building* (AISC Chicago-USA)
[2] Hasibuan P 2013 *Identifikasi Gaya Batang Tekan Baja Profil Siku for Berbagai Macam Tumpuan melalui Metode Vibrasi* Laporan Penelitian Tesis Universitas Gadjah Mada Yogyakarta (in Indonesian)
[3] Liu Y and Hui L 2008 Experimental study of beam-column behaviour of steel single angle *Journal of Constructional Steel Research.* **64** 505-514
[4] Sakla S S S 2001 Tables for the design strength of eccentrically-loaded single angle struts *Engineering Journal* (Tanta University Egypt)
[5] SNI 03-1729.1-2015 *Spesifikasi for Gedung Baja Struktural* Badan Standarisasi Nasional (in Indonesian)
[6] SNI 03-1729.1-2002 *Spesifikasi for Gedung Baja Struktural* Badan Standarisasi Nasional (in Indonesian)

The effect of ratio between rigid plant height and water depth on the manning's coefficient in open channel

M Rizalihadi, Ziana, Nina Shaskia and H Asharly
Civil Engineering Department, University of Syiah Kuala, Jl. Syekh Abdurrauf No. 7, Darussalam, Banda Aceh, Indonesia.

Email: maimunrizalihadi@unsyiah.ac.id

Abstract. One of the important factors in channel dimension is the Manning's coefficient (***n***). This coefficient is influenced not only by the channel roughness but also by the presence of plants in the channel. The aim of the study is to see the effect of the ratio between the height of the rigid plant and water depth on the Manning's coefficient (n) in open channel. The study was conducted in open channel with 15.5 m long, 0.5 m wide and 1.0 m high, in which at the center of the channel is planted with the rigid plants with a density of 42 plants/m^2. The flow was run with a discharge of 0.013 m^3/s at 6 ratios of H_{plants}/H_{water}, namely: 0; 0.2; 0.6; 0.8; 1,0 and 1,2, to obtain the velocity and water profiles. Then the value of ***n*** is analyzed using Manning's equation. The results showed that the mean velocity becomes decrease 17.81-34.01% as increase the ratio of H_{plants}/H_{water}. This results in increasing n value to become 1.22-1.52 times compared to the unplanted channel (n_o=0.038). So, it can be concluded that the ratio between the rigid plant's height and water depth in the open channel can affect the value of Manning coefficient.

1. Introduction

An important factor that greatly influences the calculation of discharge in the channel dimension planning is the Manning's coefficient. This coefficient describes the roughness of the material forming both at the base and wall of the channel. This roughness may increase the flow resistance, so that it can disrupt the flow in the channel. At present, the roughness at the base and wall of the channel is uniformly used in the Chezy or Manning equation as the basis for calculating the discharge and open channel dimensions [1]. However, another flow resistance occurring within the channel due to the presence of other obstructing objects is not included in analyzing the discharge. One of the most commonly encountered objects in the channel is the presence of plants in the channel. The presence of plants in the channel on one side can prevent scouring, but on the other side, the existence of plants in the channel can increase the flow resistance. [3] shows that the presence of plants causes an increase in flow resistance. The development of these research has been fully published in [2], [4], [5], [6], [7] dan [11]. Overall, the results of the study indicate that the presence of plants in both flexible, stiff and rigid characteristic, whether submerged or partially submerged in the water, may increase in flow resistance. However, the prediction of the vegetation resistance is very complex since there are many different species with their unique characteristics changing during the season. These plant characteristics are influencing the hydraulic resistance, which may vary significantly from place to place, and may also change in time. Beside that the inhomogeneous character of the vegetation in the field that is hard to take into account in a model of the equation. Moreover, the difference of type dan characteristic of vegetation might result in different flow resistance, although these vegetations are equally flexible, rigid or stiff vegetation. For example, mostly in irrigation channel are grown by water grass (Acorus Gramineus), water hyacinth (Eichhornia Crassipes), watercress (Ipomoea Aquatica Forsk), elephant grass (Pennisetum Purpureum) And jeringau (Acorus Calamus). These plants might be different characteristics of a different place. [9] has researched on a channel planted with jeringau (Acorus Calamus) which has flexible characteristics. Other research, [10] has also been done for elephant grass (Pennisetum Purpureum) which has rigid characteristics. The object of both researchers is the effect of plant density. The results showed that the flow resistance in the channel increased with increasing of plant density, which was indicated by the increase of Manning coefficient (n) due to the increase of plant density. Furthermore, [8] has also conducted a similar study by examining the effect of jeringau height (Acorus Calamus) on the flow resistance in the channel. The results also showed that the higher the plant in the channel the higher the increasing value of Manning roughness. Considering to the great influence of the presence of plants on the roughness in the channel, then it is necessary to do further research by studying at the different treatments and type of plants that grow in the channel to obtain the most suitable approach for general application.

This research is a continuation of previous research as mentioned in [8], [9] and [10]. This study aims to see the effect of the ratio between the height of the rigid plant and water depth in open channel on Manning coefficient (n).

2. Literature Review

2.1. Manning's equation

Flow resistance can be translated by incorporating velocity, hydraulic radius, base slope and roughness of base and wall of the channel, then formulated as called Manning's Equation [1].

$$V = \frac{1}{n} R^{2/3} S^{1/2} \tag{1}$$

where: n=Manning's coefficient; V=mean velocity (m/s); R=hydraulic radius (m); And S=the slope of the energy line.

2.2. Manning's coefficient in channel

The Manning's coefficient (n) as in the equations 1is normally derived from the roughness of materials forming channels. In fact that in nature the flow resistance is influenced by many factors, so the real Manning's coefficient can be obtained using the equation (2) as below.

$$n = (n_0 + n_1 + n_2 + n_3 + n_4)m_5 \tag{2}$$

In some cases, the values of n_1, n_2, n_3 can be ignored due to not significant, so equation (2) is dominated by the base roughness of channel and due to plants, as in equation (3).

$$n = n_0 + n_4 \tag{3}$$

where: n_0=the base roughness; n_1= the variation of channel surface; n_2=shape variation of channel section; n_3=obstacles due to the structure; n_4= due to plants; and m_5=correction factor for channel bend.

3. Research Methods

3.1 Research preparation and design

This research was conducted at Hydrotechnics Laboratory, Faculty of Engineering, Syiah Kuala University. This study used a flume with a length of 15.5 meters, a width of 0.5 meters and a height of 1.00 meters. On the upstream side of the flume is placed the 90° Thompson water gate, to control the discharge, where the water level is set to a fixed depth of 45 cm. At a center of 3 m long of the channel is planted with elephant grass (Pennisetum Purpureum), as rigid plants, with a density of 42 plants/m^2. Six ratios of plant height (H_{plant}) and water depth (H_{water}), namely: (SP-0: 0, without plant), (SP-1: 0,2), (SP-2: 0.6), (SP-3: 0,8), (SP-4: 1,0) and (SP-5: 1,2), for more details the placement of plants with 6 variations in plant height can be seen in figures 1 and 2.

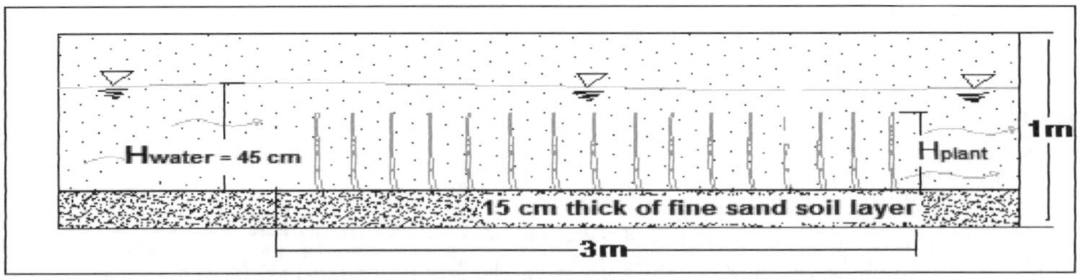

Figure 1. Side view of placement of plants in channels

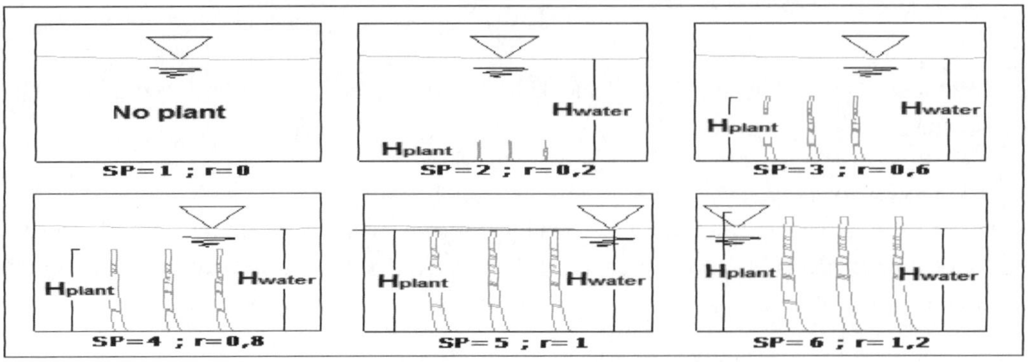

Figure 2. Variation of a ratio between plants height and water depth.

The parameters measured in this study were water level (h), flow slope (S_f) and mean velocity (V) which is measured in every ratio with a constant discharge of 0.013 m^3/s.

3.2 Data analysis

3.2.1 Distribution and mean velocity

There are 3 locations that be an object of measurements to analyze the distribution and mean velocity, namely: at the upstream, at the middle and the downstream of the channel. The velocity distribution is measured using a current meter on the flow surface, depth of 0.2h, 0.6h, 0.8h and a channel base. The mean velocity is calculated using the equation 10 as below.

$$\overline{V} = \frac{V_p + 3.V_{0,2h} + 2.V_{0,6h} + 3.V_{0,8h} + V_d}{10} \quad (4)$$

where: V = mean velocity (m/s); V_p = velocity at the surface (m/s); $V_{0.2h}$ = velocity at depth 0.2h (m/s); $V_{0.6h}$= velocity at depth 0.6h (m/s); $V_{0.8h}$ = Velocity at a depth of 0.8h (m/s); V_d = velocity at the base (m/s).

3.2.2 Slope of energy

Loss of energy can be calculated using the Bernoulli equation [1]. This value is obtained from the measurement of water depth in upstream (H_u) and downstream (H_d), as well as upstream (V_u) and downstream (V_d) velocities. The total energy losses can be written as in the equation below.

$$H_u + \frac{V^2{}_u}{2.g} = H_d + \frac{V^2{}_d}{2.g} + h_f \quad (5)$$

The slope of energy (S_f) can be calculated by the equation 6 where L is the length of the planted channel.

$$S_f = \frac{h_f}{L} \quad (6)$$

3.2.3 The Manning's Coefficient

The coefficient of Manning (n) is obtained using equation 1 above. The value of n in equation 1 is the total roughness produced by the channel and plant which can be formulated as in equation 7, so the roughness occurring due to the presence of the plant can be calculated by equation 8.

$$n = n_o + n_v \quad (7)$$

$$n_v = n - n_o \quad (8)$$

4. Result and Discussion

4.1. Profile of speed distribution to crop height

Flow velocity measurements at each depth downstream, center and upstream can be obtained from the velocity distribution profile to water depth, as shown in figures 3, 4 and 5. The graphs explain that the velocity decreases with increase in height ratio of plants compared to the unplanted channel. The presence of plants in the channel causes the flow to be blocked and cause backflow causing the increasing flow resistance, making the flow velocity becomes reduced.

Overall, this condition causes the mean velocity of each variation in the height ratio of plants in the channel to be changed, where the higher the ratio of plant height the smaller the mean velocity in the channel or it can be said that the mean velocity tends to decrease as increase the ratio of plant height, as shown in fig. 6. If compared to the unplanted channel, mean velocity becomes decreases in between 17.81-34.01% due to increasing the ratio of plant height.

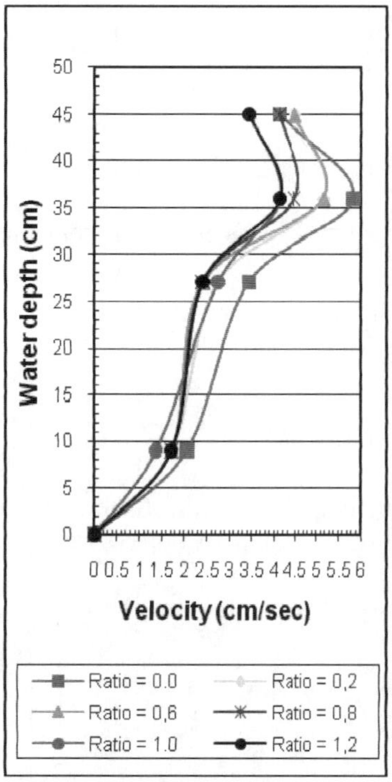

Figure 3. Velocity distribution at the downstream of channel

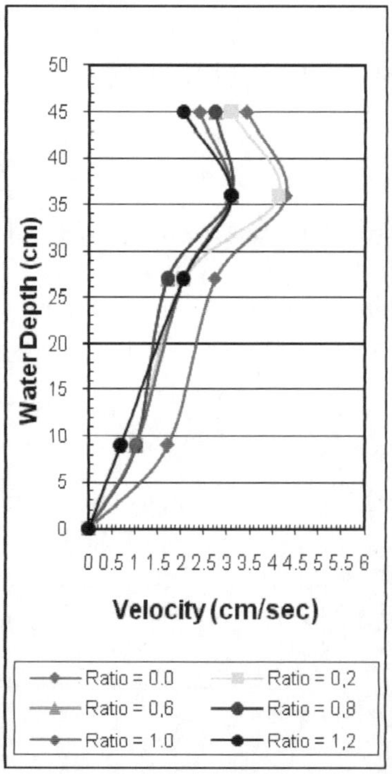

Figure 4. Velocity distribution at the central of channel

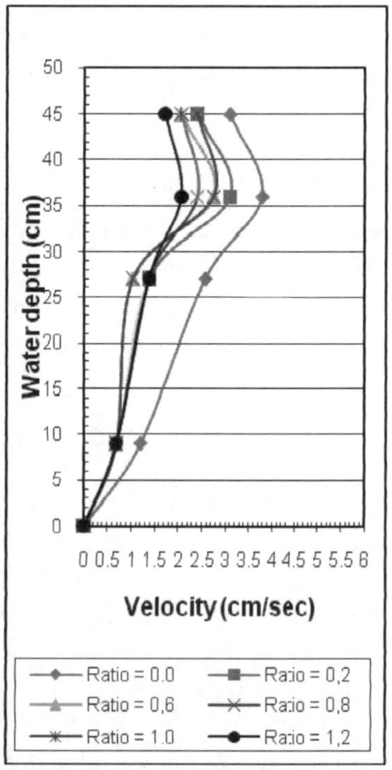

Figure 5. Velocity distribution at the upstream of channel

4.2. The relationship between the ratio of the plant height against the coefficient Manning

Using equations 1, 7, and 8, the results are shown in Table 1. The results showed that increasing the ratio of plant height result in increasing the total Manning's coefficient (n). For more details, the trend of the relationship between the height of plant and the coefficient of Manning can be seen in figure 7, where the value of n keep increasing as increasing the ratio of plant height.

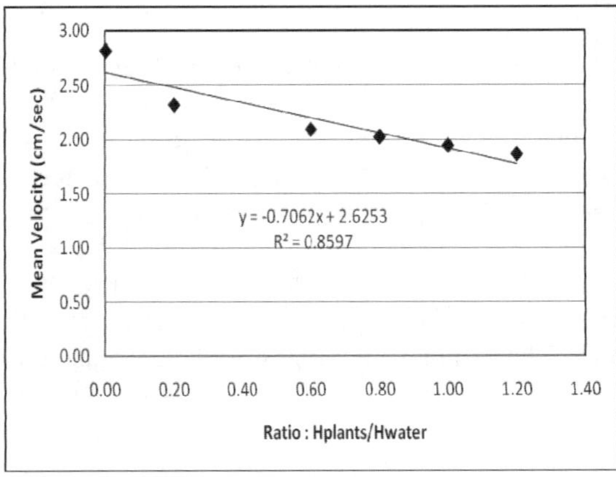

Figure 6. Mean velocity in the channel for every ratio of the plant height

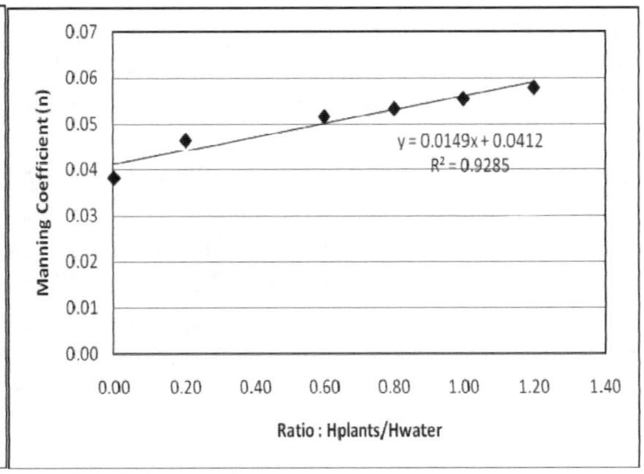

Figure 7. Relationship ratio of plant height and Manning coefficient (n)

In addition, table 1 also explains that the Manning's coefficient (n) ranges from 0.0464-0.0578, while the Manning coefficient for the unplanted channel (n_0) = 0.0381. Furthermore, by using equation 8 additional values of Manning's coefficient due to the presence of crop (n_v) is in between 0.0083-0.0196 or increase the value of Manning coefficient (n) between 21.625-51.534% by the increase Ratio of plant height. So that the total coefficient of Manning (n) is increasing 1.22 up to 1.52 times compared to the unplanted channel (n_0) = 0.038. This result shows that the Manning's coefficient obtained above is almost similar as previous research for the variation of ratio height of jeringau (acorus calamus) as published in [8], where the Manning's coefficient increase 1.30 to 1.94 times compared with the unplanted channel. This indicates that the presence of plants with increased height and type of plants can increase the value of Manning coefficient.

Table 1. Roughness of Manning (n)

Ratio Elevation of Plants	Manning Roughness		
	Total	Initial	Plant
	N	n_0	n_v
0	0,0381	0,0381	0,0000
0,2	0,0464	0,0381	0,0083
0,6	0,0514	0,0381	0,0133
0,8	0,0532	0,0381	0,0151
1,0	0,0554	0,0381	0,0173
1,2	0,0578	0,0381	0,0196

The invention is expected to be input in hydraulics science when applying in the channel planning for both irrigation and drainage. This needs to be our concern in the operational and maintenance (OM) of irrigation and drainage network systems, since the impact of increased flow resistance may affect the flow discharge to affect the irrigation system or the drainage itself.

5. Conclusion

The result of this research can be concluded that the presence of elephant grass (Pennisetum Purpureum) as rigid plant characteristics with various ratio plant height inside the channel can affect the flow velocity distribution where the greater the height of plant causing the smaller flow velocity, which is indicated by the mean velocity decrease of 17,81-34,01 %. This condition leads to an increase in the flow resistance shown by the increase in the coefficient of Manning (n) ranging from 0.0464-0.0578 compared to the unplanted channel (n_0 = 0.0381) or an increase of 1.22-1.52 times. So it can be concluded that the height ratio of rigid plants and the height of water in the open channel can affect the value of Manning coefficient. It is suggested that further research is required by including additional characteristics such as; density, diameter and leaf width of the plant or other types of plants so that the flow resistance due to the presence of plants can be modeled more perfectly for the development of the science of hydraulics in the future.

Acknowledgements

The writers would like to thank Hari Asharly and Hydrotechnic Laboratory technicians of Civil Engineering Department, Syiah Kuala University, Aceh, for carrying out the flume experiments. The writers appreciate the constructive comments and suggestions provided by AIC'S reviewers and Committee for giving us to attend the Seminar and helping in improving the quality of this article.

References

[1] Chow V T 1959 *Open-channel hydraulics* (McGraw-Hill, NY)
[2] El-Hakim O and Salama M M 1992 Velocity distribution inside and above branched flexible roughness *J. Irrig. Drain. Eng.* **118** 914–927
[3] Freeman G E, Rahmeyer W H and Copeland R R 2000 *Determination of resistance due to shrubs and woody vegetation* (Technical Report, ERDC/CHL TR-00-25, U.S.Army Engineer Research and Development Center, Vicksburg, MS)
[4] Wu F-C, Shen H W and Chou Y-J 1999 Variation of roughness coefficients for unsubmerged and submerged vegetation *Journal of Hydraulic Engineering* 934-94
[5] Järvelä J 2002 Flow resistance of flexible and stiff vegetation: a flume study with natural plants *Journal of Hydrology* **269** 44–54
[6] Kim J, Ivanov V Y and Katopodes N D 2012 Hydraulic resistance to overland flow on surfaces with partially submerged vegetation *Water Resources Res.* **48**
[7] Kouwen N and Unny T E 1973 Flexible roughness in open channels *Journal of Hydraulics ASCE* **99** 713–728
[8] Rizalihadi M and Ihsan M 2016 Pengaruh ketinggian tanaman Pandan Tikar (Acorus Calamus) Terhadap Tahanan Aliran Pada Saluran *Terbuka Prosiding Konferensi Nasional Teknik Sipil 10* (Universitas Atmajaya, Yogyakarta) (in Indonesian)
[9] Rizalihadi M and Safiana D 2015 The presence of Jeringau (Acorus Calamus) as flexible vegetation type in the channel against flow resistance *Procedia Engineering* **125** 250-256
[10] Rizalihadi M and Afrianti D 2015 Kehadiran Rumput Gajah (Pennisetum Purpureum) di saluran terhadap tahanan aliran *Prosiding Konferensi Nasional Teknik Sipil 9* (Universitas Hasanuddin, Makassar) 267-274 (in Indonesian)
[11] Tsujimoto T, Kitamura T, Fuji Y and Nakagawa H 1996 Hydraulic resistance of flow with flexible vegetation in open channel *Journal of Hydroscience, Hydraulic Engineering* **14** 47–56

IOP Conf. Series: Materials Science and Engineering **352** (2018) 012040 doi:10.1088/1757-899X/352/1/012040

The difference of drainage channel dimensions at Kopelma Darussalam on the land with and without the use of Bio Pores

Alfiansyah Yulianur, Amir Fauzi and Zaitun Humaira
Civil Engineering Department, Faculty of Engineering, Syiah Kuala University, Darussalam 23111, Banda Aceh, Indonesia

E-mail: fian_7anur@yahoo.com

Abstract. The changes of land use and diminishing of open field that persistently occur are projected to cause rates acceleration of runoff, which decreases the opportunity for rainwater to infiltrate. It has an impact on the surface runoff into the channels, which eventually may lead to overflow and inundate the surrounding area. Some efforts are required to increase the infiltration of rainfall. Thus, bio pore could be one of the most effective methods to be implemented. The objective of this study is to evaluate the effect of bio pore towards the reduction of runoff discharge into the drainage channel and to determine whether that reduction could lead to effectively lessen the channels' dimension. This study is commenced at Kopelma Darussalam in the southern part where there were several spots that submerged by inundation flood during the rainy season, namely *Sektor Timur* area. Rational modification formula is used to calculate the surface runoff discharge on the land without the use of bio pore. Meanwhile, runoff discharge on the land with the use of bio pores is calculated by the use of water balance formula. The number of bio pores that have planned in Sektor Timur area is 3350 bio pores with the diameter of each is Ø10 cm and 80 cm in depth. The result indicates that those bio pores can reduce the runoff discharge on average of 27% and its' reduction lead to the decrease of drainage channel dimension for the average of 26.9%.

1. Introduction

The major development of the cities in Indonesia today's lead to the intensification demand of land for infrastructure and residential area. Consequently, open land areas that functioned to absorb water during the rainy season are predicted to be diminished shortly. The area of Kopelma (Kota Pelajar dan Mahasiswa) Darussalam is particularly the region of Syiah Kuala University that located in Banda Aceh, is one of the locations that experiencing that case. The increasing numbers of faculties, departments as well as other academic buildings result in a land-use change from the open area to covered area. These are directly impacting the environment. As the rainy season comes, the water infiltration reduces that increases surface runoff discharge. Consequently, the drainage channel dimensions predicted will no longer enough to intercept the surface runoff and causing a flood to the surrounding areas. To avoid that, bio pore is one of the solutions to boost the infiltration of rainfall.

Base on those issues, this study is determined to find out the effect of bio pore toward the reduction of surface runoff discharge that goes into drainage channels. The study was designed in southern part of Kopelma Darussalam where there were several spots that experiencing inundation during the rainy season, for instance the area of Sektor Timur.

To achieve the objectives of this study, the methods used comprise of collected secondary data and analyze the data. The secondary data involved map of study area, data of precipitation, topographic

IOP Conf. Series: Materials Science and Engineering **352** (2018) 012040 doi:10.1088/1757-899X/352/1/012040

map, a map of land cover and use as well as data of infiltration rate on the area with and without the use of bio pore that obtained from the earlier studies by Yulia [1].

2. Literature Review

2.1. Bio pore

Biopore can intensify the ability of soil to pervade water [2]. It seeped through the wall of bio pores' surface then penetrated into the soil around the hole. Thus, it will boost water storage and prevent the surface runoff. Image of bio pore is shown in figure 1.

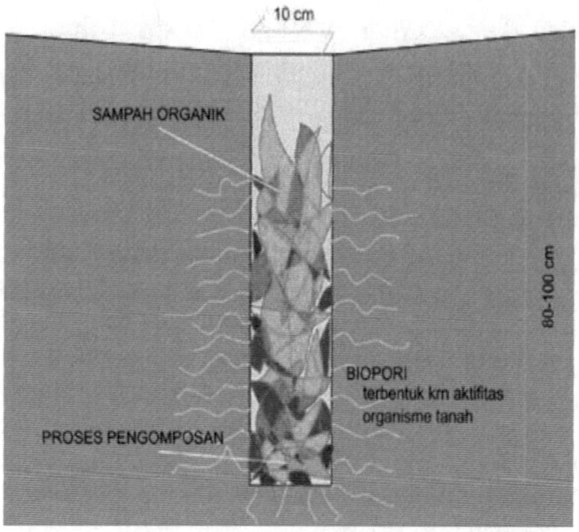

Figure 1. Bio pore

2.2. Urban Drainage

Land use change that continuously happened as an impact of the development of urban area could lead to the increment of surface runoff discharged and flood peak discharge [3]. The amount of surface runoff mostly determined by the pattern of land use that reflected by flow coefficient in the rational equation. The more rainfalls that flow to the drainage channel, the smaller water absorb to the soil. Consequently, it will cause flooding in the rainy season and the drought in the dry season.

2.3. Designed Rainfall

In hydrology, there were several probability distribution functions as listed below [4]:

- Normal and log normal distribution
- Gumbel distribution
- Log Pearson Type III distribution

There are two calibration tests that could be done to prove whether the selected distribution is appropriate with the data. Those calibration tests are Chi-quadrat test and Smirnov-Kolmogorov [5].

2.4. Time concentration

Time concentration of a watershed is the time required by the rainwater that falls to flow from the furthest point to the site of the watershed's output (control points). The formula of concentration time as given below as it is stated in [3]:

$$t_c = t_0 + t_d \tag{1}$$

2.5. Storage coefficient
According to [6], the effect of storage coefficient for the maximum flood is calculated by the equation of storage coefficient (*Cs*) as the following:

$$Cs = \frac{2t_c}{2t_c + t_d} \qquad (2)$$

2.6. Runoff Coefficient
According to [3], if watershed consists of varies land cover use with different surface flow coefficient, *C* is used as the watershed coefficient as calculated by the equation below:

$$C_{DAS} = \frac{\sum_{i=1}^{n} C_i A_i}{\sum_{i=1}^{n} A_i} \qquad (3)$$

2.7. Rainfall Intensity
The intensity of rainfall is a measurement of the amount of rain that falls over time. Rainfall intensity is calculated by Mononobe formula as given below [3]:

$$I = \frac{R_{24}}{24}\left(\frac{24}{t}\right)^{2/3} \qquad (4)$$

2.8. Designed flood discharge

2.8.1. Rational modification method.
According to[3] the method for estimating flow rate peak that commonly used is Rational Method USSCS (1973). However, the application is limited to watersheds area less than 300 ha. Accordingly, rational modification formula as the following equation is mostly used in designing urban drainage, [6]:

$$Q = 0.2778. C. Cs. I. A \qquad (5)$$

2.8.2 Water balance. According to [7], in hydrology, the relationship between input (*I*), Output (*Q*) and storage (*S*) is represented by the following equation:

$$\frac{dS}{st} = I - Q \qquad (6)$$

2.9. Efficient cross-section channel
According to [8], efficient cross-section channel is a cross-section that has the smallest wet circumference with the maximum efficiency.

A discharge of cross-section channel for any random flow can be defined by the following equation:

$$Qs = V \times As \qquad (7)$$

3. Methodology

3.1. Study location
Location for this study is in Kopelma Darussalam, particularly in the southern part with a total area 1,040 km^2. The planned drainage network is obeying the slope of the existing channel condition planned by [9].

IOP Conf. Series: Materials Science and Engineering **352** (2018) 012040 doi:10.1088/1757-899X/352/1/012040

3.2. Data collection

3.2.1. Precipitation data
Precipitation data used was the annual maximum daily rainfall of 56 years record since 1958 to 2013. These records obtained from BMKG station in Blang Bintang, Aceh Besar, Indonesia.

3.2.2. Topographic map.
The topographic map used was collected by Planning Bureau of Unsyiah. This data is used to get the information pertaining drainage networking as well as the flow direction in the region of Kopelma Darussalam.

The topographic condition of Kopelma Darussalam is relatively flat. Thus, it needs more attention regarding the preservation of drainage system as efforts toward flood-detention.

3.2.3. Map of land cover and use.
Land use in an urban area is one of the parameters that caused surface runoff. Land cover and use map are used to determined flow coefficient (C). The map was collected by the office of regional planning and development in Banda Aceh.

3.2.4. Infiltration data.
Infiltration data used obtained by earlier studies [10]. This data shows in table 1 below:

Table 1. Infiltration data of the land with and without the use of biopore in sand and silt

f_c	Sand	Silt
f_c soil (mm/minute)	0,07	0,025
f_c grass (mm/minute)	0,09	0,033
f_c thatch (mm/minute)	0,1	0,042
f_c biopori (mm/minute)	8	1

3.3. Designed flood discharged

3.3.1. Determination of designed rainfall.
Rainfall distribution type determined by calculating the distributions such as normal and log normal distribution, Gumbel distribution, and log-Pearson type III distribution. The selection of rainfall distribution is evaluated by Smirnov-Kolmogorov test.

3.3.2. The calculation of designed rainfall.
Rainfall estimation is determined based on the selected distribution type. The designed rainfall based on return period of two years. This return period is specified for tertiary channel type that designed for watershed less than 5 ha.

3.3.3. The calculation of storage coefficient.
Storage coefficient is estimated after concentration time that is determined by equation 1. Then, storage coefficient is calculated by the equation 2.

3.3.4. Runoff coefficient.
Runoff coefficient in a specific area related to its land cover and use is calculated by equation 3.

3.3.5. Rainfall intensity estimation.
Rainfall intensity estimated by Mononobe Formula as given in the equation 4.

3.3.6. Bio pore.

The bio pore is planned to be Ø10 cm in diameter and 80 cm in depth. These bio pores were dug on the land by an interval of 1 m to each other. Those determinations based on the reference, where the ideal diameter for bio pore is Ø10 cm and the depth of bio pore is depended on the water table. The depth should not exceed the water table. Based on site observation, the suitable depth of bio pore to be planned in Kopelma Darussalam is 80 cm.

3.3.7. The estimation of designed flood discharge.

The estimation of designed flood discharge is calculated in two conditions. First, flood discharge on the land without the use of bio pore by rational modification formula as in Equation 5. Second, the flood discharge on the land with the use of bio pores by water balance formula as given in equation 6.

3.4. The calculation of drainage channel dimension

The calculation of drainage channel dimension is designed with efficient cross-section channel in the shape of rectangular. Designed freeboard for this study is 30% from the water depth in the designed condition. Image of efficient cross-section channel is shown in figure 2.

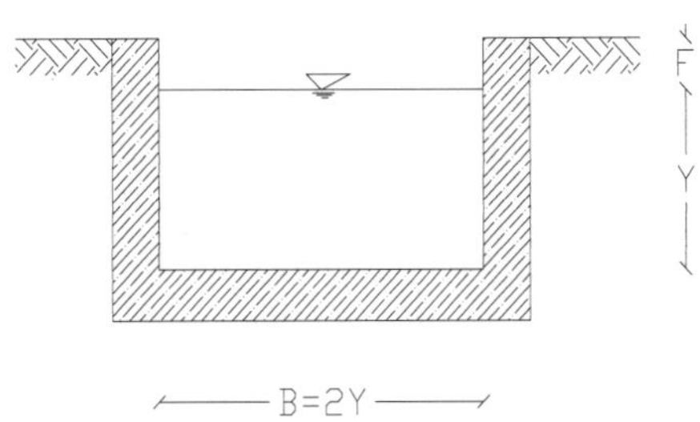

Figure 2. The efficient cross-section in the shape of rectangular.

The planned drainage channel is a tertiary channel and designed in two land conditions, with the use of bio pore and without the use of bio pore.

This study is redesigning the 13 tertiary channels with two outlets. Slope and elevation of the bottom channels for outlets are obeying the condition. Water elevation of Lamnyong river in the flood condition is projected save since it has been adapted to the condition designed by Sea Defense Consultant (SDC).

4. Result and discussion

The result and discussion involved the determination of rainfall distribution, the calculation of designed rainfall, storage coefficient, rainfall intensity, runoff coefficient, the determination of bio pore quantity, and the calculation of designed discharge as well as the calculation of dimension of drainage channel on the land with and without the use of bio pore.

4.1. Rainfall distribution

Statistical parameters used for the calculation are mean average ($\bar{x}$), standard deviation (*Sd*), an asymmetric coefficient (*Cs*), and variance coefficient (*Cv*). The calculation result of statistical parameters resumed on table 2.

Table 2. Statistical Parameters of Rainfall Distribution

Parameter	Result
Rainfall mean avg (ẋ)	111,998
Standard deviation (Sd)	40,724
asymmetric coefficient (Cs)	0,886
Varance Coefficient (Cv)	0,364
Cv3+3Cv	1,140

According to Table 2, a normal distribution is not fulfilled the rainfall distribution, since Cs≠0. It also does not match with log-normal distribution due to Cs≠ 3Cv+Cv^3. Rainfall distribution is estimated to supremely follow Gumbel distribution or log-Pearson type III distribution. Afterwards, Smirnov-Kolmogorov test is performed and results in α=0.05 as well as N=56, and $D_0 = 0{,}18$. From the overall calculation, D_{max}=0.0481 < D_0 is obtained. Meanwhile, Chi-Square testing found that X^2 = 1.5 < X^2cr = 9,488. The result of both tests shows that rainfall distribution of BMKG Blang Bintang station is obeying Gumbel Distribution.

4.2. Designed rainfall

Designed rainfall is calculated based on the selected distribution, it was Gumbel distribution. According to the calculation, R_{24} for two return periods is 105.7 mm.

4.3. Storage coefficient

Storage coefficient is determined by equation 2 that the result for storage coefficient is as it shown in table 3.

Table 3. Calculation table of time concentration and storage coefficient

No	Channel	Td(hour)	Tc(hour)	Cs
1	1-2	0,091	0,220	0,829
2	2-3	0,055	0,205	0,883
3	4-5	0,098	0,211	0,812
4	4-6	0,121	0,268	0,816
5	5-7	0,101	0,312	0,861
6	6-7	0,122	0,319	0,840
7	8-9	0,214	0,533	0,833
8	8-10	0,075	0,562	0,938
9	11-12	0,155	0,242	0,758
10	11-13	0,124	0,355	0,851
11	12-14	0,115	0,333	0,853
12	13-14	0,152	0,237	0,757
13	15-16	0,047	0,360	0,939
14	15-18	0,206	0,330	0,762
15	17-18	0,141	0,502	0,876
16	20-36	0,199	0,434	0,813
17	19-21	0,182	0,253	0,735
18	19-36	0,013	0,301	0,978
19	20-21	0,151	0,590	0,886

4.4. Runoff coefficient

Considering watersheds have varies land cover and different flow coefficient, *C* is determined by the formula that shown in the equation 3.

In this study, the region of Kopelma Darussalam separated into six watersheds. The total area for each watershed is shown in table 4.

Table 4. Watershed total areas

Watershed	Area(m^2)	Area (km^2)
1	123751,165	0,124
2	39087,801	0,039
3	143982,232	0,144
4	113793,028	0,114
5	311870,508	0,312
6	307169,400	0,307
Total	1039654,134	1,040

The calculations of runoff coefficient thoroughly provided in table 5.

Table 5. Runoff coefficient calculation (C)

Watershed	C
1	0,499
2	0,497
3	0,267
4	0,324
5	0,510
6	0,526

4.5. Rainfall intensity

Rainfall intensity is affected by daily rainfall for 24 hours (R_{24}) and concentration time (Tc) in mm/hour. Rainfall intensity is calculated by Mononobe formula as given in Equation 4. Return period used is two years with designed rainfall 105.7 mm. The rainfall intensity that obtained is illustrated in table 6.

Table 6. Rainfall intensity calculation (I)

No	Channel	Intensity (mm/hour)
1	1-2	100.404
2	2-3	105.209
3	4-5	103.351
4	4-6	88.160
5	5-7	79.653
6	6-7	78.420
7	8-9	55.770
8	8-10	53.829
9	11-12	94.303
10	11-13	73.042
11	12-14	76.244
12	13-14	95.815
13	15-16	72.405
14	15-18	76.777
15	17-18	58.047
16	20-36	63.973
17	19-21	91.689
18	19-36	81.642
19	20-21	52.113

4.6. Bio pore

The number of bio pores determined based on the circumstances where it is dug. Bio pore could be placed around the tree, in the yard, close to the fence, and in other possible locations. In this study, bio pores are placed with 1 m distance for each bio pore.

Considering a case where watershed 1 is located in a dense residential area, the bio pores were placed in the yard with varies quantities depend on the yard space. For a house with a total yard ≥ 50 m^2, eight bio pores were placed. Meanwhile, for a house with <50 m^2 total area of the yard, four bio pores is placed. Overall, the first watershed has been installed by 966 bio pores in total.

The layout of bio pore allocation in watershed 1 is shown in figure 3. Subsequently, the number of bio pore in each watershed is given in table 7.

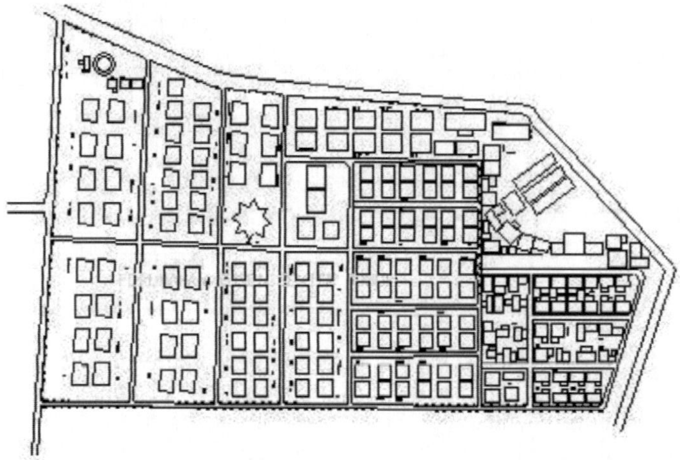

Figure 3. Layout of bio pore in watershed 1

Table 7. Number of bio pore in each watershed

Watershed	Number of bio pore	Bio pore spacious (m^2)
1	966	7,5831
2	188	1,4758
3	354	2,7789
4	388	3,0458
5	898	7,0493
6	556	4,3646
Total	3350	26,2975

4.7. Designed flood discharged

4.7.1. Designed flood discharge on the land without the use of bio pore.

Design Discharge on the land without the use of bio pore is calculated by rational modification as provided in Equation 5. For example, in the channel 2-3 with 0.0026 km^2 total area, surface runoff coefficient is 0.50, storage coefficient 0.883, and rainfall intensity is 105.4 mm/hour, so that:

$$Q_T = 0{,}278 \times 0{,}496519 \times 0{,}883 \times 105{,}4 \times 0{,}026$$

$$Q_T = 0{,}339 \text{ m}^3/\text{s}$$

Those design discharge also called inflow. The hydrograph of inflow discharge is shown in figure 4.

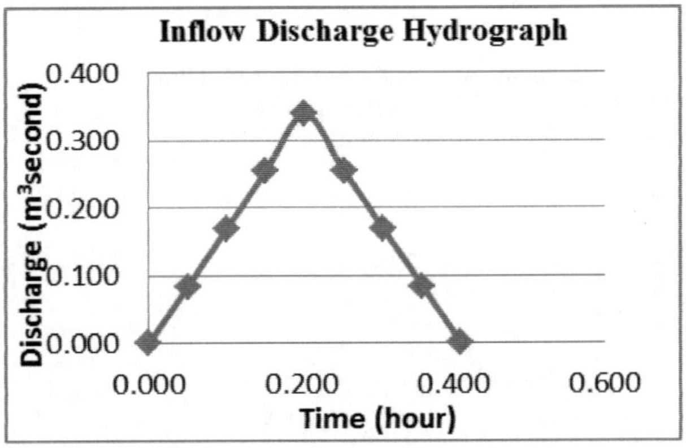

Figure 4. Inflow discharge hydrograph in channel 2-3

Figure 4 shows that time concentration is equal with rainfall duration, which is 0.205 hours.

4.7.2. Designed flood discharge on the land with the use of bio pores.

Design flood discharge on the land with the use of bio pores is calculated by water balance formula as shown in equation 6. For example, channel 2-3 has inflow discharge 0.339 m^3/s with total 102 bio pores. The bio pore is 10 cm in diameter and 80 cm in depth. The type of soil in that area is sand with the magnitude of bio pore infiltration is 8mm/minute. Outflow discharge of channel 2-3 is provided in table 8. Furthermore, combination graph of inflow and outflows discharge is illustrated in figure 5.

Table 8. The Calculation of Outflow Discharge in Channel 2-3

No.	T (jam)	Inflow (m3/s)	Vol.I (m3)	Vol.fcbiopori (m3)	ΔS (m3)	Vol.Q (m3)	Q (m3/s)
1	0,00	0,00	0,00	0,00	0,64	0,00	0,00
2	0,05	0,08	7,81	19,69	0,64	0,00	0,00
3	0,10	0,17	23,45	19,69	0,00	3,11	0,03
4	0,15	0,25	39,09	19,69	0,00	19,39	0,17
5	0,20	0,33	54,75	19,69	0,00	35,05	0,20
6	0,25	0,25	54,75	19,69	0,00	35,05	0,17
7	0,30	0,17	39,09	19,69	0,00	19,39	0,03
8	0,35	0,08	23,45	19,69	0,00	3,75	0,00
9	0,41	0,00	7,81	19,69	0,64	0,00	0,00

Table 8 shows that the implementation of bio pore has decreased the designed flood discharge by 40%. It is further illustrated in figure 5.

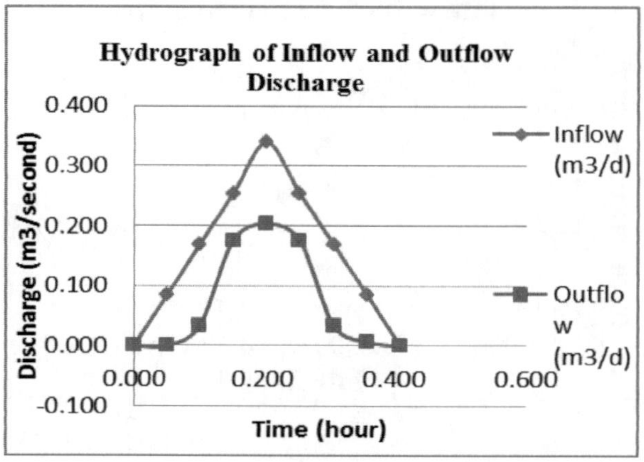

Figure 5. Inflow discharge hydrograph in channel 2-3

4.8. Channel dimension

4.8.1. Channel dimension on the land without the use of bio pores.
Manning coefficient for channel 2-3 is known 0.013, the velocity according to trial and error analysis for this channel is 0.813 m/s with slope 0.0008 (as the condition of existing channel by SDC). Then the calculation for channel dimension is:

$$V = \frac{1}{n} R^{2/3} S^{1/2}$$

$$Y = 2\left(\frac{n \times v}{\sqrt{S}}\right)^{3/2}$$

$$Y = 2\left(\frac{0{,}013 \times 0{,}813}{\sqrt{0{,}0008}}\right)^{3/2}$$

$$Y = 0{,}46 \text{ m}, \quad B = 2Y = 0{,}91 \text{ m}$$

$$F = 30\%\, Y = 0{,}14 \text{ m}$$

Afterwards, *Qs* for channel 2-3 is obtained as 0.340 m^3/s. It means that Qs =0.340 $m^3/s \geq Q_T$ = 0,339 m and dimension of drainage channel have met the requirement, where $Q_S \geq Q_T$.

4.8.2. Channel dimension on the land with the use of bio pores.
For channel 2-3, after the decrease of design discharge, the velocity is obtained as 0.716 m/s by trial and error analysis. Thus, the calculation for channel dimension is:

$$V = \frac{1}{n} R^{2/3} S^{1/2}$$

$$Y = 2\left(\frac{n \times v}{\sqrt{S}}\right)^{3/2}$$

$$Y = 2\left(\frac{0{,}013 \times 0{,}716}{\sqrt{0{,}0008}}\right)^{3/2}$$

$$Y = 0{,}38 \text{ m}, \qquad B = 2Y = 0{,}76 \text{ m}$$

$$F = 30\%\, Y = 0{,}11 \text{ m}$$

Next, Q_S is calculated and obtained $Q_S = 0.204$ m^3/s $\approx Q_T = 0{,}204$ m^3/s. This dimension of drainage channel has met the requirement, where $Q_S \geq Q_T$.

The comparison of drainage channel dimension on the land with and without the use of bio pores have been summarized in table 9, while the comparison of channel discharge on the land with and without the use of bio pores is shown in table 10.

Table 9. The difference dimensions of the land with and without the use of bio pores

No	Channel	Height (Y)			Width (B)		
		Without Biopore (m)	Biopore (m)	Difference (%)	Without biopore (m)	Biopore (m)	Difference (%)
1	2-3	0,46	0,38	17,35	0,92	0,76	17,92
2	4-6	0,51	0,40	20,69	0,92	0,81	12,05
3	5-7	0,76	0,64	15,63	1,52	1,28	15,63
4	6-7	0,40	0,33	17,28	0,80	0,66	17,50
5	8-9	0,41	0,36	12,06	0,82	0,72	12,06
6	8-10	0,89	0,73	17,38	1,78	1,46	17,73
7	11-12	0,64	0,62	2,66	1,27	1,24	2,66
8	12-14	0,74	0,68	7,46	1,48	1,36	7,94
9	15-18	0,37	0,29	20,67	0,74	0,58	21,97
10	17-18	0,40	0,35	13,54	0,80	0,70	12,67
11	19-21	1,19	1,13	5,31	2,38	2,26	4,96
12	19-36	1,02	1,02	0,01	2,04	2,04	0,01
13	20-21	0,92	0,92	0,20	1,85	1,84	0,20

Table 10. The difference of discharge in the channel on the land with and without the use of bio pores

No	Channel	Qs (m3/s)		
		Without Biopore	Biopore	Difference (%)
1	2-3	0,340	0,204	39,958
2	4-6	0,497	0,267	46,190
3	5-7	1,323	0,841	36,448
4	6-7	0,245	0,148	39,774
5	8-9	0,252	0,179	29,019
6	8-10	2,107	1,266	39,898
7	11-12	0,873	0,812	6,949
8	12-14	1,284	1,045	18,647
9	15-18	0,208	0,112	46,070
10	17-18	0,247	0,167	32,160
11	19-21	4,537	3,923	13,529
12	19-36	3,075	3,074	0,027
13	20-21	2,416	2,403	0,535

According to table 9, the comparison of the dimension of drainage channel on the land with the use of bio pores is smaller than channel dimension on the land without the use of bio pores. The average reduction of height and width are 11.5% and 11.7%. Furthermore, table 10 reveals that there is averagely 26.9% reduction of channel capacity.

5. Summary and suggestion

5.1 Summary

- Bio pore can increase the infiltration of rainfall to the soil and decrease surface runoff.
- The reduction of surface runoff discharge proves that bio pore influences over the decrease in surface water discharge.
- The dimension of drainage channel on the land with the use of bio pore is smaller compared to the dimension of drainage channel on the land without the use of bio pore. The average difference for both height (Y) and width (B) is 11.5% and 11.7%. It also reduces the average of channel capacity by 26.9%.

5.2. Suggestion

According to result and summary of this study, there are some recommendations proposed as follows.

- To compare the dimensions of the existing channel with redesign channel, it would be better to use the real existing dimensions accordance with its conditions in the field.
- To decrease more surface runoff, bio pore can be added more densely (50 cm $<x<$100 cm).

References

[1] Yulia 2014 *Studi Laju Infiltrasi Kawasan dengan Menggunakan Lubang Biopori Sebagai Upaya Penurunan Tinggi Genangan dan Upaya Konservasi Air Tanah* Thesis Magister Teknik Sipil (Banda Aceh)
[2] Brata K R and Nelistya A 2008 *Lubang Resapan Biopori* (Depok: Penebar Swadaya) (in Indonesian)
[3] Suripin 2004 *Sistem Drainase Perkotaan Yang Berkelanjutan* (Yogyakarta: Andi) (in Indonesian)
[4] Harto S 2000 *Analisis Hidrologi* (Jakarta: Gramedia Pustaka Utama) (in Indonesian)
[5] Triatmodjo B 2008 *Hidrologi Terapan* (Yogyakarta: Beta Offset) (in Indonesian)
[6] Subarkah I 1980 *Hidrologi Untuk Perencanaan Bangunan Air* (Bandung: Idea Dharma) (in Indonesian)
[7] Chow V T 1980 *Applied Hidrology* (Singapura: McGraw-Hill International Edition) (in Indonesian)
[8] Chow V T 1997 *Hidrolika Saluran Terbuka* Translated by Rosalina (Jakarta: Erlangga) (in Indonesian)
[9] Sea Defense Consultants 2004 *Proyek Drainase Package IV, Rehabilitation and Improvement of Main Drains and Structure in Banda Aceh Zona-7 and Darussalam* (Banda Aceh)

IOP Conf. Series: Materials Science and Engineering **352** (2018) 012041 doi:10.1088/1757-899X/352/1/012041

Comparative Study on the Cost of Building Public House Construction Using Red Brick and Interlock Brick Building Material in the City of Banda Aceh

Nurul Malahayati[1], Yulia Hayati[1], Cut Nursaniah[2], T. Firsa[3], Fachrurrazi[1], & Aris Munandar[1]

[1]Department of Civil Engineering Syiah Kuala University, Jln. Syech Abdurrauf No 7 Kopelma Darussalam, Banda Aceh
[2]Department of Architecture Engineering, Syiah Kuala University, Jln. Syech Abdurrauf No 7 Kopelma Darussalam, Banda Aceh
[3]Department of Mechanical Engineering, Syiah Kuala University, Jln. Syech Abdurrauf No 7 Kopelma Darussalam, Banda Aceh

[*]E-mail: malahayatijj@unsyiah.ac.id

Abstract. Red brick and interlocking brick are the building materials that are often used for wall installation work on houses construction. In the development of building materials technology and cost savings, interlocking brick can be alternative to replace red bricks. In Aceh Province, the use of interlocking bricks is less popular compared to other big cities in Indonesia. Interlocking brick is made from a mixture of clay, concrete sand and compacted cement and one of the environmentally friendly materials because it does not burn the process like red brick material. It is named interlocking brick because the installation method is locked together and it serves as a structural and partition wall of residential buildings. The aims of this study are to compare the cost of building a house in Banda Aceh City using red brick and interlock brick building materials. The data were obtained from interviews and questionnaires distributed to respondents who had built houses in Banda Aceh City. The results concluded that the house construction cost using interlock brick offer lower construction cost at comparable quality rather than using red brick.

1. Introduction

After the tsunami disaster in 2004, Aceh Province had been introduced to several building material replacements such as red brick that serves as a wall in house construction projects. One of which is called the interlocking brick. Interlocking brick was introduced to the community in Baitussalam Sub-district, Aceh Besar District, Aceh Province of Indonesia by an NGO's during Aceh and Nias tsunami reconstruction period. This interlocking brick is known as elephant brick among the Aceh people because it sizes is bigger than the sizes of ordinary red brick. The elephant brick production tool called the Soeng Thai Model BP6 compactor was used, which is capable of producing elephant bricks with the size of 300 x 150 x 100 mm (12"x 6" x4") in 9 different types of elephant bricks by adding or reducing parts of the mold. The mixture of elephant brick material used in this industry is cement, clay, and sand with a ratio of 1: 1: 2. However, production is not continuous and only produced when there is demand from consumers. These caused due to lack of promotion and knowledge-sharing to quantify and verify the benefits for the community about the use and adoption of this model.

Interlocking brick is made from a mixture of cement, clay, sand, and water compacted and is one of the environmentally friendly materials because it does not burn the process like red brick material. Interlocking brick or also called Interlocking Compressed Earth Block (ICEB) is a mixture of cement and compacted soil [1]. It is said interlocking brick because the material that serves as a wall in this system has an interlock or locked arrangement between one brick with another.

Method of installation of elephant bricks that do not use mortar (a mixture of cement, water, and sand) are possible because the bricks are shaped with projecting parts, which fit exactly into depressions in the bricks placed above and are automatically aligned horizontally and vertically. This locked system reduces labor and waiting time consumes for mortar mixing and cement drying. In addition, the attractive outward appearance of interlock brick causing unnecessary plastering and painting works to be done. The sloof position on the foundation was replaced with interlocking brick and placing vertical reinforcement at an angle, sides opening for the door, windows, on the deviation and apart, that has a large enough area. The Installation continues just like regular brickwork but does not use mortar with the addition of horizontal reinforcement on the side of the doors, windows and every 4-block layer [2].

Interlocking brick has a compressive strength of 2.6 MPa or equivalent to 7800 kg/beam with a mutual lock system (interlock), so it can function as part of building structure [2]. The use of interlocking brick as structural can eliminate the cost of reinforced concrete work such as sloof, column and ring balk. Therefore, there is no cost needed for reinforced concrete material and formwork. Interlocking brick masonry has the propensity to provide affordable, sustainable construction around the world and relatively low cost [3, 4, 5, 6].

Red brick definition according to Indonesian National Standard Number SNI 15-2094-1991 is a solid red brick made from clay with or without a mixture of other ingredients, which are burned at a sufficiently hot temperature until it has water resistance when soaking and has a cross-sectional area of less than 15% of pieces cutting flat areas [7]. Red bricks are usually traded with thick or 5 cm high, 10 cm wide, 20-24 cm long and weight less than 3 kg/pcs. The raw material requires red brick wall works are sand and cement.

Several studies have been conducted to get the cost of labor wage per unit of public house construction, type 36 house in Banda Aceh City from preparatory work until finishing by the method of wage per house. A wall work using red brick building material in 2012 is Rp. 19.500.000 with a duration of 65 days to completed [8].

This research will find out whether the use of interlocking brick as structural and filler wall can offer lower cost when compared to red brick wall work. The study also tries to obtain a cost comparison in public houses construction in the City of Banda Aceh using red brick and interlocking brick building materials. Further research for such comparisons is still required by weighting each of the aspects of the review [9].

2. Methods

The research was conducted in Aceh Besar district because respondents, who are the producer of interlocking brick, is located. The data sampling from the city of Banda Aceh were selected because it was assumed that most of the public houses construction type-36 are built in this location. The subject of the research is the cost of building public house construction type-36 from preparatory work to finishing. While the object of the research is the cost of reinforced concrete work and walls using red brick and elephant brick building materials.

The data used in this study consists of primary and secondary data. Primary data are respondent's characteristic, requirement, material cost, labor cost, duration, workforce composition per house, and implementation method of both for red brick and interlocking brick. These primary data was obtained by visiting workshop, interviewing the producer and head of worker for both red brick and interlocking brick. The head of worker are who had experienced on building public house construction using interlocking brick and red brick building material. Data about the interlocking brick is compulsory to gather because it is not included in the list of material units and wages in the Regulation of Governor of Aceh No 72 of 2016 [10]. Methods of the guided interview are selected in collecting primary data. Guided interviews are interviews conducted by the interviewer by making complete and detailed inquiries [11]. The collection of primary data is conducted for a duration of 1-2 months by guiding and explain directly to respondents.

Secondary data in the form of DED (Detail Engineering Design) of public house construction (type-36) is used as a model for quantity of work. The data for material unit price and wage per unit was refer to the Aceh Governor Regulation Year 2016 (in section of Banda Aceh area) [10],

Data analysis was conducted by steps: (1) calculating the cost of each buildings of public house construction (type-36) that using red brick. The method of cost calculation uses Unit Price Analysis. The unit price analysis refers to the Regulation of the Minister of Public Works and Public Housing. 28/ PRT/M/2016 [12]; (2) analyzing the actual price of materials and labor wages for wall installation work and reinforced concrete that using interlocking brick; (3) calculating the cost of each buildings of public house construction (type-36) that using interlocking brick. The method of cost calculation is similar to red brick in the point one. The differences of both building type are on the material of wall and material of the reinforced concrete works which are using the analysis at point (1) and (3).

3. Result and Discussion

3.1. Analysis of the Project Situation

This research assumed public housing construction project will be developed in the city of Banda Aceh for wall work using red brick building material as alternative 1 (one) and using interlocking brick as alternative 2 (two). Type of the Public housing is Type-36 that consists of 1 living room, 2 bedrooms, 1 bathroom and 1 kitchen. The specification of building uses a continuous foundation, tiled floors, wooden frame and zinc tile for roofing. The descriptions of works for both alternatives are same, the exception for concrete work, walls works, plastering and painting works. Table 1 describes the differences of the job description for both alternatives. The alternative 1 (one) is using reinforced concrete for sloof, column, ring balk, with additional plastering and painting works. While alternative 2 (two) for sloof, column, and ring balk just need placing a plain steel in horizontal and vertical directions. The attractive outer appearance on interlocking brick allows without plastering works and painting works.

Table 1. Differences of job description between alternatives 1 and 2

Job description	Alternative 1	Alternative 2
Concrete Reinforced Works		Uses interlocking brick, sloof, column and ring balk are replaced by placing plain steel in vertical and horizontal direction
1. Sloof	√	
2. Column	√	
3, Ring Beam	√	
Wall Works		
1. Brick Work 1: 2	√	Using Interlocking Brick
2. Brick Work 1: 4	√	Using Interlocking Brick
3. Plastering Work 1: 2	√	X
4. Plastering Work 1: 4	√	X
Painting Work	√	X

3.2. Unit Price of Interlocking Brick Building Material

The price per unit of material, the type, and the interlocking brick function are obtained from the literature and the results of the interviews with the respondents. The requirement of unit quantity for interlocking brick material of building type 36 houses is obtained from the analysis of quantity that refer to DED drawing. The data can be seen in table 2 below.

IOP Conf. Series: Materials Science and Engineering **352** (2018) 012041 doi:10.1088/1757-899X/352/1/012041

Table 2. The unit price of interlocking brick material

No.	Brick Interlocking Type	Picture	Price (IDR)	Function	Requirement for type-36 house (piece)
1	Building Block	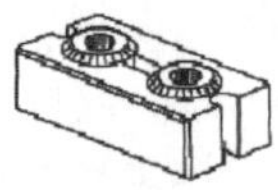	3.400	General Wall	3.297
2	Building Block corner	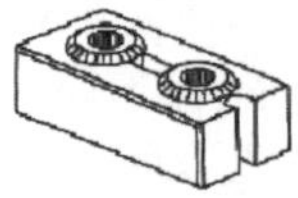	3.400	Corner Wall	553
3	Building Block Channels	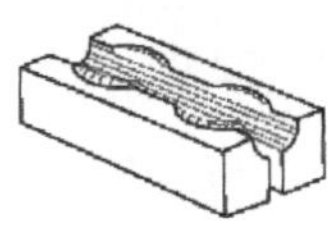	3.400	Horizontal Wall edification	744
4	Block Pillar	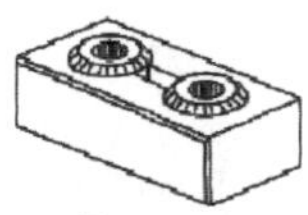	3.400	Pillars / Columns	0
5	Block Channel Angle	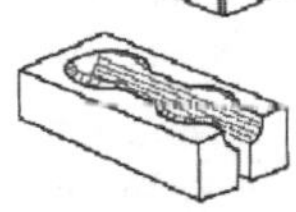	3.400	Horizontal Enlargement in Wall in Angle	74
6	Half Building Block	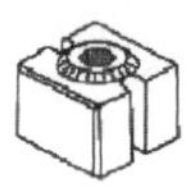	1.800	Junction Frame Not to Cut	340
7	Half Channel Block		1.800	Horizontal Wall edification	72
8	Half-Angle Block	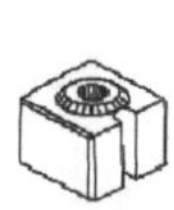	1.800	Corner Wall	82
9	Half Channel Corner Block	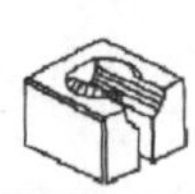	1.800	Horizontal Wall edification	82

The requirement of 1 m^2 wall works is 33,333-pieces of interlocking brick, 3.13 kg of Portland cement and 0,008 m^3 of sand. Each holes in the interlocking brick will be poured with sand. These works do not require mortar that serves as specie. Requirement for 1 m vertical iron reinforced works are 0.888 kg of plain iron, 0.25 kg of Portland cement and 0.011 m^3 of sand. Iron reinforced work on vertical direction serves as column and placed on the sides of the door and windows frames. Every column will be stick in hole of the interlocking brick with 4 pieces of iron and the other sides inserted 2 pieces of iron. For 1 m horizontal iron reinforced work requires 0.888 kg plain iron, 0.630 kg Portland cement and 0.0027 m^3 of sand. Iron reinforced work on horizontal direction function as sloof and ring balk, in addition toward this direction iron reinforced works are placed on every 4-layer block of interlocking brick with 1 piece of iron steel with dia.10. The requirement coefficient of interlocking brick material can be seen in table 3 below.

Table 3. Material requirements of interlocking brick

No.	Jobs Description	material	Unit	Coefficient
1.	1 m^2 interlocking brickwork	Interlocking brick	piece	33,333
		Portland cement	kg	3,130
		Sand	m^3	0,008
2.	1 m Vertical iron reinforced work	Plain iron	kg	0,888
		Portland cement	kg	0,250
		Sand	m^3	0,0011
3.	1 m horizontal iron reinforced work	Plain Iron	kg	0,888
		Portland cement	kg	0,630
		Sand	m^3	0,0027

The quality of the initial design in the form of the reinforced concrete frame structure and red brick works as a filler is certainly safe if it is implemented in accordance with the existing Indonesian National Standard code requirements. For interlocking brick material as structural as well as filler, the wall is having no Indonesian National Standard code that could determine the criteria for safe implementation. Buildings that have been built in Aceh Province from 2007 to 2010 still use designs with horizontal reinforcement every 8-layers.

3.3. Composition and wages of labor

According to the respondents to build public house construction such as the model in this study for the red brick and interlocking brick wall work requires Foreman, head of workers and worker in 1 day with a ratio of:

- Alternative 1 composition 1 Foremen: 3 Head of Workers: 5 workers
- Alternative 2 compositions 1 Foreman: 2 Head of Workers: 3 workers

So, for 1 m^2 area the required labor is: 1/36 X 6 = 0,167. This value is converted into the foreman and worker indexes by dividing it. The Head of workers has a larger contribution compared to workers. The ratio 60:40 then is used to generate the index needs between the head of worker and worker. 1The worker then the 60:40 ratio is compared to the worker and the builder in their index needs. The result of requirement index analysis for interlocking brickwork for every 1 m^2 horizontal/vertical works is shown in table 4. In this analysis, the head of workers requirement index is assumed 5% of the foreman requirement index.

Table 4. Labor requirements of interlocking brick

No.	Description	Labor	Unit	Coefficient
1.	1 m2 interlocking brickworks	Workers	P / day	0,100
		Head of Workers	P / day	0,067
		Foreman	P / day	0,005
2.	1 m vertical/horizontal reinforced works	Workers	P / day	0,100
		Head of Workers	P / day	0,067
		Foreman	P / day	0,005

The wages for works in alternative 1 is calculated based on the List of price per unit provided by the Aceh Governor Regulation Year 2016 for the city of Banda Aceh, while for alternative 2, in particular, the interlocking brickwork is fro, interview results with the respondents. The comparison of wages per day can be seen in table 5 below.

IOP Conf. Series: Materials Science and Engineering **352** (2018) 012041 doi:10.1088/1757-899X/352/1/012041

Table 5. Comparison of labor wages per day

No.	Type	Wages per day (IDR)		
		Foreman	Head of Workers	Workers
1	Alternative 1	130.000, -	98.000,-	87.000,-
2	Alternative 2	200.000,-	125.000, -	100.000,-

The difference in wages for both alternative are caused by the difference in actual wage. It is possible because the alternative 2 (two) is having additional skill on interlock brick and the number of worker are limited in availability that area. This constraint will affect the wages per day of workers which is higher compared to alternative 1 (one).

3.4. Comparison of Alternative Costs 1(one) and 2 (two)

This comparison utilized the Minister of Public Works and Public Housing No. 28 / PRT / M / 2016 regulation for calculation in of the unit price analysis of, except for interlocking brickworks, which used the result of coefficient need analysis based on the table below. In alternative 1 for red brick wall works the volume is 160, 58 m^2, the waged is IDR. 7,179,850, -, and the cost of materials is IDR.25.737.600,- While the wall works using interlocking brick the volume is 163,93 m^2, wages is IDR. 3,652,524, - and the cost of material is IDR. 24.673,596. Table 6 below explains the cost comparison between alternatives 1(one) and 2 (two).

Table 6. Comparison of alternative costs 1 and 2

No	Job description	Alternative Cost 1 (IDR)	Alternative Cost 2 (IDR)
1	Preparatory work	500.000,00	500.000,00
2	Land Works	3.467.642,66	3.467.642,66
3	Foundation work continuously	12.690.198,77	12.690.198,77
4	Reinforced concrete work (Sloof, Column, Ring Beam)	29.329.071,79	
5	Horizontal & Vertical Iron Works		23.898.161,87
6	Pair Brick Works	32.917.455	
7	Plastering Work	18.681.380,71	
8	Pair Interlocking Brick Work		28.326.120
9	Wall Painting work	9.771.134,16	
10	Frame Door/Window/Ventilation Painting Works	1.500.000,00	1.500.000,00
11	Frame and Door/Window/Ventilation Works	16.444.592,76	16.444.592,76
12	Floor and Ceramic Work	3.366.475,99	3.366.475,99
13	Roof and Ceiling Work	29.548.479,33	29.548.479,33
14	Sanitation Work	4.160.000,00	4.160.000,00
15	Mechanical and Electrical Work	2.685.000,00	2.685.000,00
16	Other Miscellaneous Works	4.222.115,39	4.222.115,39
	Total Cost (IDR)	**169.280.000,00**	**130.800.000,00**

In alternative 1 (one) the budget plan is IDR 169.280.000,- while in alternative 2 have a budget of IDR. 130.8 million, with the cost difference of IDR 38.480.000, - or there is a cost saving of 23% when using alternative 2. In alternative 1 the largest cost occurred in the reinforced concrete work in form of the formwork, iron and casting, while reinforced concrete work on alternative 2 is replaced with interlocking brick, where there is a hole that will be filled with plain iron with a diameter of 10 or D10, and in addition the horizontal and vertical direction is filled with a mixture of cement and sand.

The problems in this research have been answered, where building the public-house construction type-36 using interlocking brick is cheaper than using red brick. This outcome can be given to consumer, society, and public for consideration in selecting a substitute for red brick building material in for walls works used in Aceh Province.

4. Conclusion

Result of cost comparison for public houses (type-36) that are assumed to be built in Banda Aceh City using interlocking brick (elephant brick material) is cheaper than using the red brick material. Buildings that using interlocking brick (elephant bricks) in Banda Aceh from 2007 to 2010 have not collapsed and cracked despite being shaken by earthquakes several times. Research on the structural strength of elephant brick wall is underway at the Syiah Kuala University Building Materials and Structures Laboratory. Knowledge of elephant bricks to consumers will be provided by promoting the products through the event held in Aceh Province.

Acknowledgement

The research reported in this paper was supported in part by Institute for Research and Community Service (LPPM) Syiah Kuala University under the Contract No. 24/UN11.2/PM/SP3/2017. The authors wish to acknowledge the sponsor. Any opinions, findings, conclusions, and recommendations presented in this paper are those of the authors and do not necessarily reflect the views of the sponsors.

References

[1] Qu, B., Stirling, B. J., Laursen, P. T., Jansen, D. C., & Bland, D. W. (2012). *Interlocking compressed earth block walls: in-plane structural response of flexure-dominated walls.* In 15th world conference on earthquake engineering, IIAE, Lisbon.

[2] Wheeler, G. (2005). Interlocking Compressed Earth Blocks Volume II. Manual of Construction. Center for Vocational Building Technology, Thailand, 110.

[3] Adedeji, Y. M. D. (2008). *Interlocking masonry: Panacea for sustainable low-cost housing in Nigeria.* Pakistan Journal of Social Sciences, 5(8), 744-750.

[4] Calkins, M. (2008). *Materials for sustainable sites: a complete guide to the evaluation, selection, and use of sustainable construction materials.*

[5] Raheem, A. A., Bello, O. A., & Makinde, O. A. (2010). *A comparative study of cement and lime stabilized lateritic interlocking blocks.* The Pacific Journal of Science and Technology, 11(2), 27-34.

[6] Assiamah, S., Abeka, H., & Agyeman, S. (2016). *Comparative Study of Interlocking and Sandcrete Blocks for Building Walling Systems.* International Journal of Research in Engineering and Technology, vol.05(1),

[7] Departemen, P. U. (1989). *Spesifikasi Bahan Bangunan Bagian A* (SK SNI S-04-1989-F).

[8] Malahayati, N. (2013). *Durasi Proyek dan Upah Tenaga Kerja Berdasarkan Pengalaman Kepala Tukang Pada Konstruksi Rumah di Kota Banda Aceh.* Jurnal Teknik Sipil, 2(2),173-180.

[9] Fachrurrazi, F., Away, Y., & Husin, S. (2017). *The Weights Detection of Multi-criteria by using Solver.* International Journal of Electrical and Computer Engineering (IJECE), 7(2), 858-868.

[10] Dinas Pendapatan dan Kekayaan Aceh (2017). *Penetapan Standar Satuan Harga Barang Bahan Bangunan/Jasa Pemerintah*, Nomor: 028 /782 /2016, Banda Aceh, Indonesia.

[11] Nazir, M. Phd.(2013). *Metode penelitian.* Edisi. 8, Jakarta. Ghalia Indonesia.

[12] Kementerian Pekerjaan Umum (2016). *Peraturan Menteri Pekerjaan Umum dan Perumahan Rakyat No: 28/PRT/M/2016 Tentang Pedoman Analisis Harga Satuan Pekerjaan Bidang Pekerjaan Umum*, Jakarta, Indonesia.

IOP Conf. Series: Materials Science and Engineering **352** (2018) 012042 doi:10.1088/1757-899X/352/1/012042

The influence of Government's role and workforce's competence towards the construction workforce performance in Central Sulawesi

Nirmalawati, Mastura Labombang
Civil Engineering, Faculty of Engineering, University of Tadulako, Palu, Central Sulawesi, Indonesia

E-mail: nirmalawati_she@yahoo.co.id

Abstract. The performance of construction workforce should be enhanced, as it is a form of success of Central Sulawesi government in optimizing its management of construction workforces. The quality of construction workforces is when the proliferation of competence meets the standard. This research aims to identify the relationship between the role of the government and workforce competence in the advancement of construction workforce performance. This research used the quantitative method with the descriptive-correlation approach. The population in this research includes contractors, field observers, and project owner that conduct construction works in Central Sulawesi. The data collection in this research is conducted by distributing questionnaires directly to 42 respondents. The data analysis is done with statistics with the support of SPSS software. The result of the study shows that the role factor of the government has a powerful effect of 0401 on the performance of local workforce, while the competence of the construction workforce is 0.264 to the labor performance. Based on the research result, it is suggested that the policymakers, particularly in Central Sulawesi, take advantage of the research result as a reference in supervising and enhancing the performance of construction workforce in Central Sulawesi.

1. Introduction

The construction workforce's performance is one of the processes of working to achieve results. As stated by Armstrong et al. [1] on the definition of performance as work result or achievement. Work performance also has multiple meanings as not only the result of work but also the work process. When the performance of construction workforce increases, it will have a positive impact on the infrastructure construction which nowadays is developing rapidly especially in Central Sulawesi. Infrastructure construction will run smoothly and will not experience failure only when there is effective and efficient management in which one of it is to manage the construction workforce. The current construction workforce's competence is regulated by the latest construction service regulations where it is expected for the workforce to have competence in the field.

Based on the early study conducted by researchers on construction workforce in Central Sulawesi, it showed that all the construction workforce has good performance. However, most of the construction results are not as expected, experience failures, experience delays and other issues that resulted in poor construction results in Central Sulawesi.

The performance of construction workforce is a process, and the involvement of government is imperative. The role of government is critically important in increasing the performance of the construction workforce. Under the law no. 2, the year 2017 on Construction Service, it is said that the role of regional government is to be enhanced in which the regional government should form a standard of guidelines and criteria used as the basis for construction guidance in their respective regions [2]. Under this guidance, the construction training for the workforce conducted by the government shall be by the competence requirements of the workforce. As stated by Spencer-Spencer [3], competence is a part of character within an individual with regards to the effectiveness of its work performance.

The current result of field observation showed that the construction workforces in Central Sulawesi are mostly not competent in its field where they only work based on their experiences or work to fill the activities. These things obviously affect the performance of the workforce, such as Wibowo [3], including their productivity, quality, punctuality, cycle time, resource utilization and cost.

Based on the lack and weaknesses of the performance of construction workforce in Central Sulawesi, the researcher is keen to conduct further research on the performance of construction workforce in Central Sulawesi and the factors affecting it. The purpose of this research is to understand the relationship between government's role, workforce competence, and construction workforce's performances.

2. Literature Review

2.1 The Role of Government

The regional government is the head of a region and as an element to run the regional governance. Under the government regulation no. 30 the year 2000 article 7 on the Conduct of Construction Service Training, "Provincial government, regional government and the municipal government shall organize the Construction Service Training" in which one of its duties is on the development of human resources [5]. In the law no. 23 of 2014 on regional government, it is explained about the division of authority on the task of training & coaching between the local and central government with regards to the construction service training. More recently, under the construction service law no. 2 of 217 chapter VIII, the training of construction service is the responsibility of regional government. The role of government includes (a) government support, such as support of infrastructure facilities as well as fund support for the workforce. (b) the implementation of policies from the laws including the implementation of regional regulations regarding wage of construction workforce and their placement by their competence areas. (c) organizational training including cooperation with labor organizations in training the workforce.

2.2. Workforce Competence

Wibowo explained that competence is the ability to conduct a job or task based on skills and knowledge also supported with work attitude demanded by the job [4]. Under the government regulation no. 23 of 2004 on National Agency for Professional Certification (BSNP) Article 1 Paragraph 1 and 2 sub-paragraph B which stated that the Indonesian National Work Competence Standard is a formulation of workability that covers the aspects of knowledge, skills, expertise and works attitude relevant to the completion of duties and position terms set under the current regulations [6].

The workforce can be defined as any person that can work both inside and outside the working relationship to produce service or goods to fulfill their own needs and the society. Ervianto defined workforce as one of the resources difficult to manage [7]. It can be concluded that construction workforce is people that can work in a construction environment.

Based on the description above, it can be inferred that workforce competence is a set of knowledge, skills, and work attitude that must be owned by the people who works in a construction environment. The competence in this research covers the competence of knowledge, skills and work attitude.

2.3. The Performance of Construction Workforce

Rifai & Basri brought multiple definitions of work performance by many management experts [8] including: (1) work performance as a set of result achieved and refers to the execution of requested work [9]; (2) performance is one of the total set of work within a worker [10]; (3) performance is motivation function and the ability to work in which a person must have a degree of willingness and a certain label of ability [11]. Suntoro (in [12]) stated that performance is a result of work obtained by someone or a group of people in an organization to achieve the purposes and ais of the organization in a certain period. Mahsun stated that performance is a description of the achievement level of the conduct of a program or implementation of policy in achieving organization's aims, visions, and missions which are within the organization's strategic planning [12]. Rue & Bayars (in [12]) defined performance as a degree of accomplishment. Wibowo defined performance as a process f work to achieve work results [3].

Amstrong also classified few measurements of work including [1]: (1) money measurement, it is used to maximize income, minimize outcome and increasing income level; (2) time measurement, to work as scheduled; (4) influence measurement, including standard accomplishment, changes in behaviours, physical complementary work, and level of service acceptance; and (4) reaction, showing how others value the workers and therefore is objective. Wibowo also classified performance measurement including (1) productivity; (2) quality; (3) punctuality; (4) cycle time; (5) utilization of resources; and (6) cost [4].

Based on the above explanation elaborated by experts, we can conclude that the measurements of construction workforce performance that can be used are (1) work result; (2) working activities; and (3) competence possessed (certified).

2.4. Previous Researches

1. This research is titled "*Supply Demand* resources on construction in Central Sulawesi" [13]. It concludes: (1) There is still no conformity between the supply and the demand for human resources of construction supporting the implementation of construction works in Central Sulawesi Province, and there are still some of the less competent and professional road experts in their work; (2) the result of the supply and demand in road construction human resources in Central Sulawesi showed that there is only 24.6% (318 people) compared to the needs (1292 people) in 2012.Meanwhile, seeing from the graduation of the Faculty of Engineering in Central Sulawesi, the average number of graduates per year is about 25.3% (246 people) compared to the need (974 people), assuming all graduates work according to their field.
2. The research is titled "The Role of Government in Increasing Community Participation in the Development of Toundanou Village of Toluaan Sub-district of Southeast Minahasa Regency" [14]. The conclusions are: The role of government in implementing policy about increasing community participation in the implementation of development play well importance
3. Research entitled "The Influence of Public Participation and Government Participation in Land Acquisition Process for Toll Road Development Project in Surabaya" [15]. The conclusion is: the influential factor of the government's role is the timeliness of service.

3. Research Method

3.1. Research Location

The research took place in Central Sulawesi, the largest province in Sulawesi island with the land area of 68.033,00 km2.The northern part of Central Sulawesi Province is bordered by the Sulawesi Sea and Province of Gorontalo, the eastern part is adjacent to Maluku Province, the southern border with South Sulawesi and Southeast Sulawesi provinces, and the western part is limited by the Makassar Strait. The data in this research was taken from all districts including Banggai Kepulauan, Banggai, Banggai Laut, Morowali, Morowali Utara, Poso, Donggala, Tolitoli, Buol, Parigi Moutong, Tojo Una Una, Sigi, dan Palu city.

3.2. Research Design

This research is descriptive-correlation research. It is said to be descriptive because the researcher is trying to obtain information that is in accordance current condition [16].The researcher tries to describe the data. The population in this research are the contractor, field supervisor and project owner who carry out construction work in Central Sulawesi Province. The primary data collection used direct questionnaire technique with 42 respondents while the secondary data regarding the number of current construction projects and regional (district) data on construction works are obtained from the data of Ministry of Public Works. Data on the number of construction workers are taken from the Provincial Construction Services Development Agency (LPJK). Data analysis technique is done after the questionnaire collected that is as much as 42 respondents and calculated by using statistic with the support of SPSS software.

4. Results and Discussion

4.1. Instrument Validity and Reliability

To test the instruments' validity and reliability, the analysis is conducted with the help of SPSS. The following is the result of validity test. For the level of validity, the significance is tested by comparing the value of r arithmetic with r value.

Table 1. Validity & Reliability

No	Code	Questions	Correct. Item Total Correlation	r table	
1	P13	The government always supports the facilities and infrastructure of the implementation activities of the competency test of construction workers.	0.596	0,3044	Valid
2	P14	The government always supports the implementation of the training of construction workers	0.618	0,3044	Valid
3	P15	The government always provides funds for the empowerment of workers in proportion to the number of construction services work in Central Sulawesi	0.541	0,3044	Valid
4	P16	The government always implement the policy with regards to the current laws related to the placement of the competence of construction workers	0.559	0,3044	Valid
5	P17	The government always implement the policy with regards to the current laws related to the wages of construction workers in Central Sulawesi	0.627	0,3044	Valid
6	P18	The government makes local regulations for the benefit of empowerment, protection of construction workers by national laws	0.658	0,3044	Valid
7	P19	The government conducts cooperative activities with the organization of labor in carrying out guidance for construction workers	0.641	0,3044	Valid
8	P1	The construction workforce can understand	0.683	0,3044	Valid

		job descriptions			
9	P2	Construction workforce can process job directions clearly	0.753	0,3044	Valid
10	P3	Construction workers can think according to their competence	0.557	0,3044	Valid
11	P4	Construction workers have skills according to their competence	0.627	0,3044	Valid
12	P5	Construction workers have work-related experience	0.777	0,3044	Valid
13	P6	Construction workers can quickly accept the use of renewable technology in jobs	0.570	0,3044	Valid
14	P7	The construction workforce has a high discipline in working	0.570	0,3044	Valid
15	P8	Construction workers can communicate well in the field	0.556	0,3044	Valid
16	P9	Construction workers have ethics & good behavior in work	0.559	0,3044	Valid
17	P10	The construction workforce is responsible for the work	0.508	0,3044	Valid
18	P11	Construction workers never give up in work	0.680	0,3044	Valid
19	P12	Construction workers appreciate working time	0.579	0,3044	Valid
20	P20	Construction workers are skilled in carrying out work by the standards specified.	0.525	0,3044	Valid
21	P21	Workforce performs work following the result targets	0.727	0,3044	Valid
22	P22	Workers are active in carrying out work following work procedures	0.674	0,3044	Valid
23	P23	The workforce works according to the planned schedule of the work.	0.679	0,3044	Valid
24	P24	The workers have competence in accordance with the skills needed in the field work	0.606	0,3044	Valid
25	P25	Workers have competency certificates according to their skill levels.	0.631	0,3044	Valid

From the table above it can be seen that each question item has r count> of r table (0,325) and positive value. Thus, the item in question is declared valid. Also from the description of the table above, it can be seen that each variable has Cronbach Alpha> 0.60, meaning that the variable of Competency, Government Role and Performance can be said as reliable.

Table 2. Cronbach Alpha

Variable	Cronbach Alpha	Information
Role of government (X1)	0.628	Reliable
Competency (X2)	0.678	Reliable
Performance (Y)	0.645	Reliable

4.2. Correlation and Determination Coefficient

Multiple linear correlation coefficients are the index number used to measure the relationship between two variables. The coefficient of determination has a function to explain the extent to which the ability of the dependent variable (Performance) to independent variables (Role of Government and

Competence). The processed statistical results of the program assisted by SPSS 16.0 for windows shows that:

Table 3. Summary Model

Model	R	R Square	Adjusted R Square	Std. Error of the Estimate	Change Statistics					Durbin-Watson
					R Square Change	F Change	df1	df2	Sig. F Change	
1	**.911[a]**	**.830**	.821	.73046	.830	94.936	2	39	.000	1.785

The correlation coefficient (R) test between Performance (Y) variable with Competence variable (X1) and Government Role variable (X2) obtained the result of 0,830 meaning that the independent variable can explain the dependent variable equal to 83%, while the other 17% is explained by other variables that are not included in this model (not researched).

Table 4. Coefficients

Model	Unstandardized Coefficients		Standardized Coefficients	Collinearity Statistics	
	B	Std. Error	Beta	Tolerance	VIF
(Constant)	**.274**	1.871			
GOVERNMENT	**.401**	.075	.506	.484	2.066
COMPETENCE	**.264**	.053	.477	.484	2.066

The result of the regression equation is:

$$Y = 0,264X_1+0,401X_2+0,274, \quad (1)$$

the regression coefficient interpretation is as follows:

(1) The value of c = 0,274 meaning that other factors other than competency and government's role influence performance. The contribution of other unpredictable factors is 0,274.
(2) The value of A= 0,264meaning that if the competency increases by one unit then there will be a performance increase of 0,264.
(3) The value of b = 0,401meaning that if the government's role increases by one unit, then there will be an increase of 0,401 in work performance.
(4) R = 0,911, meaning that there is a strong correlation between the independent variables of X1 and X2 with the dependent variable Y.
(5) R^2 = 0,830 meaning that about 83,0 % variation of independent variables X1 and X2 can explain the variation of Y independent variable. Based on the above regression formula it can be concluded that the performance of construction workers depends on competence factors and the role of government.

4.3. Role of Government

The role of Government is the most influencing factor on the performance of construction workers. From the calculation results, indicating that if the role of government increases by unit then there will be a performance increase of 0.401. So the role of government in Central Sulawesi is very necessary for improving the performance of its workforce. This is following Act No. 2 of 2017 concerning Construction services, which states that the role of local government determines in progress in the field of construction services [14].

The results of the research indicate that the government's role is among others: government support, policy implementation of the applicable law, and organizational development. In the provision of paragraph (1) section b, the governor as the representative of the Central Government in the regions has the authority to (a) organize supervision on the selection of service providers in the implementation of Construction Services; (b) to organize supervision of construction and contract activities to conduct an orderly supervision of the implementation and orderly utilization of Construction Services in the province. In article 7, it states that the authority of the Provincial Government in the sub-affairs of Construction Services includes: (a) the holding of training of construction experts; And (b) Provision of the information system of Construction Services coverage of the provincial area.

Due to the immense importance and influence of the government, it is expected that the government will be able to support the training of construction workforce, the government provides funds for empowerment of workers in proportion to the number of construction services work in Central Sulawesi, the government implements the policies related to the placement of competence of construction workers, the government makes local regulations for the benefit of empowerment, and the government carries out cooperation activities with the organization of labor in carrying out guidance for construction workers.

4.4. Construction Workforce's Competency

Competence of construction work also influences the performance of construction workers in Central Sulawesi but less than the influence of the government's. The competence of construction workers here is measured by the components of knowledge, skills, and attitudes, and in other words, these components have a high validation in reflecting the competence of construction workers. The calculation results showed the value of $a = 0.264$ which means that if the competence increases by one unit then there will be a performance increase of 0.264. This is in line with the Construction Services Act no 2 2017 in Chapter VII section 70 stating that every construction worker working in the field of Construction Services must have a Work Competency Certificate, and each Service User and/or Service Provider shall employ the workforce Construction that has a work competency certificate as referred to in paragraph (1). Work competence certificates as referred to in paragraph (1) shall be obtained through a competency test following the standard of work competence.

It is also in line with the Regulation of the Minister of Public Works Number: 09 / PRT / M / 2013 about the Competency Requirements for Sub communication of Experts and Skilled Workers in the Ministry of Public Works Construction Services, in Article 4, Paragraph 1 states that an individual who provides construction services or an individual employed by a business entity providing construction services must have a certificate in accordance with the qualification level.

4.5. Construction workforce's performance

The performance of the construction workforce is measured by the approach of the work achieved, including the results of work, work activities and the level of competence possessed. Based on the result of the research, the role of government is very influential compared with competence factors. It can be concluded that the role of government in increasing work performance includes: (a) support in infrastructure facilities in every activity related to the conduct of labor competency test;(b) support in construction workforce training activities;(c)fund support in any construction empowerment activities; (d) it is expected that local governments can implement the current laws relating to the placement in accordance with competence, remuneration, protection of construction workers as well as in the development of appropriate regional regulations; and (e) the government may cooperate with existing labor organizations in carrying out its workforce development.

Wibowo stated that performance is a process how the work takes place to achieve the results of work [4]. Similarly, as stated by Wibowo, that competence is an ability to perform or perform a job or task based on skills and knowledge and supported by the work attitude demanded by the job [4]. So the competence of the workforce is also needed in improving the performance of the workforce, although the influence of competency is not as tight as the role of government. So it can be concluded that although a region has a lot of competent workforces but no support at all from the local government can impact to the workers seeking for appreciation outside the province or even abroad. Therefore, the local government must pay attention in handling this issue.

From the description above, it can be concluded that the workforce that has high performance are: (a) have the work results in accordance with the what has been planned; (B) having a good work activity which are the attitude and behavior in carrying out the work; and(c) and has competencies appropriate to its field of work and may be indicated by the certificate it holds.

5. Conclusion and Recommendations

Based on the results of the research analysis it is concluded that the role factor of the government has a strong influence of 0401 on the performance of local workforce if the role of government increases by unit then there will be a performance increase of 0.401. While the competence of construction workers has an effect of 0.264 on labor performance, meaning that if the competence increases by one unit then there will be a performance increase of 0.264. This shows that the role of government in Sulawesi Province is very necessary for improving the performance of its workforce.

5.1. Recommendations

- Both the central and regional government is recommended to fully support the workers' empowerment to achieve the increased level of work performance.
- It is recommended for the policymakers to take the benefits of this study as a reference in observing and increasing the performance of construction workforce.

References

[1] Amstrong M 1994 *Performance Management Translated by* Toni Setiawan 2004 (Jogyakarta: Tugu Publisher)

[2] Undang-undang Republik Indonesia No 02 Tahun 2017 About Construction Services

[3] Spencer & Spencer 1983 *Competence at work: modes for superior performance* (New York: John Willey &Sons, Inc)

[4] Wibowo 2007 *Manajemen Kinerja* (Jakarta: PT Raja grafindo Persada) (in Indonesian)

[5] Peraturan Pemerintah nomor 30 Tahun 2000 tentang *Penyelenggaraan Pembinaan Jasa Konstruksi*

[6] Peraturan Pemerintah nomor 23 Tahun 2004 tentang *Badan Nasional Sertifikasi Profesi*

[7] Ervianto WI 2002 *Manajemen Proyek Konstruksi* (Yogjakarta:Andi Offset) (in Indonesian)

[8] Rivai V and Basri F M A 2005 *Performance Appraisal* (Jakarta: PT Raja Grafindo) (in Indonesian)

[9] Stolovich & Keeps 1992 *Handbook of Human Performance Technology A Comprehensive Guide for Analysis Solving Performance Problems in Organizations* (San Fransisco: Jprney Bass Publisher)

[10] Griffin 1987 *Management* (Boston: Houghton Miffin)

[11] Hersey P & Blanchard 1995 *Management of Organizational Behavior* Translated by Agus Dharma 2004 (Jakarta: Gelora Aksara Pratama)

[12] Mahsun M 2006 *Pengukuran Kinerja Sektor Publik* (Yogyakarta: BPFE -Yogyakarta) (in Indonesian)

[13] Nirmalawati 2012 *Suppy Demand Sumber Daya di Bidang Konstruksi di Provinsi Sulawesi Tengah* (Jakarta: PU Bina Konstruksi) (in Indonesian)

[14] Steve R R 2010 *Peran Pemerintah dalam Meningkatkan Partisipasi Masyarakat dalam Pembangunan di desa Toundanau satu Kecamatan Toluanan Kabupaten Minahasa Tenggara* https://ejournal.unsrat.ac.id (Menado: Universitas Ratulangi) (in Indonesian)

[15] Zanuar Firmanto 2015 *Pengaruh Partisipasi Masyarakat dan Peranserta Pemerintah dalam Proses pembebasan lahan untuk Proyek Pembangunan Jalan Tol di Surabaya (*Surabaya: Perpustakaan ITS) (in Indonesian)

[16] Arikunto S 2002 *Prosedur Penelitian (Suatu pendekatan praktek*) (Jakarta: Rineka Cipta) (in Indonesian)

IOP Conf. Series: Materials Science and Engineering **352** (2018) 012043 doi:10.1088/1757-899X/352/1/012043

The northern tidal dynamic of Aceh waters: A 3D numerical model

M Irham[1], E Miswar[1], Y Ilhamsyah[1] and I Setiawan[1,2]
[1]Faculty of Marine and Fisheries, University of Syiah Kuala, Banda Aceh
[2]Center for Marine and Fisheries Study of Syiah Kuala University, Banda Aceh

E-mail: irham@unsyiah.ac.id

Abstract. The northern tidal dynamic of Aceh waters studied by employing three-dimensional (3D) numerical hydrodynamic model. The purpose of this study is to understand the phenomena and the characteristic of the northern tidal dynamic of Aceh waters. The research used the explicit-splitting scheme numerical model of Navier-Stokes formulation. The result displays that the vertical rotation of flow movement (vertical eddy) at a depth of 15 to 25 meter eastern part of the study area. Hence, the result also informs that the current circulation identically to the upwelling in the western region of Aceh during the wet season and vice versa. However, during the transitional season, the flow circulation depends on how the tidal dynamic occurs in the area.

1. Introduction

Aceh is one of the provinces in Indonesia, which is directly adjacent to the oceans. Geographically, the Aceh waters become the gateway or the world's sea transportation routes [1]. The vast sea area makes the Aceh waters have great potential in the marine field especially the current movement [2], [3], [4], [5]. This phenomenon is still under-researched and becomes an interesting object to be studied specifically. Aceh waters, similarly to other waters, changing the seasons and the transitional seasons are a distinct phenomenon that plays a role in reviewing the dynamics of the currents [6]. In January to March is the west monsoon, April to May is the season of transition, and June to September representing the east monsoon and November and December seasons of transition.

From the above explanation, it turns out that the waters of Aceh are interesting oceanographically to be investigated not only field study but also numerical model analysis. This is because of the complexity of topography and forces acting on these waters [2], [7], [8]. In numerical model analysis, however, there are inadequate studies about the dynamics of the current in Aceh waters. The research was conducted in a large coverage area and limited to the circulation of tidal currents in Indonesian waters, the Malacca Strait, and the Andaman Sea [6]. Therefore, to enhance the research about waters of Aceh, in this paper we try to describe the dynamics phenomenon of tidal currents in the waters of northern Aceh using the three-dimensional hydrodynamic numerical model.

2. Method

The method used in this research is as follows: literature study, secondary data collection, interpretation of satellite data, mathematical modeling of hydrodynamic model, and field data for verification. For the mathematical modeling, the discrete solution of Navier-Stokes hydrodynamic equations was used by applying the explicit-spitting finite different scheme method.

The study was conducted along the north of coastal waters of Aceh starting from north of Sabang island to east Aceh. The location of the research station is focused on the front, which is the area of the encounter of two opposite currents. Physical data such as tides, temperatures, salinity, and currents are secondary data, which then discretized and analyzed to process the data.

The basic equation used in this research can be written as [9]:

$$\frac{\partial u}{\partial t}+u\frac{\partial u}{\partial x}+v\frac{\partial u}{\partial y}+w\frac{\partial u}{\partial z}-fv=-\frac{1}{\rho}\frac{\partial p}{\partial x}+\frac{\partial}{\partial x}\left(A_H\frac{\partial u}{\partial x}\right)+\frac{\partial}{\partial y}\left(A_H\frac{\partial u}{\partial y}\right)+\frac{\partial}{\partial z}\left(A_v\frac{\partial u}{\partial z}\right) \quad (1)$$

$$\frac{\partial v}{\partial t}+u\frac{\partial v}{\partial x}+v\frac{\partial v}{\partial y}+w\frac{\partial v}{\partial z}+fu=-\frac{1}{\rho}\frac{\partial p}{\partial y}+\frac{\partial}{\partial x}\left(A_H\frac{\partial v}{\partial x}\right)+\frac{\partial}{\partial y}\left(A_H\frac{\partial v}{\partial y}\right)+\frac{\partial}{\partial z}\left(A_v\frac{\partial v}{\partial z}\right) \quad (2)$$

In z-direction, the hydrostatic equation then simplified in the form of

$$\frac{\partial p}{\partial z}=-\rho g \quad (3)$$

Equations 1, 2, and 3 used to model the dynamic of Aceh water through the process of the discretized method.

3. Result and Discussion

Based on the bathymetry map of the study area (Figure 1), the topography condition of northern Aceh varies widely. Shallow water with a depth of 100 to 200 m can be found along the east coast of Aceh from Lhokseumawe to Langsa. The gentle slope increase of topography to the depth of 1000 m from south to north shows in the region of Bireuen to Samalanga offshore. However, the northern of Aceh Besar to Sabang a drastic increase of depth occurs significantly to 1500 to 3000 m. The topographic feature in this region is complex, the trench and steep slope of the north of Sabang up to Andaman sea near the Nicobar waters varies. Geologically, the topographic variation in this area is because of the ring of fire of active subduction zone resulted in the lifting and folding ocean floor [10].

The simulation results in figure 2 show the average daily result of northern Aceh. The simulation results show the pattern of current movement following the tidal pattern, which is the existence of periodic alternating movement. Meanwhile, figure 3 shows the result of current dynamics generated by tides and winds during west monsoon from January-March 2014 showing that the current movement is highly concentrated in the tip of the island of Sumatra. This concentration changes as a wind change the direction monthly. The pattern of current movement shows that the current shifted every month during this west monsoon heading towards the northern part of Aceh.

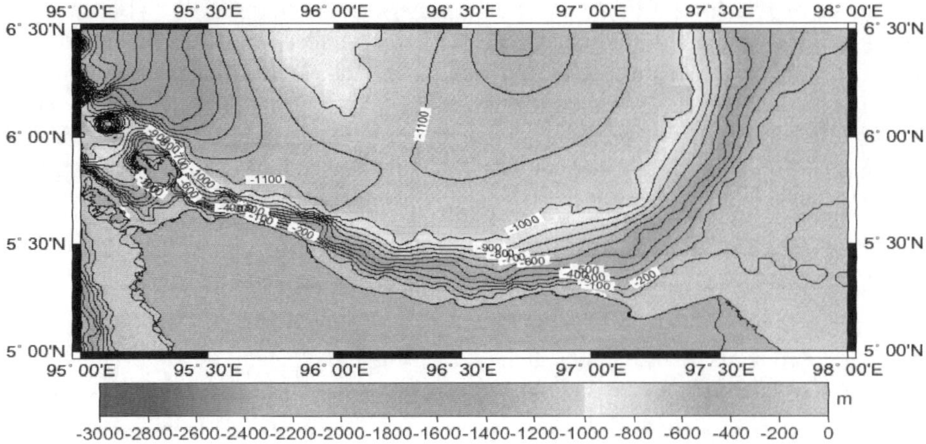

Figure 1. Bathymetry of study area showing a 2D horizontal contour of 100 contour interval.

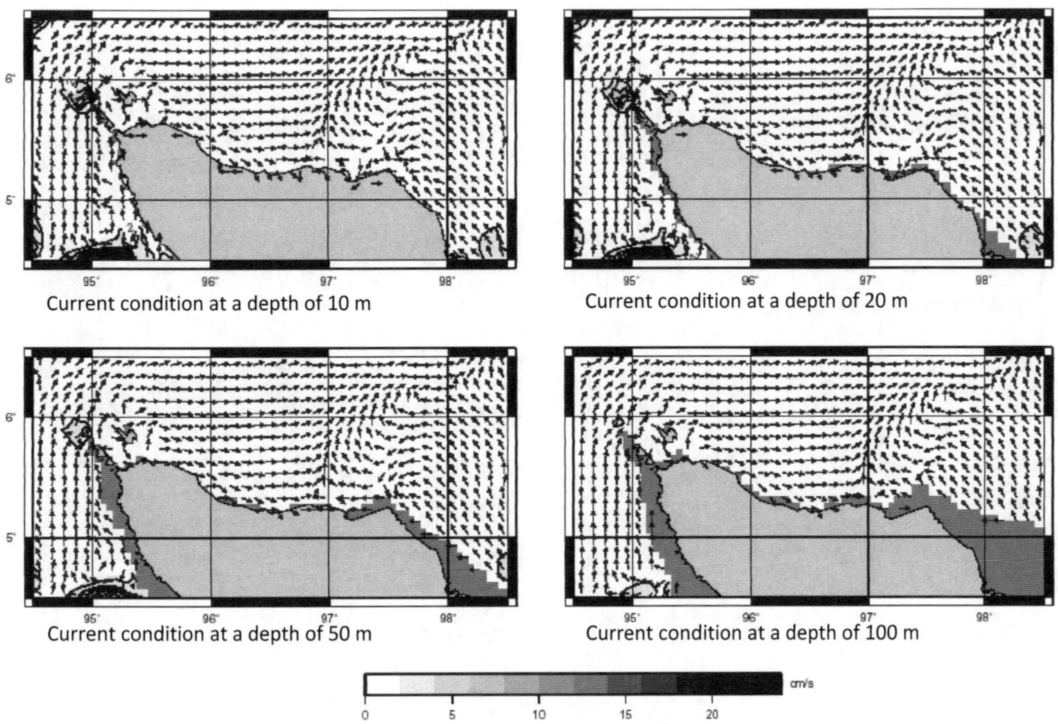

Figure 2. The tidal dynamic monthly average of 3D current on March 2014 during west monsoon

During the season of transition (April to June) in figure 3, the current condition shows almost similar everywhere. However, at the end of the transition year where the condition of the season tends to enter the east monsoon, the pattern of the current gradually increases especially in the eastern part of Aceh water.

The result of simulation in the east season from July to September (Figure 4) shows that the pattern of current movement is dominant in the east of Aceh while the western part of Aceh is dominated only in the area close to the Pulo Aceh. On the north part of Aceh, the seasonal influence for the current pattern movement is consider small which does not affect the east monsoon. This pattern occurs because of the topography of that area is a shallow region which the influence of wind is less dominant compared to another season.

During the second transition season from east monsoon to west monsoon (October to December) the current concentrated in the northern area of Aceh (figure 5). The pattern of currents moves from east to west of study area with an average ~20 cm/s. The pattern of current movement during this season steadily moves toward western part which a sign of the new season of west monsoon is approaching.

The cross-section of vertical velocity distribution in figure 6 shows the contour distribution pattern that is identical to the sinking and upwelling movement. This happened because of the massive water mass movement on the west coast of Aceh causing a vacuum of water masses on north and east coasts [1], [11]. The void of water volume was further filled by the mass of water coming from the West coast of Aceh.

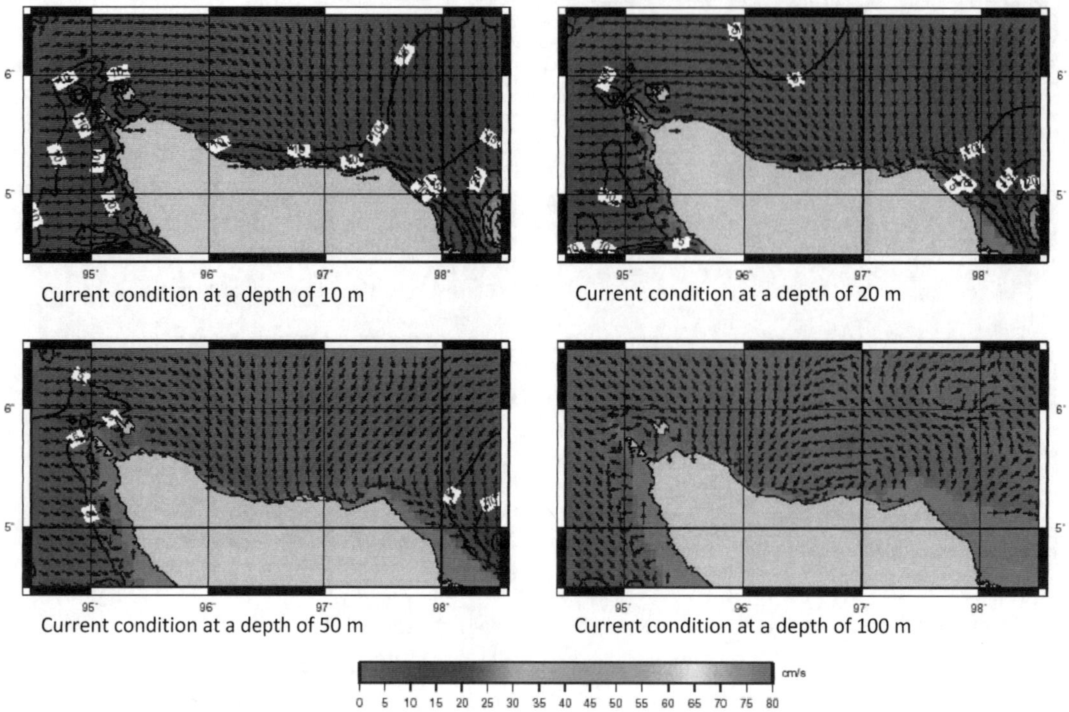

Figure 3. The tidal dynamic monthly average of 3D current on June 2014 during transition season

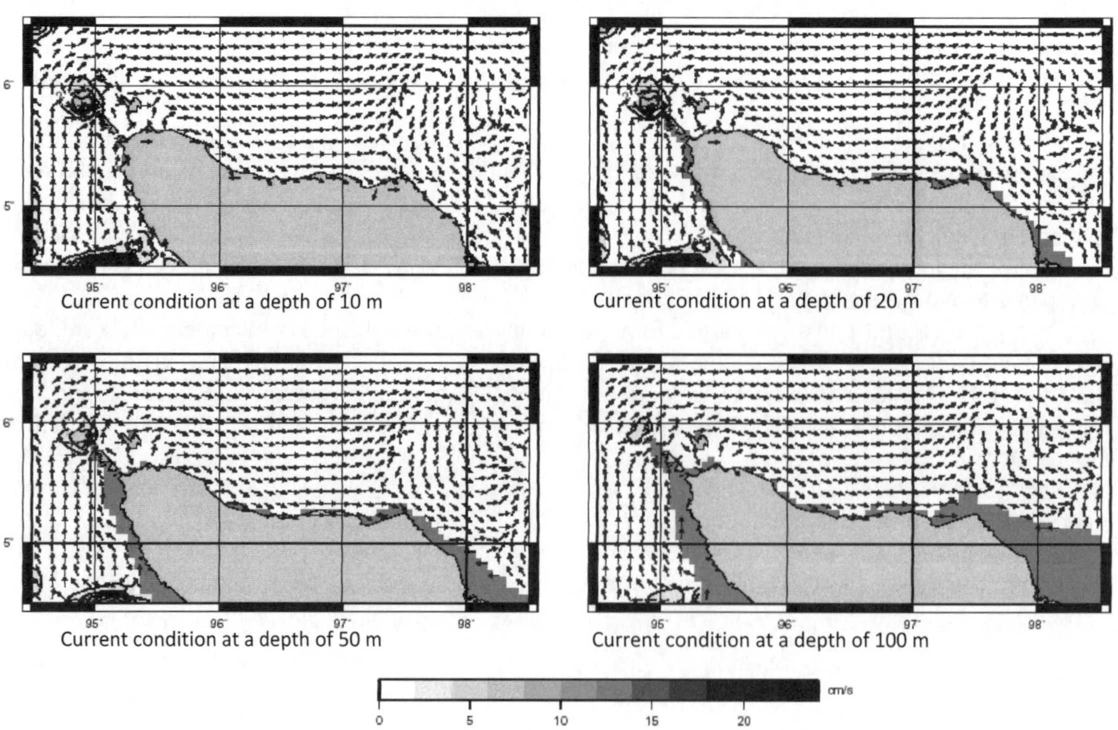

Figure 4. The tidal dynamic monthly average of 3D current on July 2014 during east monsoon

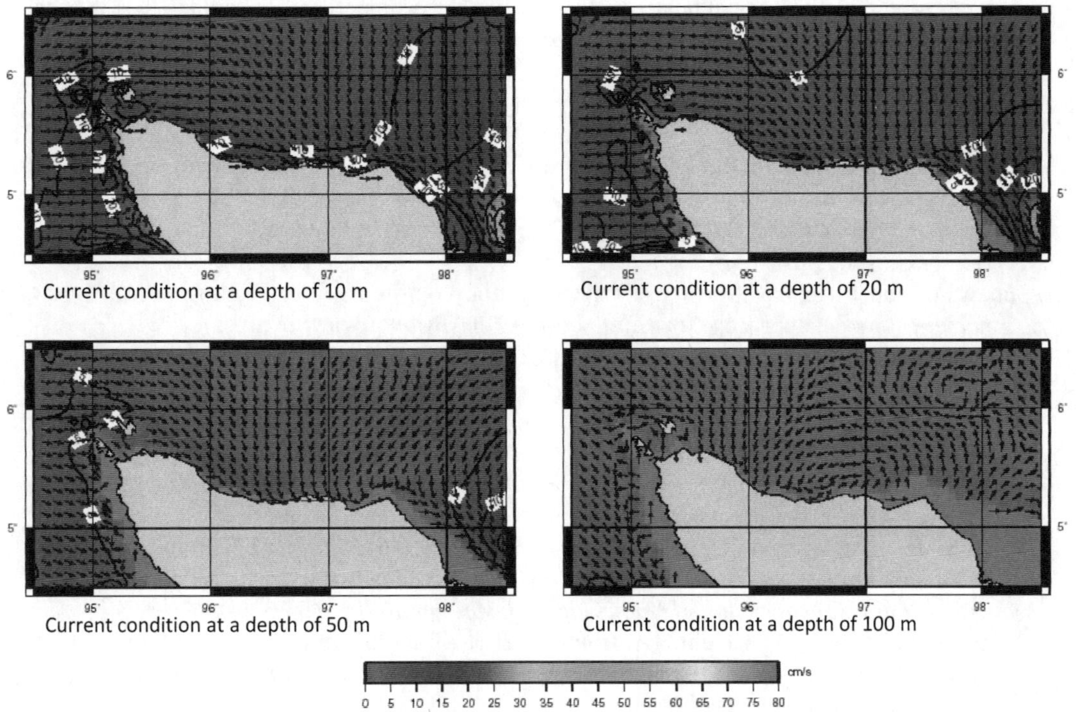

Figure 5. The tidal dynamic monthly average of 3D current on December 2014 during transition season

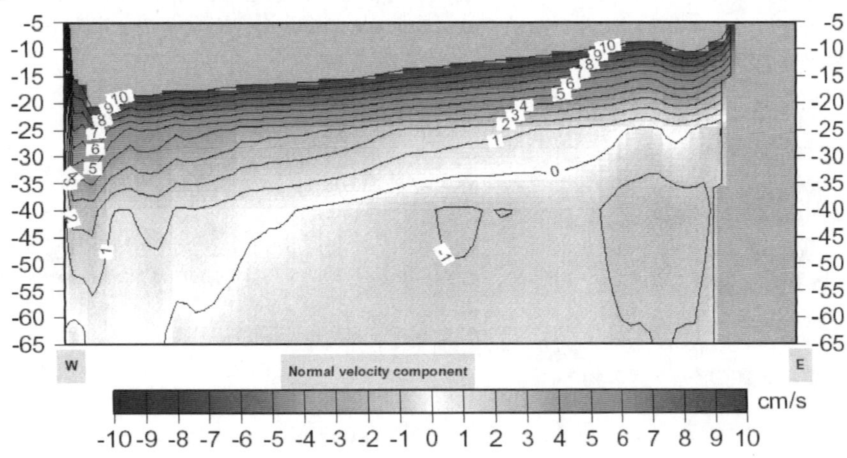

Figure 6. The pattern of cross-section vertical flow on the coast of Aceh water showing the sign of sinking and upwelling movement.

4. Conclusion

From the results, it is concluded that the seasonal wind movement profoundly dominates the dynamics of tidal currents in Aceh. During west monsoon, a dominant current occurs at the tip of the Sumatra Island, on the contrary in east monsoon, the dominant currents occur on the North and East coast of Aceh. Meanwhile, in the transitional season, the current conditions are almost similar. Furthermore, the pattern of current vertically is almost equal to the pattern of upper layer currents. This is because the stratification layer is not deep enough to occur the pattern of changing in current direction.

References

[1] Rizal S, Damm P, Wahid M A, Sundermann J, Ilhamsyah Y, Iskandar T and Muhammad 2012 General circulation in the Malacca Strait and Andaman Sea: A numerical model study *American Journal of Environmental Sciences* **8** 479-488

[2] Irham M and Setiawan I 2017 The Study of Flow Resulting from wave on Lhonga Beach, Aceh Besar *Omni Akuatika* **13** 5-12

[3] Setiawan I and Irham M 2017 Wave trajectory study on the coast of Lhoknga, Aceh Besar: A numerical model approach *Civil Engineering Dimension* (review in press)

[4] Irham M and Setiawan I 2018 The numerical model of the sediment distribution pattern at Lampulo National Fisheries Port *IOP Conference Series: Earth and Environment Science* **106** (012067) 1-6.

[5] Irham M, Fadla Y, and Setiawan I 2018 The spatial distribution of suspended sediment analysis along Krueng Cut River, Banda Aceh *IOP Conference Series: Earth and Environment Science* **106** (012066) 1-6.

[6] Rizal S, Haridhi H A, Wilson C R, Hasan A and Setiawan I 2013 Community collection of ocean current data: An example from northern Aceh province, Indonesia *SPC Trad. Mar. Res. Management and Knowledge Information Bulletin* **31** 3-11

[7] Purnawan S, Setiawan I, Haridhi HA, Irham M 2018 Granulometric analysis at Lampulo Fishing Port (LFP) substrate, Banda Aceh, Indonesia *IOP Conference Series: Earth and Environmental Science* **106** (012070) 1-6.

[8] Gomes E R, Mulligan R P, Brodie K L, and McNinch J E 2016 Bathymetric control on the spatial distribution of wave breaking in the surf zone of a natural beach *Coastal Engineering* **116** 180-194.

[9] Rizal S, Muhammad, Iskandar T, Setiawan I, Satriadi A and Radinal 2011 Simulation of Sea Surface Temperature (SST) and Sea Surface Salinity (SSS) in the Bay of Bengal *Proceedings of Annual International Conference Syiah Kuala University (AIC Unsyiah)* 153-159

[10] Setiawan I, Ilhamsyah Y, Miswar E dan Haddrevi A 2008 A prediction study of wave propagation before and after tsunami in Ulee Lheue coastal waters, Aceh, Indonesia *International Symposium Land Use after the Tsunami-Supporting Education, Research and Development in the Aceh Region* 249-254

[11] Chen G, Wang D, and Hou Y 2012 The features and interannual variability mechanism of mesoscale eddies in the Bay of Bengal *Continental Shelf Research* **47** 178-185

IOP Conf. Series: Materials Science and Engineering **352** (2018) 012044 doi:10.1088/1757-899X/352/1/012044

Characterization a binderless particleboard of coffee husk using Hydrogen Peroxide (H_2O_2) and Ferrous Sulfate ($FeSO_4$)

Milawarni[1], Nurlaili[2], Sariyadi[3]
[1] Electrical Engineering Department, Lhokseumawe State Polytechnic
[2] Civil Engineering Department, Lhokseumawe State Polytechnic
[3] Chemical Engineering Department, Lhokseumawe State Polytechnic

E-mail: milawarni@pnl.ac.id

Abstract. Binderless particleboard is particleboard that can be made of a lignocellulose material which is formed into a board only by heat pressing without the addition of adhesive or resin. The particleboard in this study was made from coffee husk (*endocarp*) using H_2O_2 and $FeSO_4$ catalyst to activate lignin coffee husk component by oxidation method. Initial treatment of coffee husk is the variation of steam then Oxidation (S + O) and Oxidation without steaming (O). In this study H_2O_2 and $FeSO_4$ catalysts were varied, including H_2O_2 levels of 10,20,30 wt% based on particle dry weight and $FeSO_4$ is 5 and 7.5 wt% based on H_2O_2 weight. From the results of the study, it can be concluded that the coffee husk particleboard whose raw material is treated oxidation without steam can improve the physical properties of binderless particleboard. Increased wt% of H_2O_2 and $FeSO_4$ catalysts in the oxidation process of coffee husk particles produce binderless particleboard with good physical characteristics such as density, water content, water absorption and swelling thickness. Therefore, considering the efficient aspects of the use of chemicals, the combination of H_2O_2 and $FeSO_4$ catalysts that can be made according to JIS A 5908 2003 standard are 20% H_2O_2 and 7.5% $FeSO_4$. The ester linkages were detected by Fourier transform infrared spectroscopy, indicated that cross-link due to the incorporation of phenoxyl radicals.

1. Introduction

Particleboards are normally fabricated by using synthetic adhesives such as urea formaldehyde, melamine formaldehyde or phenol formaldehyde. Synthetic adhesives emit formaldehyde emissions, which may pollute the environment. Therefore, in this study, we utilize the extractive substances such as lignin, hemicellulose as a natural glue. These substances are contained in coffee husk waste from the skin of coffee beans known as *endocarp*. Particleboard from coffee husk waste is normally processed using polypropylene [1]. However, the quality of the product is still low due to its modulus of elasticity (MOE) value that has not met the standard.

The utilization of natural glue substance can reduce the dependence on synthetic adhesive obtained from the processed petroleum origin, which is a non-renewable resource [2]. Based on its polyphenol structure, lignin can also function as an adhesive since its structure is similar to phenol formaldehyde resins. Both materials have almost identical chemical components of the phenolic group. Thus lignin is able to substitute phenol formaldehyde. Some works by researchers that develop a method to produce lignin directly from wood have inspired the potential of lignin as an adhesive. It allows the fibers in the wood to bind themselves without additional adhesives [3]. One method that has been successfully developed is the oxidation method. This method uses hydrogen peroxide (H_2O_2) and ferrous sulfate

($FeSO_4$) catalyst to activate lignin component of wood particles [4]. Previously, particleboard was fabricated using a natural glue based on bamboo and wood, which was obtained by oxidation method in conjunction with boiling treatment [5]. The result showed that the particleboard obtained by this method has dimensional stability and high mechanical properties compared to particleboard obtained using formaldehyde adhesive. On the other hand, the development of particleboard by using steam heat injection has also attracted attention from researchers. It was shown that the particleboard manufactured from the kenaf core using a heat steam injection method has a high adherence strength at board density of 0.65 g cm^{-3} or more [6].

In this work, we fabricate and characterize binderless particleboard using coffee husk with oxidation method. The optimal hydrogen peroxide (H_2O_2) and catalyst ferrous sulfate ($FeSO_4$) used for the oxidation process is also investigated.

2. Experimental Procedures

The experimental procedure is conducting in three stages; coffee husk treatment, preparation of percent weight of H_2O_2 and FeSO4, preparation of molder and fabrication of particleboard. Coffee husk treatment; the materials are coffee husk, hydrogen peroxide (H_2O_2) and ferrous sulfate ($FeSO_4$). The raw material is coffee husk (Arabica coffee) which was obtained from the coffee-milling factory. The material is dried in the open air for one week until water sorption less than 8% before it is grinded into a particle. The grinding process uses coffee skin cutter, which then followed by filtering to obtain the same size of 20/24 mesh. Preparation of percent weight (wt%) of hydrogen peroxide (H_2O_2) and catalyst ferrous sulfate ($FeSO_4$) and type of treatment coffee husk can show from table 1.

Table 1. Percent weight (wt%) of H_2O_2 and $FeSO_4$ from type of treatment

Type of treatment coffee husk	$FeSO_4$ (wt%)	H_2O_2 (wt%)
Steaming with Oxidation (S+O)	5	10,20,30
	7.5	10,20,30
Oxidation only (O)	5	10,20,30
	7,5	10,20,30

Preparation of molder; particle molding tool was made from iron with the size of 100 x150 x 10 mm. The coffee husk particles are then molded to fabricate particleboard using the prepared iron molding tool.

Fabrication of particleboard; the amount of H_2O_2 used varies based on the dry weight of the particles, while the $FeSO_4$ content is also determined based on the weight of hydrogen peroxide (H_2O_2). The weighted particles are steamed for 15 minutes, before they are drained and cooled. The particles are treated with further oxidation with spraying hydrogen peroxide (H_2O_2) and ferrous sulfate ($FeSO_4$) onto the coffee husk particles to complete the oxidation treatment (S+O). The treatment of coffee husk without steam (O) can be done by directly spray with hydrogen peroxide (H_2O_2) and ferrous sulfate ($FeSO_4$) catalyst. The particles, which have been sprayed by H_2O_2 and $FeSO_4$ are then conditioned for 90 minutes before being heat-treated. Then the particles are inserted into the mold. The temperature of the fitting plate used is 200°C with duration of 15 minutes, and pressure 25 $kgfcm^{-2}$. Determination of the effectiveness of the treatment method is conducted based on physical properties test result of the particleboard without adhesive produced. In this research will be done physical properties tested include: density, water content, water sorption and swelling thickness. Tests are carried out refer to JIS A 5908 2003 [7].

3. Results and Discussions

The physical properties of the particleboard such as density, water content, water sorption and the swelling thickness are investigated and discussed with and without steaming during the oxidation process. We also perform Fourier Transform Infrared (FTIR) measurement for the fabricated particleboard. Figure 1 shows the density test result indicating the effect of the addition of H_2O_2 and

$FeSO_4$ on the treatment with steam and Oxidation (S + O) and Direct Oxidation (O). As shown in the figure, the density of the particleboards with steam and oxidation (S+O) are obtained in the range from 0.37 g/cm^3 to 0.46 g/cm^3.The treatment with direct oxidation is 0.47 g/cm^3 to 0.53 g/cm3. It is observed that the treatment of oxidation without steam (O) can increase density. The highest density of 0.53 g/cm^3 was obtained with the use of 20% H_2O_2 and 7.5% $FeSO_4$. This indicates that a heat oxidation process at this level capable to activate the lignocellulose material (cellulose, hemicellulose, and lignin). It is also revealed through this research that adhesive particleboard can be made of lingo-cellulosic material formed into a board only by hot presses without additional adhesives or resins [8]. However, at 30% H_2O_2 and 5% and 7.5% $FeSO_4$, degradation of density values were observed. This indicates the degradation of coffee husks chemical components which evaporates during the oxidation process or heat-forging process. This shows that extractive substances in the coffee husks play an important role in the bonding process and the oxidation process can strengthen the bonds between the fibers and provide the chance of fiber bonds, which is getting in contact with both and become compressed. All samples met density values, it means that they met the JIS standard (2003) ranging from 0.4-0.9 gr/cm^3. However, the exception occurs only for steam with oxidation treatment (S + O) samples with 10% H_2O_2 and 5%$FeSO_4$ and 30% H_2O_2 and 5%$FeSO_4$.

The average water content value of the particleboards is shown in Figure 2. The average value of water content in the steam with dioxide (S + O) process ranges from 10.4-14.5% and in the direct oxidation process (O) ranges between 6.5-8.3%.

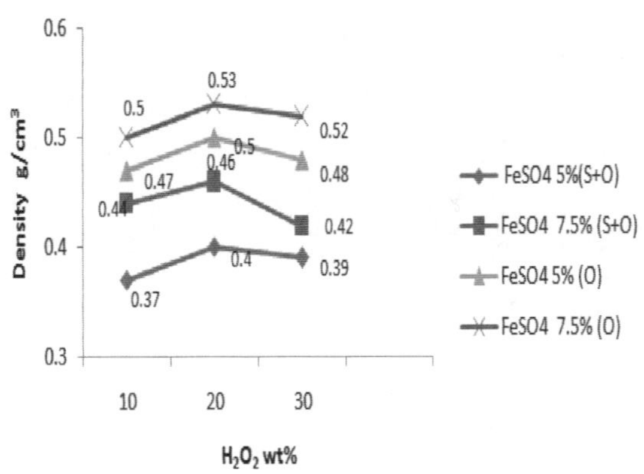

Figure 1. The density of particleboard with various wt % of H_2O_2 and $FeSO_4$.

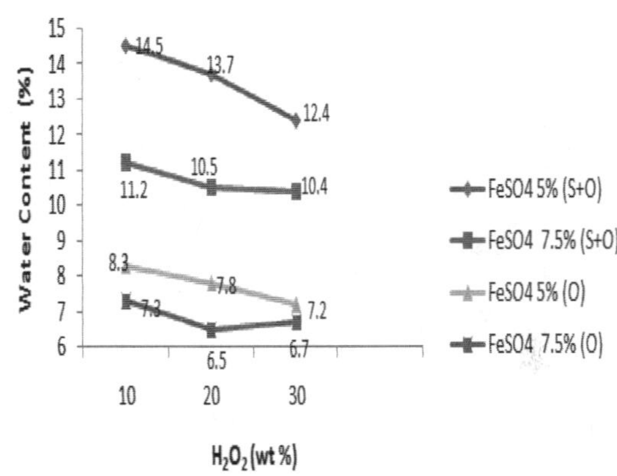

Figure 2. The water content of particleboard with various wt% of H_2O_2 and $FeSO_4$.

Figure 2 shows that there is a decline of water content for both treatments (20% H_2O_2 and 7.5% $FeSO_4$), indicating the effect of H_2O_2 and addition of $FeSO_4$, where oxidation process at this level should activate the lignocellulose material (cellulose, hemicellulose, and lignin). Although lignin levels in the coffee husks are not too large since the lignin that lies on the surface of the fiber. Lignin is a component that tends to be hydrophobic so that it inhibits the absorption of H_2O_2 and air. This condition becomes the cause so that the H_2O_2 of the particleboard equilibrium with the oxidation treatment tends to be lower. In the treatment of coffee husks directly oxidation (O) can reduce the H_2O_2 level. This indicates that the oxidation treatment tends to decrease the hydrophilic nature of the particles so that the balance of H_2O_2 becomes low. In general, the higher density value, the lower water content will be [9]. The water content of the board is very important in maintaining its dimensional stability. Low water content can improve the dimensional stability of a board, which affects the board quality as well. All samples for the water content test in this study met JIS A (2003) standard (5-13%), except in 10-20% H_2O_2 and 5 wt % FeSO4 in S + O coffee husk treatment.

Figure 3 shows the ability of particleboard in absorbing water as the ability of the particles to absorb water in its walls and cell cavities [10]. As shown in the figure, the water sorption of the particleboards with steam and oxidation (S+O) are obtained in the range from 58% to 75%.The treatment with direct oxidation is 48% to 61%. This phenomenon is usually followed by the return of the cell to its original form after previously *collapsed* due to high temperature and pressure. In addition, the inter-spaces between the particles can also be the entry of water that participates in improving the water sorption of the particleboard. However, the ability may decrease when the bonding capacity of the particles is sufficiently high. Thus, by preventing the cell from returning to its original shape and preventing water from entering empty spaces between particles, then the phenomenon of a reduced particle of water sorption along with increased H_2O_2 levels appears due to stronger bonds between particles as more oxidizing agents become available. In addition, the amount of lignin that is present on the surface of the particles due to the oxidation process and the heat crush seems to also be a barrier to the entry of water in the particle. The treatment of coffee husk without steam treatment (O) has water absorption properties that tend to decrease at 20% H_2O_2 and 7.5% $FeSO_4$. This indicates that in this level can decrease the hydrophilic nature of the particles so that it is more difficult to absorb the water caused by the bonding of the OH groups on the particles or substituting some of the OH groups into O radicals that form cross-links with other components.

Figure 4 shows the value of swelling thickness during 24 hours immersion in the steaming and oxidized (S+O) of coffee husk treatment ranges from 12-29%, while direct oxidation treatment ranges between 9-15%.

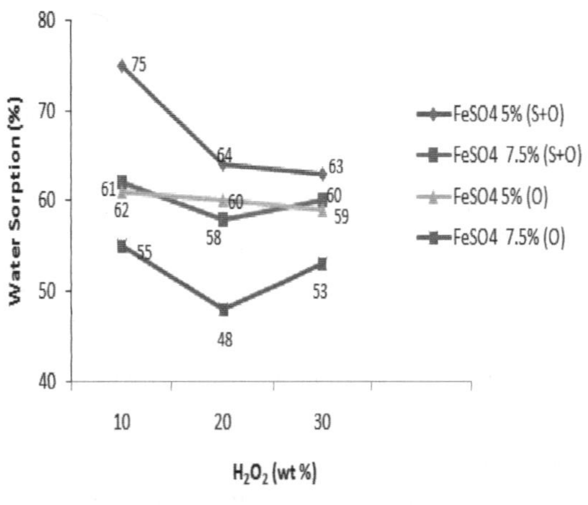

Figure 3. Sorption of particleboard with various level of H_2O_2 and $FeSO_4$.

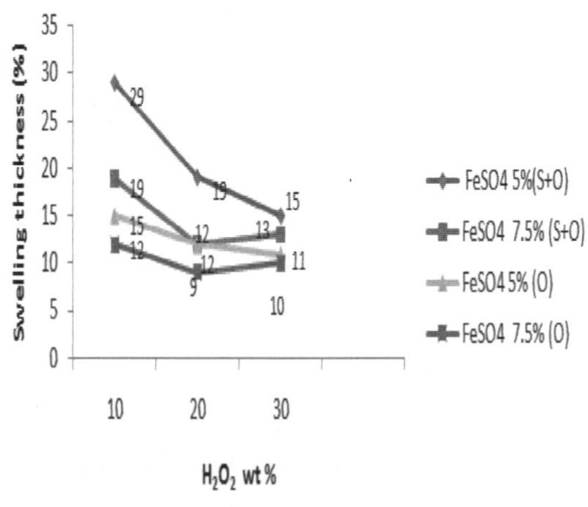

Figure 4. The swelling thickness of particleboard with various level of $H_2O_2$and $FeSO_4$.

Figure 4 also shows that H_2O_2 and $FeSO_4$ levels have an effect on the swelling thickness of particleboard. The addition of H_2O_2 decreases the swelling thickness property of particleboard. The value of swelling thickness decreases in the direct oxidation treatment, at 20% of H_2O_2 and 5% of $FeSO_4$ and also 30% $H_2O_2$7.5% $FeSO_4$ which are complied with JIS A 5908 (2003) standard is maximum 12%. Thus for application purposes, H_2O_2 and $FeSO_4$ catalyst levels in this range have cause no problems in particle quality regarding their dimensional stability. The smallest swelling thickness value is the best development since it can anticipate the absorption of water into the board through the pores of particles and empty space between the particles slowly [11]. The average value of swelling thickness in a combination of oxidation treatment tends to be lower. This phenomenon can cause by cross-linking between radical groups produced by oxidation processes and cross-linking between hydroxyl groups [12]. The radical groups generated during the oxidation process appear to form covalent bonds during the heat-pressing process so that the bonds are stronger and stable.

Figure 5 shows the infrared spectra of coffee husk binderless particleboard at different wt% H_2SO_4 and wt% $FeSO_4$ catalyst.

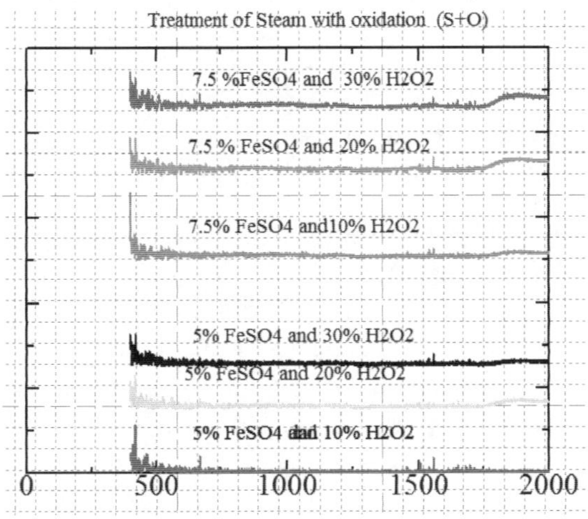

Figure 5 FTIR spectra of coffee husk binderless particleboard at different wt% H_2O_2 and $FeSO_4$ catalyst (Treatment of steam with oxidation).

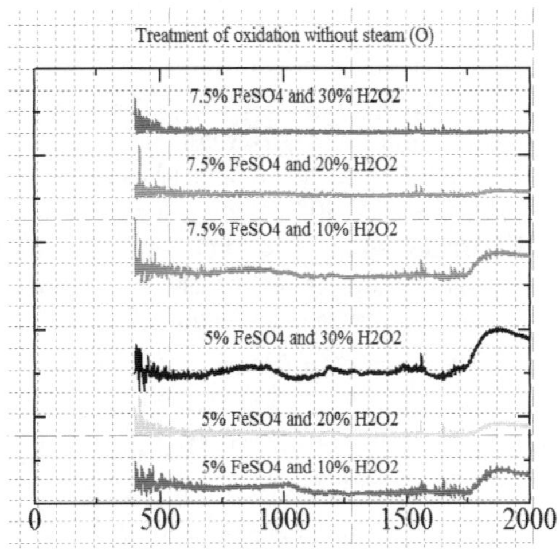

Figure 6. FTIR spectra of coffee husk binderless particleboard at different wt% H_2O_2 and $FeSO_4$ catalyst (Treatment of oxidation without steam).

Figure 5 shows the particleboard has increased intensity on the board treatment by steaming and oxidizing (S+O). The peak change occurs in the area of 1050 wave number. This area is the area of the C-H or C-O deformation absorption band. Figure 6 shows the infrared spectra in the treatment of oxidation without steam, there is a change of indentation in the 1250 cm-1 wave number which is the absorption area of ester. Based on this fact, one type of bond that plays an important role in the formation of bonds between particles on the particles without adhesive is a cross-link due to the incorporation of phenoxyl radicals.

4. Conclusions

From the results of the study, it can be concluded that the coffee husk particleboard whose raw material is treated oxidation without steam (O) can improve the physical properties of binderless particleboard. Increasing levels of H_2O_2 and $FeSO_4$ catalysts in the oxidation process of coffee husk produces binderless particleboard with good physical characteristics such as density, water content, water sorption and swelling thickness. Therefore, considering the efficient aspects of the use of chemicals, the combination of H_2O_2 and $FeSO_4$ catalysts that can be made binderless particleboard according to JIS A 5908 2003 standard are 20 wt% H_2O_2 and $FeSO_4$ catalyst 7.5 wt% in direct oxidation treatment (O). The ester linkages were detected by Fourier transform infrared spectroscopy, indicated that cross-link due to the incorporation of phenoxyl radicals.

Acknowledgment

The author would like to thank KEMENRISTEK DIKTI who has provided the financial support for this research through a competitive research grant program in 2017. The author would like also thank the students and laboratory staff that have been involved supporting the research in this study.

References

[1] Milawarni 2015 Manufacture and characterization of particleboard from coffee husk *J.Technology* **12** 20-23

[2] Pizzi A 1994 *Advanced Wood Adhesives Technology* 2nd ed Marcel Dekker inc.USA

[3] Kharlsson O and U westermark *2002 Proc.Int.Conf. Resin-Free Particleboard by Oxidation of Wood.* In: Humphrey P E, compiler. Procceding of The 6th Pacific Rim Bio-Based composites Symposium and Workshop on The chemical Modification of Cellulosics. (Portland, Oregon, USA) p 149-153

[4] Khadazipour A and Huttemann A 1998 *Bio technology Production of Wood Composites*. In: Bruce A and Palfreyman J W editors *Forest Prodict Bio technology*. (UK Taylor and Francis Ltd) p 141-150

[5] Suhasman, Massijaya, Hadi M Y and Santoso Y S A 2011 Particle Oxidation Time for the Manufacture of Binderless Particleboard. *J.Wood Research* **2** No1 27-33

[6] Kawai S, Zhang M, Okudaira Y, Xu J, Widyorini R and Han G 2002 *Proc.Int.Conf. Development of Kenaf bast fiberboard and core binderless particleboard* .In: Humphrey PE compiler. Procceding of The 6th Pacific Rim Bio-Based composites Symposium and Work shop on The chemical Modification of Cellulosics. (Portland, Oregon, USA) p 129-134

[7] [JIS] Japanese Industrial Standart 2003 Japanese Industrial for Particleboard (JIS A 5908;2003). Japanese Standard Association. Japan

[8] Okuda N, Hori K and Sato M 2006 Chemical changes of kenaf core binder less boards during hot pressing (II): effects on the binder less board properties *J. Wood science* **52** 244-248

[9] Ratri Y L 2013 Characterization binder less particleboard of oil palm empty bunches *J. Widyariset* **16** 219-226.

[10] Adi S, 2013 Utilization of lignin and tanin as subtitute alternative of glue raw material for wood composite. *Proc. National Conf. of Polymer* V ISSN 1410-8720 p 155-163

[11] Xu J, Widyorini and Kawai 2006 Development of Binder less fibreboard from kenaf core using oxidation. *J Sci* **52** 236-243

[12] Widsten P A and Kandelbauer 2008 Laccase applications in the forest products industry: A review. *J Enzyme and Microbial Technology* **42** 293-307

Influence of the apex angle of a hollow prism made from an ordinary commercial glass plate as a simple refractometer to the accuracy of the refractive index measurement of the edible oil

N Idris, Maswati and E Yusibani
Department of Physics, Faculty of Mathematics and Natural Sciences, Syiah Kuala University, Jl. Syech Abdurrauf No. 3 Darussalam, 23111 Banda Aceh, Aceh, Indonesia

E-mail: nasrullah.idris@unsyiah.ac.id

Abstract. The influence of the apex angle of a hollow prism used as a simple refractometer to the accuracy of a refractive index measurement of the edible oil samples was studied. The hollow prism was made from an ordinary commercial glass plate with a thickness of 2 mm. The apex angle of the constructed hollow prism was varied. The edible oil sample used in this study was palm oil, namely the packaged, branded oil sample and the bulk oil sample. For measuring the refractive index, the oil sample was filled in the constructed hollow prism, and then a helium-neon laser beam was passed through the oil sample at a certain angle of incidence. The angle of minimum deviation of the transmitted laser He-Ne beam was measured and then was used for calculating the refractive index of the oil sample. The refractive index measurement was made using the hollow prism with different apex angles, ranging from 30^0 to 60^0. The measurement accuracy was estimated by comparing the refractive index measured using the hollow prisms to that of obtained using a standard Abbe refractometer. It was found that the refractive index of the edible oil can be measured accurately by using the hollow prism. It was also found that the accuracy of the refractive index measurement significantly changes with the apex angle of the hollow prism. The refractive index values measured using this simple refractometer deviate up to 3,49% from the refractive index value measured using the standard Abbe refractometer, especially when the apex angle of the prism is 30°. The measurement results with high accuracies obtained when using the hollow prisms with apex angles of 45^0 and 60^0. The optimum apex angle for the present constructed hollow prism is 45^0. The refractive index obtained using the hollow prism with the apex angle of 45^0 is 1,4623 and 1,4438 for the bulk oil and the packed, branded oil samples, respectively. This result suggests that the apex angle of the prism used affects largely the accuracy of the refractive index measurement.

1. Introduction

Each food materials we use daily have a decent value of consumption by default. Determination of this standard value serves as a prevention of the abuse of use bringing about the bad impact on the health of costumers [1-2]. Thus, inspection of the quality of the food materials must always preserve. One basic important food material is edible oil. There are two kinds of edible oil including animal oil and vegetable oil and it already has a decent value of consumption. Normally along with other properties, physical and chemical parameters are used to determine the quality of cooking oil, such as chemical composition, viscosity, conductivity, reflectivity and refractive index [3]. Therefore, several instruments have been developed based on those parameters for determining the quality of vegetable oil such as refractometer, interferometer and Ostwald viscometer [4-8].

In the previous preliminary study, the present author has designed and constructed a simple refractometer for measuring the refractive index of edible oil [9]. The main components of the simple refractometer are a hollow prism as sample cavity and a He-Ne laser as a light source. A hollow prism is a prism made using glass plate with the center part of the prism is vacant. This cavity can be filled with gas or other fluids. Thus, a hollow prism is also commonly called as a fluid prism. This prism is a well-known tool used for determination refractive index of fluid and normally made from a high-quality glass for obtaining highly accurate estimation. The estimation of the refractive index of any fluid sample using the fluid prism based on the refraction of the passed light beam by the fluid sample [10-12]. The measurement of refractive index is carried out by measuring the minimum deviation angle, d_m of the light beam passed through the prism and the apex angle, A of the prism as expressed in the following equation (1).

$$n = \frac{\sin \frac{1}{2}(d_m + A)}{\sin \frac{1}{2} A} \tag{1}$$

n is the refractive index of the fluid sample; d_m is the minimum deviation angle of the light beam after passing through the prism, A is the apex angle of the prism. The minimum deviation occurs when the incident angle and refraction angle are equal. In the previous study, since there is a necessity to make a simple and cheap tool for quickly analyzing the quality of the edible oil sample, a hollow prism was made from an ordinary commercial glass easily available in market at very low cost. The constructed prism was used to make simple, cheap refractometer for determining quickly the quality of the edible oil. In the preliminary study, the thickness of the ordinary commercial glass plate used for constructing the hollow prism was 5 mm and the apex angle of the constructed hollow prism was 60°. As the results, the simple refractometer can be used to measure the refractive index of edible oil with relatively high accuracy, namely the refractive index of the new branded, packed oil sample and the new bulk oil sample is 1,5054 and 1,5152, respectively. These values differ only about 2,00-3,07% from the reference value of the refractive index of edible oil, namely 1,47. The previous results led to a conclusion that a simple refractometer consisting of a hollow prism made from an ordinary commercial glass and a He-Ne laser can be used to measure the refractive index of edible oil with a good measurement accuracy [9]. However, several physical parameters of the prism influence significantly the refractive index. As explicitly formulated in Eq. (1) Above, the apex angle, A determines the value of the refractive index. This present work is intended to study in greater detail the influence of the apex angle of the hollow prism to the measurement accuracy of the refractive index. Changing the apex angle to find the optimal apex angle of a hollow prism that allows for the highest accuracy of the refractive index measurement of the vegetable oil sample.

2. Experimental Procedure

The tool used in this work are a hollow prism and a green He-Ne laser (λ = 594 nm, 4 mW). The hollow prism constructed from the ordinary commercial glass plate (Asahi ACC) normally uses for house window and other furniture with apex angles of 30°, 45° and 60°. The thickness of the ordinary commercial glass plate is 2 mm. The dimension of each constructed prism is different depending on the intended apex angle. **Figure 1** shows a design of the hollow prism for an apex angle (A) of 60°. The dimension of both sides of the prism is 6,5 x 7,5 cm with the base length of 7,5 cm.

The edible oil sample used in this study is two kinds of palm oil samples, namely the bulk oil and the branded, packed oil samples. For measuring the refractive index, the He Ne laser light beam was passed through the hollow prism before and after the cavity of the hollow prism is filled with the oil sample. The deviation angle of the He-Ne laser beam is then measured as shown in **Figure 2**. The experimental procedure and setup are basically similar to our previous preliminary work published elsewhere [9]. Using the deviation angle and the apex angle, the refractive index of the oil sample is then calculated using equation (1). The refractive index of the edible oil samples was also measured using a standard Abbe Refractometer. To estimate the accuracy, the measurement was also carried out for a distilled water sample. The temperature of each oil samples was measured using a thermometer.

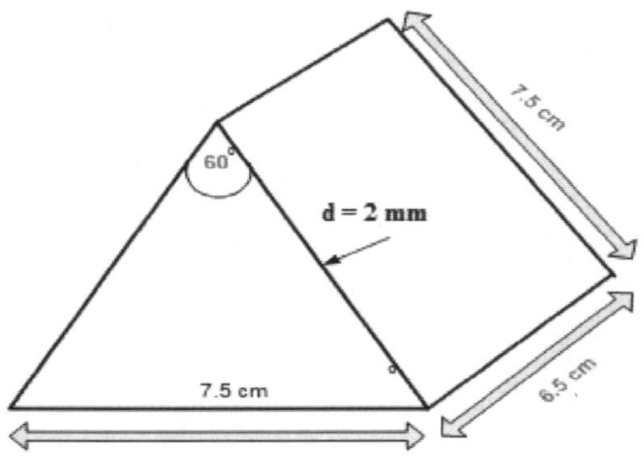

Figure 1. Design of a hollow prism with an apex angle of 60°

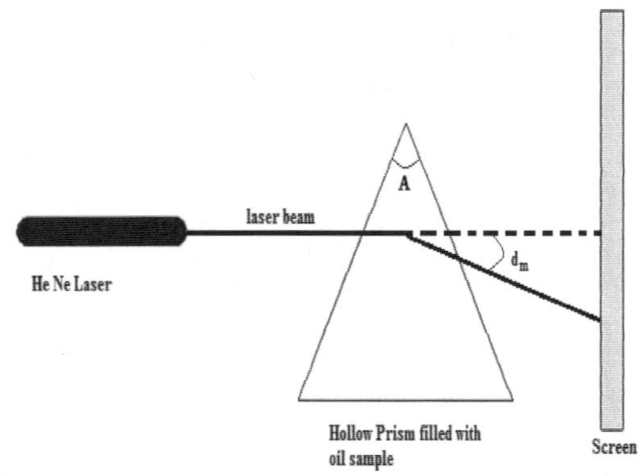

Figure 2. Experimental procedure for measuring the refractive index using the refractive index

3. Results and discussion

The hollow prism with different apex angles of 30^{0}, 45^{0} and 60^{0} has been designed and constructed from an ordinary commercial glass plate with a thickness of 2 mm. Figure 3 shows the photographs of the hollow prisms with different apex angles and the laser He-Ne were used as a simple refractometer for measuring the refractive index of edible oil sample. Table 1 shows the refractive index of the bulk palm oil sample measured using the hollow prisms with different apex angles. To estimate the measurement accuracy, the refractive index of the oil sample was also measured using a standard Abbe refractometer and found that the refractive index of the bulk oil sample is 1,4586. The refractive index measured by using the simple refractometer is then compared to the refractive index measured using the standard Abbe refractometer. It can be seen that the apex angle of the hollow prism strongly influences to the measurement of refractive index. The largest value of the refractive index was obtained when using the hollow prism with an apex angle of 30°, while the smallest is when using the hollow prism with an apex angle of 45°. The smallest, 1,4623, is the closest value to the reference value of the refractive index and it deviates only 0.25% from the refractive index value measured using the standard Abbe refractometer. The deviation is expressed in percentage by comparing the measurement results obtained by using the hollow prism to that of using the standard Abbe refractometer.

These results display that the apex angle of the hollow prism determines largely the accuracy of the refractive index measurement. The refractive index measured using the hollow prism with apex angles of 30° and 60° shows higher values, namely 1,5095 and 1,4829, respectively and also larger deviations, namely about 3,49% and 1,16%, respectively from the refractive index measured using the standard Abbe refractometer. The refractive index measured using the hollow prism with the apex angle of 45° exhibits a far lower deviation from the standard Abbe refractometer measurement result, namely 0,25%. The refractive index obtained using the hollow prism with 45° is much closer to the reference value in comparison with the results obtained using the hollow prism with apex angles 30°

and 60°. It can be observed from the results shown in Table 1 there is a general tendency that the smaller apex angle of the hollow prism the higher refractive index obtained. However, these results figure out that the change of the refractive index with a different apex angle of the hollow prism is not purely linear. Thus there is an optimum apex angle of the hollow prism to bring about measurement with high accuracy. The optimum apex angle is featured by the closest value of the measured refractive index to the reference value. As mentioned above, the reference value of the refractive index of the oil sample is the refractive index value measured by using the standard Abbe refractometer. In case of the bulk palm oil sample, the optimum apex angle of the hollow prism is 45°. This implies that the highest measurement accuracy is obtained using the hollow prism with apex angle of 45^0. Similar results were also observed for the branded, packed oil palm sample.

Table 1. Refractive index of the bulk palm oil measured employing the simple refractometer

Apex angel (A, degree)	Thickness of the glass plate, (mm)	Refractive index (n)	Deviation from Abbe refractometer (%)
30,0	2	1,5095±0,0001	3,49
45,0	2	1,4623±0,0000	0,25
60,0	2	1,4829±0,0001	1,16

Table 2. Refractive index of the branded, packed palm oil measured employing the simple refractometer

Apex angel (A, degree)	Thickness of the glass plate, (mm)	Refractive index (n)	Deviation from Abbe refractometer (%)
30,0	2	1,4993±0,0004	2,60
45,0	2	1,4438±0,0028	-1,54
60,0	2	1,4829±0.0001	1,16

Table 3. Refractive index of the distillate water measured employing the simple refractometer

Apex angel (A, degree)	Thickness of the glass plate, (mm)	Refractive index (n)	Deviation from Abbe refractometer (%)
30,0	2	1,3570 ± 0,0023	2,03
45,0	2	1,3306± 0,0002	0,04
60,0	2	1,3293±0,00017	-0,02

Table 2 displays the refractive index of the branded, packed oil samples measured by using the hollow prisms with different apex angles of 30°, 45°, and 60°. It can be seen the refractive index of the oil sample changes with apex angle. It is considered although the bulk palm oil sample and the branded, packed oil sample are originally made from the similar raw material, their qualities are different. Thus there is different in their physical and chemical properties at certain extend. The measurement results are shown in Table 1 and Table 2 confirm the difference of the refractive index from the bulk palm oil sample and the branded, packed oil sample. However, the measurement results for the branded, packed oil sample exhibited in Table 2 shares a similar tendency to that for the bulky oil sample shown in Table 1 above regarding the refractive index values and measurement accuracy. The highest value of the refractive index of the oil sample was obtained in case of using the hollow prism with the apex angle of 30° while the lowest is in case of using the hollow prism with the apex angle of 45 °. It also can be seen that in case of using the hollow prism with apex angle of 45 °, the value of refractive index measured displays the smallest deviation from the measurement result using the standard Abbe refractometer, while in case of using the hollow prisms with apex angles of 30 ° and 60 ° the deviation increases. The minus (-) in Table 2 and Table 3 imply that the present measurement gets lower value than the Abbe refractometer measurement.

To confirm the accuracy of the hollow prism and the tendency observed for the oil sample, the same measurement was carried out for distillate water sample. Distillate water was used a standard fluid since its chemical and physical properties are well known and unaltered during the experiment. As exhibited in Table 3, it can be seen that the values of the refractive index of the distillate water samples measured using the hollow prisms with different apex angles of 30^0, 45^0 and 60^0 are very close to the reference value of the refractive index of water, namely 1,33. The deviations of the results measured using the hollow prism from that of obtained using the Abbe refractometer are relatively lower compared to the cases of the bulk oil and the branded, packed oil samples, ranging from -0.02% to 2.03%. This means the measurement of the refractive index using the hollow prism for the distillate water is more accurate than that for the cases of the oil samples. This is probably due to the chemical and physical properties of the distillate water. The result shown in Table 3 displays several important points. Firstly, these results confirm that the hollow prisms made from the ordinary commercial glass plate can be used for measuring the refractive index of fluid including oil and water samples with high accuracy. Secondly, these results display a consistent tendency to the results obtained for the oil samples, namely the apex angle of the hollow prism significantly influences the refractive index values. The higher value of the refractive index observed is in case of using the hollow prism with the apex angle of 30°, the same as observed for the cases of the bulk oil and the branded, packed oil samples. From Table 3, it is observed a general tendency that the smaller apex angle of the hollow prism used yields higher value of refractive index obtained means the higher deviation from the reference value. However the tendency is also not purely linear, similar to the cases of the bulk oil and the branded, packed oil samples. The optimum apex angle of the hollow prism showing the highest measurement accuracy in this case of the distillate water is 45°, the same as the cases of the bulk oil and the branded, packed oil samples. Finally, these results also confirm that the apex angle of the constructed hollow prisms influence significantly measurement accuracy of the refractive index, since the deviation of the refractive index values of the distillate water samples measured using the hollow prisms with different apex angles from the results obtained using the Abbe refractometer changes significantly with the apex angle. The deviation pattern, in this case, shares similarity to the cases of the bulk oil and the branded, packed oil samples, namely the largest deviation observed for the case of the hollow prism with the apex angle of 30^0. These confirm the strong influence of the apex angle of the hollow prism to the accuracy of the refractive index measurement. On the other hand, the hollow prisms were constructed using glass plate with a thickness of 2 mm. These results probably will be affected by the difference in the glass plate thickness. Thus other study is being conducted to examine the influence of the glass plate thickness to the accuracy of refractive index measurement and will be published elsewhere in the near future.

4. Conclusion

The hollow prism made from the ordinary commercial glass plate and the He-Ne laser as a light source can be used for measuring the refractive index of the vegetable oil samples including the bulk oil and the branded, packed oil samples. The accuracy of the refractive index measurement using the hollow prism for the oil samples was also confirmed using the distillate water sample. Comparing the measurement results obtained using the hollow prism for the three kinds of samples, namely the bulk edible oil, the branded, packed edible oil, and the distillate water to the results obtained by means of the standard Abbe refractometer confirms that the accuracy of the refractive index measurement using the hollow prims made from the ordinary glass plate is relatively high since the refractive index of the samples, obtained from the measurement, is very close to that of measured by the standard Abbe refractometer. It was found that the apex angle of the hollow prism determine the measurement accuracy. From the present study, the measurement with high accuracies obtained using the hollow prisms with apex angles of 45 ^{o} and 60^{0} since for the three kinds of samples, the deviation of the refractive index measured using the hollow prism from that measured using the standard Abbe refractometer is much smaller in comparison with those of using the hollow prism with apex angles of 30 ^{o}. The refractive index measured using the hollow prism with the apex angle of 30^{0} shows a largest deviation, in present case up to 3.49% from the reference data of Abbe refractometer result. As the deviation for the hollow prism with apex angle of 45^{0} is in the range of 0,04 to 1,54 % for three samples, thus it is suggested that the apex angle of 45° is the optimal apex angle of the present simple hollow prism design.

Reference

[1] Sackheim J I and Lehman D D 1995 Chemistry for the Health Sciences (Macmillan Publisher Company) 5th ed.
[2] Tanaka T, Kohno H and Mori H 2001 Cemoprevention of Colon Carcinogenis by Dietery Non-Nutritive Compounds *Asian Pacific J. of Cancer* **2**(3) 165-177
[3] Shyam S 2002 Refractive Index Measurement and Its Applications *Phys. Scripta* **65** 167-180
[4] Yunus W M M, Fen YW and Lim MY 2009 Refractive Index and Fourier Transform Infrared Spectra of Virgin Coconut Oil and Virgin Olive Oil *American J. Appl. Sci.* **6**(2) 328-331
[5] Govindan G and Raj S G 2009 Measurement of Refractive Index of Liquids using Fiber Optic Displacement Sensors *J. American Sci.* **5** 13-17.
[6] Khodier S A 2001 Refractive index of standard oils as a function of wavelength and temperature *Optics Laser Tech.* **34**(2) 125–128
[7] Synowickia R A, Pribil G K, Cooney G, Craig M, Herzinger, and Green S E 2004 Fluid refractive index measurements using rough surface and prism minimum deviation techniques, *J. Vacuum Sci. Tech. B, Nanotechnology and Microelectronics: Materials, Processing, Measurement, and Phenomena* **22** 3450
[8] Yunus M M and Rahman A A 1988 Refractive index of solutions at high concentrations *Appl.Optics* **27**(16) 3341-3343
[9] Idris N, Sarina, Maswati and Susilayani D 2017. "Developing a Refractive Index Measurement Instrument Using a Hollow Prism Made from Ordinary Commercial Glass Plate and a He-Ne Laser for Examining Quality of Edible Oil". Translated from the original publication in Indonesian titled "Perancangan dan Pembuatan Sebuah Prisma Berongga dari Kaca Komersial dengan Laser Helium Neon sebagai Sumber Cahaya Sebuah Instrumen Optik Sederhana untuk Pengujian Kualitas Minyak Makan" *Risalah Fisika* **1** 39-47.
[10] Nolan P J 2001 Fundamentals of College Physics (New York. Wm. C. Brown Publishers)
[11] Southall P C J 1933 Mirrors, Prisms and Lenses -A Text-Book of Geometrical Optic- (New York. The Mac Millan Company)
[12] Hecht E 2002 Optics (New York, Addison Wiley) 4th ed.

IOP Conf. Series: Materials Science and Engineering **352** (2018) 012046 doi:10.1088/1757-899X/352/1/012046

Microtremor survey to investigate seismic vulnerability around the Seulimum Fault, Aceh Besar-Indonesia

Andrean V H Simanjuntak[1,2,3], **Umar Muksin**[1,3], **Febrina Rahmayani**[3,4]
[1]Department of Physics, Syiah Kuala University, Banda Aceh, Indonesia
[2]Mata Ie Geophysical Station of BMKG Aceh, Aceh Besar, Aceh
[3]Tsunami and Disaster Mitigation Research Center, Syiah Kuala University, Banda Aceh, Indonesia
[4]Department of Geophysics, Syiah Kuala University, Banda Aceh, Indonesia

E-mail: andreansimanjuntak@gmail.com

Abstract. The Seulimeum Fault has generated inland earthquake with magnitude larger than M 6.5 that destroyed houses in the Lamteuba and Krueng Raya Villages. Earthquakes along the Seulimeum Fault are mostly right lateral strike-slip characterizing the Fault. The understanding of the seismic vulnerability around the highly populated Banda Aceh City and the villages in Aceh Besar is therefore very important since the city, and the villages are very close to the Seulimeum Fault. A microtremor survey has been conducted to investigate seismic vulnerability in the area closed to the Seulimeum Fault. The waveforms of the microtremor have been recorded in Lamteuba and Kreung Raya villages, Aceh Besar at 20 sites for 7 days from August 14, 2017 with the interval of measurement of 1 km. The waveforms recorded for 30 minutes at each site by using one Taurus Seismometer in miniseed format. The data has been analyzing by using Geopsy to obtain the Horizontal-Vertical Spectral Ratio for each site. The seismic vulnerability is considered to be high if the value of the Horizontal-Vertical Spectral Ratio is high. The HVSR values are then interpolated to obtain the seismic vulnerability map. The preliminary result shows high seismic vulnerability in the area around the first site.

1. Introduction

Since 1892, more than 20 destructive inland earthquakes with magnitudes larger than M 6.0 have occurred along the Sumatran Fault including along the secondary faults. The Seulimeum Fault generated large earthquakes in 1964 and 1975 although the localization facilities were limited at that time to obtain an exact earthquake location. People in the Krueng Raya, the Lamteuba, and the Seulimeum regions informed that their regions were among the highly affected areas by the massive earthquake in 1964. Geoscientists have focused their studies on the Seulimeum Fault as the fault has not been activated for a quite long time by large earthquakes. The length of the Seulimeum Fault is expected to be around 120 km and moves 2.5 mm/year with the maximum magnitude of earthquake could reach M 7.5.

The Pidie Jaya earthquake with magnitude of M 6.7 is the latest earthquake which occurred along the secondary fault in the Aceh province. The Pidie Jaya earthquake could affect the activity of the Seulimum Fault because the stress could be transferred to the Seulimeum Fault (the closest fault). Although the area around the Seulimum Fault is very vulnerable, information on the seismic vulnerability index in the region is still very limited. Quantitative and qualitative data on seismic vulnerability analysis becomes very important as that information could be used for the earthquake risk reduction program.

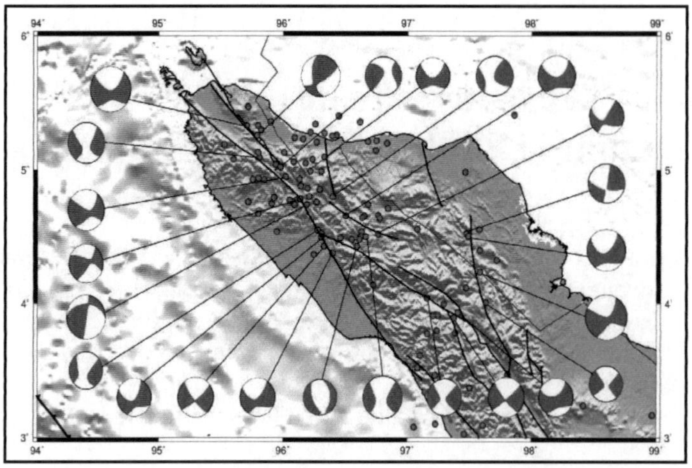

Figure 1. Historical of earthquake data and focal mechanism in Aceh Province from Global CMT, International Seismological Center (ISC), and GFZ-Potsdam.

Figure 1 explains the seismotectonic condition in the Aceh Province. The epicenters are from local fault activity and have strike-slip or oblique focal mechanism. The strikes of some focal mechanisms follow the straightness of the fault. We can see that the strike direction from NorthWest (NW) to SouthEast (SE) which is in agreement with the strike of the fault. Also, seismic susceptibility for microzonation index can be used as a reference to build safer buildings by infrastructure developer.

Earthquakes and tremors can be distinguished when we view the seismic waveforms. An earthquake waveform has several peaks indicating the arrivals of seismic energy of the earthquake such as P and S arrivals. Natural vibration (ambient vibration) of the earth comes from two main sources, nature and human activity with the frequency below 1 Hz while microtremor is natural ambient noise [1]. Microtremor is a ground vibration with an amplitude displacement between 0.1 to 1 micron with speed range from 0.001 to 0.01 cm/s which could be detected by a seismometer located in the highly amplificated region [2]. Among the natural sources of ambient vibrations are sea waves (with a frequency of ~0.2 Hz), the sea and beach waves with a frequency of ~0.5 Hz, the activity in the atmosphere (frequency below 0.1 Hz), and wind and water flow (with the frequency higher than 1 Hz).

Microtremor has been widely used to estimate the risk level and damage index caused by earthquakes and also used to investigate the geological structure of a region [3]. Nakamura [4] analyzed and found that the horizontal and vertical spectrum ratio of microtremor increased at a resonance frequency. Nakamura [4] suggested that the H/V ratio reflects the amplification level of ground motion.

2. Geology and Tectonic Settings

The geological and tectonic setting of the Seulimum Fault region has been well explained and mapped [5] as shown in figure 2. The Sumatran Fault starts to bifurcate around Tangse to create the Seulimeum fault striking NNW-SSE or around 10^0 north of the Aceh segment.

The lineation of the Seulimeum Fault is precisely in the middle of the sediment formation from Plio-pleistocene era and volcanic rocks from the Seulawah Agam volcano. The lithology of Krueng Raya, an area located in the northernmost of the Seulimeum Fault, is dominated by the same volcanic rock which spreading in the Lam Teuba area as shown in figure 2.

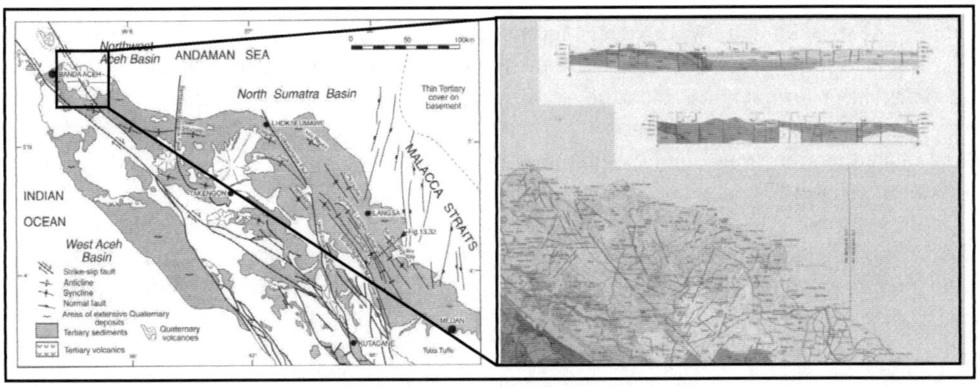

Figure 2. Structural, Geology and Tectonic Setting Map of the North Sumatran Basin [5].

The northern part of the Seulimeum Fault was created in the Cretaceous period. The rock in the Kreung Raya region is dominated by the same rocks in the Lam Teuba area which is volcanic rock caused by the Seulawah Volcano eruption [6]. The oldest rocks in this area from the pre-quartener rocks era consist of limestone, slates and phyllites [7].

3. Data and Method

Prior the survey, a literature study has been performed to better design the microtremor measurement and was conducted for 6 days from August 11th, 2017. One set low-frequency Lenard Seismograph, owned by Mata Ie Geophysical Station, was used to record 20 observation points microtremor. The locations of the measurements are shown in Figure 3. We recorded the microtremor in the area with a low level of noise and far from human activities. The duration of measurement is 30 minutes in each point which is considered to be sufficient to record the microwave around the site.

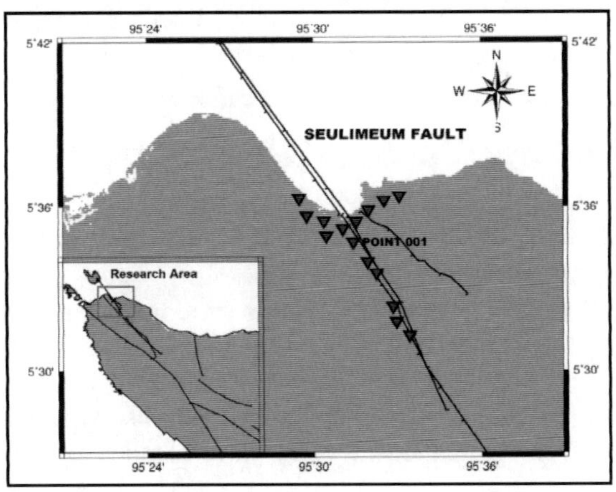

Figure 3. Some point observation with blue triangle near Seulimeum Fault with first point (001).

The microtremor data retrieval was performed by the rules set by the SESAME European research project [8]. The seismic waveforms in a MiniSeed format which were downloaded each hour through data logger and downloaded by connecting the RJ-45 cable to laptop or pc. Data retrieval was based on the measurement point that has been prepared previously using google maps and Garmin.

3.1 Method - Horizontal to Vertical Spectrum Ratio (HVSR)

The HVSR method compares the microtremor spectrum of the horizontal component to the vertical component. Nakamura [4] stated that the spectrum of H/V as a frequency function is closely related to the site transfer function of shear waves.

Nakamura [4] showed that the H/V ratio on the tape that has ambient noise is closely related to the frequency found on the ground, which is one of the amplification factors. In 1989, Nakamura [4] obtained observations indicating that the recording at a station located in hard rock region, the maximum value of the horizontal component spectrum ratio to the vertical close to the value of 1. At the station located in soft rocks, the H/V maximum is greater than 1.

Non-natural wave sources on the surface tend to trigger Rayleigh waves that propagate in the soil layer or soft sediment [9]. Rayleigh waves affect the horizontal and vertical components on the surface, but hardly affect the wave component in bedrock [1]. Nakamura [4] stated that there are two horizontal components measured in the N-S and W-E direction in the observation field. It can be expressed with:

$$SM(w) = \left[\frac{\left(HSN(w)^2 + \left(HWE(w)^2\right)^{1/2}\right)}{VS}\right] \tag{1}$$

HSN(w) = North-South horizontal spectrum component of microtremor
HWE(w) = East-West horizontal spectrum component of microtremor
VS(w) = Vertical spectrum component of microtremor

4. Result and Discussion

We conducted a proper and comprehensive evaluation of each measurement point. Evaluation is conducted by visually investigate the characteristics of the signal after the merging.

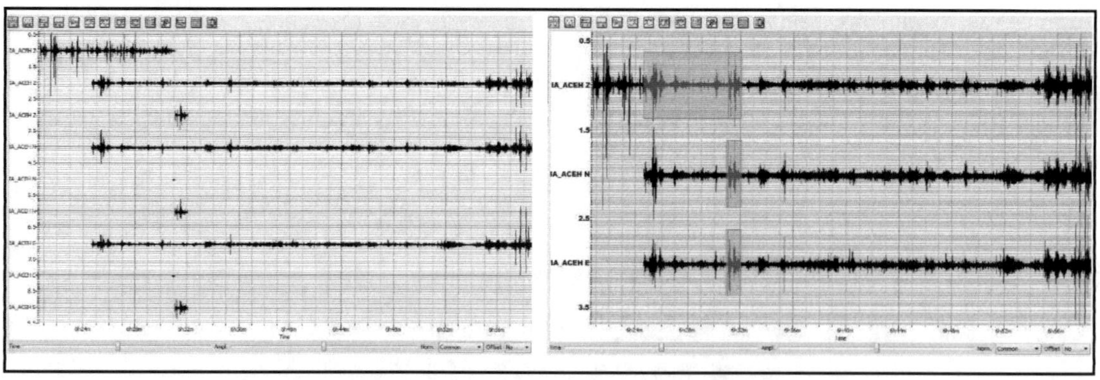

(A) **(B)**

Figure 4. **(A)** The record has not match signal and each component separated in 3 part **(B)** The record was merged to make all separate signal to be united in one part.

Figure 4 showed we combined several separated waveforms (figure4(A)) into 3 components (vertical, north, and east). After merging, the waveforms have identical origin time measurement.

4.1. Frequency/time graphic

We evaluate the frequency of the waveforms with time to see when the time have a maximum and minimum frequency amplitude values. We then perform automatic windowing to obtain the value of H/V and resonance frequency. The function of frequency and time of the Site 001 is shown in figure 5 for the Z component waveform.

From figure 5, we can see that the maximum frequency of the vertical component is the range between 40 and 60 Hz, and a minimum frequency of 0 -1 Hz. High amplitude at a certain time shown in figure 5 indicates that the waveforms contain transient noise from the surrounding environment.

IOP Conf. Series: Materials Science and Engineering **352** (2018) 012046 doi:10.1088/1757-899X/352/1/012046

From this evaluation, we could determine the characteristic of recording frequency and distinguish the natural seismic waveform and transient noise.

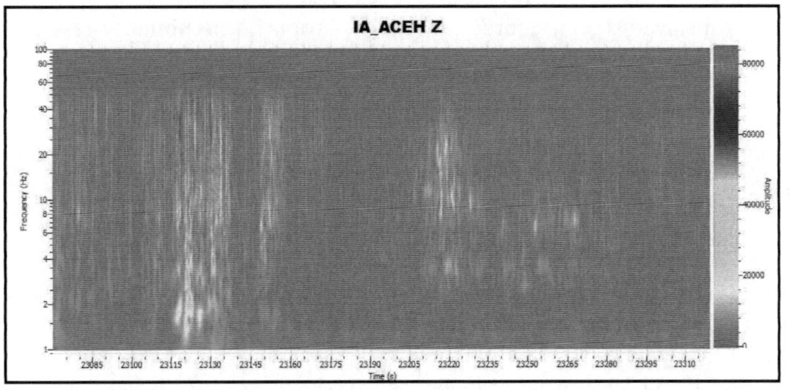

Figure 5. The comparison between frequency over time of the Vertical (Z) component of the waveform recorded at the 001 site.

The maximum frequencies of the N-S and E-W components are the same as that of the Z component between 40 and 60 Hz and a minimum frequency between 0-1 Hz. The frequencies in those ranges were used in the determination the value of H/V while noise was not included in the calculation. To determine the value of H/V, we performed a band-pass filter between 0.01 - 20 Hz, which is the characteristic of the microtremor signal. After that, the STA/LTA procedure was applied to exclude transient noise from the waveforms.

4.2. H/V Ratio

The waveform is windowed in 25 seconds and the HVSR is for each window. Based on the preliminary result, we can see that the predominant frequency value (*fo*) at the point 001 is 0.093851 Hz with the amplitude of 3.8481. The waveforms have a range of frequency between 0.01 – 20 Hz which is a typical seismic frequency. The result also shows the stable graphical trend from 1 – 20 Hz. The region of the 001 site has low amplification of seismic energy. The type of constituent rocks found are following the alleged preliminary study since the majority is composed of volcanic and tuff rocks. This makes the land in the area quite compact and not easily amplified. The preliminary results also show that recorded data is stable and could be used for the next data analysis (for other points).

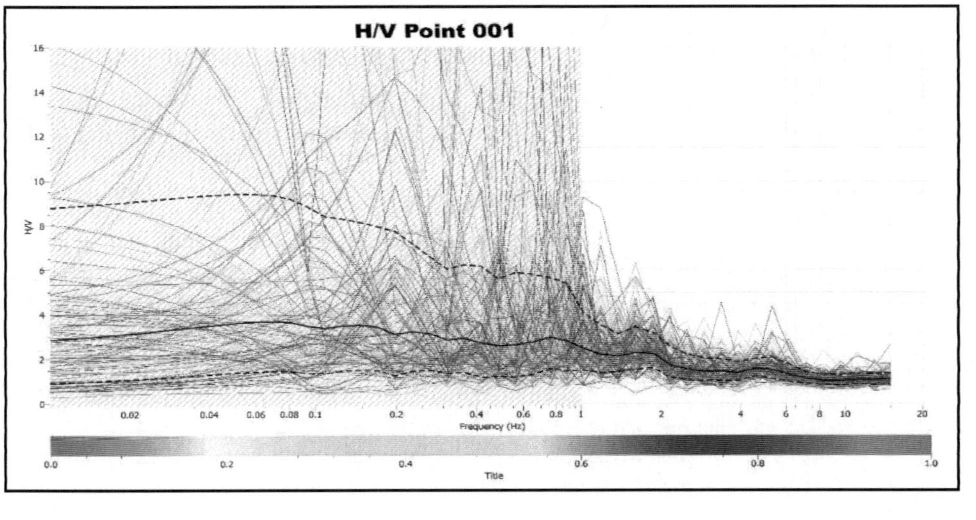

Figure 6. H/V spectrum values and resonance frequency from first point observation (001)

The study still does not cover all the points, because it is still in the process of assessment at observation points that have a good record and results. From the preliminary result, we can make a good waveform for reducing the noise, after knowing the frequency characteristic. Next, all values from each point will be mapped by interpolating all the H/V values. The seismic vulnerability index can be derived from the H/V values. All that result is very useful and important to make mitigation near the Seulimeum Fault. The microtremor survey will be continued next year by adding more observation points.

5. Conclusion

The preliminary results show the reliability waveform data we have recorded and could be used for further data analysis. The region around the first point (001) has resonant frequency f_0 of 0.093851 Hz with an amplitude of 3.8481. The values obtained by automatic windowing with the filter band pass from 0.01 - 20 Hz on the combined recordings. The recorded waveform has a maximum frequency range between 40 and 60 Hz, and a minimum frequency between 0 - 1 Hz. The value of resonant frequency and H /V spectrum value of the 001 site indicate that the region has an acceptable soil susceptibility and does not experience high amplification effect if an earthquake occurs.

References

[1] Takai dan Tanaka. 1961 On microtremors VIII. Tokyo: Bull. Earthquake Res. Inst 39 97-114

[2] Konno K, dan Omachi T 1998 Ground Motion Characteristics Estimated from Spectral Ratio Between Horizontal and Vertical Components of Microtremor Bull. Seism. Soc. Am 88 228-241

[3] Nakamura Y 2008 On the H/V Spectrum Beijing: The 14th World Conference on Earthquake Engineering NEHRP. 2000. Site Classifications Taken from Table 1615 1.1 Site Class Definitions published in 2000 International Building code International Code Council, Inc. on page 350

[4] Nakamura Y 1989 A Method for Dynamic Characteristics Estimation Od Subsurface Using Microtremor on The Ground Surface Tokyo: Quarterly Reports of the Railway Technical Research Institute 30 25-33

[5] Bennett J D, McC Bridge D, Cameron N R, Djunuddin A, Ghazali S.A, Jeffrey D H, Kartawa W, Keats W, Rock N M S, Thomson S J and Whandoyo R 1981 Geologic map of the Banda Aceh Quadrangle (Sumatra-Indonesia)

[6] Barber A J, and Crow M J 2005 Chapter 4: Pre-Tertiary Stratigraphy. In Barber A J, Crow MJ and Milsom J S Sumatra: Geology, Resources and Tectonic Evolution, Geological Society (London, Memoirs) 31 24-53.

[7] Culshaw M. G, Duncan S V and Sutarto N R 1979 Engineering geological mapping of the Banda Aceh alluvial basin, Northern Sumatra, Indonesia. Bulletin of Engineering Geology and the Environment 19 40-47

[8] Bard P Y 2005 SESAME-Team (2005) Guidelines for the implementation of the H/V spectral ratio technique on ambient vibrations: measurements, processing, and interpretations SESAME European research project

[9] Mirzaoglu M dan Dýkmen U 2003 Application of Microtremors to Seismic Microzoning Procedure Journal Of The Balkan Geophysical Society 6 143 – 156

IOP Conf. Series: Materials Science and Engineering **352** (2018) 012047 doi:10.1088/1757-899X/352/1/012047

Comparison of cadmium adsorption onto chitosan and epichlorohydrin crosslinked chitosan/eggshell composite

Rahmi, Marlina, Nisfayati
Department of Chemistry, Syiah Kuala University
Darussalam, Banda Aceh, 23111, Indonesia

E-mail: rahmi@fmipa.unsyiah.ac.id

Abstract. The use of chitosan and epichlorohydrin crosslinked chitosan/eggshell composite for cadmium adsorption from water were investigated. The factors affecting adsorption such as pH and contact time were considered. The results showed that the optimum pH of adsorption was pH = 6.0 and the equilibrium time of adsorption was 40 min. The adsorption isotherm of Cd ions onto chitosan and composite were well fitted to Langmuir equation. The maximum adsorption capacity (fitting by Langmuir model) of chitosan and composite were 1.008 and 11.7647 mg/g, respectively. Adsorption performance of composite after regeneration was better than chitosan.

1. Introduction

Heavy metal ions removal from polluted water is an important environmental issue. The existence of heavy metal ions will endanger to human health and environment. Some methods have been reported for heavy metal ions removal. Adsorption is one the methods that has much attention due to its effectivity and low cost [1-3]. The use of biopolymer as an adsorbent for heavy metal ions has many advantages such as availability, biodegradability and renewability. Most recent studies have focused on the seeking of effective adsorbent that locally available and cheap. One of heavy metal ion adsorbents that has much attention was chitosan [4-9].

Chitosan is deacetylated form of chitin which obtained in crustaceans and insects. Chitosan contains hydroxyl (-OH) and amine (-NH_2) groups that may form complexes with heavy metal ions and easy to modified. In order to improve the chitosan performance on heavy metal ions removal, some methods have been studied such as the use of crosslinking agent or filler on adsorbent based chitosan preparation. Some studies have been reported the use of filler for the improvement of chitosan performance such as the use of clay [10], sand [11], and cellulose [12]. However there is no study reporting the use of eggshell particles as a filler of chitosan. Meanwhile, Laus et al. (2010) reported the improvement of chitosan performance on heavy metal ion removal after crosslinking with epichlorohydrin [13]. Therefore, in this study we used both methods in order to improve the performance of chitosan on Cd ions removal from aqueous solutions. Epichlorohydrin was used as crosslinking agent and eggshell particles as a filler of chitosan based adsorbent. The availability, chemical compositions and low cost are the advantages of the utilization of eggshell as a filler on chitosan based adsorbent preparation.

The preparation and characterization of epichlorohydrin crosslinked chitosan/eggshell composite has been reported in our previous work [14]. The results showed that eggshell particles enhanced the mechanical properties and decreased the cristallinity, therefore in the present study we evaluated the adsorption capacity of epichlorohydrin crosslinked chitosan/eggshell composite on Cd ions removal from aqueous solutions and compared with chitosan. The adsorption was conducted with various pH, agitation time and initial cadmium concentration. Langmuir and Freundlich model were used to study isotherm adsorption and adsorption kinetics were studied by using pseudo-first order and the pseudo-second order model.

2. Materials and Methods

2.1. Materials

Chitosan (degree of deacetylation: 75.0 to 85.0%) and epichlorohydrin were purchased from Tokyo Chemical Industry Co., Ltd. Japan. Eggshell particles were produced by using ball milling machine.

2.2. Preparation of epichlorohydrin crosslinked chitosan/eggshell composite

Preparation of the composite was conducted according to our previous report [10], where chitosan (1%, w/v) was dissolved in acetic acid solution (2%). The epichlorohydrin was added to the chitosan solution under vigorous magnetic stirring at room temperature for 1 h. The eggshell particles were mixed with crosslinked chitosan solution and homogenized for 3h. The mixture was spread on the petri dish. The dish was kept at room temperature. Dried composite films were peeled off from petri dish and kept for further use.

2.3. Batch Adsorption experiments

$Cd(NO_3)_2$ was used for stock solutions (1000 ppm) of Cd ions preparation. Appropriate concentrations of standard solutions were prepared from the stock solutions. In 250 ml beakers, 25 ml aliquots of these standard solutions were placed and 50 mg of chitosan or composite was added. Batch adsorption experiment was conducted with various pH (1-9), contact time (20, 25, 30, 35, 40, 45 and 50 min) and initial Cd ions concentration (2.4, 4.8, 9.5, 16.6, 28.5, 57.0, 95.0 ppm) in order to obtain optimum condition. Cd ions concentration was analized by using UV-Vis spectrophotometer.

Adsorption equilibrium and adsorption kinetic studies were conducted under optimum condition. Adsorption capacity (Q_e) was calculated based on the difference of Cd ions concentration in aqueous solution before and after adsorption (Equation 1)

$$\text{Adsorption Capacity } (Q_e) = \frac{(C_0 - C_e)V}{W} \quad (1)$$

Where C_0 is the initial Cd ions concentration, C_e is final Cd ions concentration, V is volume of Cd ions solution and W is the weight of chitosan or composite.

3. Results and Discussion

3.1. Effect of pH

In the adsorption processes, the pH of the aqueous solution is an important parameter. Batch adsorption tests were conducted in pH range of 1-9 to study the influence of pH on the adsorption capacity of Cd ions onto chitosan and composite. From the results shown in Figure1, the adsorption capacities of the chitosan and composite were dependent on the pH of solutions. The adsorption capacity of Cd ions onto chitosan or composite increased up to pH 5.0 and tended to be constant until pH 7.0 for chitosan and pH 8.0 for composite.

In acidic condition, lower adsorption capacity was observed due to the high concentration of H^+ in the solutions compete with heavy metal ions in terms of binding with NH_2 groups of chitosan. By increasing pH, the concentration of H^+ ions decreases, facilitating the adsorption of metal ions by adsorbent. It also reported that the amine groups on chitosan surfaces are easily protonated at lower pH, inducing an electrostatic repulsion of heavy metal ions. Based on zeta potential values, Calagui, et al. (2014) reported that chitosan in less acidic solution had negative surface charge where there would be electrostatic attraction between negatively charged adsorbent surface and positively charged sorbate [15].

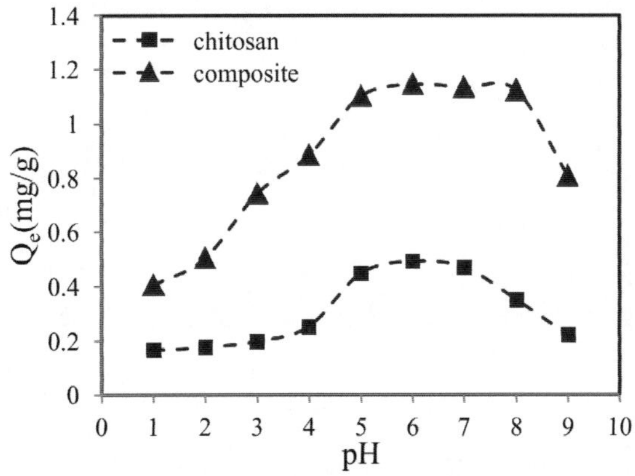

Figure 1. Effect of pH on adsorption capacity of Cd ions onto chitosan and composite

At pH 9.0, it was observed that the adsorption of Cd ions onto chitosan and composite decreased significantly. It was probably due to at alkaline solutions, amino groups of chitosan become deprotonated and allowing the shrinking of polymer chains. It reduced the possibility of heavy metal ions to contact with NH_2 groups of chitosan polymer chains and effect the decreasing of adsorption.

3.2. Adsorption Isotherms

Figure 2 shows adsorption equilibrium isotherm of Cd ions adsorption onto chitosan and composite. The adsorption capacity of Cd ions onto chitosan and epichlorohydrin crosslinked chitosan/eggshell composite increased by increasing initial concentration of Cd ions and then tended to be constant due to the saturation of the adsorbents.

Langmuir and Freundlich isotherm model were used to study the adsorption isotherm of Cd ions onto chitosan and composite. The equation for Langmuir and Freundlich model were represented in equations (2) and (3), respectively [13].

$$\frac{C_e}{Q_e} = \frac{1}{K_L Q_{max}} + \frac{C_e}{Q_{max}} \tag{2}$$

$$\log Q_e = \log K_F + \frac{1}{n}\log C_e \tag{3}$$

Where Q_{max}, K_L , K_F and n are maximum adsorption capacity (mg/g), Langmuir constant (L/mg), and adsorption intensity, respectively. Langmuir model assumes that the adsorption of adsorbate on the surface of adsorbent is monolayer adsorption. The Freundlich model does not assume monolayer adsorption, it describes the equilibrium on the heterogenous surfaces.

Table 1. Comparison of adsorption isotherm parameter for adsorption Cd ions onto chitosan and composite.

Adsorbent	Langmuir isotherm model			Freundlichisothermmodel		
	R^2	Q_{max} (mg/g)	K_L (L/mg)	R^2	K_f (mg/g)	n
Chitosan	0.996	1.008	0.197	0.815	43.954	0.417
Composite	0.952	11.765	0.141	0.822	0.404	0.553

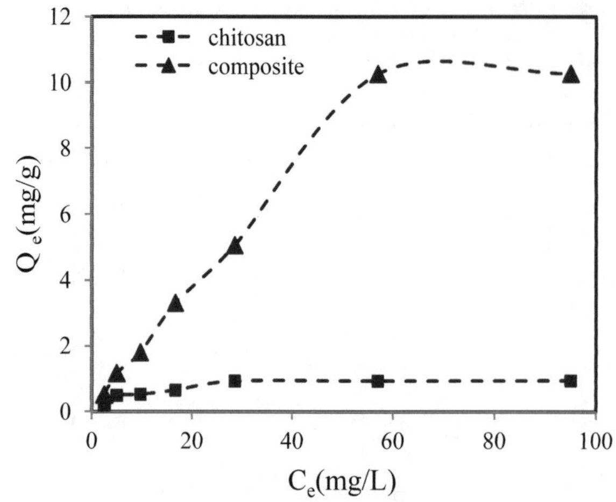

Figure 2. Adsorption equilibrium isotherm of Cd ions adsorption onto chitosan and composite.

The Langmuir and Freundlich isotherm model of Cd ions adsorption onto chitosan and composite were shown in figure 3 and 4, respectively. Comparison of adsorption isotherm parameter for adsorption Cd ions onto chitosan and composite was shown in table 1. Based on the correlation coefficient (R^2) obtained from isotherm model curves, it confirmed that the adsorption of Cd ion adsorption onto chitosan and composite were fitted to the Langmuir model. The equation of Langmuir model was employed to determine the Q_{max}. Q_{max} value of Cd ions adsorption onto composite was found higher than Q_{max} value of Cd ions adsorption onto chitosan. It indicates the improvement of chitosan performance after modification with crosslinking agent and filler.

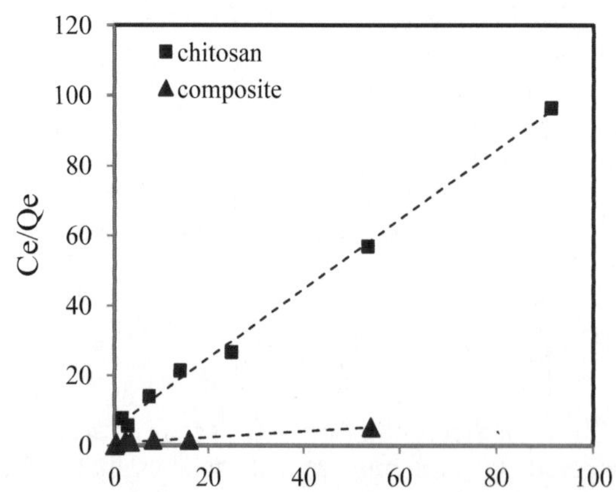

Figure 3. Langmuir isotherm model of Cd ions adsorption onto chitosan and composite.

3.3. Adsorption kinetics

Figure 5 shows the adsorption kinetics for adsorption of Cd ions onto chitosan and composite. The equilibrium time of Cd ions adsorption onto chitosan and composite were reached in the same time (40 min).

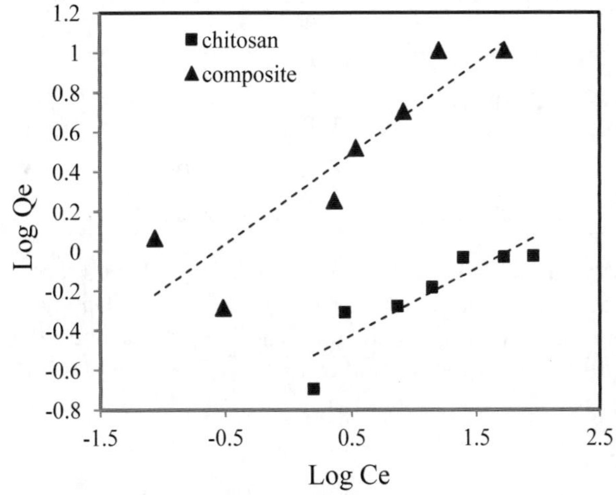

Figure 4. Freundlich isotherm model of Cd ions adsorption onto chitosan and composite.

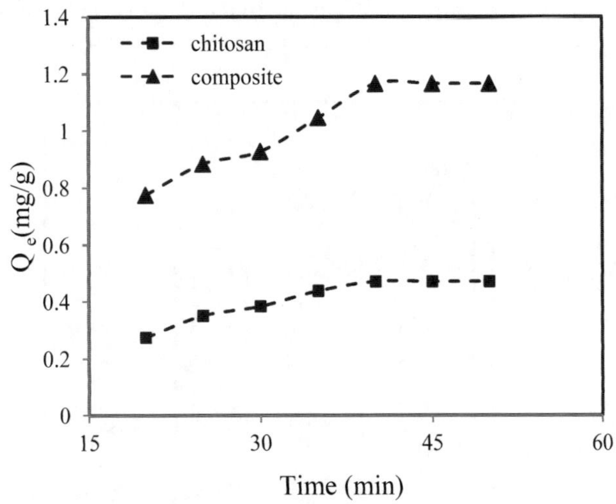

Figure 5. Adsorption kinetics of Cd ions onto chitosan and composite.

The pseudo-first order model and pseudo-second order model were used to determine the kinetic mechanism of Cd ions adsorption onto chitosan and composite. The equation for the pseudo-first order model and the pseudo-second order model are shown in Equations (4) and (5), respectively.

$$\log(Q_e - Q_t) = \log Q_e - \frac{k_1}{2.303}t \quad (4)$$

$$\frac{t}{Q_t} = \frac{1}{k_2 {Q_e}^2} + \frac{1}{Q_e}t \quad (5)$$

Where Q_e is the amount adsorbed of Cd ions (mg/g) at equilibrium. Q_t is the amount adsorbed of Cd ions (mg/g) at time t (min). k_1(min^{-1}) and k_2 (g/mg min) are the rate constant of pseudo-first order model and the pseudo-second order model, respectively. The results of parameters for the adsorption of Cd ions onto chitosan and composite are given in table 2.

Table 2. Comparison of kinetic models for the Cd ion adsorption onto chitosan and composite.

Adsorbent	Pseudo first order			Pseudo second order		
	R^2	Qe (mg/g)	k_1 1/min)	R^2	Qe (mg/g)	k_2 (g/mg min)
Chitosan	0.944	0.470	0.569	0.895	0.898	1.241
Composite	0.923	1.658	0.074	0.962	1.901	0.018

Table 2 shows the correlation coefficient (R^2) values of pseudo-first order model for adsorption of Cd ions onto chitosan was higher than pseudo-second order model. It indicates that kinetic adsorption of Cd ions onto chitosan was fitted to pseudo-first order model. On the other hand, adsorption of Cd ions onto composite agreed with pseudo-second order model. It might be influenced by the addition of crosslinking agent and eggshell particles to chitosan.

3.4. Regeneration studies

Regeneration studies are important to evaluate the probability of the adsorbent for subsequent reuse. In this study, HNO_3 solution at concentration 0.05 M was used as eluent for the regeneration of composite. In this concentration, chitosan was dissolved. Therefore, the lower concentration (0.0001 M) was used for the regeneration of chitosan. Figure 6 and 7 show experimental data of regeneration efficiency of chitosan and composite for Cd ions adsorption, respectively.

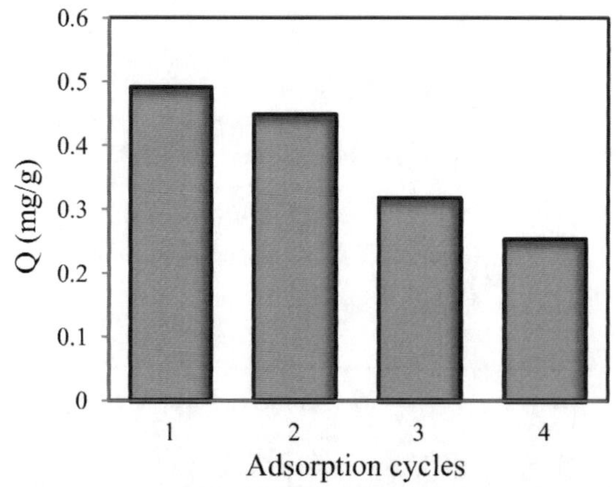

Figure 6. Regeneration efficiency of chitosan for Cd ions adsorption.

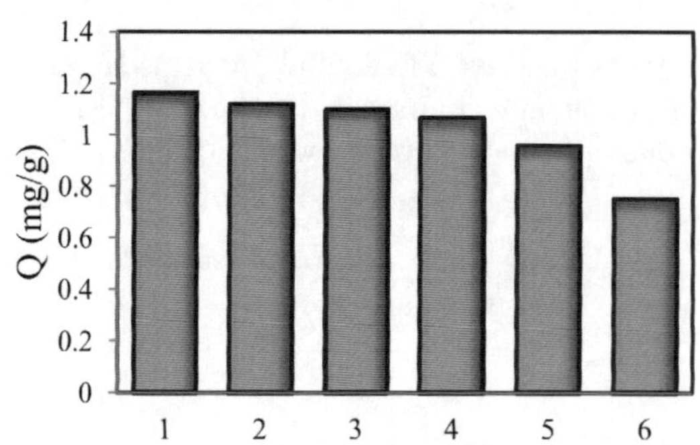

Figure 7. Regeneration efficiency of composite for Cd ions adsorption

The adsorption capacity of Cd ions onto chitosan and composite were decrease after adsorbent regeneration. The adsorption capacity of Cd ions onto composite was not significantly decrease after four times regeneration. The composite still showed good performance after fourth regeneration. While chitosan was only effective for once regeneration. It proves that the crosslinking agent and filler could improve chitosan performance on heavy metal ions adsorption.

4. Conclusions
It was observed that the crosslinking and addition of filler in chitosan could improve chitosan performance for Cd ions adsorption. Optimum condition for adsorption was found at pH 6 and contact time 40 minutes. Langmuir isotherm indicated that maximum adsorption capacity of Cd ions onto composite was 11.765 mg/g and this value was more than ten times larger than maximum adsorption capacity of Cd ions onto chitosan. Regeneration studies showed that the adsorbents can be used several times for adsorption of Cd ions.

References
[1] Mansour M S, Ossman M E and Farag HA 2011 *Desalination* **272** 301–305
[2] Barakat M A 2011 *Arab. J. of Chem.* **4** 361-77
[3] Cataldo S, Gianguzza A, Pettignano A and Villaescusa I 2013 *React.Funct. Polym.***73** 207-217
[4] Laus R,Fávere V T, 2011, *Bioresour. Technol.* **102**, 8769–8776
[5] Vieira R S and Beppu M M 2005 *Adsorption* **11**(1) 731-736
[6] Zhao F, Yu B, Yue Z, Wang T, Wen X, Liu Z 2007 *J. Hazard. Mater.* **147** 1-2 67-73
[7] Rahmi, Fathurrahmi, Irwansyah and Purnaratrie A 2015 *Orient. J. Chem.* **31**(4) 2071-2076
[8] Rahmi 2016 *Asian J. Chem.* **28**(10) 2267-2271
[9] Rahmi, Lelifajri, Julinawati and Shabrina 2017 *Carbohydrate Polymers* **170** 226-233
[10] Tirtom VN, Dincer A, Becerik S, Aydemir T, and Celik A 2012 *Chemical Engineering Journal* **197**(8) 379-386
[11] Wan M, Kan C, Rogel B D and Dalida M L P 2010 *Carbohydrate Polymers* **80**(9) 891-899
[12] Khan A, Khan R A, Salmieri S, Tien C L, Riedl B, Bouchard J, Chauve G, Tan V, Kamal MR and Lacroix M 2012 *Carbohydrate Polymers*, **90**(4) 1601-1608
[13] Laus R, Costa T G, Szpoganicz B and Favere V T 2010 *J. Hazard. Mater.* **183** 233-241
[14] Rahmi, Marlina and Nisfayati 2017 *Orient. J. Chem.* **33**(1) 478-482
[15] Calagui M J, Senoro D B Kan C-CJ, Salvacion W L, Futalan C M and Wan M W 2014 *J. Hazard. Mater.* **277** 120-126

IOP Conf. Series: Materials Science and Engineering **352** (2018) 012048 doi:10.1088/1757-899X/352/1/012048

Double layer mixed matrix membrane adsorbers improving capacity and safety hemodialysis

Saiful[1], **Z. Borneman**[2], **M. Wessling**[3]
[1]Chemistry Department, Syiah Kuala University Banda Aceh, Indoensia 23111
[2] Department of Chemical Engineering and Chemistry,
Technische Universiteit Eindhoven, Netherland
[3] Chemical Process Engineering. RWTH Aachen University, Germany

E-mail: saiful@unsyiah.ac.id

Abstract. Double layer mixed matrix membranes adsorbers have been developed for blood toxin removal by embedding activated carbon into cellulose acetate macroporous membranes. The membranes are prepared by phase inversion method via water vapor induced phase separation followed by an immersion precipitation step. Double layer MMM consisting of an active support and a separating layer. The active support layer consists of activated carbon particles embedded in macroporous cellulose acetate; the separating layer consists of particle free cellulose acetate. The double layer membrane possess an open and interconnected macroporous structure with a high loading of activated carbon available for blood toxins removal. The MMM AC has a swelling degree of 6.5 %, porosity of 53 % and clean water flux of 800 $Lm^{-2}h^{-1}bar^{-1}$. The prepared membranes show a high dynamic Creatinine (Crt) removal during hemodilysis process. The Crt removal by adsorption contributes to amore than 83 % of the total removal. The double layer adsorptive membrane proves hemodialysis membrane can integrated with adsorption, in which blood toxins are removed in one step.

1. Introduction

Blood purification is natural process, but absolutely important to maintain the life [1]. The liver, kidney are responsible for eliminating toxins from the body. When this organs stop working, can not performed their job, the condition is called liver and kidney failure or renal failure, the toxic and harmful substances will accumulate in the blood. Then we need extracorporeal blood purification to correct it. Nowadays, extracorporeal blood purification, particularly hemodialysis are very important in biomedicine. A patient is treated with hemodialysis membrane to remove accumulation of harmful subtances in blood, to control fluid balance and remove excess of water. Blood is drawn from your body then circulated to organ support system then clean blood return back to the body [2].

The disadvantage of hemodialysis is removing the toxins depending only on the molecular weight (pore size) with a lack of specificity and they cannot remove middle molecule and macromolecular poisonous substance [3]. Futhermore, the problem with dialysis are cannot remove albumin bonded toxins and needed ultrapure dialysate. Therefore, one big issue is also concern that the hemodialysis membrane is unavoidable from back-transport phenomenon, i.e. endotoxin can transfer from

contaminated dialysate fluid to blood stream leading to potential risk, that stimulate mononuclear cell cytokine production and produce a microinflammatory state. The failure of these therapies has stimulated the development of other or improvement treatments for blood purification.
Modification hemodialysis membrane properties are proposed by increasing membrane pore size and combining with adsorption properties. In fact that among extracorporeal detoxification, kidney support, liver support, and multi organs failure can not possibly work without adsorptive purification of biologic fluid. Historically, resins and charcoals were used for their adsorptive capacity and ability to remove toxic substances from blood. It removes substances of higher molecular weight than standard hemodialysis [4].

Therefore, it is essential to develop or modify a novel hemodialysis (hemofiltration) membrane which integrates filtration and adsorption [5]. Most of the extracorporeal blood detoxification treatments including kidney and liver support do not work without adsorptive purification systems [6, 7]. A new approach to obtain adsorptive hemodialysis membranes is integration of small functionalized particles (sorbents) with hemodialysis membrane. The small adsorbent particles can be putted in biocompatibility porous support may be we can integrate diffusion, convection, and in one step and then you can used for extracorporeal blood purification. The membrane properties can integrate hemodialysis with adsorption. Entrapping smaller particles into porous matrix support could improve surface area, and shorten distance diffusion of toxic compound to active sites. This concept offer possibility to protect release of small micro-particles, particle coalition and possibly fragmentation of the sorbent. Clearly, there is a room for innovations and improvements in capacity and efficacy of hemodialysis membrane as well as in the safety of sorbent systems.

In this paper, the adsorptive hemodialysis membrane offers an alternative approach for blood toxins removal by blood purification membranes meanwhile increasing the biocompatibility of hemoperfusion sorbents (activated carbon) based on mixed matrix membrane (MMM) concept [8]. A major advantage of the membrane concept is the ability to create particle-loaded membranes using any type of particles and almost any type of polymeric material. The activated membranes are prepared by incorporating activated carbon into a biocompatible macroporous polymeric matrix without affecting the activity. The adsorptive membrane are prepared by phase inversion methods. To improve the blood-compatibility, and to avoid mixing of small activated carbon particles with the blood stream, the membran is co-cast with a particle-free cellulose acetate solution with a particle free top layer preventing blood getting in contact with adsorbent particles. Creatinine (Crt) is used as model blood toxin to characterize the membrane adsorber performance.

2. Materials and Methods

Cellulose acetate with an average acetate content of 38.8 mol % was purchased from Aldrich (M.W≅30.000) and used as membrane material without further modification. Acetone (Merck) was employed as solvent. Ultra pure water, prepared using a Millipore Milli-Q plus purification unit, was used as additive in the polymer solution and as non-solvent in the coagulation bath. Activated carbon Norit A Supra EUR, kindly supplied by Norit Netherlands B.V. was used as adsorbent particles. Creatinine was used as a model blood uremic-toxin to investigate the adsorption capacity of the MMM. The buffer solutions were freshly prepared in ultrapure water. Tyrode buffer was chosen as representing adequately the ionic solutes, pH and glucose concentrations of the blood plasma. The Tyrode buffer composed of 137 mM NaCl, 5.4 mM KCl, 1.8 mM CaCl2, 0.5 mM MgCl2, 11,9 mM NaHCO3, 5.5 mM glucose (pH=7.4). The dialysate buffer contained of 140 mM NaCl, 2 mM KCl, 1.5 mM CaCl2, 0.25 mM MgCl2, 35 mM NaHCO3 and 5.5 mM glucose (pH=7.4).

a. Membrane preparation and characterization

To obtain membranes with adsorptive properties, activated carbon was dispersed into a homogenous solution containing 12 wt% cellulose acetate (CA) and 20 wt% water in acetone. The water, being a non-solvent additive was added to the casting solution in order to facilitate instantaneous demixing, which improves the membrane morphology. The dispersion was stirred over night at room temperature to break down possible particles clusters. The MMM were prepared by water vapor induced phase separation. The cast polymer solution was exposed to water vapor (temperature 30-33 oC and relative humidity of 98 %) prior to immersion in a water containing coagulation bath The polymer solutions were co-cast on a glass plate using a double casting knife with adjustable height. The height of the first and the second slit was 475 and 550 μm respectively. After the formation process, the membranes were washed with milli-Q water at room temperature to remove the residual solvent traces. The membranes were overnight dried in air and stored in a vacuum oven at 30 °C.

The membrane structures were characterized by Scanning Electron Microscopy (SEM) using a JEOL TM 220 A. The membrane top surfaces with and without co-casting layer as well as the cross sections and bottom surfaces were examined. Cross-section samples were prepared by cryogenic breaking fresh samples in liquid nitrogen. The samples were allowed to dry overnight under vacuum at room temperature and then gold-coated using a Balzers Union SCD 040 sputter coater.

b. Haemodialysis test

The diffusive transport of Crt was investigated using a two-compartment dialysis cell at room temperature. The first compartment was filled with Crt in the Tyrode buffer and the second compartment was filled with dialysate buffer. The compartments were separated by the MMM AC. The volume of each compartment was 105 ml and the active membrane area was 19.6 cm^2 (0.25 gram). Both solutions were recirculated at flow rate of 3 ml/minute using a peristaltic pump. During the experiment, 1 ml samples were taken as function of time to determine the Crt concentration in both compartments. Immediately after measuring the UV absorbance, at 230 nm the samples were returned to the compartments. The dialysis experiments were evaluated for six hours before the process were stopped.

3. Results and Discussion

3.1. Membrane preparation and characterization

Figure 1 shows SEM images of the MMM, which have been prepared with activated carbon (AC) as adsorbent. The MMM AC was prepared from a solution of 12 % CA, 20% water in acetone containing 60 wt% loading activated carbon, under exposure to water vapor environment. Regular membrane structures were obtained by means of these compositions. The prepared MMM AC shows good mechanical properties and no particle loss is observed during membrane formation. The membranes possess open and interconnected macroporous structures. The membranes display open structures both at bottom and top surface. The top layer of the membranes consists of a highly interconnected macroporous structure, which forms a continuous structure with the sub-layer. The average pore size of the membranes as determined by bubble point measurement is 0.234 μm. The MMM AC has a swelling degree of 6.5 %, porosity of 53 % and clean water flux of 800 $Lm^{-2}h^{-1}bar^{-1}$.

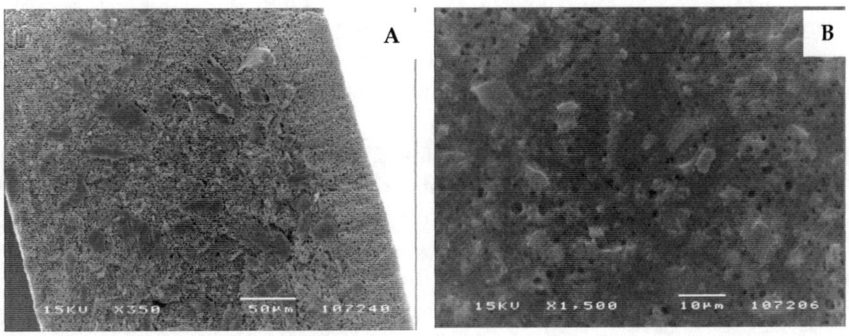

Figure 1. SEM micrographs of co-cast MMM prepared by water vapor induced phase separation and immersion precipitation.

Figure 1 indicated a cross section, magnification x350, the size bar indicates 50µm, MMM with embedded activated carbon particles (A), and a bottom surface, magnification x1500, the size bar indicates 10 µm MMM (B).

3.2. Dynamic creatinine removal

The MMM AC offers the possibility to integrate adsorption and hemodialysis treatments in one step. We investigated the performance of the MMM AC for this purpose. The adsorptive membrane was placed in a dialysis cell, which separates two compartments: one containing the Crt solution and the other containing the dialysate. The total Crt removal from the sample solution and the Crt concentration in the dialysate side were determined in time. The amount of Crt removed by adsorption onto the MMM AC can be calculated by subtraction the increase in Crt in the dialysate cell from the Crt that is removed from the sample side. Figure 2 presents the Crt removal during six hours of dialysis. Crt is significantly removed during the dialysis process by a combination of adsorption in the MMM AC and diffusion across the membrane.

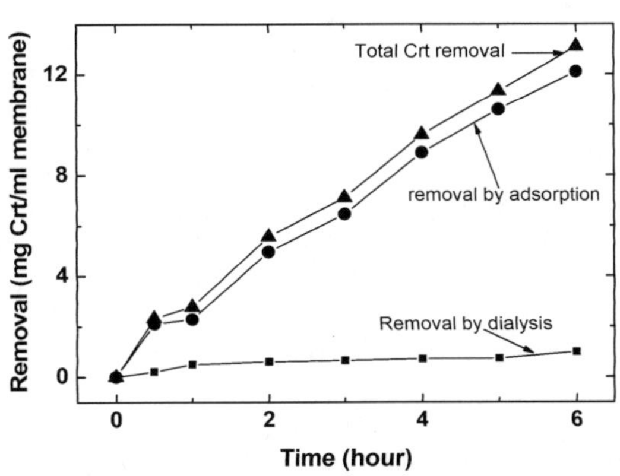

Figure 2. Creatinine removal during adsorptive dialysis. Starting concentration of Crt was approximately 0.1 mg/ml.

The relative contribution of adsorption and dialysis to the total removal of Crt can clearly be observed (figure 2). This indicates that most of Crt is captured by active carbon when diffusion through the membrane. The Crt removal by adsorption contributes to amore than 83 % of the total removal. Capacity of the hemodialysis membrane are still available for removal of contaminants after six hours hemodilaysis process. An increase in Crt removal by diffusion (dialysis) across the membrane can be expected when all of the active sites in the MMM AC are occupied (breakthrough). Activated carbon based hemoperfusion columns has recently been applied to increase the efficiency of hemodialysis, to remove toxic compounds during kidney or liver failure, autoimmune diseases and encephalopathy[1]. Furthermore, the use of activated carbon hemoperfusion columns are introduced to improve protein-bound and water soluble toxins removal, detoxification by drugs overdoses and poisonous and endotoxin removal [3, 4].

These results proved that the integration of adsorption functions in dialysis membranes significantly improves the performance of a dialysis treatment by preventing the diffusion of toxins between the two fluid streams (compartments). The adsorptive MMM can be adapted to different extracorporeal blood purification treatments, i.e. hemoperfusion, hemofiltration and hemodialysis. The total amount Crt removed during a six hour treatment depends solely on the MMM adsorption capacity. The MMM concept is ensure as safe treatment since only parts of the blood perfuses through the adsorbent and blood cellular component do not interact with the embedded sorbents.

3.3. Non-Specific Removal

Protein adsorption at blood-biomaterial interfaces is of prime importance for the biocompatibility of the material. The main factor promoting protein adsorption on surfaces is the hydrophobic interaction force between the sorbent surface and the protein molecule. BSA adsorption on adsorptive membrane is investigated to determine non-specific BSA (a bovine blood plasma protein) adsorption on entrapped active carbon particles. Figure 3 presents the BSA adsorption during six hours process hemodialysis. The BSA adsorption capacity of entrapped activated carbon particles in cellulose acetate membrane is equal to the pure cellulose acetate BSA adsorption capacity. This value is in agreement with data reported by Liu et al.[9], who measured 0.3 mg BSA/g membrane onto pure CA membranes, feed concentration 0.5 mg BSA/ml. The low BSA adsorption of the MMM AC demonstrates that entrapped activated carbon particles in the MMM are not accessible for high molecular weight BSA molecules where low molecular weight Crt molecules have free access. This makes that the MMM AC is very suitable for removal of toxic Crt out BSA containing (plasma) solutions.

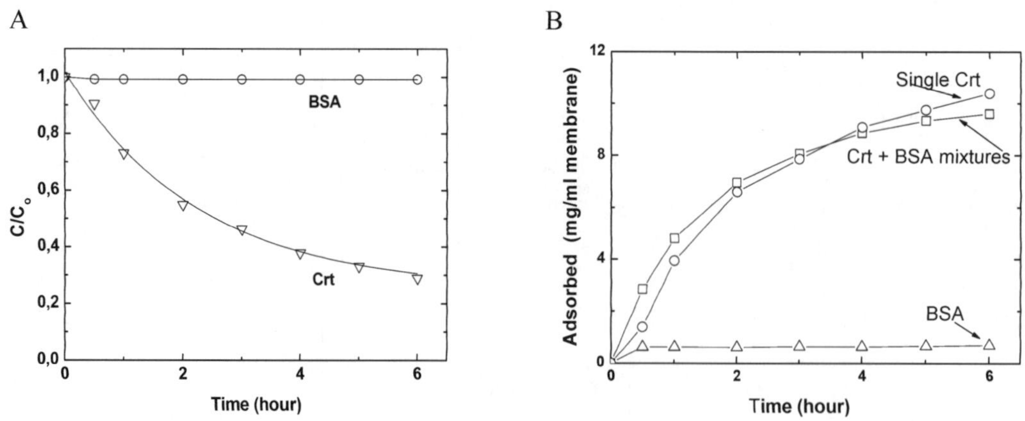

Figure 3. A. Creatinine removal during adsorptive dialysis. Starting concentration of Crt was approximately 0.1 mg/ml. B. The amount of adsorption single Crt, mixture of Crt and BSA dan single BSA on the membrane

The membrane possesses a porous structure and a very good pore interconnectivity so that harmful substances can utilize all active sites. The BSA adsorption of coated particles was approximately 0.8 mg BSA/g sorbent which is close to the non-specific adsorption value of the MMM AC. These results proof that the biocompatibility problems of sorbent hemoperfusion processes can be solved with the MMM approach. Furthermore, the biocompatibility of the blood side of the MMM adsorber can even be further improved by co-casting the MMM with a biocompatible layer.

Marlon et.al [5] proves that the combination of diffusion and adsorption in a single step is possible and paves the way for the development of more efficient blood purification devices, excellently combining the advantages of both techniques. The contribution of creatinine removal after 7 h by adsorption is over 80% of the total creatinine removal. No quick saturation occurs for the developed MMMs under these experimental conditions. The MMM combines uremic retention solute removal via both diffusion as well as adsorption in one single step. A polyethersulfone (PES)/polyvinylpyrrolidone (PVP) polymer blend is used for the preparation of the macro-porous membrane matrix and activated carbon is used as adsorptive particle.

4. Conclusion

The MMM double layers have been prepared by incorporating activated carbon into macroporous cellulose acetate membranes by phave inversion method. . To improve the blood-compatibility, and to avoid mixing of small activated carbon particles with the blood stream, the MMM is co-cast with a particle-free cellulose acetate solution with a particle free top layer preventing blood getting in contact with adsorbent particles. The MMM AC has a swelling degree of 6.5 %, porosity of 53 % and clean water flux of 800 $Lm^{-2}h^{-1}bar^{-1}$. The prepared membranes show a high dynamic Creatinine (Crt) removal during hemodilysis process. The Crt removal by adsorption contributes to amore than 83 % of the total removal. The MMM technology is suitable for integration of dialysis with adsorption in one step.

Acknowledgements

We want to acknowledge the financial support of this work through the Membrane Technology Group of the University of Twente and the Ministry of Education of the Republic of Indonesia.

References

[1] Stamatialis D F, Papenburg B J, Girones M, Saiful, Srivatsa N M, Bettahalli, Schmitmeier S and Wessling M 2008 *Journal of Membrane Science* **308** (1-2) 1-34

[2] Armignacco P, Lorenzin A, Neri M, Nalesso F, Garzotto F and Ronco C 2015 *Semin Dial* **28** 125–130

[3] Chan C T, Covic A, Craig J C, Davenport A, Kasiske B L, Kuhlmann M K, Levin N W, Li P K, Locatelli F, Rocco M V, Wheeler D C 2013 Kidney International **83**(3) 359-371

[4] Winchester F J, Ronco C, Salsberg J, Yousha E, Brady J A, Cowgill L D, Choquette M, Albright R, Clemmer J, Davankov V, Tsyurupa M, Pavlova L, Pavlov M, Cohen G, Horl W, Gotch F and Levin N W 2002 *Contributions to Nephrology* **137** 170-180

[5] Marlon S.L. Tijink, Maarten Wester, Junfen Sun, Anno Saris, Lydia A.M. Bolhuis-Versteeg, Saiful Saiful, Jaap A. Joles c, Zandrie Borneman , Matthias Wessling, Dimitris F. Stamatialis 2012 Acta Biomaterialia **8** 2279 –2287

[6] Ash S R, T A Sullivan, and D J Carr 2006 *Therapeutic Apheresis and Dialysis* **10**(2)145-153

[7] Mikhalovsky S V, Gun'ko V M, Turov V V, Leboda R and Betz W R 2005 *Adsorption* **11** 657

[8] Tijink M S L, Kooman J, Wester M, Sun J, Saiful S, Joles J A, Borneman Z, Wessling M and Stamatialis D F 2014 *Blood Purification Journal* **37** (1) 1-3

[9] Liu C and R Bai 2005 *Journal of Membrane Science* **267**(1-**2**) 68-77

IOP Conf. Series: Materials Science and Engineering **352** (2018) 012049 doi:10.1088/1757-899X/352/1/012049

Chitosan-stabilized Silver Nanoparticles for Colorimetric Assay of Mercury (II) Ions in aqueous system

Fitri Zarlaida [1], M. Adlim[1], M. Syukri Surbakti[2], Ahmad Fairuz Omar[3]
[1]Chemistry Department, FKIP, Syiah Kuala University, Darussalam Banda Aceh, 23111 Indonesia.
[2]Physics Department, FMIPA, Syiah Kuala University, Darussalam Banda Aceh, 23111 Indonesia
[3]School of Physical Sciences, Universiti Sains Malaysia, Minden, Penang 11800, Malaysia.

E-mail: zarlaida.fitri@unsyiah.ac.id

Abstract. Mercury is considered as dangerous pollutant. Among the many form of mercury, the most stable and soluble in water is mercury (II) ions which it cause threat to human health and surroundings. Silver nanoparticles (AgNPs) used in this method were prepared by chitosan (chi) which act as stabilizing agent. The Chi-AgNPs has good dispersity with size ranging from 2.50 to 6.00 nm as shown by transmission electron microscopy (TEM) analysis and it is stable for 3 months. Color of Chi-AgNPs fades from brownish-yellow to colorless only with Hg^{2+} ions, but it shows no significant changes upon addition of other metal ions such as Al^{3+}, Ba^{2+}, Ca^{2+}, Cd^{2+}, Cr^{3+}, Co^{2+}, Cu^{2+}, Fe^{2+}, K^{+}, Mg^{2+}, Mn^{2+}, Na^{+}, Ni^{2+}, Pb^{2+}, and Zn^{2+}. The detection limit for Hg^{2+} ions by bare-eye is estimated to be ~1μM. This method can be used for sensing mercury(II) ions in numerous water samples.

1. Introduction

Nobel metal nanoparticles such as silver nanoparticles (AgNPs) and gold nanoparticles (AuNPs) have attracted researchers' attention because of their unique chemical and optical properties, which are very much dependent on particle shape and size. [1,2]. Noble metal nanoparticles commonly synthesized through chemical reduction method because of its convenient procedure, easy in controlling, and use simple equipment [3]. However, nanoparticles are generally unstable and tend to aggregate. One way to solve this problem is to disperse them into natural or synthetic polymer matrices such as chitosan, cellulose, and PVA. Chitosan is one of the most abundant natural, renewable, biocompatible, biodegradable, non-toxic, having ability to form film, has high mechanical strength, and highly adsorptive polymer. It can be used as stabilizing agent and immobilization matrix due to the existence of hydroxyl and amino functional groups [4,5].

The prepared metal nanoparticles have been reported to be applied for catalysis, sensors, antibacterial, and numerous other areas [4,6,7,8]. Nobel metal nanoparticles can be used as colorimetric assay because of their high absorption coefficient, light stability, functionality, extent of surface-to-volume ratio, changing optical properties [9] and absorbing colors in visible light areas [10].

Mercury is considered as dangerous pollutant. Among the many form of mercury, the most stable and soluble in water is mercury (II) ions, which it cause threat to human health and surroundings through food chain [11]; hence, the development of methods for its detection at very low concentration is important. Detection of mercury(II) ions can be done using instrument like *Atomic Absorption/Emission Spectrometry* (AAS/AES), Inductive Coupled Plasma-Mass Spectrometry (ICP-MS), and Fluorescence Spectrometry AFS) [12]. Nevertheless, these methods rather complicated, expensive, and take a long time to analysis. An alternative method such as colorimetric sensing offers advantages over other methods due to its simplicity, low cost, and bare eye application [13,14].

In present work, focus of study was to prepare and characterize a material comprised of chitosan-stabilized nanoparticle (Chi-AgNPs) and use it for colorimetric assay of mercury (II) ion. As commented earlier, AgNPs are used as catalysts for many reactions and chitosan is used to stabilize metallic nanoparticles. However, there has been very limited use of materials comprised of chitosan-stabilized silver nanoparticles for the fabrication of sensor to detect mercury (II) ions.

2. Materials and Methods

All chemicals used in the present study were of the highest purity (p.a). Chitosan of medium molecular weight and sodium borohydride purchased from Aldrich (USA). Silver nitrate, acetic acid, methanol, salt of additional cation tested and all other chemicals bought from Merck (Germany). Characterization of size and shape of the synthesized silver nanoparticles was carried out using Transmission electron micrograph (TEM) recorded on JEOL JEM-1400 operating at 100 kV.

2.1. Synthesis of Silver Nanoparticles

A colloidal silver was prepared by reduction of $AgNO_3$ using sodium borohydride as our reported method with slight modification [15]. All glassware was washed with a mixture of nitric acid and hydrochloric acid (1:3) before use. Distilled water was used throughout the experiment. Silver (I) stock solution was prepared 10^{-4} mol/mL from silver nitrate as a precursor. In typical experiment, 5.6 x 10^{-4} mol (0.0900) was dissolved in 88.8 ml acetic acid (1.5%) and the mixture was stirred overnight to obtain homogeneous solution. Next, 90 ml of methanol slowly added with vigorous stirring. Afterwards, 1.12 ml (1.12 x 10^{-4} mol) of silver stock solution was added little by little to the mixture. Then 0.0265 g of sodium borohydride was added slowly while kept stirring vigorously to reduce the silver ions to silver particles to yield brown yellowish colloid. This silver nanoparticles were characterize with TEM.

2.2. Colorimetric assay of mercury (II) ions using Chi-AgNP as a probe

For colorimetric assay of mercury (II) ions, 3.0 mL of Hg^{2+} ions with various concentration were mixed with 0.5 mL of the prepared chi-AgNPs. The mixture were shaken and settled at room temperature for 15 minutes and then UV-vis absorption spectra were recorded. Colorimetric color change of the sensor after addition with mercury (II) ion were also observed by bare eye and the photograph was taken by camera. In the experiment of selectivity, all samples containing metallic ions were tested in same way. This method was also applied in similar manner for detection of mercury ions using tap, river, and bottle water from nearby Syiah Kuala University campus.

3. Result and Discussion

The chi-AgNPs based chemical sensors are synthesized by reduction of silver nitrate as precursor with sodium borohydride. Silver nitrate is added to a solution of chitosan kept under continuous stirring, dispersion of Ag^+ ions is enabled by the action of chitosan as a stabilizing agent/surfactant. The Ag^+ ions is reduced to Ag^0 (metallic silver) after addition of sodium borohydride and remnants embedded in the chitosan matrix [16]. The presence of $-NH_2$ and –OH groups in the chitosan can assist in possible adsorption interaction between chitosan and heavy metal ions including Ag^+ [17]. Besides as stabilizing or capping agent, chitosan also acts as a reducing agent [15]. Figure 1(a) shows TEM image of silver nanoparticle stabilized with chitosan. It is revealed that chi-AgNPs had good dispersity with size ranging from 2.50 to 6.00 nm. After addition of mercury(II) ions to the chi-AgNP shows its aggregated nature as exhibited in Figure 1(b). The interaction between Hg^{2+} ions with chitosan as stabilizing agent to form the larger size of nanoparticles that lead to agglomeration.

IOP Conf. Series: Materials Science and Engineering **352** (2018) 012049 doi:10.1088/1757-899X/352/1/012049

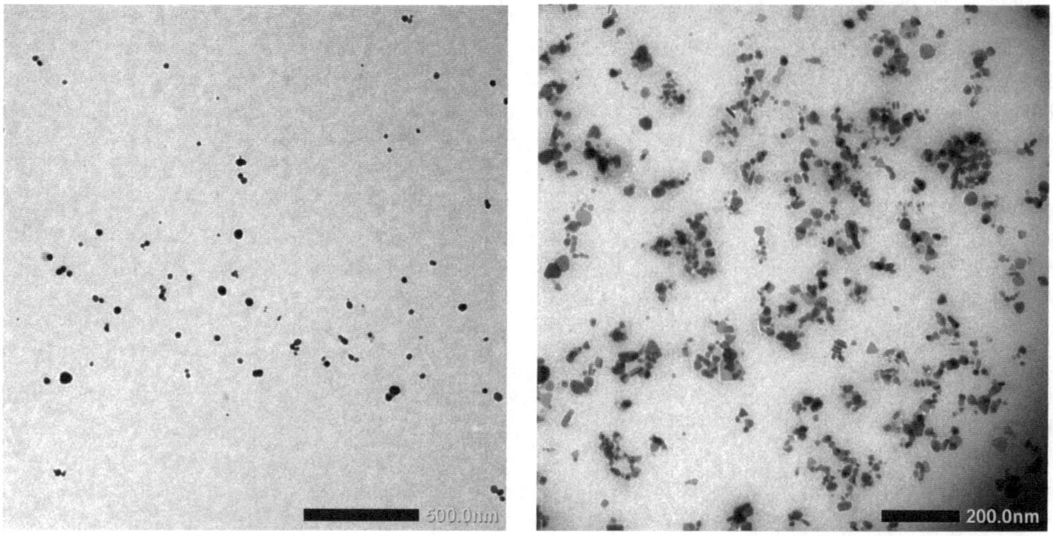

(a) (b)

Figure 1. (a) TEM image of non-aggregated AgNPs stabilized by chitosan and (b) TEM image of the aggregated AgNPs after addition of Hg^{2+} ions.

3.1. Selectivity and sensitivity of assay

The selectivity of this probe for Hg^{2+} ions has been evaluated through testing the response of the assay to different metal ions Al^{3+}, Ba^{2+}, Ca^{2+}, Cd^{2+}, Cr^{3+}, Co^{2+}, Cu^{2+}, Fe^{2+}, K^{+}, Mg^{2+}, Mn^{2+}, Na^{+}, Ni^{2+}, Pb^{2+}, and Zn^{2+} at concentration of 100 μM after 10 minutes of mixing. As depicted in Figure 2, only the Hg^{2+} sample shows a significant color change to colorless relative to that of the blank whereas all others remain brownish-yellow color without any eye-distinguishable change. It means that the probe highly selective to detect mercury(II) ions.

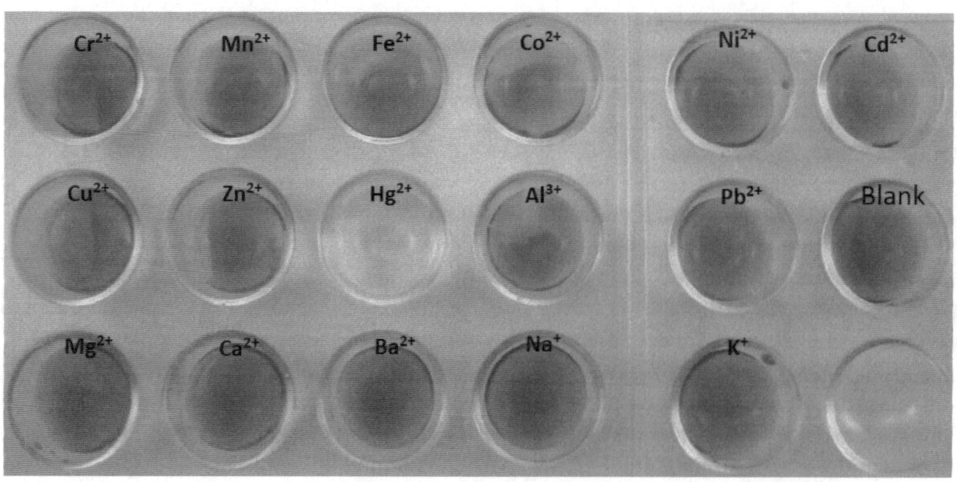

Figure 2. Color change of Chi-AgNPs in the existence of various metal ions

To estimate sensitivity of chi-AgNPs for detecting Hg^{2+} ions, a series of various concentration of Hg^{2+} ions were mixing with the probe. Figure 3 demonstrated the digital image of the AgNPs after addition with various concentraion of Hg^{2+} ions and the limit detection is estimated to be ~1 μM.

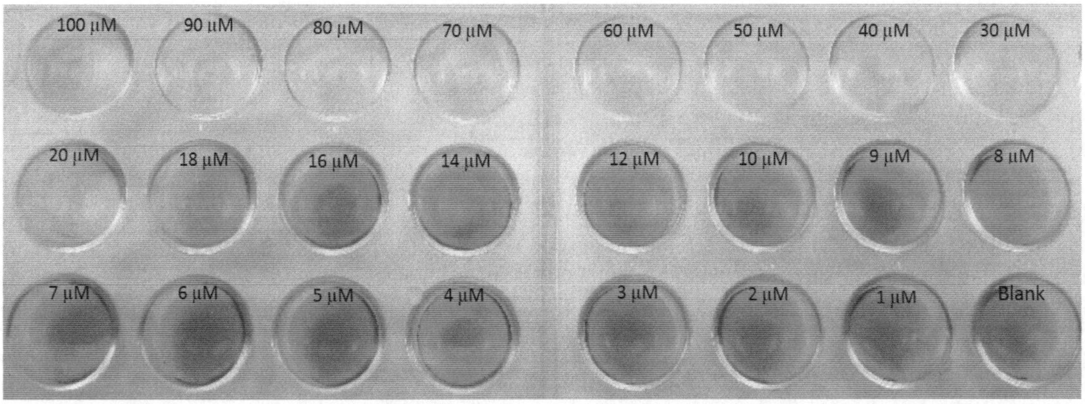

Figure 3. Sensitivity of Chi-AgNPs toward different concentration of Hg^{2+} ions.

3.2. Detection of Hg^{2+} ions in real water samples
Practicability of this colorimetric method is evaluated for various water sample application. It is found that none of the water samples caused visible color change in chi-AgNPs probe (Figure 4a), indicating that concentration of Hg^{2+} ions of these water samples are below 1 µM. On the contrary, when the water samples spiked with Hg^{2+} 100 µM, the color of the water change to colorless (Figure 4b). These results confirm that the detection method established here can be used for the detection of Hg^{2+} ios in real water samples.

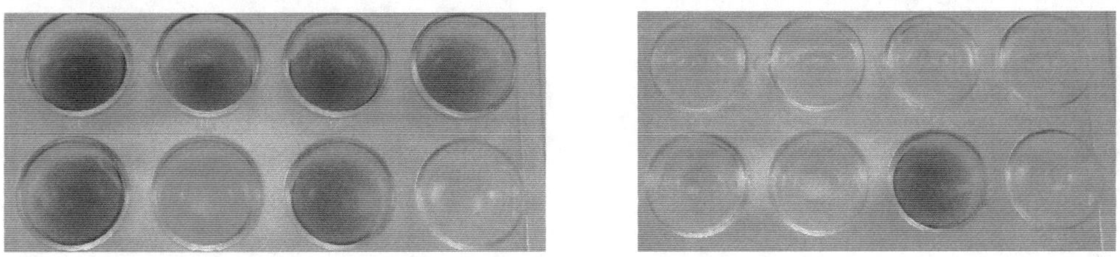

a) b)

Figure 4. a) Color of water samples after addition of the AgNPs
b) Color of water samples that spiked with Hg^{2+} after addition of the sensor. (Top line from left to right: local water refills, well water, tap water, Lamnyong river water. Bottom line, from left to right: bottled water (Cleo)), mercury (II) ions solution and blank.

4. Conclusions
In summary, AgNPs stabilized by chitosan can be used as a colorimetric assay for mercury (II) ion with limit of detection 1µM. Several other commonly encountered metal ions present in the environment does not affect detection of Hg^{2+} ions. The practical applicability of AgNPs stabilized by chitosan exhibits selective sensing of Hg^{2+} ions in various water samples such as bottled, well, tap, and river water.

References
[1] Oliveira E, Nunez C, Santos HM , Fernandez-Lodeiro J, Fernandez-Lodeiro A, Capelo J L and Lodeiro C 2015 *Sens. Actuators B* **212** 297-328
[2] Huber J and Leopold K 2016 *Trends Anal. Chem.* **80** 280-292
[3] Tran Q H, Nguyen V Q and Le A-T 2013 *Adv Nat Sci Nanosci Nanotechnol.* **4**(3) 33001-20
[4] Adlim 2006 *Indo. J.Chem.* **6**(1) 1-10
[5] Dutta P K, Duta J and Tripathi V S 2004 *J. Sci. Ind. Res.* **63**(1) 20-31

[6] Fisher E, Kenisgberg L, Carreira M, Fernández-Gallardo J, Baldwin R and Contel M 2016 *Polyhedron* **120** 82-87
[7] Bindhu M R and Umadevi M 2014 *Spectrochim Acta - Part A Mol Biomol Spectrosc.* **128** 37-45
[8] Zarlaida F and Adlim M 2017 *Microchim. Acta* **184**(1) 45-58
[9] Chen Z, Zhang C, Tan Y, Zhou T, Ma H, Wan C, Lin Y and Li K 2015 *Microchim. Acta* 1**82**(3-4) 611-616
[10] Vilela D, González M C and Escarpa A 2012 *Anal Chim Acta.* **751** 24-43
[11] Bernhoft R A 2012 *J. Environ. Public Health. 2012*
[12] Rastogi L, Sashidhar R B, Karunasagar D, and Arunachalam J 2014 *Talanta* **118** 111-117
[13] Choi Y W, You GR, Lee M M, Kim J, Jung K D and Kim C 2014 *Inorg. Chem. Commun.* **46** 43-46
[14] Duan J, Yin H, Wei R and Wang W 2014 *Biosens Bioelectron.* **57** 139-142
[15] Adlim 2006 *J Sains Tek.* **12**(3) 185-191
[16] Nivethaa E A K, Narayanan V and Stephen A 2015 *Spectrochim Acta Part A Mol Biomol Spectrosc.* **143** 242-250.
[17] Mohammadi S and Khayatian G 2017 *Spectrochim Acta Part A Mol Biomol Spectrosc.* **185** 27-34

Characterization of corn starch-based edible film incorporated with nutmeg oil nanoemulsion

Y Aisyah[1], L P Irwanda[2], S Haryani[1], N Safriani[1]
[1]Lecturer of Agricultural Product Technology Department, Syiah Kuala University, Banda Aceh 23111, Indonesia
[2]Undergraduate student of Agricultural Product Technology Department, Syiah Kuala University, Banda Aceh 23111, Indonesia

E-mail: yuliani.aisyah@unsyiah.ac.id

Abstract. This study aimed to formulate corn starch-based edible films by varying concentrations of nutmeg oil nanoemulsion and glycerol. Furthermore, the resulted edible film was characterized by its mechanical properties and antibacterial activity. The edible films were made using corn starch, nutmeg oil nanoemulsion, and glycerol. Concentrations of nutmeg oil nanoemulsion were 1%, 2%, and 3%, and glycerol were 10%, 20%, and 30%. Results indicated that the increase of nutmeg oil nanoemulsion concentration could increase the film thickness. However, the nutmeg oil had no effect on the film tensile strength and elongation. Glycerol had no effect on the film tensile strength. The best treatment of the corn starch-based film was obtained by adding 1% of nutmeg oil and 30% of glycerol, yielding a tensile strength of 18.73 Kgf/mm^2, elongation of 69.44% and thickness of 0.0840. The addition of 1% nutmeg oil nanoemulsion has been able to inhibit the growth of two types of the bacteria tested (*Staphylococcus aureus* and *Escherichia coli*).

1. Introduction

Food products like fruits, vegetables, meat, and processed products are easily damaged and degraded. Therefore, the packaging is used to reduce food degradation. One of the primary packaging types is an edible film, a thin layer coating the food that can be consumed and naturally degraded [1]. Edible films are, in fact, a relatively new food preservation technique; yet, they have been widely researched. Several studies have shown that edible films can extend food shelf life and improve its quality.

One of the potential and widely researched polymer materials for edible films is starch. However, starch-based films have some weaknesses i.e. low resistance to water and low water vapor barrier due to its hydrophilic nature that affects its stability and mechanical properties [2]. However, the physical and functional characteristics of starch films can be improved by adding biopolymers or other materials that are hydrophobic and/or have antimicrobial properties such as essential oils and glycerol (plasticizers) [3].

Essential oils contain natural preservative compounds because of their antimicrobial properties [4]. These properties prevent microbial growth by interfering the microbial cytoplasmic membranes, disrupting cell constituents, and reacting with cell membranes; and thus, increasing permeability and causing cell component loss and/or coagulating cell contents [5]. Nutmeg oil is one of the essential oils containing phytochemical compounds e.g. sabinene (25.4%), â-pinene (15.8%) and myristicine (14.8%) [6].

The antibacterial compounds in nutmeg oil are non-polar, thus having a low solubility rate. Therefore, the nanoemulsion system is required to increase solubility, bio-accessibility, and bio-availability, and to maintain stability as well. The use of nanocomposites in edible film production could enhance the film tensile strength, storage modulus, glass transition temperature, and water vapor barrier properties [7].

This study used corn starch in making the edible films, reinforced with nutmeg oil nanoemulsion and glycerol. Corn starch contains a high amount of amylose, at approximately 25%, thus it can be utilized in developing the film matrix [8]. The addition of nutmeg oil nanoemulsion and glycerol into the edible film formulation could increase the mechanical and functional properties of the films. The aims of this study were to find the best formula out of several compositions of nutmeg oil nanoemulsion and glycerol in the edible films and to find out the effects of nutmeg oil nanoemulsion and glycerol addition on the film characteristics.

2. Materials and Methods

2.1. Materials
Materials used in this research were the essential oil of nutmeg oil (food grade) from PT. Djasula Wangi, Jakarta, Indonesia. Tween 80 (polyoxyethylene-20-sorbitan monooleate) and Glycerol analytical standard (Sigma-Aldrich, US). Corn starch was purchased from local market in Banda Aceh, Indonesia.

2.2. Preparation of Nutmeg Oil Nanoemulsions
Nanoemulsion was produced by using *High-Pressure Homogenization* (HPH) method. The type of emulsion was oil in water (O/W) with the nutmeg oil as the dispersed phase and water as the dispersing phase. The coarse emulsion was prepared by dispersing 15% of nutmeg oil into the double distilled water and tween 80 as a surfactant. The concentration of tween 80 was 20% of the nutmeg oil (w/v). The mixture then homogenized by a high shear homogenizer (IKA T25 digital Ultra Turrax) at 12000 rpm for 5 minutes. The coarse emulsion was then passed into High-Pressure Homogenizer (HPH) at 500 bar with 3 cycles [9].

2.3. Preparation of Edible Films
Edible films were prepared by dissolving 3% of corn starch and glycerol (10%, 20%, 30%) into the aqueous solution. The suspension was stirred at 85°C for 30 minutes. Afterward, the suspension was cooled to 45°C and then the nutmeg oil nanoemulsion at the concentration of 1%, 2%, 3% was added to the suspension. The mixture was cast onto 30 cm x 20 cm glass plates and was dried at ± 50°C for 12 hours and subsequently cooled at room temperature for 24 hours. Then the films could be easily removed out from the glass plates. The films were then analyzed [10].

2.4. Characterization of Edible Film
The characteristics of the resulted films were observed by measuring those mechanical properties; tensile strength, thickness, elongation [11]; and the antibacterial as well [12].

3. Results and Discussions

3.1. Morphology of edible film
The analysis results of the edible film surface morphology can be seen in Figure 1. Based on the SEM test results in Figure 1, It can be seen that the film morphology with no addition of nutmeg nanoemulsion, 20% of corn starch, and 30% of glycerol shows viscous, smooth, and non-porous surface of the film molecular structure (Figure 1a). On the other hand, the surface of the film

molecular structure produced by adding 1% of nutmeg oil nanoemulsion, 20% of corn starch, and 30% of glycerol is quite porous (Figure 1b).

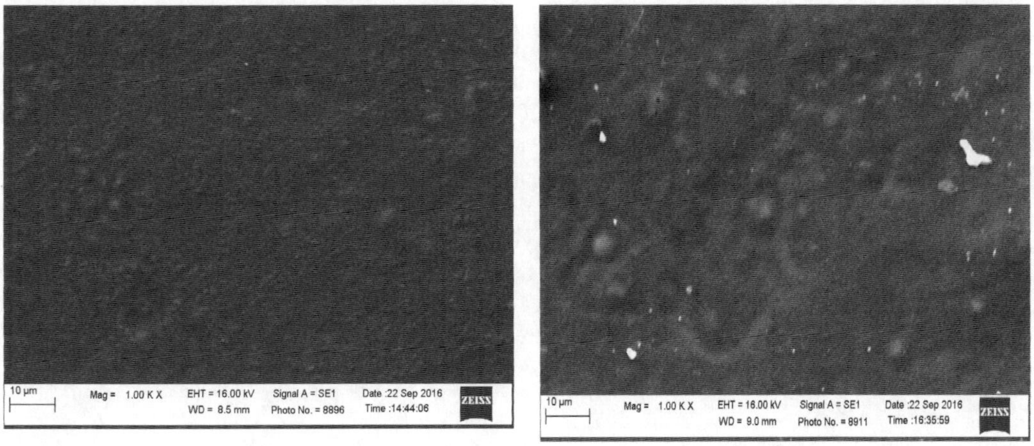

(a) (b)

Figure 1. The film surface with no addition of nutmeg oil nanoemulsion, 20% of starch, and 30% of glycerol (a) and with the addition of 1% nutmeg oil nanoemulsion, 20% starch, and 30% glycerol (b) using 100x magnification

3.2. Thickness (mm)

The film thickness was measured by using a screw micrometer at five random points with the area of 7x7 cm each. The average measurement data of the thickness were collected. The thickness data showed that the films had a range of thickness from 0.0643 – 0.1013 mm with an average of 0.0830 mm.

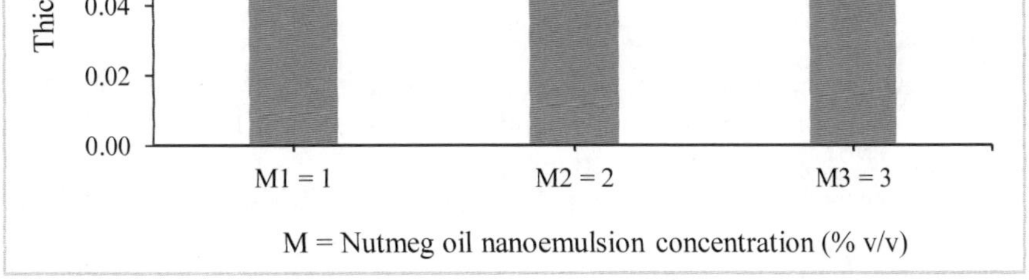

Figure 2. The effects of nutmeg oil nanoemulsion concentration (M) on edible film thickness

Figure 1 shows that nutmeg oil nanoemulsion at 3% of concentration had the highest thickness while at 1% of that resulted in the lowest thickness. It indicates that the higher the nutmeg oil concentration, the thicker the film. Then, after the LSD test between treatments was carried out, there was a significant difference on the film thickness between the concentration of nutmeg oil nanoemulsion 1%, 2%, and 3%, shown from different notation. The nutmeg oil nanoemulsion consisted of 3 (three) components: nutmeg oil, water, and surfactant. The addition of nanoemulsion could enhance the viscosity of the solution leading an increase of the edible film thickness. The increase in thickness occurs due to differences in the concentration of the filmmaker, while the volume of solution poured on each plate is the same. Furthermore, both the total solid in the dried films and the polymers that make up the film matrix increased. Bertuzzi (2007) showed that the addition of glycerol concentration increased the viscosity of the solution thus increasing the film thickness as well. The increased concentration of the materials used will lead to an increase in film thickness (Mc Hugh, 1993).

Figure 2 shows that the glycerol concentration at 30% produced the highest thickness the edible film, while the glycerol concentration at 10% yielded the lowest thickness. It means that the higher of the glycerol concentration, the thicker of the films. Then, after the LSD test between treatments was carried out, there was a significant difference on the film thickness between the concentration of glycerol at 1%, 2%, and 3%, shown from different notation.

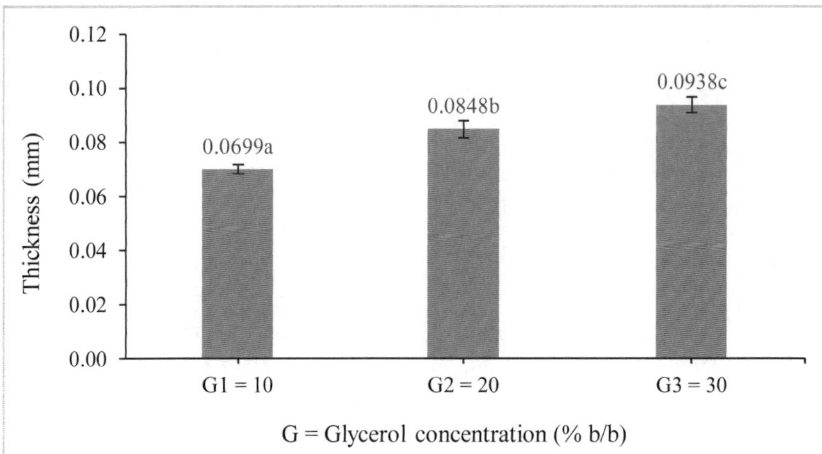

Figure 3. The effects of glycerol concentration (G) on edible film thickness

According to Zhang and Han [13], the film thickness elevates as the plasticizer increases, this is because glycerol can bind quite a lot of water, thereby raising viscosity. In addition, the viscosity of a solution has an impact on water evaporation in the drying process [14], thus, when the viscosity is higher, the film will be thicker.

3.3. Tensile Strength (Kgf/mm^2)

Tensile strength is the mechanical property of an edible film which reflects the maximum stress that the film sustains before it eventually breaks. Tensile strength shows the film strength in resisting the mechanical damage. The device used here was an autograph with the measurement in Kgf/mm^2. The value of tensile strength obtained in this study ranged from 15.07 – 18.93 Kgf/mm^2 with an average of 16,90 Kgf/mm^2.

The addition of nanoemulsion at the concentration of 1%, 2%, 3% and glycerol of 10%, 20%, 30% was not notable in the film tensile strength. However, generally, it seems that the higher the concentration of nanoemulsion (M), the lower the film tensile strength, as depicted in Table 1. The decrease is mainly caused by the ability of glycerol to enhance flexibility and lessen intermolecular forces along the polymer chains [15].

Table 1. The effect of nutmeg oil concentration on the mechanical properties of the edible films

Concentration of nutmeg oil nanoemulsion	Tensile strength (Kgf/mm^2)	Elongation at break (%)
1%	17.38	35.57
2%	16.56	33.71
3%	15.78	32.89

Table 2. The effect of glycerol concentration on the mechanical properties of the edible films

Concentration of glycerol	Tensile strength (Kgf/mm^2)	Elongation at break (%)
10%	16.00	11.52
20%	16.67	28.93
30%	18.04	52.84

The decrease in tensile strength and increase in elongation are common results of essential-oil incorporation and have been broadly discussed in research into other biopolymer films. Maizura et al. (12) reported a decrease in the TS of the film prepared from a starch-alginate film to which different concentrations of lemongrass oil had been added. They reported that this was due to the effect of the lipid on the starch chains and that the phase rich in polysaccharide had a higher tensile strength than lipid phase. Furthermore, the addition of glycerol led to the formation of free volumes, a result of broken bonds between polysaccharides [16]. The tensile strength keeps declining as molecular interactions decreases [11]. Glycerol can penetrate into polysaccharide molecules and remove hydrogen bonds.

3.4. Elongation (%)
Elongation is the maximum extension percentage that an edible film can achieve before it is finally broken/torn. The observation data in this study showed that the elongation values obtained ranged from 4.81% to 69.44% with an average of 31.10%.

Table 2 shows that the largest percentage of elongation of 52.84% was obtained at 30% of glycerol concentration and the lowest of 11.52% was at 10% of glycerol concentration. This suggests that the increase of glycerol concentration tends to elevate the percentage of the film elongation. The increase of glycerol may decrease the intermolecular force of the edible film matrix structure, increase flexibility, and lower the number of hydrogen bonds; thus, reducing fragility and not easily breaking.

3.5. The antibacterial activity of edible films
The results of the study showed that the edible films with the addition of nutmeg oil nanoemulsion were able to inhibit the growth of *Staphylococcus aureus* and *Escherichia coli.* The film antimicrobial activity with 3% of nutmeg oil nanoemulsion addition was capable of inhibiting *Staphylococcus aureus* at 18.27 mm and *Escherichia coli* at 14.89 mm. Reinforcing nutmeg oil nanoemulsion to the films causes the obstruction of the growth of tested microbes (Table 3). When the nanoemulsion was added to the films, it would diffuse into the agar medium and produce clear zones in the microbial growth medium.

Table 3. The inhibition zone diameter of edible film with the addition nutmeg oil nanoemulsion against *S. aureus* and *E. coli*

Type of bacteria	Nutmeg oil nanoemulsion concentration	Inhibition zone (mm)
	1%	10.04
Staphylococcus aureus	2%	14.62
	3%	18.27
	1%	10.76
Escherichia coli	2%	12.22
	3%	14.89

The Table 3 describes that the higher of nutmeg oil nanoemulsion concentration, the greater the ability to inhibit the growth of tested bacteria. Factors affecting the size of the inhibitory area include the sensitivity of the organism, the culture medium, the incubation conditions, and the speed of the agar diffusion. Factors affecting the speed of agar diffusion are the concentration of microorganisms, the composition of the media, the incubation temperature, and the incubation time. According to [17], the diameter of the inhibitory zone does not always elevate in corresponding with the rise of the bacterial concentration. It is highly likely that different diameters of the inhibitory zones over a period occur due to the differences in the diffusion speed of antibacterial compounds on the agar medium and the different types and concentrations of antibacterial compounds.

It can be concluded that the addition of nutmeg oil nanoemulsion of 1% has been able to inhibit the growth of two types of the bacteria tested (*Staphylococcus aureus* and *Escherichia coli*). Therefore, the addition of 1% nutmeg oil nanoemulsion is chosen as the minimal concentration of the inhibitory ability.

4. Conclusion

The concentrations of nutmeg oil nanoemulsion and glycerol have had an effect on the characterization of corn starch-based edible films. The best treatment of edible film was produced by adding 1% of nutmeg oil nanoemulsion and 30% of glycerol, resulting in the film tensile strength of 18,73 Kgf/mm^2, elongation of 69.44% and thickness of 0.0840 mm. The addition of 1% nutmeg oil nanoemulsion was able to inhibit the growth of two bacteria tested (*Staphylococcus aureus* and *Escherichia coli*).

Acknowledgment

The authors thank the Ministry of Research, Technology and Higher Education of The Republic of Indonesia for funding this research through Research Institution of Syiah Kuala University.

References

[1] Guilbert S, Gontard N and Gorris L G M 1996 Prolongation of the shelf-life of perishable food products using biodegradable films and coatings *Lebensm. Wiss. Technol.* **29** 10-17

[2] Garcia N L, Ribbon L, Dufresne A, Aranguren M and Goyanes S 2011 *Carbohydrate Polymers.* **84** 203-210

[3] Kester J and Fennema O 1989 Resistance of lipid films to water vapor transmission *J. Amer. Oil Chem. Soc.* **66** 1139-1146

[4] Burt S 2004 Essential oils: their antibacterial properties and potential applications in foods a review *Intl. J. Food Microbiol.* **94** 223-253

[5] Politeo O, Jukić M and Miloš M 2006 Chemical composition and antioxidant activity of essential oils of twelve spice plants. croatica chemica acta *Croatica Chemica Acta CCACAA.* **79** 545-552

[6] Chang P R, Jian R, Yu J and Ma X 2010 Starch-based composites reinforced with novel chitin nanoparticles *Carbohydrate polymers.* **80** 420-425

IOP Conf. Series: Materials Science and Engineering **352** (2018) 012050 doi:10.1088/1757-899X/352/1/012050

[7] Sandhu K S, Singh N 2007 Some properties of corn starches II: Physicochemical, gelatinization, retrogradation, pasting and gel textural properties *Food Chem.* **101** 1499-1507
[8] Aisyah Y, Safriani N, Muzaifa M and Fakhrurrazi 2015 Optimasi proses emulsifikasi minyak pala (Myristica fragrans houtt) *Prosiding Seminar Agroindustri dan Lokakarya nasional FKPT-TPI* Universitas Trunojoyo Madura Surabaya (In Indonesian)
[9] Kusumawati D H and Putri W D R 2013 Karakteristik fisik edible film dari patijagung yang diinkorporasi dengan perasan temu hitam *Jurnal Pangan dan Agroindustri.* **1** 90-100 (In Indonesian)
[10] Bourtoom T 2008 Edible films and coatings: characteristics and properties *International Food Research Journal.* **15** 237-248
[11] Madigan M T, Martinko J M and Parker J 2003 *Brock Biology of Microorganisms* Prentice Hall (New Jersey USA: Prentice Hall-Pearson Education Inc)
[12] Zhang V and Han J H 2006 *Jurnal Food ist.* **71** E253-E260
[13] Mali S, Grossmann M V E, García M A, Martini M N and Zaritzky N E 2004 Barrier, mechanical and optical properties of plasticized yam starch films *Carbohydrate Polymer.* **56** 129-135.
[14] Rodriguez M, Oses J, Ziani K and Mate J I 2006 Combined effect of plasticizers and surfactants on the physical properties of starch based edible films *Food Res. Int.* **39** 840-846
[15] Intan D H and Aizan A 2011 Tensile and Water Absorption Properties of Biodegradable Composites Derived from Cassava Skin/Polyvinyl Alcohol with Glycerol as Plasticizer *Sains Malaysiana.* **40** 713-718
[16] Matan N 2012 Antimicrobial activity of edible film incorporated with essential oils to preserve dried fish (Decapterus maruadsi) *International Food Research Journal.* **19** 1733-1738

IOP Conf. Series: Materials Science and Engineering **352** (2018) 012051 doi:10.1088/1757-899X/352/1/012051

The effect of poly ethylene glycol additive on the characteristics and performance of cellulose acetate ultrafiltration membrane for removal of Cr(III) from aqueous solution

S Mulyati[1], S Aprilia[1], Safiah[2], Syawaliah[3], M A Armando[1], H Mawardi[1]
[1]Syiah Kuala University, Banda Aceh, Indonesia, 23111
[2]Graduate School of Engineering Science, Syiah Kuala University, Banda Aceh Indonesia, 23111
[3]Doctoral School of Engineering Science, Syiah Kuala University, Banda Aceh, Indonesia,23111

E-mail: sri.mulyati@unsyiah.ac.id

Abstract. The effect of polyethylene glycol (PEG) additive on the characteristics and performance of the cellulose acetate ultrafiltration membrane to chromium metal removal has been studied using some variation of concentration in the casting solution. The concentration of cellulose acetate polymer was 17.5%, whereas the variations of PEG concentration were regulated at 0, 2.5, 5, 7.5 and 10% by weight. Dimethyl formamide (DMF) was used as a solvent. Pure water flux, membrane morphology test, functional group analysis, and molecular weight cut off (MWCO) were investigated to characterize of the prepared membranes. Membrane performance was tested against Cr(III) metal removal. The results confirmed that the pure water flux improved with the increasing of additive concentration. The maximum improvement occurred at membrane with modification using 7.5% PEG. At this PEG concentration, the pure water flux elevated from 49.5 $L/m^2.h$ to 62.2 $L/m^2.h$. The addition of PEG successfully improved the membrane flux because the role PEG plays as a pore-forming agent. Membrane with addition of 7.5% PEG showcased rejection result for chromium metal of 31.89%. This value is lower than that of pure CA membrane which rejection value against Cr(III) metal amounted to 35.72%.

1. Introduction

The scarcity of clean water has become an extremely serious problem with the increasing of population. Approach such as degradation by advanced oxidation process, distillation, and filtration have been performed to address this issue. The membrane process has gained a particular attention as it can separate or eliminate bacteria, viruses and ions based on pore size and pressure differences as driving forces [1]. This process is used to remove particles, turbidities and microorganisms in water and wastewater [2]. This technology also produces high quality water, eliminates toxicity or recovers valuable components from the output of various industrial processes [3]. Hydrophilicity and pore structure of the membrane play an important role in membrane separation processes. For separation process, a membrane has to have a good porosity, as well as high permeability, hydrophilicity and resistance to impurities.

Phase inversion is one of many important processes used to fabricate asymmetric porous membranes such as microfiltration, ultrafiltration, reverse osmosis and support for composite membranes. In the process of membrane structure formation, besides thermodynamic process, kinetic process also plays an important role. The structure and properties of membranes made with phase inversion techniques are influenced by many factors. Additives are one of those that plays a role in enlarging or preventing macrovoid formation, increasing pore size and enhancing membrane hydrophilicity [4].

A considerable number of researchers has investigated the influence of additive types on morphology and membrane performance. Some researchers have studied the effects of natural additives on membrane performance. The hydrophilic structure was obtained by addition of additives such as glicerol, poly ethylene oxide, LiCl and $ZnCl_2$[5] Polyvinylpyrrolidone (PVP) is the most commonly used additive for membrane modification [6-7] . Polyethylene glycol (PEG) has the same role as PVP in membrane formation, PEG acts as a macrovoid formation barrier and increases membrane hydrophilicity [4]. In addition, PEG additive can also improve membrane characteristic in term of pore size, flux and rejection to protein [8]. The addition of additives on the ultrafiltration polymer membrane made by phase inversion has been done by some researchers including Poly Ethylene Glycol [9], Aminated polysulfone [10], and Polyvinylpyrrolidone [11]. Intensive investigation on the effect of several molecular weight of PEG has been done to improve the performance of polyethersulfone membrane [5]. So far, the modification on cellulose acetate membrane by PEG as additive has been done only using one molecular weight of 600 Da [3].

In this study, cellulose acetate (CA) membrane was prepared using poly ethylene glycol PEG additive with molecular weight of 20000 Da. The aim was to investigate the best PEG loading in order to obtain an excellent modified membrane. The effect of additive concentration is reviewed on membrane characteristics and performance. The evaluated membrane characteristics include pure water flux, membrane morphology, and functional group analysis. Membrane performance was studied against chromium metal removal in solution samples.

2. Experimental

2.1 Materials

Cellulose Acetate obtained from Wako Chemical Japan with an average molecular weight of 30,000 g/mol was used as the main polymer in the casting solution, Dimethyl Formamide (DMF) purchased from Merck used as a solvent. Polyethylene Glycol (PEG Mw_20000 Da) obtained from Merck was used as an additive. The distilled water is used as a non-solvent. Chromium nitrate, $Cr(NO_3)_3.9H_2O$ was used as a sample of artificial solution used for the solute rejection test.

2.2. Membrane Preparation

Celulosa Acetate flat sheet membrane was prepared by phase inversion technique. The CA Polymer of 17.5%, a certain amount of PEG additive (0, 2.5, 5, 7.5 and 10%) and the DMF solvent were stirred at room temperature. The solution was stirred until homogeneous. The solution was then allowed for 24 hours without stirring to remove air bubbles present in the casting solution. Afterwards, the membrane was casted on top of the glass plate using a casting knife. Furthermore, layer that are still stuck on the surface of the glass was dipped into a coagulation bath containing precipitant media in the form of pure water that serves as a non-solvent (precipitation process). This process was allowed until the membrane layer detached from the glass plate. Following that, the membranes underwent annealing process for 15 minutes at temperature of 70°C. After annealing, the membranes were ready for use. An illustration of the experimental conditions for the membrane composition is shown in Table 1.

2.3. Membrane Characterization

Pure water flux was investigated through filtration experiment using a set of dead-end ultrafiltration module as shown in Figure 1. The distillate water was flowed through a membrane that has been compacted with Trans Membrane Pressure (TMP) in the pressure range of 1, 1.25, 1.5, 1.75 and 2 bar. The membrane surface area was 0.00152 m^2. After reaching a constant state, pure water flux was calculated using Equation 1.

$$J_w = \frac{Q}{A.t} \quad (1)$$

Where J_w represents pure water flux ($Lm^{-2}h^{-1}$), Q is generated permeate (L), A is effective surface area of the membrane (m^2) and t is permeation time (h).

Table 1. Composition of CA/PEG casting solution.

Membrane	CA (%W)	PEG (%W)	DMF (%)
M-1	17.5	0	82.75
M-2	17.5	2.5	80
M-3	17.5	5.0	77.5
M-4	17.5	7.5	75
M-5	17.5	10	72.5

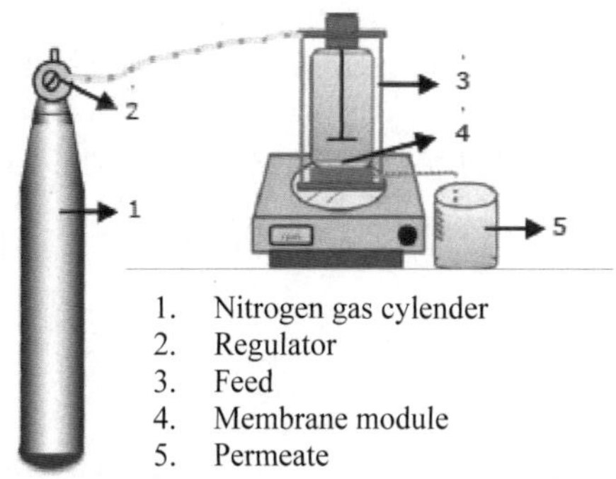

Figure 1. Ultrafiltration Equipment with Dead End Process

Membrane morphology was observed using Scanning Electron Microscopy (SEM). Observations were made on surface, and cross-sectional structure, which prepared by breaking the membrane with liquid nitrogen and then coated with gold. The compositions of M-1, M-3, M-4 and M-5 membranes were analyzed using ATR-FTIR spectrometer (UATR, Perkin Elmer) in the wavenumber range of 500-4000 cm^{-1}.

2.4. The removal of Cr(III)

Chromium-containing solution was prepared by dissolving Chromium nitric $Cr(NO_3)_3.9H_2O$ in distillate water. A total of 200 mL solution containing Cr(III) metal with a concentration of 20 ppm was passed through the membrane at a pressure of 2 bar. The filtration experiment was conducted for 2 hours. The concentration of Cr(III) ion in permeate was analyzed by means of Atomic Absorption Spectroscopy (AAS). The percentage of rejection was calculated using Equation 2

$$\text{Rejection} = \frac{C_F - C_p}{C_F} x100\% \quad (2)$$

Where C_F and C_p are dextran concentrations in the sample (feed) and permeate

3. Results and Discussion

3.1. Pure Water Flux (Jw)

The pure water flux is directly related to the number and size of the pore on the membrane surface [4] and is an important characteristic for membranes when applied to the industry [8]. Pure water flux is also used to confirm membrane stability [4]. The effect of additive concentration on pure water flux is shown in Figure 2.

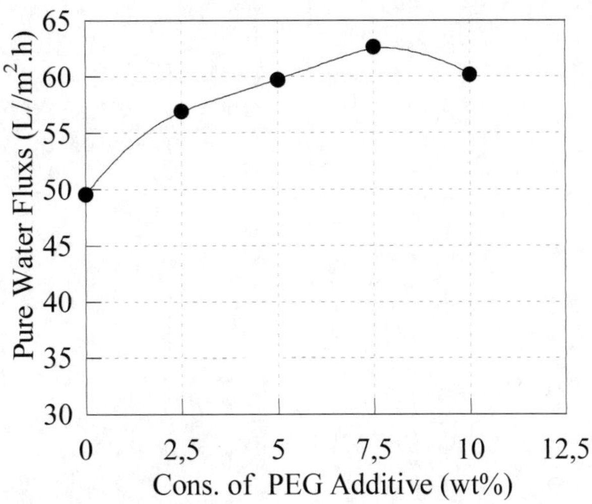

Figure 2. Effect of PEG Additive Concentration on Pure Water Flux

Figure 2 shows that the pure water flux is affected by the concentration of PEG additives added to the casting solution. M-1 membrane produces a pure water flux of 49.53 L/m^2.hour. The addition of 7.5% PEG increases the pure water flux up to 62.63 L/m^2.hour. PEG is a pore forming agent and its presence in the casting solution assists the formation of more pores on the membrane surface as it facilitates the formation of the nucleus after the casted film is immersed into the coagulation bath [3]. Increased loading of PEG declines the flux of pure water, this is because the possibility of PEG additives added to the casting solution is detached from the polymer solution and moves into the coagulation bath when casted film is immersed into a non-solvent [2]. Addition of PEG additive with concentration of 0-7.5% increase thermodynamic instability in casting film causing the occurrence of instantenious demixing. The addition of additives above the concentration makes the casting solution more viscous thus the diffusion rate of solvent and non-solvent exchange in the membrane formation process in the coagulation batch decreases resulting in the occurrence of delayed demixing event. This phenomenon caused the top-layer to become denser and suppress its pores.

3.2.Membrane Morphology

Scanning Electron Microscopy (SEM) was used to analyze the changes in modified membrane structure with variations performed on the casting solution. The membranes are code named as M-1, M-4 and M-5. The membrane with PEG concentration of 7.5% (M-4) gives a fairly good pore distribution on the membrane surface, so does cross sectional structure, the pore structure of membrane sublayer seen larger. If the PEG concentration is continuously increased and reaches 10% (M-5) the membrane surface structure looks almost identical to the membrane without the PEG additive (M-1).

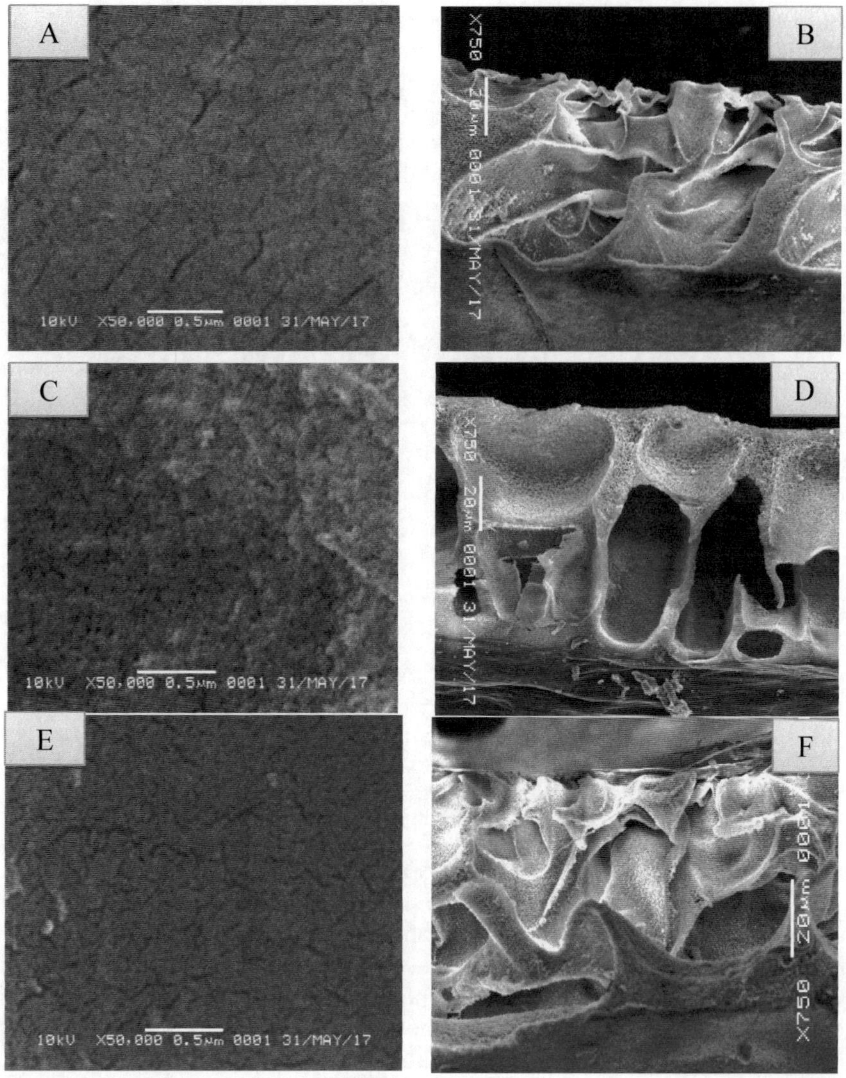

Figure 3. SEM images of CA membranes. (A)The surface of M-1, (B) The cross section of M-1 (C) The surface of M-4, (D) The cross section of M-4 (E), The surface of M-5, (F) The cross section of M-5

3.3. Functional Groups Analysis

Spectroscopy is a method used to identify the basic content of a polymer. Attenuated total reflectance infrared spectroscopy (ATR-FTIR) is the most commonly employed method for analysing functional groups of a polymer [12]. The analysis was conducted on pure CA membrane (M-1) and that with addition of 2.5% (M-2), 7.5% (M-4) and 10% (M-5) of PEG. FTIR analysis for pure and PEG-modified cellulose acetate membrane is shown in Fig. 4.The spectrum for all samples shows that the pure and modified membranes contain a hydroxyl group (O-H). The group is present at wavenumber of 3376-3377 cm^{-1}. The alkane group (C-H) appears at wave number of 1374 cm^{-1}. The C-O group also appears at 1044.90 and 1233 cm^{-1}. The C=O group is seen at the wave number of 1641 cm^{-1}. For membranes with the addition of PEG the appearance of little peak at wave number of 619.33 cm^{-1} indicates the presence of O-H torsional vibrations.

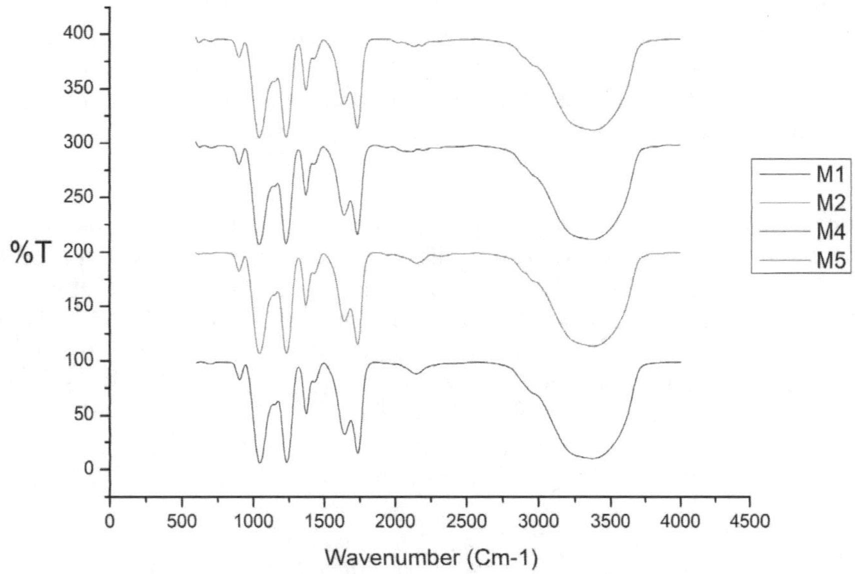

Figure 4. The spectrum of CA membranes with and without additive; M1 is unmodified CA membrane, M2, M4 and M5 are modified membranes with PEG concentration of 2,5%, 7,5%, and 10% respectively

3.4. Membrane Performance

Membrane performance was determined by measuring rejection of Cr(III) metal removal by passing a 100 ml metal sample through the prepared membrane. The membrane having the best permeability of pure water is a membrane with a 7.5% of additive concentration. The rejection of Cr(III) ion for each of fabricated membranes is shown in Figure 5.

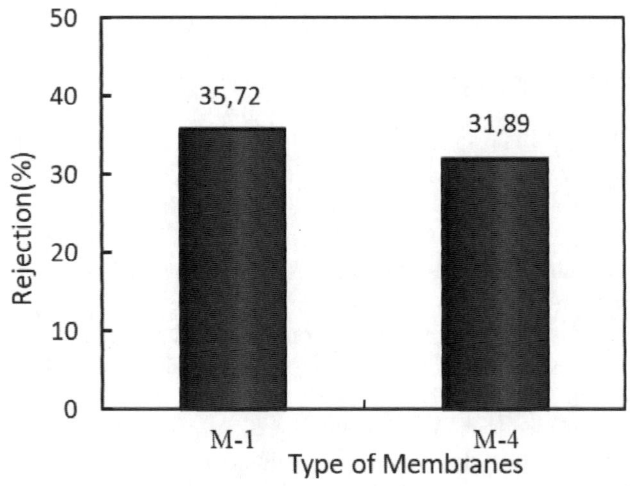

Figure 5. Cr(III) removal of pure CA membrane (M1) and that with 7,5 % PEG addition (M4)

From Figure 5 presents the removal performance of membrane without PEG additive (M1) and that with 7,5% PEG addition (M4). Only M4 is used for performance test as it showcased the best characteristic amongst all the other PEG-modified membranes. It can be observed that the Cr(III) metal removal for pure CA membrane (M-1) is 35.72% while the rejection for this metal in the modified membrane (M-4) is 31.89%. In comparison to pure CA membrane (M-1), the rejection of Cr(III) metal on modified membrane is lower. The addition of PEG additive that acts as a pore-forming agent causes the membrane to pass the solute more easily. This leads to the decline of rejection. The residue of Cr(III) in the permeate is still beyond the permissible threshold. The removal using a membrane process can reduce 35.72% of Cr(III) metal in maximum.

4. Conclusions

Cellulose acetate membranes have been prepared by preparing a casting solution containing cellulose acetate as the main polymer and Dimethyl Formamide (DMF) as solvent. PEG with molecular weight of 20000 Da of varying concentration was used as an additive to enhance the characteristic and performance of the membrane. In general, it is confirmed that the presence of PEG in the membrane system did improve the characteristics such as morphology and pure water flux. This is because PEG plays a role as a pore-forming agent which facilitates the pore distribution of the membrane. In addition, PEG enhanced the hydrophilicity of the membrane as it is a hydrophilic material, which can be observed by the appearance of O-H functional group from the FTIR analysis result. However, the removal performance of the PEG-modified membranes declined with increasing concentration used.

References

[1] Lee J S, Heo S A, Jo H J and Min B R 2006 *React. Funct. Polym.* **99** 701
[2] Sun W N, Chen C X, Li J D and Lin Y Z 2009 Ultrafiltration membrane formation of pes-c, pes and ppesk polymers with different solvents *Chin. J. Polym. Sci.* **27** 165-172
[3] Ehsan S and Toraj M 2009 Cellulose accetate (CA)/polyvinylpyrrolidone (PVP) blend asymetric membranes: Preparation, morphology and performance *Desalination* **249** 850-884
[4] Yuxin M, Fengmei S, Jun M, Miaonan W, Jun Z and Congjie G 2011 Effect of PEG additive on the morphology and performance of polysulfone ultrafiltration membranes *Desalination* **272** 51-58
[5] Idris A, Mat Zain N and Noordin M Y 2007 Synthesis, characterization and performance of asymmetric polyethersulfone (PES) ultrafiltration membranes with polyethylene glycol of different molecular weights as additives *Desalination* **207** 324-339
[6] Shockravi A, Vatanpour V, Najjar Z, Bahadori S and Javadi A 2017 A new high performance polyamide as an effective additive for modification of antifouling properties and morphology of asymmetric PES blend ultrafiltration membranes *Microporous Mesoporous Mater.* **246** 24-36
[7] Ping Y Z, Yan L W and Zhen L X and Hu Y 2011 Preparation of poly (vinyl butyral) hollow fiber ultrafiltration membrane via wet-spinning method using PVP as additive *Desalination* **278** 186-193
[8] Arthanareeswaran G and Starov V M 2011 Effect of solvents on performance of polyethersulfone ultrafiltration membranes: Investigation of metal ion separations *Desalination* **267** 57-63
[9] Rajagopolan M, Ramamoothy M and Doraiswamy R M 2004 Cellulose acetate and polyethersulfone blend ultrafiltration membranes. Part I. Preparation and characterizations *Polym. Adv. Technol.* **15** 149-157
[10] Arockiasamy D. L, Nagendran A, Shobana K H and Mohan D 2009 Preparation and Characterization of Cellulose Acetate/Aminated Polysulfone Blend Ultrafiltration Membranes and their Application Studies *Sep. Sci. Technol.* **44** 398-421
[11] Arahman N, Mulyati S, Lubis MR, Razi F, Takagi R, Matsuyama H 2016 Modification of polyethersulfone hollow fiber membrane with different polymeric additives *Membr Water Treat.* **7** 355-365
[12] Riaz T, Ahmad A, Saleemi S, Adrees M, Jamshed F, Hai A. M. And Jamil, T 2016 Synthesis and characterization of polyurethane-cellulose acetate blend membrane for chromium (VI) removal *Carbohydr. Polym.* **153** 582-591

IOP Conf. Series: Materials Science and Engineering **352** (2018) 012052 doi:10.1088/1757-899X/352/1/012052

Characterization of Polydopamine-Coated Polyethersulfone (PES) membrane for water purification

Syawaliah[1], S Mulyati[2], Muzaitun[2], R Mulyasari[2]
[1]Doctoral School of Engineering Science, Syiah Kuala University, Banda Aceh, Indonesia 23111
[2]Chemical Engineering Department, Syiah Kuala University, Banda Aceh, Indonesia, 23111

E-mail: sri.mulyati@unsyiah.ac.id

Abstract. The polyethersulfone (PES) membrane has been prepared by phase inversion method using N-methyl-2-pyrolidone (NMP) as solvent and polydopamine (PDA) as additive. The fabricated membrane was modified by coating with PDA of 0.5 g/l concentration and 180 minutes immersion time. The characteristic of the PES membranes before and after the modification was studied in this paper. The result of the pure water permeation experiment showed that the PDA-coated PES membrane showcased a higher flux than that of pure PES membrane. Scanning Electron Microscopy (SEM) analysis confirmed that the membrane had an asymmetric structure consisting of two layers. There was no significant influence on the addition of PDA to the morphology of the pore matrix because the modification was done by surface coating. Fourier Transform Infrared Spectroscopy (FTIR) analysis showed that PDA was successfully introduced on the surface of PES membrane with the appearance of peak O-H from catechol at wavenumber of 3348 cm^{-1}. Modification with PDA increased the mechanical strength of the membrane which affirmed by the results of the tensile and elongation at break evaluation.

1. Introduction

The shortage of clean water continues to increase and has become a pivotal issue. Water sources such as rivers, lakes and ground water have been contaminated directly or indirectly by industrial waste disposal [1, 2]. Many approaches have been done to remove humic acid contained in water, such as by coagulation method [3], flotation [4, 5], plasma [6], magnetic carbon [7] and ultrafiltration membranes [8].

Membrane technology is one of the most effective and widely applied methods for water purification process. Membranes have been widely used in the food industry, pharmaceutical, biotechnology, water treatment and other fields due to their excellent performance, low energy consumption, good selectivity, no phase change and operation can be done at room temperature [9]. Ultrafiltration process is generally employed for water purification process because this process has proven effectively in removing organic macromolecules in water [8, 10, 11]. Polyethersulfone (PES) is one of the most commonly applied membrane in membrane separation due to its outstanding chemical resistance, as well as thermal and mechanical stability [12]. However, fouling on the membrane is a major constraint in the application of this process [13], especially on hydrophobic polymer based like PES. Fouling has a negative effect on membrane performance by decreasing the flux value, membrane selectivity, and also increasing the operating cost [13].

In order to reduce the occurrence of fouling, modification is done to the membrane by adding another substrate be it inside or on the surface of the membrane [14, 15]. Coating is a very simple and most used technique to modify the membrane surface. Surface coating using polydopamine has been really popular nowadays [15-17]. PDA attracted the attention of many researchers because of its adhesiveness to any substrate[18]. In addition, PDA coating can reduce the hydrophobicity of membrane due to its superhydrophilic nature [19]. More hydrophilic surface of membrane improves the antifouling as well as enhancing the membrane performance in terms of flux and selectivity.

However, modifications must be designed in such a way as to obtain the right conditions. Excessive addition of additive will also negatively impact the characteristics and performance of the membrane. Such as that reported by [20], long immersion time in a dopamine solution caused the coating layer to become thicker on the membrane surface which leading to pore blockage and resulted in the decrease of flux of the membrane. Similar case was also proposed by [21] on the concentration of PDA.

In this present study, the research team has been working on observing the ideal condition for PDA coating by varying concentrations and immersion times. However, in this paper, only the characterization results of the modification experiment performed using polydopamine at a concentration of 0.5 g/l with immersion technique for 180 min will be discussed. The base membrane used is Polyethersulfone (PES) prepared using phase inversion method. Membranes were characterized by conducting pure water permeability tests, functional group tests, membrane morphological structure tests and tensile tests.

2. Experimental

2.1. Materials

Polyethersulfone (PES) was used as the main polymer. N-methyl pyrrolidone (NMP) was utilized as a solvent and aquadest as a non-solvent. Dopamine hydrochloride and tris-HCl buffer with pH 8.8 was used to make polydopamine (PDA) solution to coat the membrane surface. UV-VIS 1700 Shimadzu Spectrophotometer, Scanning Electron Microscopy (SEM), and Fourier Transform InfraRed (FTIR) Shimadzu 8400 and a dead-end ultrafiltration module were used to characterized the prepared membranes.

2.2. Preparation and Modification of PES Membrane

Polyethersulfone (PES) of 17.5 wt% was dissolved into 82.5 wt% NMP solvent. This mixture was stirred using a magnetic stirrer until homogeneous. The result of this stirring is called Dope PES. Dope was then stored into the freezer to remove air bubbles that are still present in the solution. Dope was removed from the freezer and allowed to reach room temperature. Dope was then casted by pouring it on a glass plate (casting process) and flattened onto the entire surface of the glass plate. The casted film on the glass plate was afterward immersed in a coagulation bath containing the aquadest that serves as a non-solvent (precipitation process). The formed membrane was then stored in a container containing aquadest.

Membrane modification was done by dissolving dopamine hydrocloride into 15-ml tris-HCL buffer solution (pH 8.8). Following that, the solution turned into a brownish color indicating the formation of polydopamine solution. The fabricated UF PES membrane was then immersed in 0.5 g/l polydopamine (PDA) solution for 180 minutes. The membranes were denoted as M1 for Pure PES and M2 for PES immersed in PDA for 180 minutes.

2.3 Membrane Characterization

The characterization of membranes was carried out in terms of pure water permeability, membrane morphology, functional group, and tensile test. Determination of pure water flux was done by dead-end filtration flow method. The test was performed by means of an ultrafiltration module. Aquadest was passed through the module and the pressure was given (1, 1.5, 2, 2.5 and 3 bar) to produce permeate. The permeate was collected and measured every five minutes.

The structural morphology of the membranes was analyzed using Scanning Elcetron Microscopy (SEM). The SEM test aims to determine the surface structure and pore size of the membrane. The sample was dried prior analysis, then immersed in liquid nitrogen for a few seconds until it solidifies and the membrane was then broken. The specimen was then placed on a sample holder and its structure was observed. Membrane functional groups were investigated using the Fourier Transform InfraRed Spectroscopy (FTIR) instrument. Type of equipment used was FTIR Shimadzu 8400. Membrane samples were analysed at wavelength 500 cm^{-1} - 4000 cm^{-1}. The mechanical properties evaluation was conducted in Laboratory of Physics, Syiah Kuala University, Banda Aceh, Indonesia. Specimens of membrane were stretched to the breaking point, and in doing so, the tensile strength and elongation were determined.

3. Results and Discussion

3.1. Pure Water Flux (Jw)

Flux is the volume of permeates that passes through each unit of membrane area in a given time. Figure 1 shows the value of pure water flux on pure PES membrane and PES/PDA membrane. The pure water flux of PDA-coated membrane is greater than that of pure PES membrane. In the unmodified membrane (M1), the fluxes obtained at the operating pressures of 1, 1.5, 2, 2.5 and 3 are 10.15, 23.6, 29, 41.7, and 48 L/m^2.hour. Whereas on the membrane after modification, the pure water fluxes obtained at the same operating pressure are slightly higher which are 15.5, 24.9, 35, 45.5, and 55.6 L/m^2.hour, respectively. An increase in the flux of pure water is due to increased hydrophilicity of the membrane with the addition of PDA as additive. Hydrophilic membranes tend to pass through the membrane pores as opposed to hydrophobic membranes [22]. Thus, water diffuses faster which confirms the enhance of pure water flux.

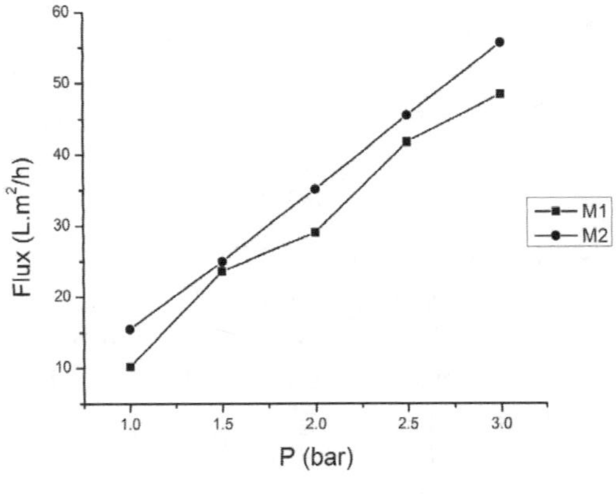

Figure 1. Pure water flux of M1 and M2 membranes at each operating pressure.

3.2. Membrane Morphology

The results of surface and cross-sectional structure of the membrane observed using Scanning Electron Microscopy (SEM) with 10 kV voltage can be seen in Figure 2. PES/PDA membrane surface structure is seen denser and has smaller pore size than that of pure PES membrane. This result is in accordance with the research done by [23] which reported that the addition of PDA can decrease the pore size of

IOP Conf. Series: Materials Science and Engineering **352** (2018) 012052 doi:10.1088/1757-899X/352/1/012052

the membrane. In addition, PDA coating can cause the membrane surface to become rough. This can occur due to the formation of PDA nanoagregates on membrane surfaces [24].

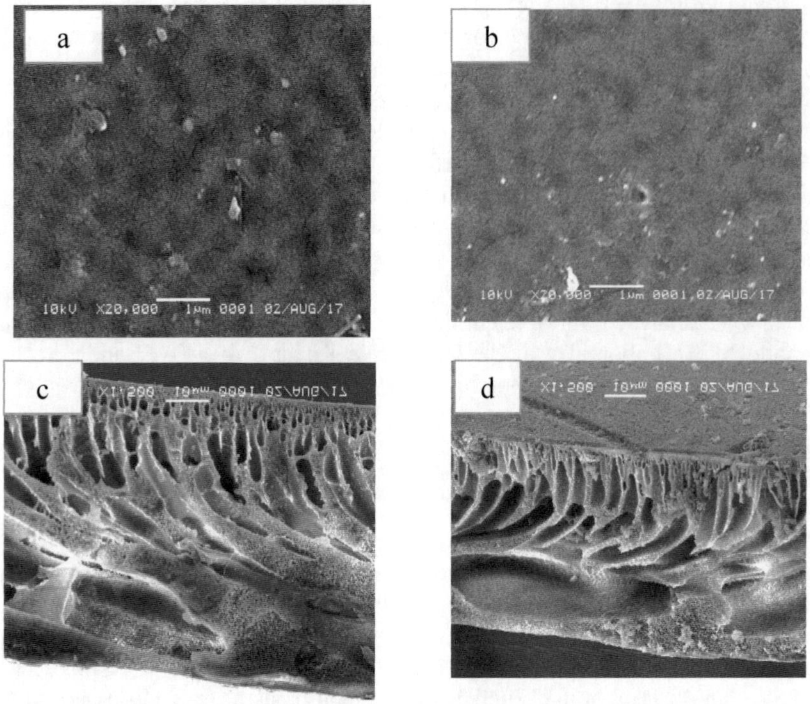

Figure 2. SEM imaging of (a) surface structure of uncoated PES membrane, (b) surface structure of PDA-coated PES membrane, (c) Cross sectional structure of uncoated PES membrane, and (d) Cross-sectional structure of PDA-coated PES membrane

The SEM observation results of cross sectional structures on both membranes evidencing that the prepared membrane has an asymmetrical structure with two layers, where the top layer is thinner with tight pores, while the lower layer has a larger pore size which serves as a support and provides resistance to membrane. The upper layer of the PES/PDA membrane has a greater thickness than the pure PES membrane. There is no noticeable change in the matrix portion of the membrane after modification. This is because the modification was done by coating on the surface. Therefore, only membrane surface is affected.

3.3. Functional Groups Analysis

The investigation of functional groups presence on the membrane was analyzed by FTIR device, and the result is shown in Figure 2. Regarding to Fig. 2, in M2 membrane, there is a new peak appearence at spectrum range of 3300-3700 cm^{-1}. This peak is confirmed to be O-H and N-H bands of catechol group from PDA [25]. This can be taken as indication that PDA has successfully been deposited on the membrane surface. However, another specific indolequinone peak should be emerging on the 1700 cm^{-1} wavenumber [26]. However, it is not detected in the analysis, could be due to the low of concentration and length of immersion time.

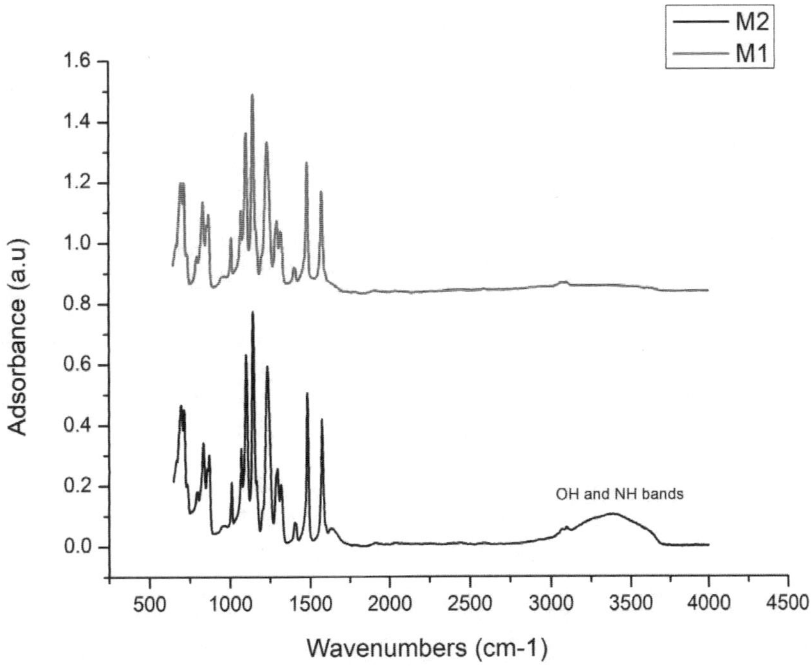

Figure 3. Chemical composition of PES membrane with (M2) and without (M1) PDA-coating as observed by FTIR

3.4. Mechanical Properties

Table 1 shows the results of mechanical strength measurement of membranes before and after modification. Mechanical strength tested is tensile strength and elongation at break. From the results, it can be observed that both tensile and elongation at break improved in the membrane after coating with polydopamine. Pure PES membrane had tensile value of 11.80 kgf/mm^2 and elongation at break of 36.62%. While on the membrane coated with polydopamine with concentration of 0.5 g/l for 180 minutes, tensile strength and elongation at break reached up to 19,30 kgf/mm^2 and 73,50%, respectively. This is because the PDA particles reduce the fragility of the membrane through the occurrence of energy absorption during tensile deformation [27]. High tensile strength and elongation means the membrane is not brittle or more ductile and not easily broken or damaged despite given a heavier workload. These characteristics have an effect on the length of membrane life-time

Table 1. Mechanical properties of pure and modified PES membranes

Membrane	Tensile Strength (kgf/mm^2)	Elongation at break (%)
M1	11.80	36.62
M2	19.30	73.50

4. Conclusions

Polyethersulfone (PES) ultrafiltration membrane has been prepared and modified using 0,5 g/l polydopamine by dip-coating for 180 minutes. The aim was to improve the antifouling of the PES membrane by enhancing the hydrophilicity. The effect of modification on the PES membrane was observed through a pure water flux test, composition analysis using FTIR, morphological test by means of SEM and mechanical strength. Based on the result of characterization it can be concluded that polydopamine has been successfully coated on membrane surface which confirmed by the appearance of N-H and O-H peak from catechol at wavenumber of 3348 cm^{-1}. PDA coating result in a narrowing of the pore on the membrane surface. The presence of PDA has an effect on the increase of pure water permeation due to the hydrophilic nature of the PDA. In addition, the PDA also improved the mechanical properties of the membranes in terms of tensile and elongation at break.

References

[1] Li J, Liu Z, He C, Yue H, and Gou S 2017 Water shortages raised a legitimate concern over the sustainable development of the drylands of northern China: Evidence from the water stress index *Sci Total Environ* **590** 739-750

[2] Lim DI, Jung SW, Choi MS, Kang, Sm, Jung SH, and Choi JY 2013 Historical record of metal accumulation and lead source in the southeastern coastal region of Korea *Marine Poll Bull* **74** 441-445

[3] Xu Y, Chen T, Cui F, and Shi W 2016 Effect of reused alum-humic-flocs on coagulation performance and floc characteristics formed by aluminum salt coagulants in humic-acid water *Chem Eng J* **287** 225-232

[4] Zouboulis A I, Jun W, and Katsoyiannis I A 2003 Removal of humic acids by flotation *Colloids Surf,. A* **231** 181-193

[5] Brum M C and Oliveira J F 2007 Removal of humic acid from water by precipitate flotation using cationic surfactants *Miner. Eng.* 2007 **20** 945-949

[6] Li Y, Qu G, Zhang L, Wang T, Sun Q, Liang D and Hu S 2017 Humic acid removal from micro-polluted source water using gas phase surface discharge plasma at different grounding modes *Sep. Purif. Technol.* **180** 36-43

[7] Li S, He M, Li D and Pan Z 2017 Removal of humic acid from aqueous solution by magnetic multi-walled carbon nanotubes decorated with calcium *J. Mol. Liq.* **230** 520-528

[8] Chu KH, Huang Y, Yu M, Heo J, Flora JRV, Jang A, Jang M, Jung C, Park CM, KIM DH and Yoon Y 2017 Evaluation of graphene oxide-coated ultrafiltration membranes for humic acid removal at different pH and conductivity conditions *Sep.Purif. Technol.* **181** 139-147

[9] Liu J, Tian C, Xiong J and Wang L 2017 Polypyrrole blending modification for PVDF conductive membrane preparing and fouling mitigation *J. Colloid Interface Sci.* **494** 124-129

[10] Huang Y, Du JR, Zhang Y, Lawless D and Feng X 2015 Removal of mercury (II) from wastewater by polyvinylamine-enhanced ultrafiltration *Sep. Purif. Technol.* **154** 1-10

[11] Acero JL, Benitez FJ, Real FJ and Teva F 2017 Removal of emerging contaminants from secondary effluents by micellar-enhanced ultrafiltration *Sep. Purif. Technol.* **181** 123-131

[12] Zhao C, Xue J, Ran F, Sun S 2013 Modification of polyethersulfone membranes – A review of methods *Prog. Mater. Sci.* **58** 76-150

[13] Guo W, Ngo HH and Li J, 2012 A mini-review on membrane fouling *Bioresour. Technol.* **122** 27-34

[14] Zhu K, Zhang S, Luan J, Mu Y, Du Y and Wang G 2017 Fabrication of ultrafiltration membranes with enhanced antifouling capability and stable mechanical properties via the strategies of blending and crosslinking *J. Membr. Sci.* **539** 116-127

[15] Arahman N, Mulyati S, Lubis MR, Razi F, Takagi R, Matsuyama H 2016 Modification of polyethersulfone hollow fiber membrane with different polymeric additives Membr Water Treat. 7 355-365

[16] Wang C, Li Z, Chen J, Yin Y and Wu H 2018 Structurally stable graphene oxide-based nanofiltration membranes with bioadhesive polydopamine coating *Appl. Surf. Sci.* **427** 1092-1098

[17] Vaselbehagh M, Karkhanechi H, Takagi R and Matsuyama H 2016 Effect of polydopamine coating and direct electric current application on anti-biofouling properties of anion exchange membranes in electrodialysis *J. Membr. Sci.* **515** 98-108

[18] Liebscher J, Mrówczyński R, Scheidt HA, Filip C, Hădade ND, Turcu R, Bende A and Beck S 2013 Structure of Polydopamine: A Never-Ending Story? *Langmuir* **29** 10539-10548

[19] Liu Y, Ai K, and Lu L, 2014 Polydopamine and Its Derivative Materials: Synthesis and Promising Applications in Energy, Environmental, and Biomedical Fields *Chem Rev.* **114** 5057-5115
[20] Zhang W, Ying Y, Ma J, Guo X, Huang H, Liu D and Zhong C 2017 Mixed matrix membranes incorporated with polydopamine-coated metal-organic framework for dehydration of ethylene glycol by pervaporation *J. Membr. Sci.* **527** 8-17
[21] Vaselbehagh M, Karkhanechi H, Mulyati S, Takagi R and Matsuyama H 2014 Improved antifouling of anion-exchange membrane by polydopamine coating in electrodialysis process *Desalination* **332** 126-133
[22] Arena J T, Manickam S S, Reimund K K, Freeman B D and McCutcheon J R 2014 Solute and water transport in forward osmosis using polydopamine modified thin film composite membranes *Desalination* **343** 8-16
[23] Fan J, Luo J, Chen X and Wan Y 2016 Polydopamine meets porous membrane: A versatile platform for facile preparation of membrane adsorbers *J. Chromatogr. A* **1448** 121-126
[24] Zhang R, Su Y, Zhao X, Li Y, Zhao J and jiang Z 2014 A novel positively charged composite nanofiltration membrane prepared by bio-inspired adhesion of polydopamine and surface grafting of poly(ethylene imine) *J. Membr. Sci.* **470** 9-17
[25] Guo H, Deng Y, Tao Z, Yao Z, Wang J, Lin C, Zhang Tong, Zhu B and Tang CY 2016 Does hydrophilic polydopamine coating enhance membrane rejection of hydrophobic endocrine-disrupting compounds? *Environ. Sci. Technol. Lett.* **3** 332-338
[26] Zangmeister R A, Morris T A, and Tarlov M J 2013 Characterization of Polydopamine Thin Films Deposited at Short Times by Autoxidation of DopamineLangmuir **29** 8619-8628
[27] Jiang J, Zhu L, Zhu L, Zhang H, Zhu B and Xu Y 2013 Antifouling and antimicrobial polymer membranes based on bioinspired polydopamine and strong hydrogen-bonded poly(N-vinyl pyrrolidone) *ACS Appl Mater Interfaces* **5** 12895-904
[28] Arahman N, Mulyati S, Lubis MR, Razi F, Takagi R, Matsuyama H 2016 Modification of polyethersulfone hollow fiber membrane with different polymeric additives *Membr Water Treat.* **7** 355-365

IOP Conf. Series: Materials Science and Engineering **352** (2018) 012053 doi:10.1088/1757-899X/352/1/012053

Economic base determination and influence of several variables against contributions percentage of the GDRP in Aceh Besar district

Keumala Andayani, Miftahuddin
Department of Statistics, Faculty of Mathematics and Sciences, Syiah Kuala University, Banda Aceh 23111, Indonesia

E-mail: keumala.andayani@students.stat.unsyiah.ac.id, miftah@unsyiah.ac.id

Abstract. The percentage contribution of Gross Regional Domestic Product (GRDP) in Aceh Besar district is influenced by several leading sectors, such as agriculture, building sector, trade, hotel and restaurant sector, transport and communications, financial sector, leasing and business services, and services sector. Based on the use of Location Quotient (LQ) method and multiple regression model, the effect of labor variables and population to Gross Regional Domestic Product by 2000 constant prices for agriculture and trade. For each addition of one workforce in the trading sector, the trade sector contribution will increase by 0.000014157%. Thus, the trade sector contribution will increase by 0.0000013786% in every addition of one soul of the population. Whereas, for every addition of one human resource in the agricultural sector will be reduced by 0.0002%. In other words, for each addition of one soul of the population will reduce the contribution of the agricultural sector by 0.00008611%.

1. Introduction

The success of economic development of a region can be seen from the economic growth achieved by the region. Economic growth is calculated from the value of the gross domestic product (GDP) or gross regional domestic product (GRDP). The increase in income per capita reflects an excellent economic development of a region and demonstrates the welfare of the people. However, it does not mean that income per capita should increase continuously, because the per capita income may decrease in any time, real growth, and society economy indicator [1]. The income per capita also related to technological improvement. For example, the agricultural sector, formerly land processing still using animal power but now changed into a tractor machine. Generally, GRDP has dominant contribution of sector for each region, for example in India and Malaysia [2,3]. The econometric model is popular to develop GRDP model.

During 2004 - 2013, GRDP of Aceh in Indonesia continues to increase. In 2007, the growth of GRDP was in a large number; 13.87% or from 1.8 trillion to 2.1 trillion. Economic growth is inseparable from the added value generated by economic units grouped by the field of business [4]. The magnitude of each sector's role describes the economic structure of the region. The economic development of Aceh Besar Regency is quite good. It is indicated by GRDP, and the high growth rate of it is higher compared to other districts in the province of Aceh.

Research on the development of the mainstay area in Aceh Besar District using data for five years shows a result that Aceh Besar district should give top priority to the construction sector, trade, hotel and restaurant sector, and services as it shows significant value in supporting the increase of GRDP

[5]. Agricultural sector which is the basic sector began to decrease its role in the economy of Aceh Besar District needs to be considered and given more support. Also, research [6] in the case study Kedungsepur states that the variable of labor negatively affects the variable of GRDP. When there is an increase of labor by 1%, it will lower the GRDP by 1.29%.

Determination of the mainstay area by base sector is a government policy that is effective enough to develop the potential of economic resources area. However, sometimes in the implementation, the concept of the development of the mainstay area is not implemented efficiently so its success can't be measured. Therefore, this study was conducted in addition to looking at the base sector in Aceh Besar district, and to see the effect of labor and other variables on the contribution of GRDP. So the government policy can be more effective and efficient in taking the step to develop economic development.

This research is intended to determine the leading sector of Aceh Besar Regency from time series data of GRDP in 2006 to 2013. And also to see the effect of several variables on the contribution of the highest percentage GRDP in Aceh Besar district. The result of this research is expected to utilized by the government and stakeholders to analyze the policy on GRDP. The data used in this study is secondary data derived from the Badan Pusat Statistik (BPS-Statistics Indonesia) years 2006 - 2013. This study uses time series data on GRDP of Aceh Besar District and Aceh Province, sector labor of GRDP, and the population of Aceh Besar district.

2. Literature Review

2.1. Methods Location Quotient (LQ)

To identify leading sectors in Aceh Besar district Location Quotient (LQ) analysis will be used. LQ analysis measures the concentration of economic activity within a region by comparing its role in the regional economy with the role of similar activities or sectors in the national economy. LQ calculation formula is often called Static Location Quotient (SLQ) because it only sees a single point in time. The LQ formula is:

$$LQ = \frac{e_{si}/e_i}{e_s/e} \tag{1}$$

where:

e_{si} = value of production sector *i*th at the district / city
e_i= total gross regional domestic product district / city
e_s= value of production sector *i*th in the province
e = total GRDP province of

The criteria measuring LQ [7] namely:

- LQ > 1: The rate of growth subsector *i*th in region *k* has a more rapid pace as compared to the same sector at the regional level reference *p*. Thus, the sector is the basis and potential for development.
- LQ <1: The growth rate subsector *i*th in the *k* has a slightly slower pace compared to the same sector at the regional level reference *p*. Thus, the sector is not a basis.
- LQ = 1: Indicates if the rate of growth of the sector *i*th in the region *k* is equal to the growth rate of the same sector at the regional level reference *p*.

2.2. Multiple Regression Model

Regression analysis is used to model the relationship between dependent variable and explanatory variable [8]. This analysis aims to estimate the value of one variable concerning other variables known through the regression line equation. A regression model consisting of two independent variables is called multiple regression. Multiple linear regression is a linear combination analysis that explains the relationship between response variables and the explanatory factors affecting more than one predictor.

The purpose of multiple linear regression analysis is to load prediction Y over X. The form of multiple linear equations is as follows [9]:

$$Y = \beta_0 + \beta_1 X_{1i} + \beta_2 X_{21} + \cdots + \beta_k X_{ki} \tag{2}$$

where:
Y is the value of the response variable in the *i*-th observation
X_{ki} is the value of the independent variable in the *i*-th observation
β_0 is an intercept parameter
β_1, β_k are parameters regression of coefficient predictors

The significance test is a procedure used to test the right or error of the results of the null hypothesis of the sample [10]. F test is conducted to find out whether the independent variables as a whole has a significant value in influencing the dependent variable. If the value of F arithmetic on the simultaneous test is greater than the value of F table or value significantly smaller than α, then the independent variables as a whole affect the dependent variable. The multi-collinearity test is performed to find out whether there is a correlation between independent variables found in the regression model. If there is a correlation, there is a multi-collinearity problem to be overcome. Multi-collinearity can be seen from the value of VIF, where if the value of VIF > 10 then there is multi-collinearity [10,11]. The heteroscedasticity test is performed to see the variance of the residue between observations one with the other observations. The assumption of heteroscedasticity can be detected by using the Glejser Test that regressing Independent variable with its residual absolute value. Autocorrelation test is done to see whether there is a correlation between independent variables by looking at Durbin Watson value.

3. Results and discussion

3.1. Location Quotient

This study uses data GRDP at constant prices in 2000, as it can be to analyze the seed sector and examines the economic conditions in Aceh Besar district.

Table 1. Analysis *Location Quotient*

No.	Business Sector	2006	2007	2008	2009	2010	2011	2012	2013	Average
1	Agriculture	1.53	1.28	1.15	1.03	0.97	0.95	0.94	0.94	1.10
2	Mining & Quarrying	0.12	0.14	0.18	0.31	0.33	0.33	0.34	0.35	0.26
3	Processing Industry	0.24	0.24	0.25	0.25	0.27	0.28	0.29	0.32	0.27
4	Electricity, Gas & Water Supply	1.26	0.90	0.74	0.83	0.76	0.74	0.73	0.73	0.84
5	Building	2.04	2.46	2.42	2.28	2.27	2.27	2.27	2.18	2.27
6	Trade, Hotels & Restaurants	1.29	1.19	1.17	1.07	1.07	1.09	1.09	1.08	1.13
7	Transportation & Communication	1.61	1.27	1.13	1.00	0.95	0.91	0.86	0.91	1.08
8	Finance, Real Estate, & Business Services	1.60	1.40	1.28	1.12	1.11	1.09	1.06	1.04	1.21
9	Offices	1.55	1.43	1.32	1.21	1.20	1.19	1.17	1.12	1.27

From the calculation of LQ in table 1. it can be seen that Aceh Besar Regency has six sectors of comparative advantage (LQ > 1) which are agriculture sector, building sector, trade sector, hotel and restaurant, transportation and communications sector, financial sector, and the services sector. While the other three sectors are non-base sector (LQ < 1) mining and quarrying sector, processing industry sector, and electricity, gas and drinking water sector. This indicates that in those six sectors, Aceh

Besar Regency has been able to fulfill their own needs and it is possible to export out of the area. For example, agricultural and plantation products in Jantho (Aceh Besar district) can be sold to Banda Aceh and other districts.

3.2. Descriptive of the Sectors

Here is presented a descriptive analysis graph of time series data GRDP in Aceh Besar district 2004 to 2013 as seen in Figure 1:

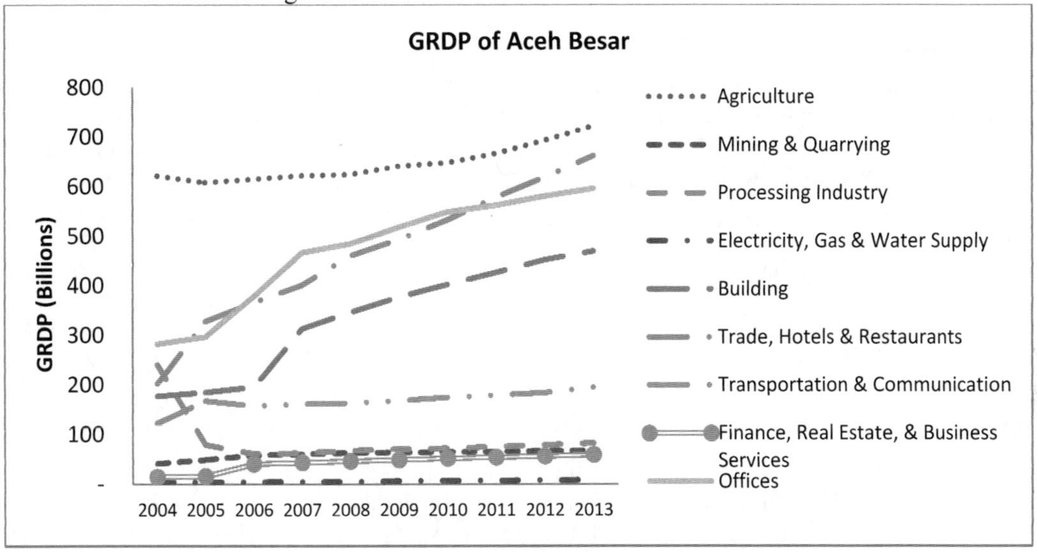

Figure 1. GRDP of Aceh Besar

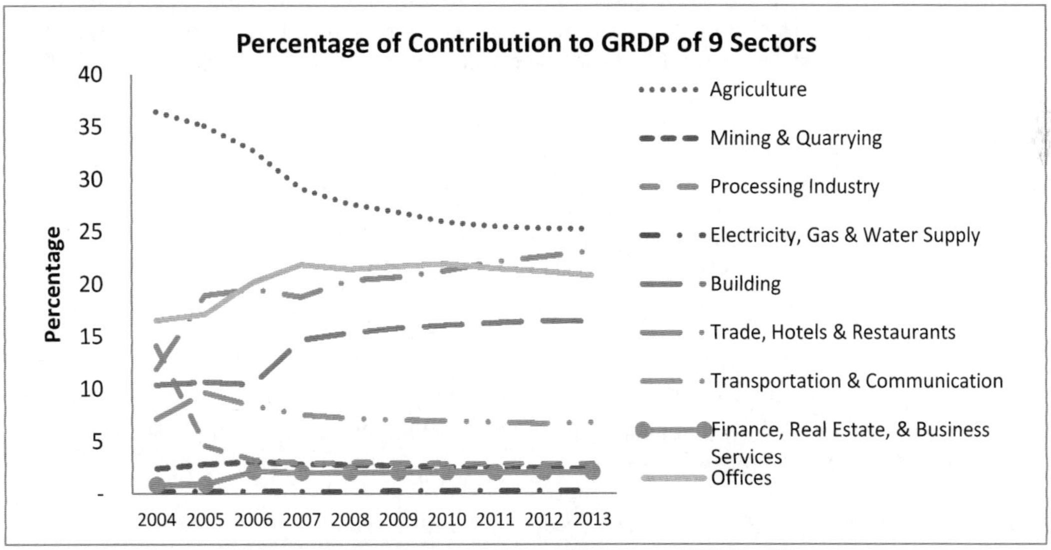

Figure 2. Percentages of contribution to GRDP of 9 sectors

There are nine sectors in GRDP: agriculture sector, mining and quarrying sector, manufacturing industry sector, electricity, gas and water sector, building sector, trade, hotel and restaurant sector, transportation and communications sector, financial sector, and company services, and the services sector. From 9 sectors, there are eight sectors which increased from 2004 to 2013, while the manufacturing industry sector decreased drastically in 2005 from Rp. 240.5 billion to Rp. 79.2 billion. Furthermore, increase from 2006 to 2013 to Rp. 82.3 billion. This decline related to the disaster that hit the coastal area of Aceh in 2004 that tsunami. The PT SAI factory (renamed Lafarge Cement

Indonesia) became one of the main contributors of the industrial sector in the cement subsector, destroyed by the tsunami waves, resulting in a decline in the manufacturing sector.

The agricultural sector and trade sector are the two largest sectors that contribute to GRDP. In 2013 the agricultural sector contributed Rp. 721.6 billion or about 25.23% of the formation of GRDP Aceh Besar District. The contribution decreased if compared to 2004 which is about 36.44%. While the trade sector in 2013 contributed Rp. 661.4 billion or about 23.13%. The contribution of the trade sector increased rapidly when compared to the year 2004, i.e., 11.88%.

3.3. Multiple Regression Model

The results of the descriptive analysis show that the percentage contribution of agricultural sector decreased every year and the trade sector increased. We found the case interesting to do further analysis such as regression analysis. The goal is to see the effect of some variables on contribution percentage of both sectors. Results of the assumptions that must be met in the regression analysis are:

Table 2. Test the assumptions on regression

No	Sector	Normality Test	Multi-collinearity Test	Heteroscedasticity Test	Autocorrelation Test	
		Kolmogorov Smirnov	VIF	Test Glejser	Durbin Watson	RunTest
1	Agriculture	0453	1,312	0.2629	2,397	1
2	Trade, hotels, and restaurants	0.64	3365	0065	2677	0703

For normality test, agriculture sector and trade, hotel, and restaurant sector should have a significant value > α (0.05) to fulfill assumption and can be concluded as a normal distribution. In the multi-collinearity test, the two sectors have VIF value < 10, so there is no multi-collinearity. In the heteroscedasticity test, the two sectors have significance value > α (0.05), so it can be concluded that there is no heteroscedasticity. The autocorrelation test of both sectors has Watson Durbin value in the area so these can't be concluded. Therefore, it is necessary to do a further test with a run test to see the residual randomness. Because the value of run test > α (0.05) then the data used is quite random so there is no problem autocorrelation.

The data of agricultural sector and trade sector, hotel, and restaurant have fulfilled the assumption so the regression analysis can be done. Parameter test conducted in this research is by using simultaneous test (F test). Here is ANOVA table for simultaneous test:

Table 3 Table ANOVA model agricultural sector

Model	Sum of Square	df	Mean Square	F	P-value
Regression	41.637	2	20.818	18.476	0005
Residual	5,634	5	1,127		
Total	47 271	7			

Labor variable and population have significant value (0.005) < α (0.05). This means that labor in the agricultural sector and the number of people (population) affect the contribution of the agricultural sector. The formulation obtained by agriculture sector:

$Y = 64.1123 - 0.0002 \text{ (labor)} - 0.00008611 \text{ (population)}$

This means that every addition of 1 labor force in the agricultural sector then the contribution of the agricultural sector will be reduced by 0.0002%. For each addition of 1 soul of the population then the contribution of the agricultural sector will be reduced by 0.00008611%.

Table 4. Table ANOVA on the model of trade, hotels, and restaurants

Models	Sum of Square	*df*	Mean Square	F	P-value
Regression	15 274	2	7637	46 889	0001
Residual	0814	5	0163		
Total	16 089	7			

IOP Conf. Series: Materials Science and Engineering **352** (2018) 012053 doi:10.1088/1757-899X/352/1/012053

The variable of labor and population have significant value (0.001) < α (0.05) This means that the labor force in the trade, hotels, and restaurants sector and the number of people influences the contribution of trade, hotel, and restaurant sector. The formulation of the contribution of trade, hotel, and restaurant sector are:

$$Y = 13.7571 + 0.000014157\ (\text{labor}) + 0.0000013786\ (\text{population}).$$

It means that every addition of 1 labor force in the trade, hotel and restaurant sector then the contribution of trade, hotel, and restaurant sector will increase by 0.000014157%. For each addition of a person of the population then the contribution of trade, hotel, and restaurant sector will increase by 0.0000013786%. If we compared to agriculture sector and the trade, hotel and restaurant sector then both has different direction. However, both sectors have potency to build and explored to get large contribution to increase GRDP in Aceh Besar in the future.

4. Conclusion

Leading sector in Aceh Besar is agriculture, building sector, trade sector, hotel and restaurant sector, transportation and communications sector, financial sector, leasing and business services, and services sector. So, the local government should give more attention and contribute to mentioned the six sectors.

Labor variable in the agriculture sector gives negative effect to the contribution of the agriculture sector. It means that every additional labor and population then the contribution of the agricultural sector will decline. Therefore, the government should contribute more to the agricultural sector, given the agricultural sector is the largest field for society as farmer and gives largest contribute to GRDP for Aceh Besar district.

Whereas, the labor variable in the trade, hotels, and restaurants sector and the number of residents positively affect the trade, hotels, and restaurants sector. Each workforce and price then the trade, hotel and restaurant sector will go up. It shows that this sector can become a leading sector in Aceh Besar district.

References

[1] Martin Feldstein. 2017. Underestimating the Real Growth of GDP, Personal Income, and Productivity. Journal of Economic Perspectives, Vol. 31, No. 2, Spring 2017, p 145–164.

[2] Akbar Khodabakhshi. 2011. Relationship between GDP and Human Development Indices in India. International Journal of Trade, Economics and Finance, Vol. 2, No. 3.

[3] Raja Nurul Aini Raja Aziz and Amalina Azmi. 2017. Factor Affecting Gross Domestic Product (GDP) Growth in Malaysia. International Journal of Real Estate Studies, Vol. 11 No. 4.

[4] Badan Pusat Statistika. 2006-2013. *Aceh Dalam Angka* Jantho, Aceh Besar (in Indonesian)

[5] Mursidah, Abubakar Hamzah, dan Sofyan. 2013 Analisis Pengembangan Kawasan Andalan di Kabupaten Aceh Besar *Jurnal Ilmu Ekonomi Universitas Syiah Kuala* 1 43-45 (in Indonesian)

[6] Utami Ratri Heningtyas. 2013 Pengaruh Tenaga Kerja, Upah Minimum Regional (UMR), Pendapatan Asli Daerah (PAD) Terhadap PDRB Perkapita Kabupaten/Kota di Kawasan Kedungsepur (Kendal Demak Ungaran Semarang Grobogan Salatiga) Universitas Negeri Semarang (in Indonesian)

[7] Kuncoro M. 2002 Evaluasi Penetapan Kawasan Andalan: Studi Empiris di Kalimantan Selatan 1993-1999 *Jurnal Ekonomi dan Bisnis Indonesia* 17 (in Indonesian)

[8] R.Carter Hill, William E. Griffiths, and Guay C. Lim. 2011. Principle of Econometrics. Fourth Edition. John Wiley and Sons, Inc.

[9] Badi H. Baltagi. 2005. Econometric Analysis of Panel Data. Third Edition. John Wiley and Sons, Ltd.

[10] William H. Greene. 2012. Econometric Analysis. Seventh Edition. Prentice Hall.

[11] Ronald E. Walpole, Raymond H. Myers and Sharon L. Myers. 2013. Probability & Statistics for Engineers & Scientists 9th Edition. Pearson Higher Education. Inc.

IOP Conf. Series: Materials Science and Engineering **352** (2018) 012054 doi:10.1088/1757-899X/352/1/012054

Description of the supporting factors of final project in Mathematics and Natural Sciences Faculty of Syiah Kuala University with multiple correspondence analysis

Asep Rusyana, Nurhasanah and Maulizasari
Study Program of Statistics, Syiah Kuala University, Banda Aceh, 23111, Indonesia

Email: asep.rusyana@unsyiah.ac.id

Abstract. Syiah Kuala University (Unsyiah) is hoped to have graduates who are qualified for working or creating a field of work. A final project course implementation process must be effective. This research uses data from the evaluation conducted by Mathematics and Natural Sciences Faculty (FMIPA) of Unsyiah. Some of the factors that support the completion of the final project are duration, guidance, the final project seminars, facility, public impact, and quality. This research aims to know the factors that have a relationship with the completion of the final project and identify similarities among variables. The factors that support the completion of the final project at every study program in FMIPA are (1) duration, (2) guidance and (3) facilities. These factors are examined for the correlations by chi-square test. After that, the variables are analyzed with multiple correspondence analysis. Based on the plot of correspondence, the activities of the guidance and facilities in Informatics Study Program are included in the fair category, while the guidance and facilities in the Chemistry are included in the best category. Besides that, students in Physics can finish the final project with the fastest completion duration, while students in Pharmacy finish for the longest time.

1. Introduction

The process of education in universities begins with the introduction of basic knowledge in the field of interest and ends with research in the form of a final project. The final project is a combination of cognitive, psychomotor and affective competencies in students by the field of science that is learned. The final project is a compulsory course that must be taken by the student as the main requirement to obtain a Bachelor of Science degree. National Accreditation Board for Higher Education (2008) stated that the best time to the completion of the task is less or equal to six months. Since 2013, the number of undergraduate graduates in FMIPA of Syiah Kuala University has continued to increase. FMIPA has a Faculty Quality Assurance Unit (SJMF) team that acts as a quality assurance institution. In 2016 the SJMF MIPA team conducted an academic evaluation of the completion of the students' final assignment. The goal was to know student constraints in doing the final project. The scope of research conducted by SJMF includes characteristics of students, duration of final project completion, guidance process, facilities, final project seminar, and general and quality effect. This research provides qualitative data information consisting of several variables with various categories. Each variable will be analyzed firstly through the chi-square test, then the related variables will be analyzed further using correspondence analysis.

IOP Conf. Series: Materials Science and Engineering **352** (2018) 012054 doi:10.1088/1757-899X/352/1/012054

2. Literature review

2.1. Chi-square test
In the correspondence analysis, the strength of relationships among categories is determined based on the value of chi-square. The chi-square value indicates the relationship between variables [2].

The hypothesis as follows:
H_o: There is no relationship between rows and columns
H_1: There is relationship between rows and columns
The formula for calculating the chi-square test:

$$X^2 = \Sigma_{ij} \frac{(x_{ij}-E_{ij})^2}{E_{ij}} \quad (1)$$

Where :
x_{ij} = The observed frequencies on the i-th row and the jth columns
E_{ij} = The expected frequencies on the i-th row and the jth columns
Where :

$$E_{ij} = \frac{x_{i.}x_{.j}}{x_{..}} \quad (2)$$

[1].

With the following criteria:
H_o is rejected, if X^2 value > X^2 table with [α, df = (a-1) (b-1)], where α is an error of type I, a and b are the number of rows and columns, respectively.
H_o can not be rejected, if X^2 value ≤ X^2 table with [α, df = (a-1) (b-1)].
Or
P-value < α then H_o is rejected, and *P-value* > α then H_o cannot be rejected

2.2. Multiple Correspondence Analysis
The calculation of multiple correspondence analysis is a looping application of a simple correspondence analysis algorithm using the Burt matrix [5]. The Burt matrix is a combination of the following matrices:

$$\mathbf{Z'Z} = n\begin{bmatrix} \mathbf{P}_r & \mathbf{P} \\ \mathbf{P} & \mathbf{P}_c \end{bmatrix} \quad (3)$$

where **P** is the correspondence matrix.
The matrix $\mathbf{P_r}$ and matrix $\mathbf{P_c}$ are diagonal matrices with diagonal elements marginal columns or marginal row matrix P, respectively.

2.2.1. Correspondence matrix
The relative frequency of matrix is:

$$\mathbf{P}_{axb} = (p_{ij}) = (\frac{n_{ij}}{n}) \quad (4)$$

With $n = \sum_i \sum_j n_{ij}$, $\mathbf{P}_{axb}$ is called the correspondence matrix, n_{ij} is the number of the object on row i and column j, and p_{ij} is the proportion of object on row i and column j.
The diagonal matrix of row number element elements is the matrix $\mathbf{D_r}$ of size (a x a), as follows:

$$\mathbf{D}_r = diag(r) = \begin{bmatrix} p_{1.} & 0 & \cdots & 0 \\ 0 & p_{2.} & \cdots & 0 \\ \vdots & \vdots & \ddots & \vdots \\ 0 & 0 & \cdots & p_{a.} \end{bmatrix} \quad (5)$$

The diagonal matrix of column number element elements is the matrix $\mathbf{D}_c$ of size (b x b), as follows:

$$\mathbf{D}_c = diag(c) = \begin{bmatrix} p_{.1} & 0 & \cdots & 0 \\ 0 & p_{.2} & \cdots & 0 \\ \vdots & \vdots & \ddots & \vdots \\ 0 & 0 & \cdots & p_{.b} \end{bmatrix} \quad (6)$$

[4].

2.2.2. Decomposition of singular value
Singular decomposition is done by reducing the dimensions of data based on the greatest diversity of data (eigenvalues /inertia) while maintaining optimum information. The matrix of decomposition value yields matrix **A** with size *a x m* and matrix **B** with size *b x m* with m is the best low rank for **Λ**. The coordinate matrix $(\mathbf{P} - \mathbf{rc}')$ is **AΛB'** where **r = P 1** and **c = P` 1**. Where **Λ** diagonal matrices have the diagonal elements of the singular value of the matrix $(\mathbf{P} - \mathbf{rc}')$.

2.2.3. Inertia
The value of inertia shows the contribution of the i-th row in total inertia. The column profiles are represented by the dimensional spaces d ≤ m, the coordinates of rows of the matrices formed by taking the first d columns of $\mathbf{F} = \mathbf{D_r^{-1}A\Lambda}$ for d = 2 or 3, and the coordinates of the column profiles are *j* rows of the matrices formed by taking *k* first columns of $\mathbf{G} = \mathbf{D_c^{-1}B\Lambda}$. Total inertia will explain all the information in all spaces $\boldsymbol{tr(E) = n\sum_{i=1}^{m}\lambda_i}$ where $\boldsymbol{E = D_r^{-\frac{1}{2}}(P - rc`)D_c^{-\frac{1}{2}}\left(D_r^{-\frac{1}{2}}(P - rc`)D_c^{-\frac{1}{2}}\right)`}$] and $\lambda_i$ are the singular values obtained from $\boldsymbol{D_r^{-\frac{1}{2}}(P - rc`)D_c^{-1/2}}$ [4].

3. Data and Methods

This research uses data from the research conducted by Quality Assurance Unit Team of Faculty of Mathematics and Natural Sciences of Syiah Kuala University in 2016. The respondents are 62 students consisting of students of Mathematics, Physics, Biology, Chemistry, Pharmacy and Informatics Study Program who have completed the final project in the academic year 2015/2016. The variables are defined as in Table 1.

Stages performed in the study are as follows:

1. Describe the characteristics of the respondent and the general description of the completion of the final project of FMIPA students of Syiah Kuala University.
2. Perform a chi-square test to know the relationship between the duration of the completion of the final task, the assessment of the students in completing the final task based on the guidance, facilities, the final seminar and the general impact and the quality of the study program at FMIPA.
3. Perform data processing using multiple correspondence analysis. The importance of each variable must be found in correspondence plot [7].

IOP Conf. Series: Materials Science and Engineering **352** (2018) 012054 doi:10.1088/1757-899X/352/1/012054

Table 1. Variables and Categories in this research.

No	Variable	Category
1	Study Program (Y)	1 = Biology, 2 = Pharmacy 3 = Physics 4 = Informatics 5 = Chemistry 6 = Mathematics
2	Completion duration of final project (X_1)	1 = ≤ 6 month 2 = 7-8 month 3 = 9-10 month 4 = 11-12 month 5 = >12 month (BAN-PT, 2008)
3	Guidance (X_2)	1 = Poor; 2 = Fair; 3 = Good; 4 = Excellent
4	Facilities (X_3)	1 = Poor; 2 = Fair; 3 = Good; 4 = Excellent
5	Final project seminar (X_4)	1 = Poor; 2 = Fair; 3 = Good; 4 = Excellent
6	Quality and general impact (X_5)	1 = Poor; 2 = Fair; 3 = Good; 4 = Excellent

4. Results and Discussion

4.1. Descriptive Analysis

Respondents in this research are students from five study programs at Faculty of Mathematics and Natural Sciences Syiah Kuala University. Some information about characteristics of respondents is in Table 2.

Based on Table 2 most of the respondents are women that is equal to 76%. Almost all students in good health condition when completing the final task that is equal to 97%. Only 11% of the student's finance is from their own while preparing the final task, while the rest is financed by others. The final assignment theme is determined by the students themselves 58% and 40% are determined by their supervisor. Furthermore, the determination of supervisor by the study program is as much as 77%.

Table 2. Percentage of respondent characteristics.

No.	Variabel	Category	n	Percentage
1	Gender	Male	15	24
		Female	47	76
2	Health condition	Healthy	60	97
		Not Healthy	2	3
3	Cost of education	Themselves	7	11
		Others (parent, scholarship, etc.)	55	89
4	Determination of topic of final project	Themselves	36	58
		Supervisor	25	40
		Others	1	2
5	The supervisor is determined by the Study Program	Yes	48	77
		No	14	23

4.2. Chi-square test

Table 3. The result of the chi-squared test between the variable of the study program and supporting variable of final project completion.

No.	Variable	Chi-Square	df	*p-value*
1	Guidance (X_1)	21.069	10	0.021*
2	Facility (X_2)	20.512	10	0.025*
3	Final project seminar (X_3)	16.310	10	0.091
4	Quality and general impact (X_4)	10.725	5	0.057
5	Duration (X_5)	36.853	20	0.012*

Note : * = Significant with α 0.05

Chi-square test is performed to find out the variables that are related to the study programs. More can be seen in the following Table 3.

By using the α value of 0.05, then the rejection criterion is used if p-value < α then reject H_0. Based on these criteria, guidance activities, facility and duration have a relationship with study program, the other words each of study programs have characteristics.

4.3. Multiple correspondence analysis

Based on the Chi-Square test, the duration of completion of the final project, guidance activities and facilities are available variables. Correspondence analysis results obtain the singular value, the value of inertia, the inertia proportion, and the cumulative proportion presented in the following Table 4.

Table 4. The singular value, the value of inertia, the proportions, and the cumulative proportions.

Dimension	Singular value	The value of Inersia	The Proportion of Inertia	Cumulative proportion
1	0.7109	0.5054	0.1555	0.1555
2	0.6745	0.4550	0.1400	0.2955
3	0.6080	0.3697	0.1138	0.4093
...	...	...	...	...
13	0.2827	0.0799	0.0246	1.0000
Total	6.263	3.25	1.00	-

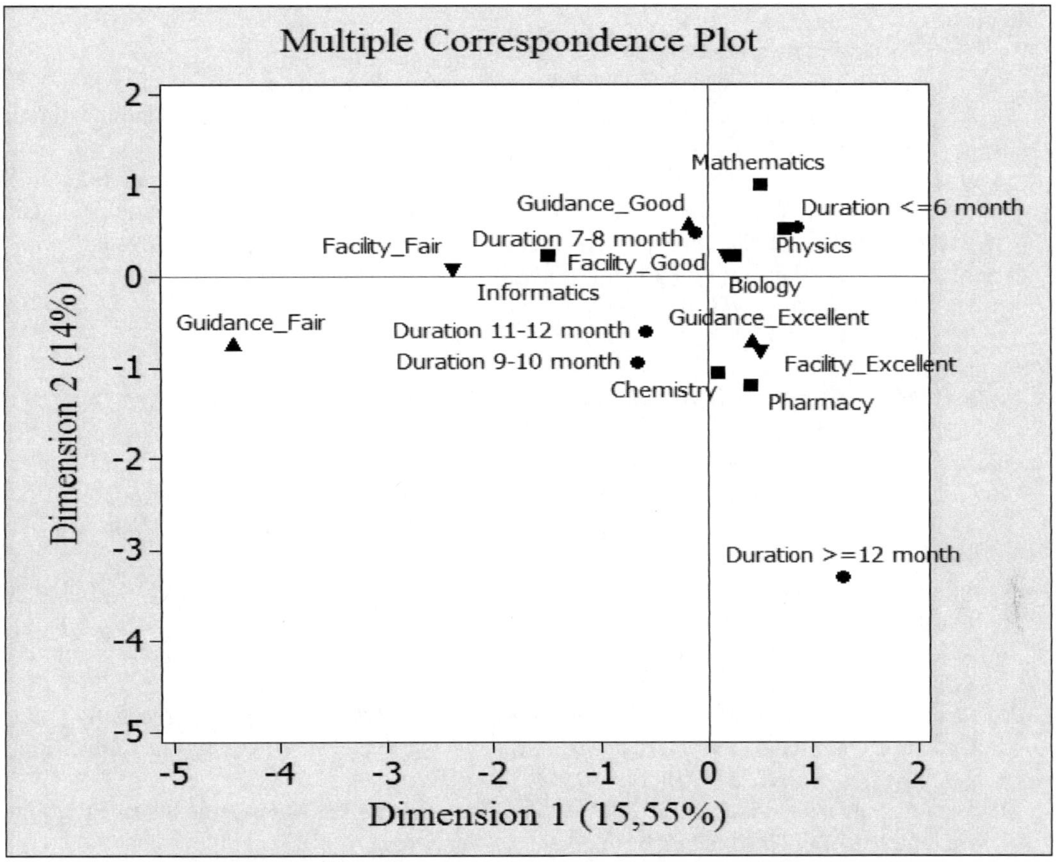

Figure 1. The plot of multiple correspondence analysis.

Table 4 shows that there are 13 dimensions of the results of multiple correspondence analysis. The data consist of 4 variables with 17 categories, so that this research can produce 17 – 4 = 13 dimensions. There are some criteria used to determine the representative model [6]. In the study, multiple correspondence analysis is only capable of displaying 2-dimensional graphics so that the first two root characteristics describe 29.55% from all information. They are 15.55% from the first dimension and 14% from the second dimension.

The result of multiple correspondence analysis in the form of correspondence plot can be seen in figure 1.It can explain about the scoring of students as respondents that:

1. Study Program of Mathematics has good facility and guidance, and ideal final project completing duration. These can be caused by computer laboratories and many Ph.D. lecturers.
2. Study Program of Physics has a good facility, good guidance, and the fastest duration. These can be caused because lecturers provide much time to guide student research.
3. Study Program of Biology has the good facility, good guidance, and duration for completing the final task between 7 and 8 months. They can be influenced by complete biology laboratories and tools.
4. Study Programs of Chemistry and Pharmacy consist of an excellent facility, excellent guidance, and greater than 7-month duration. These can be caused by advanced laboratories and also professor and Ph.D. lecturers.
5. Study Program of Informatics has a fair facility, good guidance activities and 7 to 8-month duration. These can be caused because this study program has not had a permanent building yet.

5. Conclusions

The conclusions of this study are as follows:

1. The factors that have a relationship with the completion of the final task in each study program at FMIPA Syiah Kuala University are facilities available, guidance activities, and the duration of completion of the final task.
2. Guidance activities and facilities available in Chemistry and Pharmacy are included into the excellent category. Whereas the fastest duration is owned by Physics Study Program. Guidance activities and available facilities in Informatics are included in the category fair while Pharmacy completed the final task for the longest time.
3. Chemistry and Pharmacy Study Programs have similar characteristics.

References

[1] Abdi H and Williams L J 2010 *Correspondence Analysis Encyclopedia of Resign Desain* (ed: Neil Salkind) (United States of America: Thousand Oaks: Sage Publications) 11

[2] Hair J F, Black W C, Babin B J and Anderson R E 2010 *Multivariate Data Analysis A Global Perspective 7th Edition* (United States of America: Pearson Education International)

[3] Johnson A R and Wichern D W 2007 *Applied Multivariate Statistical Analysis 6th Edition* (United States of America. Pearson Education International)

[4] Mattjik A A and Sumertajaya I M 2011 *Sidik Peubah Ganda dengan Menggunakan SAS* (Bogor: Statistics Departement of IPB) (in Indonesian)

[5] Nenadic O and Greenarce M 2005 *Computation of Multiple Correspondence Analysis with code in R*

[6] Rusyana A, Nurhasanah, Marzuki and Flancia M 2016 SARIMA Model for Forecasting Foreign Tourists at the Kualanamu International Airport. *Proc. Int. Conf. Mathematics, Statistics, and their Applications (ICMSA)* Banda Aceh IEEE p 155

[7] Rusyana A, Salwa N and Muzamil 2009 Analisis konjoin untuk menilai pembukaan Program Studi Statistika di Universitas Syiah Kuala *J. Pythagoras* **5** 14-23 (in Indonesian)

IOP Conf. Series: Materials Science and Engineering **352** (2018) 012055 doi:10.1088/1757-899X/352/1/012055

Selection for the best ETS (error, trend, seasonal) model to forecast weather in the Aceh Besar District

Chesilia Amora Jofipasi, Miftahuddin, and Hizir
Department of Statistics, Faculty of Mathematics and Science, Syiah Kuala University, Banda Aceh, Indonesia

Email: chesilia.amora@students.stat.unsyiah.ac.id; miftah@unsyiah.ac.id;

Abstract. Weather is a phenomenon that occurs in certain areas that indicate a change in natural activity. Weather can be predicted using data in previous periods over a period. The purpose of this study is to get the best ETS model to predict the weather in Aceh Besar. The ETS model is a time series univariate forecasting method; its use focuses on trend and seasonal components. The data used are air temperature, dew point, sea level pressure, station pressure, visibility, wind speed, and sea surface temperature from January 2006 to December 2016. Based on AIC, AICc and BIC the smallest values obtained the conclusion that the ETS (M, N, A) is used to predict air temperature, and sea surface temperature, ETS (A, N, A) is used to predict dew point, sea level pressure and station pressure, ETS (A, A, N) is used to predict visibility, and ETS (A, N, N) is used to predict wind speed.

1. Introduction

Weather and climate is component of the natural ecosystem that affects the life processes of human, animal and even plants. Based on data collected by the Intergovernmental Panel on Climate Change (IPCC),. the earth temperature has increased by 0,8 °C measured from the beginning of the 20th and estimated at the end of the 21st the earth's temperature will be increased to 2,4°– 6,4 °C [1]. Weather can be predicted using data in previous periods with studying trend data. Forecasting for time series data can be done with a time series univariate and can also be multivariate.

In this study, ETS (Error, Trend, Seasonal) was chosen as the method to be used to forecast weather elements (air temperature, visibility, wind speed, dew point, station pressure, sea level temperature and sea surface temperature). The purpose of this research is to get the best ETS model for every weather element in Aceh Besar.

2. Literature Review

2.1. Weather Elements

Weather is a physical phenomenon from the atmosphere at the place and at a shorter time. While, the climate is the average of the weather on a year that investigation in a long time (minimum 30 years) and covers a large area [2].

The air temperature is the degree of heat of the molecular activity in the atmosphere, expressed in Celcius (C), Reamur (R) and Fahrenheit (F) [3]. Air pressure is the power to drive air masses within each area, expressed in Celcius (C), Reamur (R) and Fahrenheit (F) [4]. Wind speed is the speed of air that moves horizontally or vertically at 2m above ground level, expressed in knots [3]. Visibility is the ability to see an object horizontally and expressed in miles [5]. Sea surface temperature is one of the parameters that determine the quality of the waters, expressed in Celcius (C). SPL is an important factor influencing the dynamics of regional climate and global climate [6].

2.2. ETS (Error, Trend, Seasonal)

ETS (Error, Trend, Seasonal) method is an approach method for forecasting time series univariate. This ETS model focuses on trend and seasonal components [7]. The flexibility of the ETS model lies in its ability to trend and seasonal components of different traits.

The components are as follows:

Table 1. ETS Components [8]

Trend Component	Seasonal Component		
	N (None)	A (Additive)	M (Multiplicative)
N (None)	NN	NA	NM
A (Additive)	AN	AA	AM
Ad (Additive Damped)	AdN	AdA	AdM
M (Multiplicative)	MN	MA	MM
Md (Multiplicative Damped)	MdN	MdA	MdM

The combination of ETS models obtained from the above components:

Table 2. Combination ETS model [8]

Model	Model	Model
ETS (M, M, N)	ETS (A, M, A)	ETS (M, N, M)
ETS (M, A, N)	ETS (A, Md, N)	ETS (M, N, A)
ETS (M, A, M)	ETS (A, Md, M)	ETS (M, N, N)
ETS (A, M, N)	ETS (A, N, A)	ETS (M, A, A)
ETS (A, N, N)	ETS (M, Ad, M)	ETS (A, Ad, M)
ETS (A, A, M)	ETS (M, Ad, N)	ETS (M, M, A)
ETS (M, M, M)	ETS (M, Md, M)	ETS (A, A, A)
ETS (A, N, M)	ETS (A, Ad, N)	ETS (A, Ad, A)
ETS (A, A, N)	ETS (M, Md, A)	ETS (M, Ad, A)
ETS (A, M, M)	ETS (M, Md, N)	ETS (A, Md, A)

From the ETS model above, there are 15 models with additive errors and 15 models with multiplicative errors. In time series analysis it is possible to have some models that meet the assumptions. Related to that, to determine the best model of 30 ETS model can be used several criteria such as Akaike's Information Criterion (AIC), Akaike's Information Criterion correction (AICc) and Bayesian Information Criterion (BIC) [9].

Akaike's Information Criterion (AIC) can be calculated using the following equation:

$$AIC = -2\left(\frac{LL}{T}\right) + \frac{2t_p}{T} \quad (1)$$

with:

AIC = Akaike Information Criterion
LL = log likelihood
t_p = Total Parameters
T = Number of observations

Akaike's Information Criterion correction (AIC_c) can be calculated using the following equation:

$$AIC_c = AIC + \frac{2k(k+1)}{n-k-1} \tag{2}$$

with $\frac{2k(k+1)}{n-k-1}$ is bias correction.

Bayesian Information Criterion (BIC) can be calculated using the following equation:

$$BIC = -2LL + k \ln(n) \tag{3}$$

with:

LL = log likelihood
k = Estimation of parameter model
n = Number of observations

3. Research Methods

3.1. Data

This study uses data obtained from the website www.noaa.gov. The data is a monthly data of weather elements in Aceh Besar (Satellite Station Blang Bintang). The data used were 132 observations from January 2006-December 2016 for each variable, but the SST variable used data from September 2006 to August 2016 with missing data from January 2013-April 2015. Data analysis was done with software R 3.0.3. Here are the variables:

Table 3. Weather elements

No	Variables	Unit
1	temp (Air Temperature)	°C
2	dewp (Dew Point)	°C
3	slp (Sea Level Pressure)	mb
4	stp (Station Pressure)	mb
5	visib (Visibility)	miles
6	wdsp (Wind Speed)	knot
7	sst (Sea Surface Temperature)	°C

3.2. Analytical Procedures

The data analysis procedure in this study consists of:

1. Data of weather elements (air temperature, visibility, wind speed, dew point, station pressure, sea surface pressure and sea surface temperature) are inputted into software R 3.0.3.
2. To descriptive analysis by looking at the characteristics of data through data summary.
3. To inferencing analysis using 30 combinations of ETS models for each weather element.
4. Choose the best ETS model based on the smallest AIC, AICc and BIC values for each weather element.

IOP Conf. Series: Materials Science and Engineering **352** (2018) 012055 doi:10.1088/1757-899X/352/1/012055

4. Results and Discussion

4.1. Des criptive analysis

To know the characteristics data of weather elements can be seen in the following table:

Table 4. Summary of weather element in Aceh Besar district on 2006-2016

Variable	Characteristics	Value	Variable	Characteristics	Value
Air temperature	Maximum	29,28	Visibility	Maximum	6,25
	Minimum	25,72		Minimum	4,00
	Mean	27,29		Mean	5,86
	Median	27,31		Median	6,00
	Quartile 1	26,72		Quartile 1	5,75
	Quartile 3	27,81		Quartile 3	6,00
Dewpoint	Maximum	24,59	Wind speed	Maximum	8,00
	Minimum	21,01		Minimum	0,67
	Mean	22,87		Mean	3,76
	Median	22,90		Median	3,75
	Quartile 1	22,24		Quartile 1	3,19
	Quartile 3	23,50		Quartile 3	4,30
Sea level pressure	Maximum	1012	Sea surface temperature	Maximum	30,71
	Minimum	1007		Minimum	28,26
	Mean	1010		Mean	29,25
	Median	1010		Median	29,22
	Quartile 1	1009		Quartile 1	29,00
	Quartile 3	1010		Quartile 3	29,33
				Data lost	34
Station pressure	Maximum	1009			
	Minimum	1005			
	Mean	1007			
	Median	1007			
	Quartile 1	1007			
	Quartile 3	1008			

Based on Table 4. we can know the characteristics of weather elements in Aceh Besar 2006 to 2016 as follows:

1. Water temperature obtained an average of 27,29 °C with a minimum temperature of 25,72 °C and a maximum temperature of 29,28 °C. The water temperature obtained the mean, median and quartile three which ranges from 27 °C.
2. Dewpoint, obtained an average of 22,87 °C with a minimum temperature of 21,01 °C and a maximum temperature of 24,59 °C. At dew point the mean and median values at the same temperature that is 22 °C.
3. Sea level pressure obtained the average of 1010mb with the lowest pressure of 1007mb and the highest pressure is 1012mb. The sea level pressure obtains the mean, median and quartile three values are precisely the same value of 1010mb.
4. Station pressure, obtained an average of 1007mb with the lowest pressure of 1005mb and the highest pressure is 1009mb. Also the station pressure also obtained the same value for the mean, median and quartile 1 is equal to 1007mb.
5. Visibility obtained an average of 5,86 miles with the closest distance of 4,00 miles and the farthest distance of 6,25 miles. On visibility also obtained the same value for median and quartile three that is equal to 6,00 miles.

6. Wind speed obtained an average of 3,76 knots with a late speed of 0,67 knots and the fastest speed of 8.00 knots. Besides, wind speed has a more significant range value when compared with other weather elements that are equal to 7.
7. Sea surface temperature, obtained an average of 29,25 °C with a minimum temperature of 28,26 °C and a maximum temperature of 30,71 °C. Also, the correct values of the mean, median, quartile 1 and quartile 3 are in the same value of 29 °C. As well as on variable sea surface temperature there are lost data as much as 34 periods.

4.2. Inference Analysis

Inference analyzes were performed using 30 combinations of ETS models, and only a few model can be used to predict the weather elements. From the model used the best model is chosen by looking at the smallest AIC, AICc and BIC values.

Here are the combinations ETS model for air temperature in Aceh Besar in 2006-2016:

Table 5. ETS model air temperature

Model ETS	AIC	AICc	BIC
M, M, N	500,5896	500,9046	512,1208
M, A, N	500,7238	501,0387	512,2550
M, Ad, M	439,5239	444,8978	488,5370
A, N, N	497,6485	497,7415	503,4141
M, Md, M	442,1614	477,5299	491,1691
A, A, N	501,9201	502,2351	513,4513
A, N, A	433,8487	437,4385	474,2080
M, N, M	438,6933	442,2830	497,0525
M, N, A	432,5406	436,1304	472,8998
M, N, N	496,5606	496,6536	502,3262
M, Ad, A	437,5158	443,1233	486,7625
A, A, A	439,5158	444,2462	485,6406

From the Table 5 obtained the best model is ETS (M, N, A) to predict water temperature in Aceh Besar in 2017-2018 with AIC of 432,5406, AICc of 436,1304 and BIC of 472,8998.

Here is the combinations ETS model for dew point in Aceh Besar in 2006-2016:

Table 6. ETS model dew point

Model ETS	AIC	AICc	BIC
M, Md, N	547,9749	548,4511	562,3889
M, A, N	545,5890	545,9039	557,1202
M, Ad, M	476,1028	481,4712	525,1104
A, N, N	541,3037	541,4003	547,0729
M, Md, M	475,3339	480,7023	524,3415
A, A, N	545,3121	545,6270	556,8433
A, N, A	468,0783	471,6681	508,4376
M, N, M	470,4469	474,0366	510,8061
M, N, A	468,2307	471,8205	508,5899
M, N, N	541,7574	541,8504	547,5230
M, A, A	475,8386	480,5691	521,9635
A, A, A	473,5571	478,2876	519,6820

From the Table 6 obtained the best model is ETS (A, N, A) to predict dew point in Aceh Besar in 2017-2018 with AIC of 468,0783, AICc of 471,6681 and BIC of 508,4376.

Here is the combinations ETS model for sea level pressure in Aceh Besar in 2006-2016:

IOP Conf. Series: Materials Science and Engineering **352** (2018) 012055 doi:10.1088/1757-899X/352/1/012055

Table 7. ETS model sea level pressure

Model ETS	AIC	AICc	BIC
M,M,N	655,1631	655,4781	666,6943
M,A,N	654,9047	655,2196	666,4359
M,Ad,M	595,0740	600,4424	644,0816
A,N,N	651,5644	651,6574	657,3300
M,M,M	491,5557	496,2861	537,6805
A,A,N	565,4636	565,7786	576,9948
A,N,A	484,7299	488,3196	525,0891
M,M,N	485,2665	488,8563	525,6258
M,N,A	485,1053	488,6951	525,4646
M,N,N	561,3579	561,4509	567,1235
M,Ad,A	492,3699	497,7383	541,3775
A,A,A	492,4421	497,1725	538,5669

From the Table 7 obtained the best model is ETS (A, N, A) to predict sea level pressure in Aceh Besar in 2017-2018 with AIC of 484,7299, AICc of 488,3196 and BIC of 525,0891.

Here is the combinations ETS model for station pressure in Aceh Besar in 2006-2016:

Table 8. ETS model station pressure

Model ETS	AIC	AICc	BIC
M,Md,N	565,7366	566,2128	580,1506
M,A,N	563,7944	564,1094	575,3256
M,A,M	489,2061	493,9365	535,3309
A,N,N	559,8000	559,8930	565,5656
M,M,M	489,5715	494,3019	535,6963
A,A,N	563,8278	564,1428	575,3591
A,N,A	481,8738	485,4636	522,2331
M,N,M	482,4092	485,9990	522,7684
M,N,A	481,9825	485,5722	522,3417
M,N,N	559,7663	559,8593	565,5319
M,A,A	491,1983	495,9287	537,3231
A,Ad,A	490,1326	495,5011	539,1403

From the Table 8 obtained the best model is ETS (A, N, A) to predict station pressure in Aceh Besar in 2017-2018 with AIC of 481,8738, AICc of 485,4636 and BIC of 522,2331.

Here is the combinations ETS model for visibility in Aceh Besar in 2006-2016:

Table 9. ETS model visibility

Model ETS	AIC	AICc	BIC
M,M,N	341,4413	341,9175	355,8553
M,A,N	335,6049	355,9199	347,1361
M,Ad,M	350,9936	356,3621	400,0013
A,N,N	334,6036	334,6966	340,3692
M,M,M	361,7591	367,1275	410,7667
A,A,N	333,5488	333,8638	345,0080
A,N,A	341,3315	344,9213	381,6908
M,N,M	381,2393	384,8291	421,5986
M,N,A	339,6880	343,2777	380,0472
M,N,N	335,0598	335,1528	340,8254
M,A,A	354,9128	360,2812	403,9205
A,A,A	344,7034	350,0718	393,7110

IOP Conf. Series: Materials Science and Engineering **352** (2018) 012055 doi:10.1088/1757-899X/352/1/012055

From the Table 9 obtained the best model is ETS (A, A, N) to predict visibility in Aceh Besar in 2017-2018 with AIC of 333,5488, AICc of 333,8638 and BIC of 3345,0800.

Here is the combinations ETS model for wind speed in Aceh Besar in 2006-2016:

Table 10. ETS model wind speed

Model ETS	AIC	AICc	BIC
M,M,N	677,9229	678,2379	689,4541
M,A,N	674,6547	674,9697	686,1859
M,A,M	687,8231	692,5517	733,9461
A,N,N	666,8980	666,9910	672,6636
M,M,M	682,0401	686,7705	728,1649
A,Ad,N	667,8871	668,3633	682,3011
A,N,A	676,8035	680,3932	717,1627
M,N,M	678,1830	681,7727	718,5422
M,N,A	675,8602	679,4500	716,2195
M,N,N	673,3369	673,4299	679,1025
M,Ad,A	682,6962	688,0647	731,7039
A,A,A	683,3426	688,0730	729,4674

From the Table 10 obtained the best model is ETS (A, N, N) to predict wind speed in Aceh Besar in 2017-2018 with AIC of 666,8980, AICc of 666,9910 and BIC of 672,6636.

Here is the combinations ETS model for sea surface temperature in Aceh Besar in 2006-2016:

Table 11. ETS model sea surface temperature

Model ETS	AIC	AICc	BIC
M, Md, N	314,9518	315,5401	328,3625
M, A, N	314,6889	315,0773	325,4175
M, Ad, M	315,7892	322,5892	361,3854
A, N, N	312,5315	312,6458	317,8957
M, Md, M	314,4385	321,2385	360,0347
A, A, N	316,3125	316,7009	327,0410
A, N, A	305,2927	309,8088	312,8425
M, N, M	304,9535	309,4696	312,5033
M, N, A	304,7735	309,2896	312,3233
M, N, N	311,0270	311,1413	316,3912
M, A, A	312,8845	318,8626	355,7986
A, A, A	313,6343	319,6123	356,5484

From the table 11 obtained the best model is ETS (M, N, A) to predict sea surface temperature in Aceh Besar in 2017-2018 with AIC of 304,7735 AICc of 309,2896 and BIC of 312,3233.

5. Conclusion and Suggestion

5.1. Conclusion

Based on AIC, AICc and BIC the smallest values obtained the judgment that the ETS (M, N , A) is used to predict air temperature and sea surface temperature, ETS (A, N, A) is used to predict dew point, sea level pressure and station pressure, ETS (A, A, N) is applied to predict visibility, and ETS (A, N, N) is used to predict wind speed.

5.2. Suggestion

This study focuses on trend and seasonal components for weather forecasting in Aceh Besar with variable air temperature, dew point, sea level pressure, station pressure, visibility, wind speed and sea

IOP Conf. Series: Materials Science and Engineering **352** (2018) 012055 doi:10.1088/1757-899X/352/1/012055

surface temperature, for the next researchers to forecast the elements weather with more attention to the error.

References

[1] Solomon S, Qin D, Manning M, Chen Z, Averyt K B, Tignor M and Miller H L 2007 Contribution of working group i to the fourth assessment report of the intergovernmental panel on climate change intergovernmental panel on climate change *Climate Change 2007: The Physical Science Basis*
[2] Wuryatno 2000 *Klimatologi* (Bandung: ITB Press) (in Indonesian)
[3] Tjasjono B 2004 *Klimatologi* (Bandung: ITB) (in Indonesian)
[4] Sarjani 2004 Cuaca dan Iklim www.scribd.com (Accessed on 5 August 2017) (in Indonesian)
[5] Nugroho F 2009 Visibility, RVR dan Konversi RVR ke visibility *Penerbangan* (in Indonesian)
[6] Qu T, Du Y, Strachan J, Meyers G and Slingo J 2005 Sea surface temperature and its variability in the Indonesian region *Oceanography* **18**(4) 50
[7] Sindhanuru Hemanth 2016 Exponential Smoothing (ETS) Framework www.latenview.com/idealab/exponential-smoothing-ets-framework (Accesed on 7 April 2017)
[8] Hyndman R J and Khandakar Y 2008 Automatic time series forecasting: The forecast package for R *J. Statistical Software* **27** (Monash University)
[9] Hyndman R, Koehler A B, Ord J K and Snyder R D 2008 *Forecasting with Exponential Smoothing the State Space Approach* (Germany: Springer)

IOP Conf. Series: Materials Science and Engineering **352** (2018) 012056 doi:10.1088/1757-899X/352/1/012056

Tax revenue and inflation rate predictions in Banda Aceh using Vector Error Correction Model (VECM)

Eva Maulia[1], Miftahuddin[2] and Hizir Sofyan[2]
[1] Department of Statistics, Faculty of Math and Sciences, University of Syiah Kuala, Banda Aceh 233111, Indonesia
[2] Lecturer, Department of Statistics, Faculty of Math and Sciences, University of Syiah Kuala, Banda Aceh 233111, Indonesia

E-mail: eva.maulia@students.stat.unsyiah.ac.id, miftah@unsyiah.ac.id, hizir@unsyiah.ac.id

Abstract. A country has some important parameters to achieve the welfare of the economy, such as tax revenues and inflation. One of the largest revenues of the state budget in Indonesia comes from the tax sector. Besides, the rate of inflation occurring in a country can be used as one measure, to measure economic problems that the country facing. Given the importance of tax revenue and inflation rate control in achieving economic prosperity, it is necessary to analyze the relationship and forecasting tax revenue and inflation rate. VECM (Vector Error Correction Model) was chosen as the method used in this research, because of the data used in the form of multivariate time series data. This study aims to produce a VECM model with optimal lag and to predict the tax revenue and inflation rate of the VECM model. The results show that the best model for data of tax revenue and the inflation rate in Banda Aceh City is VECM with 3rd optimal lag or VECM (3). Of the seven models formed, there is a significant model that is the acceptance model of income tax. The predicted results of tax revenue and the inflation rate in Kota Banda Aceh for the next 6, 12 and 24 periods (months) obtained using VECM (3) are considered valid, since they have a minimum error value compared to other models.

1. Introduction

A country has several important parameters to achieve economic prosperity. The economic field includes important factors in realizing it, such as tax revenue and inflation rate. One of the largest revenues of the State Budget in Indonesia comes from the tax sector. Also, the rate of inflation occurring in a country can be used as one measure, to measure both the bad economic problems facing a country. Given the importance of tax revenue and inflation rate control in achieving economic prosperity, it is necessary to analyze the relationship and forecasting tax revenue and inflation rate.

Taxes are dues to the country payable by those obliged to pay according to the applicable regulations [1]. Inflation is a condition where prices have increased steadily [2]. Tax receipts and inflation rates have a substantial impact on the economic development of a region. To that end, the government needs to consider policies that can affect both of them, such as monetary policy and fiscal policy. Monetary policy aims to stabilize the economy through regulating the money supply. While the fiscal policy concerns the regulation of government expenditure as well as government revenue (tax) to direct the economy of a country [2]. Thus, forecasting of tax revenue and inflation rate is needed as one of the considerations of policy decision making, which may affect economic development.

This study aims to obtain the VECM model from tax revenue and inflation rate, to know the relationship between tax revenue and inflation rate and to forecast the tax revenue and inflation rate of the VECM model. After knowing the relationship between tax revenue and inflation rate, it is expected to be an input for local government in planning monetary policy and fiscal policy. Also, it can also build models used to forecast tax revenue and inflation rate. In this article, VECM (Vector Error Correction Model) was chosen as the method to be used to look at the relationship, as well as predict tax revenues and inflation. VECM has chosen for this research because of the research using multivariate time series data.

2. Literature review

2.1. The definition of income tax, VAT and inflation rate

Income tax is one of the sources of state revenues derived from people's income, are a form of state obligation and people's participation in financing and national development [7]. Income tax is levied on the subject of tax in respect of the income it receives in the tax year. The tax year may be the calendar year or fiscal year that is not the same as the calendar year, throughout the year the book covers 12 months [8].

Value Added Tax (VAT) is a tax imposed on the consumption of, within the customs area [9]. The general VAT rate is 10%, but according to government regulations, tariffs can be changed as low as 5% and as high as 15%. The nature of the VAT is the tax on consumption, which makes it one of the taxes that has a vast range of tax objects. Also, the VAT characteristic which is an objective tax base the imposition of tax on the object. If there is no tax object, there is no tax that can be charged. VAT Objects include delivery, utilization, export/import of taxable goods or taxable services [7].

The inflation rate is the rate of general price level from year to year and followed by the price increase in certain year from the previous year [10]. With the inflation of the price of goods increased, so people's purchasing power will decline. An increase in the price of one or two items alone cannot be called inflation unless the increase extends or results in price increases in other goods. [11].

2.2. VECM (Vector Error Correction Model)

In general, the econometric model of time series is a structural model because it is based on existing economic theory. VAR models (Vector Autoregression) was introduced by Christopher A. Sims in 1980, as an alternative to the macro-economic analysis. VAR models have a simple model structure with many variables that bit where all the variables are considered as an endogenous variable with the independent variable is the lag. VAR models are designed for stationary variables that do not contain the trend. The VAR model cannot be used to model a co-integrated time series data (having a long-term relationship) [12].

In the time series data, the stochastic trend indicates that the long-term component (long-run) and short-term (short-run) in the data. Research on stochastic trends in economic variables had been developed, so Granger developed the concept of co integration in 1981. Then along Engle-Granger developed the concept of co-integration and error correction (error correction) in 1987. Subsequently, in 1990, Johansen and Juselius VECM developed concepts (Vector Error Correction Model). VECM Model is an econometric analysis model, which can be used to determine the short-term behavior of a variable to the long-term, due to a permanent shock [13]. The model offers an easy working procedure, to separate long-term components and short-term components from the data generation process [14]. Thus, the VECM model differs from the VAR model, where the VECM model can be used to model the time series data which is cointegrated and not stationary. The VECM model is often referred to as terrestrial VAR form [15]. The general model of VECM is as follows.

$$\Delta Y_t = \varphi_1 + \delta_1 t + \lambda_1 e_{t-1} + \gamma_{11}\Delta Y_{t-1} + \ldots + \gamma_{1p}\Delta Y_{t-p} + \omega_{11}\Delta X_{t-1} + \ldots + \omega_{1q}\Delta X_{t-q} + \varepsilon_{1t} \qquad (1)$$

and

$$\Delta X_t = \varphi_2 + \delta_2 t + \lambda_2 e_{t-1} + \gamma_{21}\Delta X_{t-1} + \ldots + \gamma_{2p}\Delta X_{t-p} + \omega_{21}\Delta Y_{t-1} + \ldots + \omega_{2q}\Delta Y_{t-q} + \varepsilon_{2t} \quad (2)$$

where $e_{t-1} = Y_{t-1} - \alpha - \beta X_{t-1}$.

Here is a theoretical analysis of VECM analysis.

1. Stationary Test
   The assumption that must be met in the VECM analysis is that all independent variables must be stationary. It is characterized by all the remnant is white noise, which has a zero mean, constant variance and among the dependent variable was no correlation [13]. There are several ways that can be done to measure the data stationary, one of which is by using Augmented Dickey-Fuller test (ADF). Here are the ADF test statistics.
   $$\Delta Y_t = \alpha_0 + \gamma Y_{t-1} + \beta_i \sum_{i=1}^{P} \Delta Y_{t-i+1} + \varepsilon_t \quad (3)$$

2. Cointegration Test
   Cointegration is a long-term relationship between the variables, which although individually are not stationary, but the linear combination of these variables can be stationary [16]. Cointegration is one way to avoid the problem of spurious regression. One way to test cointegration is by using Johansen test as follows [17].
   $$LR_{tr} = (r \mid k) = -T \log(1- \lambda_i) \quad (4)$$

3. Residual Assumption Test
   The residual assumption test on the VECM model must satisfy the residual criteria. These criteria, namely, residual normal distribution (normality assumption test), there is no serial correlation in the residuals (autocorrelation assumption test), and the residual is homoscedasticity (heteroscedasticity assumption test) [13].

4. Optimal Lag Test
   The optimal lag test is the determination of the length of the lag to be used in estimating the VECM. An optimal lag determination is important in the analysis using the VECM method. This is because if the lag set is too long or too short, it will result in the wrong model specification. The criteria can be used to determine the optimal lag in between Final Prediction Error Correction (FPE), Akaike Information Criterion(AIC), Schwarz Information Criterion (SIC) and Hannan-Quinn Information Criterion (HQ), which has a minimum value among Various lags submitted. Mathematically, the values of AIC, SIC and HQ can be found using the following formula.
   $$AIC = -2\left(\frac{1}{T}\right) + 2(k + T) \quad (5)$$
   $$SIC = -2\left(\frac{1}{T}\right) + k\frac{\log(T)}{T} \quad (6)$$
   $$HQ = -2\left(\frac{1}{T}\right) + 2k\log\left(\frac{\log(T)}{T}\right) \quad (7)$$

5. Granger Causality Test
   The Granger causality test can indicate whether a variable has a two-way or one-way relationship. This test is conducted to determine the influence of one variable with another variable. It should be remembered that in this Granger causality test, what is seen is the effect of the past on the present condition so that the data used is time series data [18].

6. Prediction
   Prediction is an objective calculation and by using past data, to determine something in the future [19]. To determine the accuracy of the prediction of the model can use the value of the error follows.
   a. *Mean Absolut Percentage Error* (MAPE)

$$\text{MAPE} = \frac{\Sigma|(y_t - \widehat{y_t})/y_t|}{n} \text{ X } 100 \tag{8}$$

   b. *Mean Absolute Error* (MAE)

$$\text{MAE} = \frac{\Sigma|y_t - \widehat{y_t})|}{n} \tag{9}$$

   c. *Root Mean Square Error* (RMSE)

$$\text{RMSE} = \sqrt{\frac{\Sigma\ (y_t - \widehat{y_t})^2}{n}} \tag{10}$$

3. Method

The research method used in this research is VECM analysis. The research steps are carried out are as follows.

1. Input data into the R Studio software.
2. Perform the stationary test of the data. If the data is stationary, the process of data analysis using Vector Autoregression (VAR) analysis. If the data is not stationary, then differencing process will be conducted to obtain data that is stationary.
3. Data were stationary after differencing; co-integration tests need to be done first. If there is no co-integration (long-term relationship) between variables, then the data analysis using VAR analysis. However, if there is co-integration between variables, then the next stage of analysis uses VECM analysis.
4. The model fit test that is: normality test, autocorrelation test, and heteroscedasticity test.
5. Granger causality test was performed on stationary data.
6. Next, the determination of the length of the lag to be used in the VECM model.
7. At this stage, the VECM model is formulated.
8. Do predictions based on the VECM model Obtained.

4. Results and discussion

4.1. Descriptive analysis

The descriptive analysis aims to show a picture of the data that has been collected. Here is the plot of tax revenue and the inflation rate in January 2013 - December 2015.

From Figure 1 it can be seen that the pattern of the seven variables used have similarities. This can be seen from the fluctuations that occur, where the pattern of increase and decrease of each variable occurs at almost the same time. Tax receipts and inflation rates are increasing at the end of each year, while declines occur at the beginning of each year. This illustrates that there is a cointegration relationship between the seven variables. To see more clearly the cointegration relationship for each variable will be tested cointegration Johansen, but previously will be checked stationarity of the seven variables.

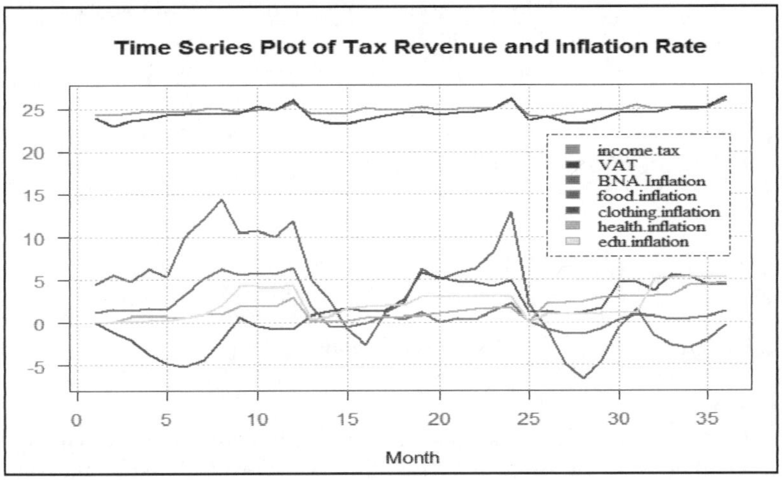

Figure 1. Time series plot of tax revenue and inflation rate for January 2013 – December 2015.

4.2. Stationarity test

The assumption that must be met in the VECM analysis is that all independent variables must be stationary. There are several ways that can be done to measure the data stationary, one of which is by using Augmented Dickey-Fuller test (ADF). The following table of ADF test results on tax revenue data and inflation rate.

Table 1. ADF test results.

No	Variable	Level		1st differencing		2nd differencing	
		Stat. ADF	P-value	Stat. ADF	P-value	Stat. ADF	P-value
1	Income.tax	-2.830	0.250	-4.312	0.01	-5.7	0.01
2	VAT	-2.790	0.266	-3.668	0.042	-4.265	0.011
3	BNA.inflation	-2.448	0.398	-2.696	0.303	-4.445	0.01
4	food.inflation	-3.112	0.140	-3.419	0.071	-4.753	0.01
5	clothing.inflation	-2.276	0.466	-3.321	0.085	-3.964	0.023
6	health.inflation	-1.353	0.825	-3.434	0.069	-4.475	0.01
7	edu.inflation	-2.831	0.250	-3.082	0.153	-4.047	0.020

* Reject H_0 if p-value < α (0.05) then H_0 is rejected.

Based on Table 1 it can be seen that all variables used are not stationary at the level. It can be seen at the level, ADF statistical p-value for each variable is greater than α (0.05). That is, can not reject H_0 or data is not stationary. Meanwhile, the results of the first differencing can be seen that there are several variables that have not been stationary or ADF statistic p-value is greater than α (0.05). While the results of the second differencing can be seen that the ADF statistic p-value of each variable is smaller than α (0.05). That is, reject H_0 or data is stationary. Thus the variable is a non-stationer variable of second order.

4.3. Cointegration test

Presence or absence of cointegration based on the value of the test statistic trace (LRtr). If the value of the test statistic trace (LR_{tr}) greater than the critical value (τ_{σ}), then there is cointegration on many variables.

Table 2 shows that all endogenous variables are cointegrated. It can be seen from the $LR_{tr} > \tau_{\sigma}$. Thus, it can be concluded that there is a long-term relationship between these variables, so the analysis is continued by using the VECM model.

Table 2. Results of the Johansen cointegration test with trace test.

Null hypothesis (H_0)	0,05 Trace Critical Value (T_σ)	Trace Test (LR_{tr})
r = 0*	131.70	362.50
r = 1*	102.14	244.91
r = 2*	76.07	155.02
r = 3*	53.12	88.87
r = 4*	34.91	48.32
r = 5*	19.96	27.55
r = 6*	9.24	11.58

* Reject H_0 (trace test value> 0.05 trace critical value)

4.4. *Residual assumption test*

In the VECM model, there is three residual assumptions test that must be met, namely normality test, autocorrelation test and heteroscedasticity test. The third test of residual assumption is:

Table 3. Residual assumption test.

Assumption	Test Name	Chi-squared	df	p-value
Normality	Jarque Bera test	2.66	7	0.915
Autocorrelation	Portmanteau test	665.69	644	0.269
Heteroscedasticity	White's Test	868.00	1176	1.000

* Reject H_0 if p-value < α (0.05) then H_0 is rejected.

The results of the tests Jarque Bera, Portmanteau dan White's in Table 3, showed that the p-value > α (0.05) or cannot reject H_0. That is, the residual of the data used in normal distribution, there is no autocorrelation and homoskedasticity. The three assumption test results indicate that the data used has fulfilled the assumption.

a. Granger causality test

The Granger Causality Test was conducted to determine the short-term relationships between observed variables. The results of Granger causality test show that the inflation rate of health as the dependent variable, income tax, VAT and the inflation rate of food have p-value < α (0.05), so reject H_0. Thus, it can be concluded that there is a short-term relationship between income tax, VAT and the inflation rate of food with the inflation rate of health. While other variables have no short-term relationship.

b. Optimal lag test

The optimal lag test is the determination of the length of the lag to be used in estimating the VECM. An optimal lag determination is important in the analysis using the VECM method. If the lag set is too long or too short, it will result in the wrong model specification. The following table determines the optimal lag length.

Table 4. Optimal lag length determination results.

Lag	AIC	HQ	SC	FPE
1	1.222	2.067	3.813	3.691
2	0.593	2.176	5.450	3.358
3	**-4.784**	**-2.462**	**2.339**	**0.099**

The VECM approach is sensitive to the determination of the amount of lag used, so it is necessary to determine the optimal number of lags. Determination of the number of lags is to determine the duration of the influence of a variable, to the variable past or other variables. Determination of lag length can be seen through minimum FPE, AIC, SIC and HQ values. Table 4 shows that the optimal lag for the VECM estimate is the 3rd lag or VECM (3). This is because the values of AIC, HQ, SC and FPE show the minimum value in the 3rd lag.

IOP Conf. Series: Materials Science and Engineering **352** (2018) 012056 doi:10.1088/1757-899X/352/1/012056

c. VECM model estimation

Based on the model specification using optimal lag analysis, then obtained the best model for data of tax revenue and the inflation rate that is VECM (3). From the seven models created, only one model that is significant, that is the model for income tax, so it can be concluded that there is a long-term relationship between the VAT, the inflation rate of Banda Aceh, the inflation rate of food, the inflation rate of clothing, the inflation rate of health and the inflation rate of education to income tax. Here is the model of income tax with VECM (3).

$$\begin{aligned}\Delta \text{income.tax} = {} & -0.877 + 0{,}004 - 0.748\Delta \text{income.tax}_{t-1} - 0.690\Delta \text{income.tax}_{t-2} + 1.123\Delta \text{income.tax}_{t-3} - 0.992\Delta VAT_{t-1} - 0.491\Delta VAT_{t-2} + 0.174\Delta VAT_{t-3} + \\ & 0.485\Delta \text{BNA.inflation}_{t-1} - 0.076\Delta \text{BNA.inflation}_{t-2} - 0.394\Delta \text{BNA.inflation}_{t-3} - 0.017\Delta \text{food.inflation}_{t-1} - 0.034\Delta \text{food.inflation}_{t-2} - 0.137\Delta \text{food.inflation}_{t-3} + \\ & 0.119\Delta \text{clothing.inflation}_{t-1} + 0.152\Delta \text{clothing.inflation}_{t-2} + 0.094\Delta \text{clothing.inflation}_{t-3} - 0.159\Delta \text{health.inflation}_{t-1} + 0.867\Delta \text{health.inflation}_{t-2} + 0.196\Delta \text{health.inflation}_{t-3} + \\ & 0.230\Delta \text{edu.inflation}_{t-1} - 0.038\Delta \text{edu.inflation}_{t-2} + 0.026\Delta \text{edu.inflation}_{t-3}\end{aligned}$$

From this model it is known that the magnitude of the absolute value of the variable coefficient ECT (Error Correction Term) amounted to 0.877 who explained that about 88% discrepancy between the actual value of income tax revenue in the short term and the value of the balance of income tax revenue in the long term, will be corrected each month. This shows that there is a long-term relationship from VAT, the inflation rate of Banda Aceh, the inflation rate of food, clothing inflation rate, inflation rate and the inflation rate health education on the income tax.

d. Predicted results

Furthermore, based on the model specification using optimal lag analysis, the best model for tax revenue variable and the inflation rate in Banda Aceh is VECM with 3rd optimal lag or VECM (3). The plot of predicted tax revenue and the inflation rate in Banda Aceh using VECM (3), for the next 6 periods (months) are as follows:

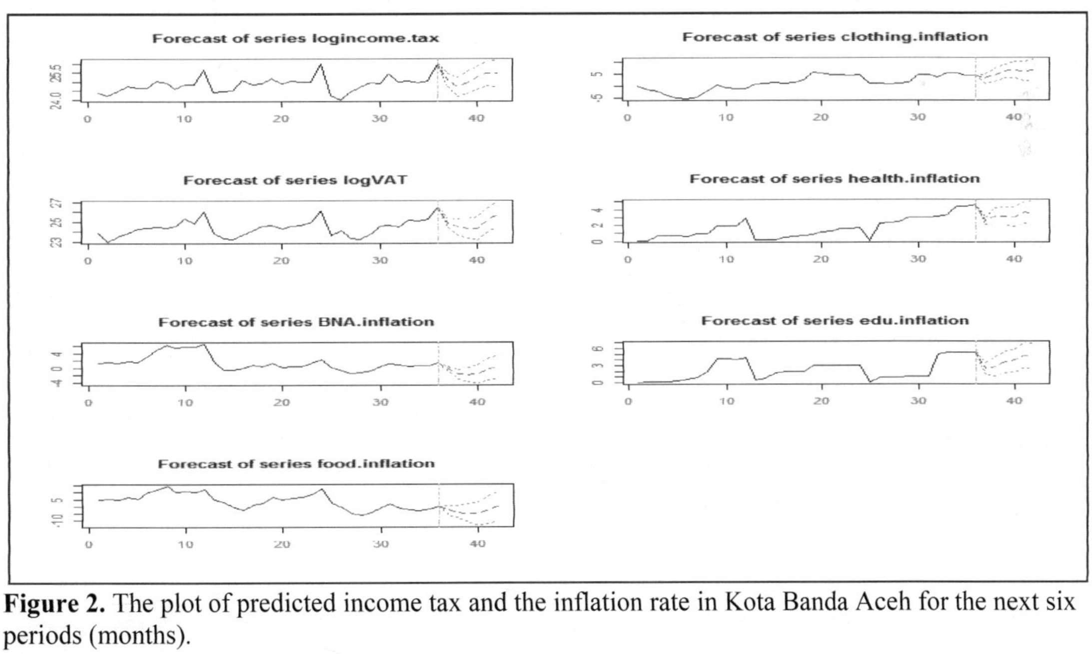

Figure 2. The plot of predicted income tax and the inflation rate in Kota Banda Aceh for the next six periods (months).

Table 5 below presents the MAPE, MAE and RMSE values of the six periods (months) predictions. In the table can be seen that the value of MAPE, MAE and RMSE minimum is the variable acceptance of income tax. That is, the result of the acceptance of income tax obtained by using VECM (3) is considered valid, because it has minimum error value that is, MAPE (2.497%), MAE (0.613) and RMSE (0.480).

Table 5. The accuracy of predicted results six periods (months).

Error Value	Income tax	VAT	Inflation BNA	Inflation BM	Inflation SDG	Inflation SHT	Inflation PDK
MAPE	2.497	5.564	1473.289	344.917	9261.81	12550.92	23944.32
MAE	0.613	1.320	23.372	19.072	27.975	24.669	24.910
RMSE	0.480	1.094	23.090	19.075	24.835	24.215	23.234

The plot of predicted tax revenue and the inflation rate in Banda Aceh city using VECM (3), for the next 12 periods (months) are as follows:

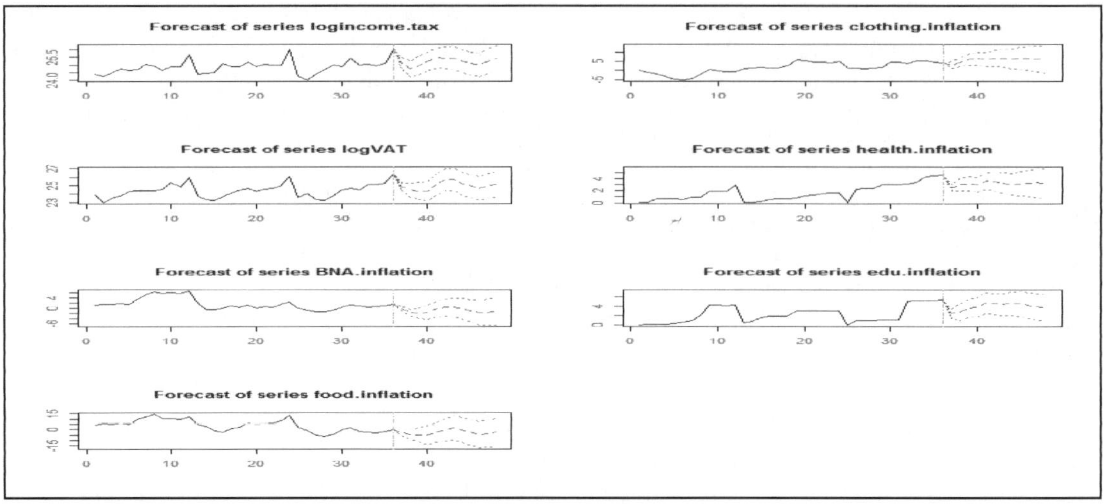

Figure 3. Plots of predicted tax revenue and the inflation rate in Kota Banda Aceh for the next 12 periods (months).

Table 6 below presents the MAPE, MAE and RMSE values of the 12 periods (months) predictions. In the table can be seen that the value of MAPE, MAE and RMSE minimum is the variable acceptance of income tax. That is, the result of the acceptance of income tax obtained by using VECM (3) is considered valid, because it has minimum error value that is, MAPE (1.882%), MAE (0.464) and RMSE (0.503).

Table 6. The accuracy of predicted results 12 periods (months).

Error Value	Income tax	VAT	Inflation BNA	Inflation BM	Inflation SDG	Inflation SHT	Inflation PDK
MAPE	1.882	4.033	905.285	233.038	6196.906	7080.919	12448.45
MAE	0.464	0.970	21.399	16.364	27.259	24.076	23.419
RMSE	0.503	1.057	21.496	16.663	27.336	24.089	23.488

The plot of predicted tax revenue and the inflation rate in Kota Banda Aceh using VECM (3), for the next 24 periods (months) are as follows:

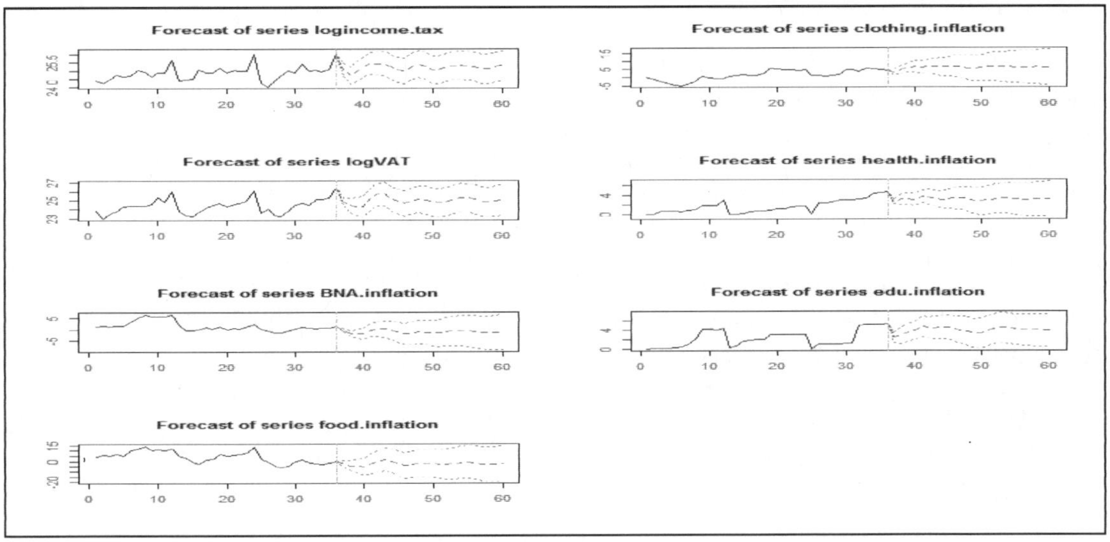

Figure 4. The plot of predicted income tax and the inflation rate in Banda Aceh City for the next 24 periods (months).

The following table 7 presents the values of MAPE, MAE, and RMSE from 24 periods (months) prediction results. In the table can be seen that the value of MAPE, MAE and RMSE minimum is the variable acceptance of income tax. That is, the result of the acceptance of income tax obtained by using VECM (3) is considered valid, because it has minimum error value that is, MAPE (1.714%), MAE (0.424) and RMSE (0.480).

Table 7. The accuracy of predicted results 24 periods (months).

Error Value	Income tax	VAT	Inflation BNA	Inflation BM	Inflation SDG	Inflation SHT	Inflation PDK
MAPE	**1,714**	4.129	4468.999	539.829	6196.906	3655.223	6587.806
MAE	**0.424**	0.992	22.982	18.602	24.662	24.205	23.189
RMSE	**0.480**	1.094	23.090	19.075	24.835	24.215	23.234

5. Conclusion

In this paper we concluded that: the specifications of the model using the analysis of optimal lag, we obtained that the best model for data obtained tax revenues and the inflation rate is VECM with 3rd optimal lag or VECM (3). From the seven models created, there is one model that significant is a model of income tax.

The result of predictions tax revenues and inflation in Banda Aceh for 6, 12 and 24 periods (months) ahead obtained by using VECM (3) to be valid, due to it has the minimum error compared with other models in this research.

References

[1] Sukardji U 2002 *Value Added Tax* (Jakarta: PT. Raja Grafindo Persada)
[2] Nopirin 2009 *Monetary Economics* (Yogyakarta: BPFE-Yogyakarta)
[3] Sinaga A R 2010 Effect of macroeconomic variables against tax receipts in Indonesia *Thesis* (Jakarta: UI)
[4] Herdiana E and Nalendra R 2013 Effect of Gross Domestic Product (GDP) and inflation rate Against Tax Receipts (Bandung: University Computer Indonesia)
[5] Wahyudi E, Sanim B, Siregar H and Nuryatono Nunung 2009 The Effect of Economic Shock on Tax Receipts in Regional Tax Offices in Indonesia *J. Development Economics* 10 68-83
[6] Renata A H, Hidayat K and Kaniskha B 2016 Influence of Inflation, Rupiah exchange rate and number of taxable entrepreneurs on value added tax *J. Taxation (TRACES)* 9 (Universita Brawijaya, Malang)

[7] Purnomo H. 2010 *The Basics of Taxation and Tax Accounting* (Jakarta: Erlangga)
[8] Diana A and Setiawati L 2014 *Taxation: Theory and Latest Regulations* (Yogyakarta: Andi)
[9] Casavera 2009 *Taxation* (Yogyakarta: Graha Ilmu)
[10] Pure A 2006 *Macro Economics* (Bandung: PT. Refika Aditama)
[11] BPS 2016 Inflation Banda Aceh (Banda Aceh: Central Bureau of Statistics)
[12] Gujarati D 2006 *Econometrics 4 th ed* (New York: Mc. Graw Hill)
[13] Ajija, S R, Sari D W, Setianto R H and Primanti M R 2011 *The Intelligent Mastering Eviews* (Jakarta: Salemba Four)
[14] Lutkepohl H 2006 *New Introduction to Multiple Time Series Analysis* (Berlin: Springer-Verlag)
[15] Juanda B. and Junaidi 2012 *Time Series Econometrics* (Bogor: IPB Press)
[16] Rosadi, D 2012 *Analysis Econometrics and Applied Time Series with Eviews* (Yogyakarta: Andi Offset)
[17] Maddala G S 2004 *Introduction to Econometrics 3 rd ed* (UK: Prentice-Hall, Inc.)
[18] Ariefanto M D 2012 *Econometrics essence and applications by using eviews* (Jakarta: Erlangga)
[19] Sumayang L 2003 *Fundamentals of Production and Operations Management* (Jakarta: Salemba Four)
[20] Makridakis S G, Wheelwright S C and Hyndman R J 1999 *Forecasting: Methods and Applications* (USA: John Wilet and Sons Inc)
[21] Miftahuddin 2008 Inflation Analysis Through Multiple Autoregression Vectors (VAB) with Five Variable Information *JMSK Journal* 5 ISSN: 1858-1382
[22] Achsani N A, Holtemöller O and Sofyan H 2005 Econometric and Fuzzy Modeling of Indonesian Money Demand *Statistical Tools for Finance and Insurance* (Berlin: Springer)
[23] Dickey D A and Fuller W A 1981 Likelihood ratio statistics for autoregressive time series with a unit root *Econometrica: J. Econometric Society* 1057-1072
[24] Engert W and Hendry S 1998 *Forecasting Inflation with the M1-VECM: Part Two* (Bank of Canada)
[25] Shibata R 1976 Selection of the Order of an Autoregressive Model by Akaike's Information Criterion. *Biometrics* 117-126
[26] Case K E and Fair R C 1999 *Principles of Macroeconomics 5th ed* (USA: Prentice-Hall, Inc.)

IOP Conf. Series: Materials Science and Engineering **352** (2018) 012057 doi:10.1088/1757-899X/352/1/012057

User acceptance of E-Government Services Based on TRAM model

K Adiyarta[1], D Napitupulu[2,1], H Nurdianto[3], R Rahim[4], A Ahmar[5],
[1]Program Studi Magister Ilmu Komputer, Universitas Budi Luhur
Ciledug Raya, Petukangan Utara, Kebayoran Lama, Jakarta Selatan, Indonesia
[2]Research Center for Quality System & Testing Technology,
Indonesian Institute of Sciences, Tangerang Selatan, Banten, Indonesia
[3]Department of Teknik Informatika, STMIK Dharma Wacana, Lampung, Indonesia
[4]School of Computer and Communication Engineering, Universiti Malaysia Perlis, Malaysia
[5]Department of Statistics, Universitas Negeri Makassar, Makassar, Indonesia

E-mail: krisna.adiyarta@gmail.com, darm003@lipi.go.id, herinurdiyanto@gmail.com, usurobbi85@zoho.com, ansarisaleh@unm.ac.id

Abstract. Developing country mostly left behind in adopting e-Government system. Miss-interpretation is assuming e-Government only about the application of technology made fail implementation. It is a whole philosophy that explores a human-citizen centered aspect in organized societies. Further, successful information system determined by user acceptance. Therefore, the evaluation of the performance of e-Government is one of the basic objectives of current government reformation. The research focus is to identify and explore the extent of user acceptance toward e-Government system. The research methodology used in this research is survey based on a questionnaire with TRAM approach and distributed to 230 respondents as customers that had been using e-Government services provided by X institution. The result of the study showed that the personality traits of TR significantly influenced cognitive dimensions of TAM. Through empirical demonstrations indicated that the TRI was said to be antecedents to TAM. The optimism dimension had higher coefficient among others, exhibit the technology users convince that the new technology will improve their productivity, while discomfort had no impact since the users have no doubt or hesitate to use technology.

1. Introduction

The purpose of developing e-Government system is to provide an electronic public service well and meet the needs of society. Yildiz pointed out the importance of formulating the e-government concept model as a reference for the transformation needs of the electronic service-based bureaucratic process by examining each dimension and orientation of public administration that became the focus of e-Government [1]. Some concept or definition of e-government that is well known such as World Bank stated that E-Government is about the optimization of ICT (Information and Communication Technology) in government agency [2]. UNESCO said that e-government refers to the use of ICT by users or community to obtain information and services needed. E-Government is neither the installation of systems and implementation of software applications nor the mechanisms to promote products and services on the market [3]. Thus, e-Government is not about ICT or new technology that could bring benefit and good relation with internal and external stakeholder. It is a basic principle that includes all of social and technology aspects.

The governance of the country had been made significantly straightforward due to the technological usage, one such technique being thee-government. Information and services availability could be accessed from website or portals of government. The essence of the e-government system is to support ICT infrastructure and strong regulation or policy in the finite environment. The primary purposes of e-government system are to provide public services that needed by the citizen. The website for information dissemination is still most developing country focus to build. From the view of world continent that could be seen from e-Government world ranking, Asia is lagging from other region developed countries like America and Europe in implementing e-Government system [4]. These conditions show that slow development and user adoption of e-Government services especially in Asia region [5]. According to Heeks (2008), it is estimated that total 85% of e-government implementation in developing countries failed. It remains only 15% of the projects to be considered as successes [6]. These indicated including Indonesia also one of developing country still suffered in e-Government implementation [7]. In nation level, PeGI had captured the usage of ICT by government agencies in Indonesia. PeGI (2015) reported government institution in province level had the total average score of 2.50, which is a poor result. Despite the potential benefit of ICT for the government agency, e-Government failure was mostly caused by low level of usage. User acceptance could predict the level of usage of the information system. Since user acceptance determines the success of information system including e-Government, thus a study of factors that affect user acceptance become important to conduct.

The main problem discussed in many pieces of research related to adoption or diffusion of technology is the different rate of technology diffusion itself. Furthermore, new technology could be accepted by a market and rejected by another market or user. An effort knows the progress of e-Government implementation for internal and external stakeholder become the critical point of organization. Therefore, it is necessary to develop ways to measure the success and failure of e-government projects [8][9]. The importance of evaluation is much greater in developing countries, where the resources are scarcer, and the opportunity costs are much higher [10]. In such countries, lack of resources does not become a huge problem since it used optimally based on strategic planning of the organization. Therefore, this study would like to explore the extent of user acceptance for e-Government services. The user acceptance could predict the level of utilization that is crucial being success indicator of technology including e-Government system [11].

2. Literature Review

Based on Roger (1995), the adoption rate of technology differs among users. Some technology or innovation may have a rapid rate of adoption, on the other hand, some innovation adopted more slowly. This important issue is being addressed in adoption or diffusion research especially to explore what factors that influence the users to adopt new technology. A behavioral theory needed to explain the user's attitude and behavior in adopting technology [12]. This theory could be used to predict whether innovation or technology will be a success or fail adopted by potential users or organizations. In another word, a behavioral approach is used to evaluate the technology being implemented [13]. There are some behavioral theories that widely used as best practice model to measure the extent of technology adoption such as TRI (Technology Readiness Index), TAM (Technology Acceptance Model) and UTAUT (The Unified Theory of Acceptance and Use of Technology). All these best practice model were parsimony but very general. Thus it may not be designed for any particular system since each system had unique characteristics that could influence the behaviour of technology adoption [14].

2.1. TRI (Technology Readiness Index)

Technology readiness was introduced by Parasuraman (2000). Definition of technology readiness is a “customers’ desire to adopt new technologies for improving their productivity in life and business.

The one is a promising theory that could be used all parties to know the difference process of people to adopt new technology [15][16]. User readiness is an important indicator that influences user acceptance to any technology offered [15]. Throughout the process of technology adoption, the user had a personal opinion about the technology-based product or service. Thus, this opinions according to Parasuraman (200) would be evaluated under dimensions as personality traits: Optimism, Innovativeness, Discomfort, and Insecurity.

Two main dimensions such as optimism and innovativeness were called contributors that could enhance the readiness of adopting the technology, while Discomfort and Insecurity stating negative feeling are inhibitors in TR pressing the preparedness as shown in figure 1. Optimism is about individual's tendency to believe that technology would bring good results in life and business, Innovativeness is an individual's lead about technological products, Insecurity is about individual's trust to technology security or privacy, while Discomfort represents consumer's anxiety in technical terms [15]. In practice, each dimension illustrated person with high optimism will convince that the new technology will be useful to increase the productivity of users. The person with high innovativeness is happy to try experiments with new technology. The person with high insecurity feels doubt about the capability of new technology to fulfill the complete transaction and the person who discomfort think that the system was not suitable for them and thus they are unpleasant [15].

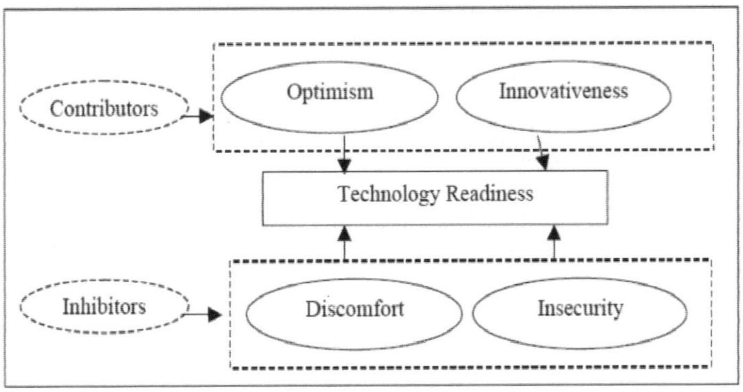

Figure 1. Technology Readiness Index (TRI) Model [15]

Figure 1 shows that people readiness of technology affected by personality traits. TRI indicates people faith or conviction toward technology and not determine the capacity of people to adopt it. Based on this concept, the people overall belief about technology could be theorized which also indicate their usage of technology-based product and services [15]. Based on technology readiness scores, there are five categories of user: explorers, pioneers, skeptics, paranoids, and laggards. Type explorers are dominant in technology readiness because they have optimism and innovativeness and on the other hand, they only have little discomfort and insecurity. Explorers are quickly attracted to the existence of the latest technology and usually being a leader to try the new technology. In contrast, laggards are the last group of adopters of latest technology because they tend inhibitors and weakness in contributors factors.

Other categories of user like pioneers, skeptics, and paranoids show more variant in technology perceived. The type of pioneers with dominant optimism and innovation like explorers, but they are weak in discomfort and insecurity. Sceptics indicated low confidence and innovativeness and showed small inhibiting level that needs to be convinced in advance about the benefits of using technology. On the other hand, paranoids with high optimism and interested in new technology but in the same they feel discomfort and insecure [17].

Table 1. Characteristic of Technology Segmentation

Technology Segments	Optimism	Innovativeness	Discomfort	Insecurity
Explorers	High	High	Low	Low
Pioneers	High	High	High	High
Sceptics	Low	Low	Low	Low
Paranoids	High	Low	High	High
Laggards	Low	Low	High	High

The Parasuraman & Colby shows that types of explorers and pioneers liable to use new technologies prior than other types [16].

2.2. TAM (Technology Acceptance Model)

Another behavioral theory that well known to predict user behavior in adopting technology is TAM. TAM was extensively used especially in the domain of Information Systems (IS/IT) in purpose to obtain a more comprehensive perspective and a better explanation of the process of user acceptance of technology [18]. TAM concept offers a simple yet powerful explanation related to technology acceptance and usage behavior [11].

The main factor in the model of TAM have perceived usefulness and perceived ease of use of technology as a reasonable act in the context of technology users, Thus, the strong reason why people use technology influenced by perceived usefulness and perceived ease of use. Based on Davis (1989), the level of IT utilization by the user will be largely determined by the level of user acceptance, while user acceptance can be predicted from perceived usefulness and perceived ease of use [11]. Based on Davis (1989), both of factors could explain the behavioral aspects of users and found significantly influence user acceptance of the technology. These findings were also supported by other studies [19][20][21].

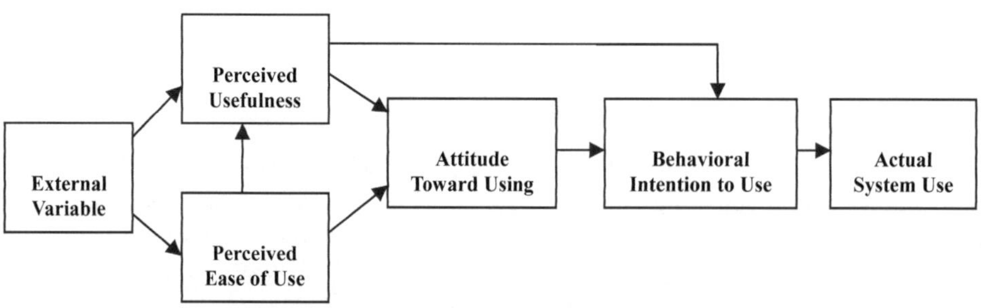

Figure 2. Original Model of TAM [11]

According to figure 2 above, it could be seen that there is five construct in original TAM model including perceived usefulness, perceived ease of use, attitude toward using, behavioral intention and actual system use. Initially, TAM incorporated attitude toward using variable, but later expelled it because of its weak effect in mediating the relation between behavioural intention to use and user belief [22]. In its development, TAM has become a widely used model for predicting attitudes, desires, and behaviors in the use of new technologies. This model explains about 40% of the variance of individual willingness to use information technology [22]. Perceived usefulness is influenced by perceived ease of use while both beliefs factor determined by external variables, such as people trust, facilitating condition, social influence, gender, age, etc. [22][23].

Although the perceived of usefulness and perceived of ease of use has been widely proven as a factor affecting behavior intention, but the relative strength of the two variables is different where the perceived of usefulness is considered as the primary factor determinant of behavior and has higher explanatory power than the perceived of ease of use. It can be possible because the primary consideration of individuals to use a system/technology is usefulness, while the degree of comfort in using it will be considered later [11][23]. Nowadays, TAM had been widely used in many fields to explore user behavior in using new technology. In practically, TAM could measure the level of user's technology adoption [23][24][25].

3. Research Methodology

Model TRI evaluates people readiness to adopt technology based on personality traits factor; they are optimism, innovativeness, discomfort and insecurity factor. Parasuraman (2000) said that a person with explorer type tend to adopt the latest technology since they have high TR. In other words, this kind of people has high optimism and innovativeness and low discomfort and insecurity [15]. Surprisingly evidence shows users that have current technology readiness did not use the kind of technology offered [25][26]. Based on evidence from the fieldwork indicates that TRI model could not satisfactorily explain why do specific individual adopt new technologies whereas other's don't? In other words, TR is incapable of defining user adoption process of technology, and it is vital for the organization that provides technology-based products or services. Thus, an effort had been made by some researchers to improve TRI model based on hypothesizing that the personality traits in TRI model were antecedents to technology acceptance model. In other words, psychological characteristics influenced the cognitive dimensions of TAM [23][27]. Lin (2007) try to construct an integrated TRI and TAM model called TRAM (Technology Readiness Acceptance Model) to explain the user behavior in adopting technology-based e-services by combining TR with the two dimension of TAM, perceived usefulness and perceived ease of use [24]. Basgoze (2015) also connected or integrated TR into TAM in the context of mobile shopping intention [25]. Walczuch (2007) incorporate TRI and TAM to measure the relation between both models [28]. Among the existing acceptance model, TAM (Davis, 1989) become the most popular best practice model since it had robust and parsimony measurement model [24][25].

According to Davis (1989), TAM model is straightforward but powerful parsimony model. If we observed further, unfortunately, TAM model also has some limitation such as not taking into account of social influence and condition of facilities that encourage user behavior in using technology [11]. In another word, TAM does not take into consideration external variables of the system such as social influence, facilitating condition, etc. Also, TAM model used to measure user behavior only in the voluntary environment, not a mandatory one. In the optional use of information technology, the measure of success is based on user acceptance. The end user has full freedom whether the user will use or leave the technology [29][30]. On the other hand, the indicator of success in the mandatory environment is user satisfaction. Thus, TAM model could not explain user behavior of mandatory technology. The other limitation of TAM models is explicitly designed only for a particular system (system-specific) not for technology beliefs (individual-specific) likely in TR model. Therefore, Susskind (2004) suggested supporting the finding to this research and extending TAM by considering TR model into TRAM model. Further, TRAM model indicated why people who have high TR do not certainly use new technology because they influenced two main constructs in TAM that determine adopting technology behavior [24].

Based on the previous research, the conceptual model used in this research refer to TRAM (Technology Readiness Acceptance Model) had been developed by other researchers as follow [24][25][28]. Thus, there are nine hypotheses proposed in this research as follows:

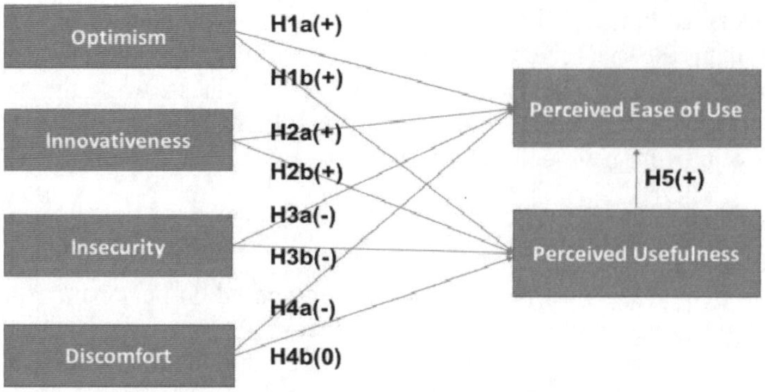

Figure 3. Conceptual Research Model (TRAM)

H1a : Optimism (OPT) significantly positive affects the Perceived Ease of Use (PEOU)
H1b : Optimism (OPT) significantly positive affects the Perceived Usefulness(PU)
H2a : Innovativeness (INN) significantly positive affects the Perceived Ease of Use (PEOU)
H2b: Innovativeness (INN) significantly positive affects the Perceived Usefulness(PU)
H3a : Insecurity (INS) significantly negative affects the Perceived Ease of Use (PEOU)
H3b : Insecurity (INS) significantly negative affects the Perceived Usefulness(PU)
H4a : Discomfort (DIS) significantly negative affects the Perceived Ease of Use (PEOU)
H4b : Discomfort (DIS) does not significantly affect the Perceived Usefulness(PU)
H5 : Perceived Ease of Use (PEOU) significantly positive affects the Perceived Usefulness (PU)

The online survey was conducted to distribute the questionnaire to consumers or users that have been using single technology in the context of e-Government services. Our experiment was for individual's own choice of a software system or technology. Thus the participant was not asked to rate a specific technology. Instead, the participants were asked to select technology they use most related to e-Government services. The questionnaire consist of total 48 item of measurement, where TRI has 36 issues (10 questions for optimism; 7 pieces for innovativeness; 10 items for discomfort; 9 items for insecurity) adapted from Parasuraman (2000) and TAM (6 issues for perceived usefulness; 6 items for perceived ease of use) has 12 things [21]. The questionnaire was designed in 5 Likert scales, where 1=" strongly disagree" and 5="strongly agree". This study collected a total 230 respondents, but the completed questionnaire only 208 data that could be analyzed further (respond rate 90,4%). The respondents were public as e-Government service users that chosen by purposive sampling.

4. Result and Discussion

Initially, the instrument was examined in term of their validity and reliability. According to Sugiyono (2007), an instrument is said to be valid means the measuring tool used to get the data (measure) is valid. Reliable instruments are instruments which, when used multiple times to measure the same object, will produce consistent data. In this research, the validity test was conducted by calculating Pearson Product-Moment Correlations between measurements of each variable.

The Pearson Correlation coefficient showed that all item met the minimum requirement (>0.396 for 208 respondent). Thus the object is said to be valid. Regarding reliability, this study used Cronbach's alpha for each variable of item's scale (Ghozali, 2002). The result of reliability test showed that each variable also met the requirement where Cronbach's alpha coefficient exceed 0.60 (>0.60). Thus, the instrument had good validity and reliability in their scale.

The second step is hypotheses testing based on regression analysis using SPSS tool of IBM 22 version 2013. As explain in research methodology, there are nine hypotheses related to TRAM model being tested in the context of e-Government services. These hypotheses divided into 3 model of regression as follows:

$$Y1 = a + b1X1 + b2X2 + b3X3 + b4X4 + e \quad (1)$$

$$Y2 = a + b1X1 + b2X2 + b3X3 + b4X4 + e \quad (2)$$

$$Y2 = a + b1Y1 + e \quad (3)$$

Where :

Y1 = Perceived Ease of Use (PEOU)
Y2 = Perceived Usefulness (PU)
X1 = Optimism (OPT)
X2 = Innovativeness (INN)
X3 = Insecurity (INS)
X4 = Discomfort (DIS)
a = The direct link between independent and dependent variables
e = Error Residue

The first model tested was regression model of Y1, where each variable of TR (Optimism, Innovativeness, Insecurity, and Discomfort) toward one of TAM variable, Perceived Ease of Use, as follows:

Table 2. Regression Model of Perceived Ease of Use (Y1)

		Unstandardized Coefficients		Standardized Coefficients		
Model		B	Std. Error	Beta	t	Sig.
1	(Constant)	1.459	.559		2.611	.011
	Optimism	.550	.074	.589	7.465	.000
	Innovative	.298	.091	.296	3.281	.001
	Insecurity	-.194	.090	-.193	-2.167	.033
	Discomfort	-.020	.089	-.019	-.224	.823

a. Dependent Variable: PEOU

Based on table 2 above can be produced by multiple linear regression equation:

$$Y1 = 1.459 + 0.550X1 + 0.298X2 - 0.194X3 - 0.20X4 \quad (4)$$

Based on the t-Test conducted, we could see that hypothesis H1a is supported in this study because Optimism had significantly positive influenced Perceive Ease of Use. The result showed positive coefficient regression (0.550), and the probability of significant was 0.000 (<0.05) with t-value 7.465 (>t table). Hypothesis H2a is also supported in this study since Innovative had significantly positive influenced Perceived Ease of Use. The coefficient regression obtained was positive (0.298) and the probability of significant 0.001 (<0.05) with t-value 3.281 (>t table).

Hypothesis H3a is accepted in this study because Insecurity had significantly negative influenced Perceived Ease of Use. The result showed that negative coefficient regression (-0.194) and the probability of significant was 0.033 (<0.05) with t-value -2.167 (>t table). On the other hand, Hypothesis H4a is not supported in this study since Discomfort had no significantly influenced Perceived Ease of Use. The coefficient regression obtained was negative (-0.20) and the probability of significant 0.823 (>0.05) with t-value -0.224 (<t table).

The second model tested was regression model of Y2, where each of TR (Optimism, Innovativeness, Insecurity, and Discomfort) toward TAM variable, Perceived Usefulness, as follows :

Table 3. Regression Model of Perceived Usefulness (Y2)

	Unstandardized Coefficients		Standardized Coefficients	t	
Model	B	Std. Error	Beta		Sig.
1 (Constant)	1.195	.686		1.744	.089
Optimism	.761	.110	.693	6.909	.000
Innovative	.312	.135	.310	2.310	.026
Insecurity	-.240	.116	-.256	-2.074	.044
Discomfort	-.097	.105	-.103	-.919	.363

a. Dependent Variable: PU

Based on Table 3 above can be produced by multiple linear regression equation:

$$Y2 = 1.195 + 0.761X1 + 0.312X2 - 0.240X3 - 0.097X4 \quad (5)$$

Based on the t-Test conducted, we could see that hypothesis H1b is supported in this study because Optimism had significantly positive influenced Perceive Usefulness. The result showed positive coefficient regression (0.761), and the probability of significant was 0.000 (<0.05) with t-value 6.909 (>t table). Hypothesis H2b is also supported in this study since Innovative had significantly positive influenced Perceived Usefulness. The coefficient regression obtained was positive (0.312) and the probability of significant 0.026 (<0.05) with t-value 2.310 (>t table). Hypothesis H3b is supported in this study because Insecurity had significantly negative influenced Perceived Usefulness. The result showed that negative coefficient regression (-0.240) and the probability of significant was 0.044 (<0.05) with t-value -2.074 (>t table). Besides, hypothesis H4b is also supported in this study since Discomfort had no significantly influenced Perceived Usefulness. The coefficient regression obtained was negative (-0.097) and the probability of significant 0.363 (>0.05) with t-value -0.919 (<t table).

The result indicated that the Optimism also had higher coefficient regression than another dimension correlated to Perceived Usefulness. It means Optimism variable was the most dominant variable in TR model among others. The last model tested was regression model of Y1 toward Y2 as follows :

Table 4. Regression Model of Perceived Ease of Use Toward Perceived Usefulness

	Unstandardized Coefficients		Standardized Coefficients	t	
Model	B	Std. Error	Beta		Sig.
1 (Constant)	1.843	.610		3.020	.004
PEOU	.563	.145	.506	3.889	.000

a. Dependent Variable: PU

Based on Table 4 above can be produced by multiple linear regression equations:

$$Y2 = 1.843 + 0.563Y1 \quad (6)$$

Based on the t-Test conducted, we could see that hypothesis H5 is supported in this study because Perceived Ease of Use (PEOU) had significantly positive influenced Perceive Usefulness (PU). The result showed positive coefficient regression (0.563), and the probability of significant was 0.000 (<0.05) with t-value 3.889 (>t table).

Besides, the result indicated the personality traits in TR significantly influenced technology use or adoption. It can be proved by empirically that the TRI is an antecedent to TAM model. The result also indicates that the optimism had a higher coefficient than other dimensions. It means optimism variable is the most dominant variable in TR model among others. In practical implication, high confidence indicates user's belief about technology was positive that the new technology would increase the productivity of them. They convince that the technology could offer more control, flexibility, and efficiency in their lives. Innovativeness also significantly positive influenced the perceived ease of use and perceived usefulness. It means the users in institution X was critical to the new technology. They are aware of the newest development of technology and is happy to try experiments with new technology first than others. Insecurity negatively influenced the perceived ease of use and perceived usefulness, as predicted. Insecure users have low trust to technology security or privacy. Discomfort had no impact on perceived ease of use, not expected finding. The possible explanation is that the technology users in X Government Institution has no doubt or hesitate to use technology. They are not feeling of being overwhelmed by it. Finally, perceived ease of use also positively affected perceived usefulness, as predicted. This finding was supported by previous research [21]. It indicated that technology users would consider the ease of use factor in adopting new technology.

5. Conclusion

Based on the result and analysis obtained that the personality traits of TR (Technology Readiness) significantly impacted the cognitive dimension of TAM (Technology Acceptance Model), the perceived ease of use and the perceived usefulness. The hypotheses were supported that TRI was antecedents to TAM Model. Therefore, TRAM model is proven by empirically could indicate user adoption in technology satisfactorily. Furthermore, TRAM model could be used to explain why do certain individual adopt new technologies whereas other's don't because TRAM not only considers for a particular system (system-specific) but also include technology beliefs (individual-specific). According to Parasuraman and Colby (2001), optimism and innovativeness represent technological facilities inducers; they induce people to embrace new technologies. On the other hand, discomfort and insecurity work as inhibitors, they demotivate or postpone new technologies embracement. Even though they coexist inside us, the inducing and inhibiting dimensions of technological facilities act separately, and each user may show different inducing or inhibiting combinations. Thus, the manager should learn and improve the TR of users as a public customer to gain technology adoption successfully in an organization.

Acknowledgments

We would like to thank the research institution that has supported the research activities was being carried out properly.

References

[1] Yildiz M 2007 E-Government research: Reviewing the literature, limitations, and ways forward *Gov. Inf. Quarterly* **24** 646-65
[2] Dhami B 2010 E-Governance in Nepal: Prospects and challenges in implementation *Master Thesis* (Corvinus University of Budapest)
[3] Ghimire B E-Governance and its role in service delivery 2013 *The Nepalese J. of Public Administration* (Ministry of General Administration)
[4] UN United Nations E-Government Survey 2014: E-Government for The Future We Want 2014 United Nations Department of Economic And Social Affairs
[5] Ibrahim J S and Taib C A and Shahzad A 2016 The behavioral Intention of citizen of Nigeria on E-Government Service by Utilization of The Unified Theory of Acceptance and Use of Technology (UTAUT) *Sains Humanika* **8** 41-6
[6] Heeks R 2008 ICT4D 2.0: The next phase of applying ICT for international development 2008 *Computer* **41** 26-33
[7] Munyoka W and Manzira F M 2013 Alignment of E-Government policy formulation with practical implementation: The Case of sub-saharan Africa *Int. J. of Social, Behavioural, Educational, Economic, Business and Industrial Eng.* **7** 3034-39

[8] Mohamadali N K and Garibaldi J M 2010 A novel evaluation model of user acceptance of software technology in health-care sector *Proc. from the Int. Conf. on Health Informatics* 392–97

[9] Seddon P and Kiew M Y 1996 A partial test and development of DeLone and Mclean's model of IS success *J. Inf. Syst* **4** 90–109

[10] Nguyen L and Bellucci L 2014 Electronic health records implementation: an evaluation of information system impact and contingency factors *J. Med. Inf.* **83** 779–96

[11] Davis F D 1989 Perceived usefulness, perceived ease of use, and user acceptance of information technology *MIS Quarterly* **13** 319-39

[12] Irani Z, Love P E, Elliman T, Jones S and Themistocleous M 2005 Evaluating E-Government: Learning from the Experiences of Two UK Local Authorities *Inf. Syst. J.* **15** 61-82

[13] Hakim S 2009 Evaluasi Tingkat Penerimaan Sistem Informasi Manajemen Kepegawaian (SIMPEG) Pada Lembaga Ilmu Pengetahuan Indonesia (LIPI) Universitas Gadjah Mada

[14] Esmaeilzadeh P, Sambasivan M and Nezakati H 2012 The limitations of using the existing tam in adoption of clinical decision support system in hospitals: an empirical study in Malaysia *Int. J. Res. Bus. Soc. Sci.* **3** 56–68

[15] Parasuraman A 2000 Technology readiness index (TRI): a multiple-item scale to measure readiness to embrace new technologies *J. of Service Research* **2** 307–20

[16] Parasuraman and Colby C L 2001 Techno-ready marketing: how and why your customers adopt technology *Free Press, New York*

[17] Angelidis V P and Chatzoglou P D 2009 Using a modified technology acceptance model in hospitals *Int. J. Med. Inf.* **78** 115–26

[18] Legris P, Ingham J and Collerette P 2003 Why do people use information technology? A critical review of the technology acceptance model *Inf. & Manag.* **40** 191–204

[19] Kim J 2005 User acceptance of web-based subscription database *Dissertation of Doctor, Tallahassee Florida*: Information Studies Department Florida State University

[20] Hong W, Thong J Y, Wong W and Tam K Y 2004 Determinants of user acceptance of digital libraries: An empirical examination of individual differences and system characteristics *J. of Manag. Inf. Sys.* **18**

[21] Putra G 2008 Kajian minat berperilaku aparatur Pemerintah Kota dalam menggunakan E-Government: Aplikasi Technology Acceptance Model (TAM) *National Conf. on Management Research* ISBN 979-442-242-8

[22] Venkatesh V, Morris M G, Davis G B and Davis F D 2003 User acceptance of information technology: toward a unified view *MIS Quarterly* **27** 425-78

[23] Karahanna E and Straub D W 1999 The psychological origins of perceived usefulness and ease-of-use *Inf. & Manag.* **35** 237–50

[24] Lin C H, Shih H Y and Sher P J 2007 Integrating technology readiness into technology acceptance: The TRAM model *Psychology & Marketing* **24** 641-57

[25] Basgoze P 2015 Integration of Technology Readiness (TR) into the Technology Acceptance Model (TAM) for m-shopping *Int. J. of Scientific Research and Innovative Tech.* **2** 26–35

[26] Tsikritis N 2004 A technology readiness-based taxonomy of customers *J. of Service Research* **7** 42–52

[27] Agarwal R and Prasad J A 1998 conceptual and operational definition of personal innovativeness in the domain of information technology *Information Systems Research.* **9** 204–15

[28] Walczuch R, Lemmink J and Streukens S 2007 The effect of service employees' technology readiness on technology acceptance *Inf & Manag* **44** 206-15

[29] Napitupulu D 2017 Validity testing of technology acceptance model based on factor analysis approach *Indonesian Journal of Electrical Engineering and Computer Science* **5** 697-704

[30] Napitupulu D 2016 Evaluasi kualitas Website Universitas XYZ dengan pendekatan webqual *Bulletin Pos & Telekomunikasi* **14** 51-64

IOP Conf. Series: Materials Science and Engineering **352** (2018) 012058 doi:10.1088/1757-899X/352/1/012058

Content validity of critical success factors for e-Government implementation in Indonesia

D Napitupulu[1,2], M Syafrullah[2], R Rahim[3], A Amar[4], YG Sucahyo[5]
[1]Research Center for Quality System & Testing Technology,
Indonesian Institute of Sciences, Tangerang Selatan, Banten, Indonesia
[2]Program Studi Magister Ilmu Komputer, Universitas Budi Luhur
Ciledug Raya, Petukangan Utara, Kebayoran Lama, Jakarta Selatan
[3]School of Computer and Communication Engineering, Universiti Malaysia Perlis, Malaysia
[4]Department of Statistics, Universitas Negeri Makassar, Makassar, Indonesia
[5]Computer Science Faculty, Universitas Indonesia, Depok, Jawa Barat

E-mail: darm003@lipi.go.id, mohammad.syafrullah@budiluhur.ac.id, usurobbi85@zoho.com, ansarisaleh@unm.ac.id, yudho@cs.ui.ac.id

Abstract. The purpose of this research is to validate the Critical Success Factors (CSFs) of e-Government implementation in Indonesia. The e-Government initiative conducted only to obey the regulation but ignoring the quality. Defining CSFs will help government agencies to avoid failure of e-Government projects. A survey with the questionnaire was used to validate the item of CSF based on expert judgment through two round of Delphi. The result showed from 67 subjects in instrument tested; there are 11 invalid items deleted and remain only 56 items that had good content validity and internal reliability. Therefore, all 56 CSFs should be adopted by government agencies in Indonesia to support e-Government implementation.

1. Introduction

Nowadays, many organizations including government agencies have chosen to utilize Information and Communication Technology (ICT) to succeed in an increasingly dynamic, demanding and complex world. Furthermore, they have seen expectations for value and quality of services rise from their various stakeholders [1]. In response, they have invested in ICT solutions in term of e-Government, with the hope of fulfilling the expectations of their stakeholders. Successful implementation of e-Government system is believed in gaining many advantages such as (1) efficiency in the form of cost reduction; (2) enhanced service quality to stakeholder; (3) transparency, accountability, democracy; and (4) competitive advantage gaining, many government agencies around the world are expending a significant amount of resource [2][3][4][5][6]. As an illustration, the central government in the United Kingdom (UK) spent on ICT in 2007/2008 had achieved £3.2 billion and estimated to expend approximately £4.2 billion in year 2010/2011 [7]. Other researcher reported that about US$4.2 billion in 2004 and US$5.8 billion in 2009 was spent by United States (US) government. Almost the same as that, about US$2.3 billion invested for ICT by Russian Federation to develop their e-government in 2006 [8].

Despite the massive resource allocated for government's ICTs expenditures, a different disparity of e-Government implementation occurs among countries. Through the last decade, the universal movement for E-government was supported. Since 2003, the world has gradually enhanced their e-government development where European countries are increasing in global classification especially till 2012. Asia as a complete progressed at a leaner rate till 2010 – approximately according to the advances with the global average – and after that go ahead. Based on the progress of e-Government for the period 2003-2012, e-government in Africa was slow implemented and said to be lack of ICT readiness among others [8][9].

Table 1. World Top Ten of UN e-Government Survey [8]

No	Country	Index Value
1.	Republic of Korea	0.9283
2.	Netherlands	0.9125
3.	United Kingdom	0.8960
4.	Denmark	0.8889
5.	United States	0.8687
6.	France	0.8635
7.	Sweden	0.8599
8.	Norway	0.8593
9.	Finland	0.8505
10.	Singapore	0.8474

There are 192 countries in the global classification of e-Government survey [8]. In fact, some countries have high position significantly more than others. South Korea is the best world country which has e-government index in (0.9283), at second rank comes the Netherlands with (0.9125), the United Kingdom with (0.8960) at third position, followed by Denmark (0.8889) at fourth position, United States with (0.8687) followed by France, Sweden Norway, Finland, Singapore, Canada and other as illustrated at Table 1 above. These top rank developed countries have achieved ideal maturity of e-Government development according to UN survey [8]. What about developing countries? It is known that there is significant gap between countries around the world in adopting e-Government such as infrastructure, human resources, culture, e-participation, computer literacy, etc. [10]. For example, the developed country had already excellent ICT infrastructure and supporting high-speed internet access for the public domain. On the other hand, many developing countries do not have adequate infrastructure to build e-government system in their area. Low ICT infrastructures and internet access became problems classically in most developing countries. Another big challenge in e-Government implementation is human resources, where developed countries had many qualified HR or expert staff who able to define requirements or needs in e-Government system. Further, the one is also lack of well-skilled and trained people that could bring the initiative of e-Government implementation [11][12].

According to Heeks (2008), level of failure implementation of e-Government was quite high until 85%, especially in developing countries. The rate of successful e-Government projects only 15% while 85% of one was failed [13][36]. Similarly, Kettani & Moulin (2014) also reported that e-Government projects tend to fail especially for developing countries (i.e. India, China and Bangladesh) [14]. The two main characteristics or criteria that must be fulfilled in the e-Government system, they are availability and accessibility [15]. The information and services could be provided every time user need it. Users are free to choose at any time concerned want to deal with the government to carry out various transactions or mechanisms of interaction. It allows communities and businesses with the flexibility to access government services outside working hours. Second, e-Government success is highly determined by the accessibility of services available on the website. If the e-Government services or transactions could not be accessed or available, it could be said of e-Government will not succeed or fail.

In the initial stage of e-Government implementation, people awareness become a most important thing [4]. According to Delon & McLean (2003), the success of technology depends on how far a technology being adopted by users [16]. In another word, e-government is said to be successful when it has been utilized optimally by its users. Many factors influence users adoption to achieve success e-Government. In this research, that kind of elements called Critical Success Factors (CSFs). Al-Kaabi (2010) had proposed some CSFs derived from best practice, specially developed country. Defining CSFs particularly during e-Government development could enable organizations to avoid of a fail e-Government implementation [17]. Heeks (2008) also argued that is critical to identify the factors of e-Government success. In this paper, we will validate a previous research a list of CSFs derived from literature study. This research's purpose is to get a big picture of CSFs that should be accommodated by government agencies in developing country especially Indonesia [13][18].

2. Literature Review

2.1. E-Government in developing country

The gap of e-Government field between developed and developing countries is vast. Most developing countries have limited resources, i.e., bad ICT infrastructure, and must encounter so many challenges as inhibit factors such as social & cultural factors, human resources factors, economic factors, political factors, etc. [10]. Thus, developing countries should overcome the barriers during e-Government development both internal and external factors.

UN (2001) reported the impediments factors of e-Government implementation as shown in Table 2 below [19]:

Table 2. Factors impeding e-Government in developing country [19]

Core Factors	Symptoms	Consequences
Institutional Weakness	Insufficient planning Unclear objectives	Inadequately design system Cost over-runs
Human Resources	Shortage of qualifies personnel Lack of professional training	Insufficient support Isolation from sources of technology
Funding arrangements	Underestimated project cost Lack of recurring expenditure	Unfinished projects Higher maintenance costs
Local environment	Lack of vendor representation	Lack of qualified technical support
	Lack of backup system/parts	Implementation problem
Technology and Information changes	Limited hardware/software Inappropriate software	System incompatibility Over-reliance on customer applicatiom

The e-Government success issue is determined by ICT capability and readiness of government [20]. One of the capabilities of government should be supported their affordability to have ICT budgets and expenditures for e-Government development. To bring good illustration of budgeting, Table 3 show estimation of ICT budgets in few developed and developing countries as follows:

Table 3. Government ICT budgets for developed and developing countries [21]

Government	India	Philippines	Hongkong	USA	UK
IT Spending in Million US $	556	120	250	52.000	10.000
IT Spending as % of Budget	2-3	0.8	1.2	13	2.4
E-Government Spending as Part of IT Budget (%)	34	75	-	0.08	5.2

To gain the large advantages of e-Government adoption, financial and budgeting issues should be concerned. Developing e-governments system should be in line with ICT infrastructure development that functions to provide and increase the performance of e-Government. In contrast, lack of ICT infrastructure to develop the initiative of e-government is still become the main problem in developing country. One of researcher had studied the conditions of ICT infrastructure in some developing countries as shown in Table 4 [22].

Table 4. Overview ICT Infrastructure in developing countries [22]

Government	India	Bangladesh	Thailand	Malaysia	China	Philippines
Fixed Lines and Mobile Telephones (per 1000 people)	35.5	5.0	142.6	412.3	177.6	124.4
Personal Computers (per 1000 people)	4.5	1.5	24.3	103.1	15.9	19.3
Internet Users	5.0 M	100.000	2.3 M	3.7 M	22.5 M	2.0 M

Table 4 shows the conditions of main ICT infrastructure such as fixed lines and mobile telephones, PC and also penetration of Internet. It can be seen that there is kind of difference of ICT Infrastructure among developing countries. The relationship of Internet Access and its Penetration including access cost in some developed and developing countries could be illustrated in Table 5.

Table 5. Internet Access and Penetration [23]

Correlation of Internet VS Penetration (based on 20 hours of monthly access)		
Country	Access Cost (% of GDP per capita)	Internet Users per 1000
India	16.82	6
South Africa	5.26	35
Malaysia	4.85	90
Singapore	0.84	302
UK	0.91	554

Table 5 shows that internet access cost has a positive correlation to the rate of internet users in developing countries. There are gaps between countries that have a lower cost of Internet usage and the countries that have a higher cost of Internet resulted in the number of Internet users [23].

3. Research Methodology

This research evaluated the significance of success factors (CSFs) proposed through its validity [24]. The efficacy will be focused in this study is content validity, defined as the representation of content that should be to evaluate [25]. Content validity should be conducted in the instrument development stage and also judgmental one [26]. Developing instrument had the purpose of understanding the construct that is being measured. The constructs could be obtained from qualitative ways such as literature reviews, interviews, and focus groups. Selecting the domain of constructs will bring undoubtedly research variables, scope and elements of the subject could be obtained.

In the other hand, the instrument judgment stage is based on expert opinion that surveyed with a questionnaire in quantitative ways [27]. In this case, during testing, the validity of an item, an expert gave opinion or agreement according to the measurement item that is being assessed. Content validity is determined by professional or expert and assessed through an expert agreement, not by the researcher [28]. In another word, content validity shows the extent of experts agreement toward item or construct in the instrument being assessed. Involvement of at least five to ten experts in the same domain was valuable to evaluate each item or construct of the instrument [27]. Conducting the content validity become the main activity in testing instrument had been designed. Therefore, a survey method needed to explore expert agreement or opinion about CSFs of e-Government as an item proposed in the instrument. A total number of ten experts suggested had been involved in evaluating the significance of each CSFs of e-Government in Indonesia [27]. The expert's criteria determined was having expertise and experiences in e-Government field. They made their rating on a five-point scale (Likert) from 1 means strongly disagree to 5 means strongly agree for each item in the instrument. The demography of respondents can be shown in Table 6 as follows:

Table 6. The Demographic of Experts

No	Background	Gender	Education Level
1.	Academician	Male	Professor
2.	Academician	Male	Ph.D
3.	Academician	Male	Ph.D
4.	Academician	Male	Ph.D
5.	Academician & Practician	Male	Ph.D
6.	Academician & Practician	Male	Ph.D
7.	Practician	Female	Master
8.	Academician & Practician	Male	Master
9.	Practician	Male	Master
10.	Practician	Male	Master

Based on expert comment or rating quantitatively, the statistical approach used to measure the extent of instrument validity. In this research, we calculated the index following Aiken model (1980, 1985) that widely used in validating items of the instrument [28][29][30][31][33].

The extent of agreement between experts indicated the significance of items, and it was calculated symbolized by V coefficient. V coefficient based on Aiken was formulated [31][32]:

$$Vj = Sj / [\, n\,(c-1)\,] \tag{1}$$

$$Sj = \Sigma rij - 1 \tag{2}$$

With n is the number of experts. The coefficient of V had a value from lowest 0 to maximum of 1. According to Aiken Table [31][33], content validity index (V) required of the item is significant if above the cut off value 0.70 (V>0.70). It means if the validity index of the item below 0.70 (V<0.70), an item doesn't have a good content validity or not significant [29][30][31][34]. In the other hand, V Coefficient was confirmed by the extent of consistency of items namely homogeneity reliability. Homogeneity reliability was calculated symboliezed by H coefficient. H coefficient based on Aiken was formulated [29][30][31][32]:

$$Hj = 1 - 4S_j/(c-1)(n^2-k) \tag{3}$$

With n is some experts. In another side, some researchers proposed Cohen's kappa introduced by Jacob Cohen (1960), as a robust method for nominal scale collecting data comparing to Aiken [34]. In this study, we use Aiken since we collect and analyze data based on five points scale.

IOP Conf. Series: Materials Science and Engineering **352** (2018) 012058 doi:10.1088/1757-899X/352/1/012058

4. Result And Discussions

Based on Aiken's formula described above, the content validity coefficient (V) measured to indicate how significant each item in the instrument. As said, the content validity is conducted by expert judgment and rely on an expert agreement about significant of items in the instrument proposed. In this research, exploring the expert judgment was conducted using Delphi method. The purpose of Delphi is to combine experts opinion to seek convergence on the specific problem [35]. The result of item validity from Delphi 1st round could be shown in Table 7 as follows:

Table 7. Content Validity of Item (Delphi 1st Round)

No	Item (CSF)	V Coefficient
1.	Participation of User & Stakeholder	0.925
2.	Project Plan	0.825
3.	System Accessibility	0.800
4.	Regular Training	0.675
5.	Ease of Use	0.875
6.	Website Promotion	0.525
7.	Pilot Project	0.675
8.	Skills and Expertise	0.750
9.	E-Leadership	0.850
10.	Project Coordination	0.900
11.	Clear Guidance	0.725
12.	Funding Continuity	0.850
13.	Business Process Reengineering	0.675
14.	E-Government Policy and Regulation	0.825
15.	Stable Government	0.650
16.	Outsourcing strategy	0.525
17.	Basic Infrastructure of ICT	0.850
18.	ICT Literacy	0.725
19.	Organizational Structure	0.675
20.	International Cooperation	0.375
21.	Privacy & Security	0.775
22.	Usefulness	0.825
23.	Monitoring and evaluation	0.900
24.	Private Partnership	0.500
25.	Change Management Strategy	0.800
26.	Socio-Cultural	0.675
27.	System Modeling	0.600
28.	Top Management Support	0.950
29.	System Actual Usage	0.775
30.	Citizen Relationship Management	0.675
31.	Compatibility	0.775
32.	Project Management	0.725
33.	Information Quality	0.825
34.	System Quality	0.825
35.	Service Reliability	0.900
36.	Trust	0.850
37.	Awareness	0.800
38.	ICT Governance	0.850
39.	Public Satisfaction	0.775
40.	Methodology or Approach	0.700
41.	E-Transaction and E-Payment	0.650
42.	User-Friendly	0.750
43.	Gradual Implementation	0.725
44.	Re-Usable	0.700
45.	Continuous Improvement	0.825
46.	Service Innovation	0.800
47.	Loyalty	0.750
48.	Acknowledgement	0.650
49.	Public Intention to Use	0.700
50.	Sustainable Revenue	0.475
51.	E-Participation	0.675
52.	Roadmap	0.725
53.	Market Sinergy	0.600
54.	Political Pressure	0.675
55.	Inter-Governmental Relationship	0.800

56.	Tools and Equipment	0.625
57.	E-Initiative	0.700
58.	Vision	0.875
59.	Citizen Empowerment	0.725
60.	Knowledge Management	0.750
61.	Service Guarantee	0.825
62.	Personalization of Service	0.575
63.	Empathy	0.600
64.	Flexibility of Technology	0.750
65.	Alignment of Organization Goal and ICT direction	0.800
66.	Good Responsiveness	0.750
67.	E-Democracy	0.675

There are 21 items had validity coefficient (V) below the cut off value (V<0.70) so those items (highlighted) was removed from the list and would be included in Delphi 2nd round to obtain the conformity from experts about the significance of items.

Table 8. Content Validity of Item (Dephi 2nd Round)

No	Item (CSF)	V Coefficient
1.	Regular Training	0.825
2.	Website Promotion	0.600
3.	Pilot Project	0.575
4.	Business Process Reengineering	0.725
5.	Stable Government	0.625
6.	Outsourcing strategy	0.600
7.	Organizational Structure	0.750
8.	International Cooperation	0.425
9.	Private Partnership	0.525
10.	Socio-Cultural	0.725
11.	System Modeling	0.575
12.	Citizen Relationship Management	0.750
13.	Acknowledgement	0.725
14.	Sustainable Revenue	0.750
15.	E-Participation	0.775
16.	Market Sinergy	0.600
17.	Political Pressure	0.675
18.	Tools and Equipment	0.575
19.	Personalization of Service	0.600
20.	Empathy	0.750
21.	E-Democracy	0.725

Based on Table 8 from 21 items asked, there are 11 items was still invalid based on expert judgment. Therefore, it already showed a stable result of 11 items had no significance of content validity based on Delphi 1st and 2nd round. Thus, total 11 item eliminated from the instrument because they could not meet the minimum requirement (V<0.70). The total 11 items are website promotion, pilot project, stable government, outsourcing strategy, international cooperation, private partnership, system modeling, market synergy, political pressure, tools & equipment and personalization of service.

The content validity will be supported by the level of the constancy of the items called homogeneity reliability. After eliminating 11 invalid items, we got only 56 items that already had good content validity. Therefore, all 56 items would be tested and computed into the homogeneity reliability coefficient (H) as shown in Table 9.

Table 9. Homogeneity Reliability of Item

No	Item (CSF)	H Coefficient
1.	Participation of User & Stakeholder	0.63
2.	Project Plan	0.67
3.	System Accessibility	0.68
4.	Regular Training	0.73
5.	Ease of Use	0.65
6.	Skills and Expertise	0.70
7.	E-Leadership	0.66
8.	Project Coordination	0.64
9.	Clear Guidance	0.71

10.	Funding Continuity	0.66
11.	Business Process Reengineering	0.73
12.	E-Government Policy and Regulation	0.67
13.	Basic Infrastructure of ICT	0.66
14.	ICT Literacy	0.71
15.	Organizational Structure	0.73
16.	Privacy & Security	0.69
17.	Usefulness	0.67
18.	Monitoring and evaluation	0.64
19.	Change Management Strategy	0.68
20.	Socio-Cultural	0.73
21.	Top Management Support	0.62
22.	System Actual Usage	0.69
23.	Citizen Relationship Management	0.73
24.	Compatibility	0.69
25.	Project Management	0.71
26.	Information Quality	0.67
27.	System Quality	0.67
28.	Service Reliability	0.64
29.	Trust	0.66
30.	Awareness	0.68
31.	ICT Governance	0.66
32.	Public Satisfaction	0.69
33.	Methodology or Approach	0.72
34.	E-Transaction and E-Payment	0.74
35.	User-Friendly	0.70
36.	Gradual Implementation	0.71
37.	Re-Usable	0.72
38.	Continuous Improvement	0.67
39.	Service Innovation	0.68
40.	Loyalty	0.70
41.	Acknowledgement	0.74
42.	Public Intention to Use	0.72
43.	Sustainable Revenue	0.81
44.	E-Participation	0.73
45.	Roadmap	0.71
46.	Inter-Governmental Relationship	0.68
47.	E-Initiative	0.72
48.	Vision	0.65
49.	Citizen Empowerment	0.71
50.	Knowledge Management	0.70
51.	Service Guarantee	0.67
52.	Empathy	0.76
53.	Flexibility of Technology	0.70
54.	Alignment of Organization Goal and ICT direction	0.68
55.	Good Responsiveness	0.70
56.	E-Democracy	0.73

Based on table 8 above, all 56 items seems had good homogeneity reliability because they had obtained significant standard (H>0.51). Thus, no more items deleted from the list. In other words, this study got 56 significant items that already valid and reliable based on expert judgment. All the CSFs obtained are the contribution of this study and being state of the art on CSFs research in the domain of e-Government among other researches.

5. Conclusion

Based on the result, the study had obtained the significant CSFs in the e-Government implementation. The content validity and homogeneity reliability of instruments had been tested through expert judgment. The Delphi used in this study eliminated 11 invalid items and resulted in only 56 items that already had good content validity and internal consistency. Those 56 items obtained to bring the picture of CSFs of e-Government that should be adopted by government institutions in Indonesia.

Acknowledgments

This research could be finished well because of full support from the research institution.

References

[1] Devadoss P R, Pan S L and Huang J C 2002 Structurational analysis of e-government initiatives: A case study of SCO *Decision Support Systems* **34** 253–269

[2] Kim S, Kim H J and Lee H 2009 An institutional analysis of an e-government system for anti-corruption: The case of OPEN *Government Information Quarterly* **26** 42–50

[3] Ndou V 2004 E-government for developing countries: Opportunities and challenges *The Electronic Journal on Information Systems in Developing Countries (EJISDC)* **18** 1–24

[4] Sang S and Lee J D 2009 E-government adoption in ASEAN: The case of Cambodia *Internet Research* **19** 517–534

[5] Srivastava S C and Teo T S 2007 E-government payoffs: Evidence from cross-country data *Journal of Global Information Management* **15** 20–40

[6] Haldenwang C V 2004 Electronic government (e-government) and development *The European Journal of Development Research* **16** 417–432

[7] Kable Central government ICT expenditure forecast 2004/05 to 2007/08 2006 http://www.kable.co.uk/

[8] UN E-government for the people 2012 http://www.unpan.org/egovkb/global reports/08report.htm/

[9] Hasan A 2013 Status of E-Government in Iraq and what the challenges of development and implementation *International Journal of Science and Research* **5** 1511-1516

[10] Nawafleh S A, Obiedat R F and Harfoushi O K 2012 E-Government between developed and developing countries *International Journal of Advanced Corporate Learning* **5** 8-13

[11] Chen Y, Chen, Ching R and Huang W 2006 E-Government Strategies in Developed and Developing Countries: An Implementation Framework and Case Study *Journal of Global Information Management* **14** 23-46

[12] Furuholt B and Wahid F 2008 E-Government challenges and the role of political leadership in Indonesia: the case of Sragen *Proceeding of the 41st International Conference on System Sciences*

[13] Heeks R 2008 Causes of e-Government Success and Failure: Factor Model (Institute for Development Policy and Management University of Manchester)

[14] Kettani D and Moulin B 2014 *E-government for Good Governance in Developing Countries: Empirical Evidence from the EFez Project* Anthem Press

[15] Napitupulu D and Sensuse D 2014 The Critical success factors study for e-Government implementation *International Journal of Computer Applications* **89** 23-32

[16] Delone W H and McLean E R 2003 The DeLone and McLean model of information systems success: A ten- year Update *Journal of Management Information Systems* **19** 9-30

[17] Al-Kaabi R 2010 Critical success factors of e-government: A proposal model for e-Government implementation in Kingdom of Bahrain *The 6th International Conference one-Government (ICEG)* 1-9

[18] Napitupulu D and Sensuse D 2014 Toward maturity model of e-Government implementation based on success factors *International Conference on Advanced Computer Science and Information Systems (ICACSIS)* 108-112.

[19] ASPA 2001 *Benchmarking e-government: a global perspective* United Nations, Division of Public Economics and Public Administrations, and American Society of Public Administration

[20] Basu S 2004 E-Government and developing countries: an overview *International Review of Law Computers* **18** 109-132

[21] Harvard Policy Group Improving budgeting and financing for promising IT initiatives 2001 Network- Enabled Service and Government Report

[22] Harris R 2003 Government policies and strategies on ICT for rural poverty reduction in the Asia and Pacific region: review and analysis *Ad Hoc Expert Group Meeting on ICT in Rural Poverty Reduction* ESCAP Bangkok

[23] CIOL Seven years of internet in India – reliving the incumbency factor 2002 www.ciol.com/content/news/trends/102090501.asp
[24] Wui-Chui M Patient Satisfaction with Anesthesia Care in Taiwan: Development and validation of a Pilot Version of a Psychometric Questionnaire 2009 *Acta Anaesthesiol Taiwan* **47** 180-88
[25] Napitupulu D 2015 Kajian Empiris Faktor Sukses Implementasi e-Government Studi kasus: Pemerintah Kota Bogor *J. Sisfo.* **5** 229-36 (in Indonesian)
[26] Burns N and Grove S 1993 The practice of nursing research: conduct, critique, and utilization (2nd ed) W.B.Saunders: Philadelphia, Pennsylvania, USA
[27] Yaghmaie F 2003 Content validity and its estimation *J. Med. Educ* **3** 25–27
[28] Yang W 2011 Applying content validity coefficient and homogeneity reliability coefficient to investigate the experiential marketing scale for leisure farms *J. Global Business Management* **7**
[29] Yu M N 1993 The computation of content validity coefficient and homogeneity reliability coefficient in ordinal data *Psychological* **40** 199-214
[30] Lai S K and Chang C M 2007 The Application of aiken content validity coefficient and homogeneity reliability coefficient in enterprise business risks scale for hot-spring hotel industries *Journal of Management Practices and Principles* **1** 64-75
[31] Aiken L R 1985 Three coefficients for analyzing the reliability, and validity of ratings *Educational and Psychological Measurement* **45** 131-142
[32] Azwar S 2012 *Reliabilitas and Validity, 4.0 ed* (Pustaka Pelajar) (in Indonesian)
[33] Aiken L R 1980 Content Validity and Reliability of Single Items or Questionnaires *Educational and Psychological Measurement* **40** 955-59
[34] Cohen J 1960 A coefficient of agreement for nominal scales *Educ. Psycol. Meas.* **20** 37-46
[35] Boar B H 2001 *The art for strategic planning for information system* Second Edition ed John Wiley and Son's Inc
[36] Silalahi M, Napitupulu D and Patria G 2015 Kajian Konsep dan Kondisi E-Government di Indonesia *Jurnal Penerapan Ilmu-Ilmu Komputer* **1** 10–16 (in Indonesian)

IOP Conf. Series: Materials Science and Engineering **352** (2018) 012059 doi:10.1088/1757-899X/352/1/012059

Marketing practitioner's tacit knowledge acquisition using Repertory Grid Technique (RTG)

Afdhal Azmi[1], Ramzi Adriman[2]
Department of Electrical and Computer Engineering
Faculty of Engineering, Syiah Kuala University
Banda Aceh, Indonesia

E-mail: afdhal@unsyiah.ac.id, ramzi.adriman@unsyiah.ac.id

Abstract. The tacit knowledge of Marketing practitioner's experts is excellent resources and priceless. It takes into account their experiential, skill, ideas, belief systems, insight and speculation into management decision-making. This expertise is an individual intuitive judgment and personal shortcuts to complete the work efficiently. Tacit knowledge of Marketing practitioner's experts is one of best problem solutions in marketing strategy, environmental analysis, product management and partner's relationship. This paper proposes the acquisition method of tacit knowledge from Marketing practitioner's using Repertory Grid Technique (RGT). The RGT is a software application for tacit acquisition knowledge to provide a systematic approach to capture and acquire the constructs from an individual. The result shows the understanding of RGT could make TKE and MPE get a good result in capturing and acquiring tacit knowledge of Marketing practitioner's experts.

1. Introduction

Marketing management is the process of delivering the product and services to consumers. Marketing management process needs concepts and strategies to communicate and develop the business process to identify and keep the customer, get their interest, and strive to reach their satisfaction. Customer satisfaction will be successfully achieved when employing all its resources effectively. It includes how to generate the underlying strategies sales techniques, business communication, and business development. It also consists of the integration process through which a company can build an active communication link with its customers and can create value for their clients and themselves.

Typically, in the development of the Marketing process, the essential things is the creativity and skill of Marketing practitioners in advertising, distribution, and selling. It consists of concept and strategy, environmental analysis, product management, organizes a partnership with producer, and build relationships with customers and create dependencies. Those things are call marketing knowledge. Sometimes, that experience does not appear in manual, guidebooks, documents, technical report, business rules, procedures, databases, etc. This knowledge called tacit knowledge. Tacit knowledge is non-articulated knowledge. Tacit knowledge does not manifest as rules; instead, it exists as the domain expert's skill, common sense, and intuitive judgment while solving the problem [1].

[1] afdhal@unsyiah.ac.id
[2] ramzi.adriman@unsyiah.ac.id

Tacit knowledge is expert experiences that difficult to transfer using writing it down or verbalizing it. Many enterprises excel at collecting data and information, but few have systematic processes, supported by the right culture and technologies, for capturing tacit knowledge owned by their employees. Tacit knowledge has deemed as the knowledge that is unexpressed yet implied or indicated. The act of an expert imparting their consciousness is merely an external expression of in-depth, experimental, intuitive and inculcated knowledge that make up their being [1].

Tacit knowledge of Marketing practitioner's experts is excellent resources and priceless. It is including the experiential, skill, ideas, belief systems, insight and even involves their speculation to decision-making. This expertise also could be an individual intuitive judgment and personal shortcuts to complete the work efficiently. Even more importantly, tacit knowledge of Marketing practitioner's experts always is problem solutions in marketing strategy, environmental analysis, product management and partner's relationship, thus making the organization get more advantages from it. This paper focuses only on tacit knowledge of Marketing practitioner's and how to capture and acquire it to reach and present more benefits to their enterprise.

We propose the acquisition framework and method of tacit knowledge from Marketing practitioner's using repertory grid technique (RGT) to eliciting personal constructs. Computational WebGrid 5 RGT tool is using to employ the framework and method. The rest of this paper is organized as follows: In section 2 literature reviews, and in section 3 propose the framework tacit knowledge acquisition of the Marketing practitioner's experts, while in part 4 suggested the methodology to capture and acquire implicit knowledge from the Marketing practitioner's expert. In section 5, we will discuss and analyze the process of tacit knowledge acquisition of the Marketing practitioner's experts. Finally, conclude and summarize the work in section 6.

2. Literature Review and Related Works

Capital knowledge is divided into tacit and explicit knowledge[2]. The four keys to knowledge creation lie in the way of Socialization, Externalization, Combination, and Internalization (SECI) [3,4]. Polanyi also stated that the dimension of knowledge consists of tacit and explicit knowledge [5]. Tacit's knowledge is the personal knowledge that is difficult to formalize. This knowledge consists of the subjective knowledge, insight, and intuition that exist in a person because he has been immersed in activities for a long time. Explicit knowledge is a formal knowledge that is easily transmitted to individuals and groups. It is often articulated in the form of mathematical formulas, rules, specifications, and so on. According to the SECI Model, the interaction between two forms of knowledge is critical to the dynamics of knowledge creation in business and organizations. It produces four meaningful modes of knowledge conversion that require a special learning process and as a whole is the creation of knowledge [6]. The mode of externalization is the process of tacit knowledge into explicit knowledge; it holds the key to the creation of knowledge as it produces a new explicit concept of tacit knowledge. The internalization mode is the process of explicit knowledge into tacit knowledge; it is closely related to learning by doing and leading to operational/procedural knowledge. While the combination mode is a process for gathering explicit knowledge from multiple sources; Socialization, from tacit to tacit knowledge, is a common mode of instructing, training and supervising employees [6].

Tacit knowledge represents knowledge based on individual experience of human action in the form of evaluation, attitude, the point of view, commitment, and motivation [7]. Polanyi expressed the essence of tacit knowledge in the phase of "*We know more than we can tell*".

The acquisition of knowledge in computer science is practiced earlier manually and then automated. The acquisition of manual knowledge is carried out in several ways: protocol analysis, interviews, and observations into general acquisition activities. Protocol analysis involves examining verbal records of experts who describe their thought processes as they solve a typical problem. Interviews are processes in which a knowledge engineer formally discusses with an expert. Observation is a process of acquisition where the observer determines useful information by observing an expert completing a unique task. All of these processes are labor-intensive and error-prone. Automating knowledge acquisition offers a solution to the problem [8].

In the first part of the 20th century, American psychologist George Kelly formulated Personal Constructs Psychology, a psychological theory that explains why people have different views and attitudes toward events in the world. Kelly [10] claims that people during their lifetime use very personal criteria. Kelly devised a method for obtaining personal construction, The Role Construct Repertory Test. This method has been redefined and self-developed and the other and is known as Repertory Grid Technique (RTG) [9]. According to Kelly, the extent to which we understand others - or ourselves - is measured insofar as we understand how they perceive their experiences. The term personal construction in Kelly's theory refers to a set of models, or hypotheses, or representations, which everyone has made about their world. Kelly finds interviews as a way to get people to express their models.

RGT is a representation of how the experts to see a particular problem. The grid is a bipolar scale or construction whose elements are placed in gradation. The knowledge engineer gets the construction and then asks the domain expert to provide a set of examples, called elements. For a more accurate assessment of the required elements assessed by the construction provided and gradations to solve the scale [10], as shown in figure 1.

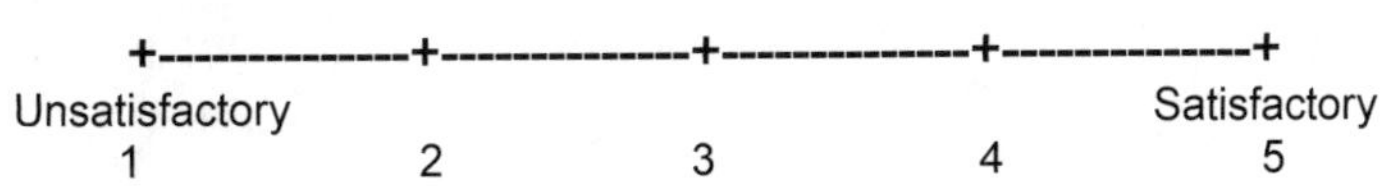

Figure 1. The conceptual framework [10].

One of the benefits of RGT is that it may prompt the expert to think more seriously about the problem and how to solve it [11]. The RGT provides a systematic approach to capture and acquire such constructs from an individual. The RTG first stage has defined the purpose of the project. The next stage is the setup of elicited elements from that person. The subject must be knowledgeable about all elements. The final stage in RGT is the elicited construct.

3. Proposed Framework

The conceptual framework for tacit knowledge acquisition of the Marketing practitioner's expert in this paper focuses on the second phase of Nonaka's model of knowledge creation and transformation, called Externalization [4] and knowledge management processes and activities [12], as shown in Figure 2. The first step in this conceptual framework of tacit knowledge acquisition process is determined domain area. The second step is decomposing the knowledge acquisition task. When attempting to solve any significant scale problem, we need to break the Marketing practitioner's expert professional activity into some smaller tasks to build a knowledge base and breaking the task into sub-tasks. By breaking the tasks to subtasks could be recognizing that knowledge inherent in well-developed skills are complex and expected to help and manage the expert's focus of attention in building a knowledge base to acquire a reliable and get an accurate result. Structuring the task into some distinct sub-steps will ease the expert into the task building a knowledge base gradually and also enable the expert to maintain a global perspective that prevents confusion when detailed knowledge is specified. Decomposition of knowledge acquisition task also helps the expert to focus their attention on one aspect of expertise at a time and expected could be increasing the availability of pertinent information in a timely fashion.

The third step is determining interdependencies. By determines, interdependencies between these tasks will facilitate identifying the missing pieces of knowledge, determine the number of knowledge are related and determine any inconsistencies with the knowledge gathered for that domain. The next step is focused on pattern recognition as the basis of expertise [13]. When the expert challenged to solve most of their problem in a project case, they expected can recognizing an aspect of the similar problem has encountered before in the past as the experiential knowledge of them that ever has been the personal shortcut to complete work easily. The methodology of tacit knowledge acquisition from Marketing practitioner's expert will be discussed in section 4.

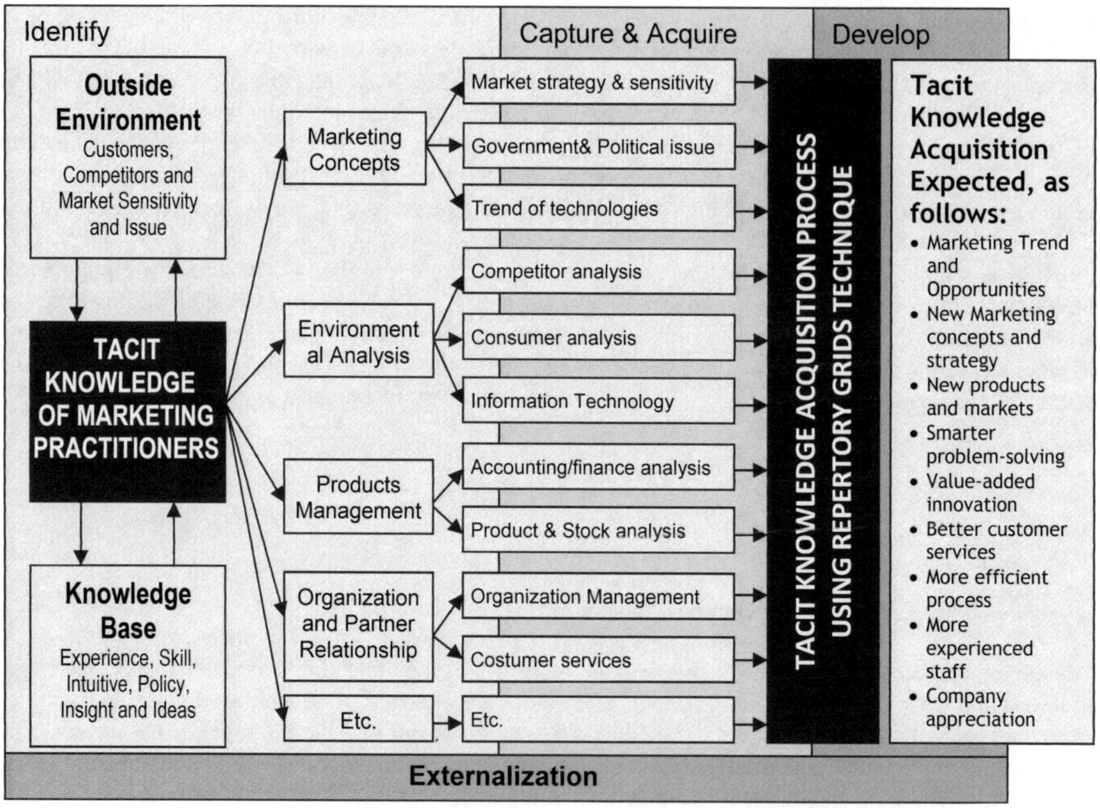

Figure 2. Marketing Practitioner's Tacit Knowledge Acquisition framework.

4. Tacit Knowledge Acquisition Methodology

The acquisition methodology suggested in this paper using RTG, as shown in figure 3.

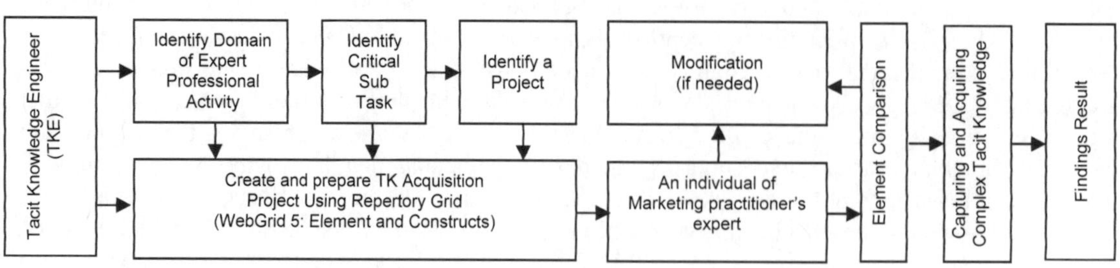

Figure 3. Tacit knowledge acquisition methodology using RTG.

The sequences of tacit knowledge acquisition methodology using repertory grid technique can describe into multistep as shown in figure 3, as follows:

Step 1: Identify domain, critical Sub-Task and scenario of Marketing practitioner's professional activity, as proposed in framework in figure 2;

Step 2: Create and prepare a tacit knowledge acquisition scenario mimicking a real market situation of Marketing practitioner's professional activity sub-task using repertory grid technique, as follows:

1. Define the purpose of the scenario;
2. Build element selection; choose a set of elements which are consistent with the objective of a scenario and targeted subsystem of constructs to be elicited from Marketing practitioner's experts;
3. Make constructs elicitation; elicit the content and hierarchical structure of the subjective meanings.

Step 3: Establish an individual of Marketing practitioner's expert to depicting their professional activity (sub-task).

Step 4: Select a scenario with element selection and constructs elicitation for an individual of Marketing practitioner's expert with a knowledge base on his/her experiential.

Step 5: Element comparison; this stage involves Marketing practitioner's expert in completing a repertory grid scenario that created by TKE to capturing and acquire the complex tacit knowledge from his/her.

Step 6: Modification (if needed); suggest to an individual of Marketing practitioner's expert to make some modification of their constructs.

Step 7: Findings result from a repertory grid scenario has been completed as an output of tacit knowledge acquisition process from Marketing practitioner's expert, involves display, cluster, map, cross-plot, and matches. Save and put into the repository as the knowledge to analyze the same case.

The result of tacit knowledge acquisition methodology to capture and acquire the expertise Marketing practitioner's expert of using RGT will discuss in section 5.

5. Result and Discussion

The implementation of the RGT involves agreement on a purpose is the identification or provision of a series of scenarios. There are many software packages that administrate the eliciting process and supply the researcher with different statistical tools. We discuss the implementation of tacit knowledge acquisition methodology to capture and acquire the expertise Marketing practitioner's expert of using RGT.

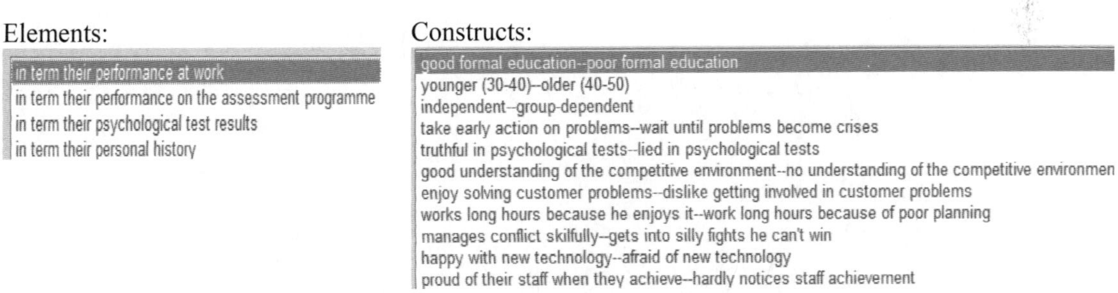

Figure 4. The elements and construct within WebGrid 5

We take one of the existing sub-task within the framework of *organizational management* is a routine activity of a Marketing expert who has the ability, the skills, and experience. TKE set up a scenario as described in the methodology and select the scenario mimicking a real market situation with *the topic and purpose to analyze the differences between successful and unsuccessful managers in the new environment.* Create elements with the two against one comparison were: regarding their performance at work; regarding their performance on the assessment programme; regarding their psychological test results, and regarding their personal history, and create the constructs, as shown in figure 4.

5.1. The rating of elements and the resulting grid of MPE-1

TKE set up a scenario in the default form with element selection and constructs elicitation, without gradation or rating elements and establish an individual of Marketing practitioner's expert (MPE-1) make element comparison. This stage involves Marketing practitioner's expert to completing a repertory grid scenario that created by TKE to capturing and acquire the complex tacit knowledge. To get better results, TKE can suggest to MPE-1 to do the modification by his experience in running their daily routines. Modified scenarios by MPE-1 can serve as a model for capturing and gaining tacit knowledge from other individual Marketing experts (MPE-n)later. The final step is the findings result from a repertory grid scenario has been completed as an output of tacit knowledge acquisition process from Marketing practitioner's expert, involves display, cluster, map, cross-plot, and matches, as shown in Figure 5.

5.2. Analysis of the Grid of MPE-1

The program noted two poles of the construct, and all the elements were rated accordingly by an individual Marketing practitioner's expert, belonging to a different extent to one or the other of poles. The construct and the rating of all the elements noted in the program on the first row of the grid. The resulting grid as shown in figure 5.(a). The elicited constructs forms row and the elements columns in the grid. Statistical methods may be used to find similarities or differences in the data. The resulting grid as shown in figure 5.(b), called hierarchical clustering analysis, the grid has been sorted and rearranged to bring closely matching elements together and closely matching construct together. The similarity scores of adjacent elements or construct are provided in a numeric form and graphically in a tree structure. In the figure 5.(c), the principal component analysis identifies a distinct pattern of variance on figures in a grid. In figure 5.(d) a cross-plot of the principal component analysis identifies the finding result.

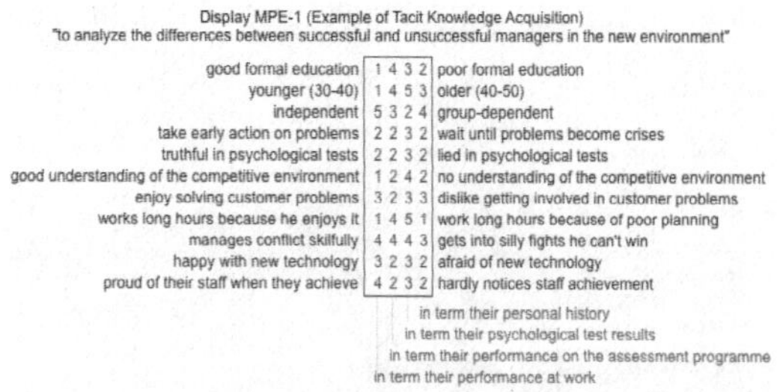

(a) Display of MPE-1 constructs

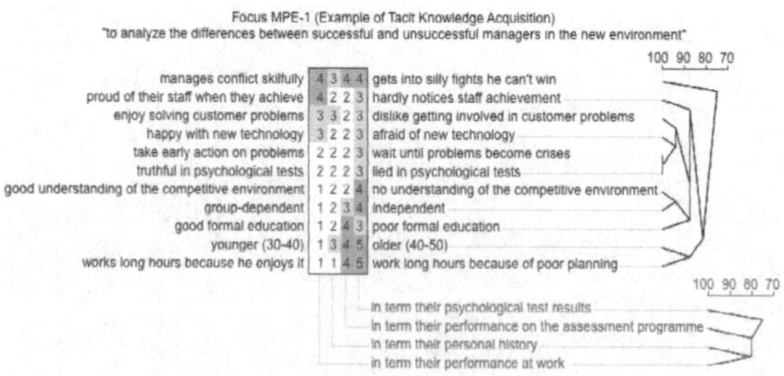

(b) Focus display of MPE-1

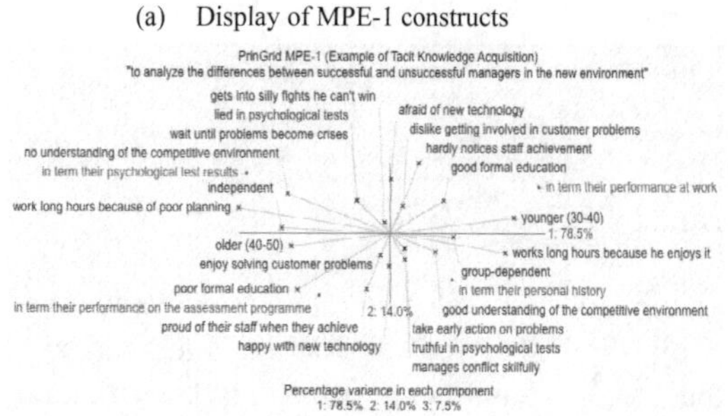

(c) Map of principal component analysis by MPE-1

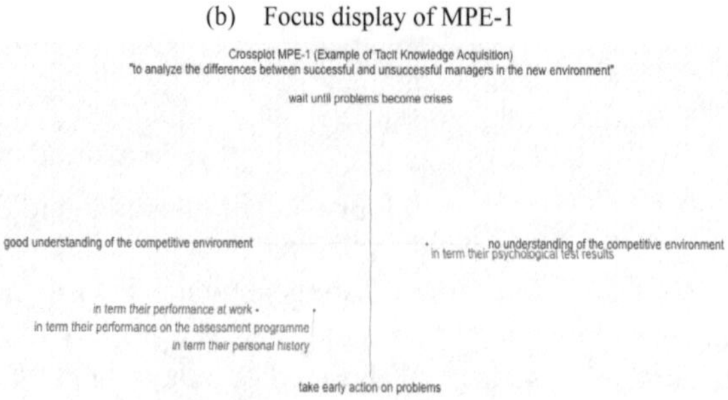

(d) Cross-plot of principal component analysis by MPE-1

Figure 5. The result of RGT construct the display, focus display, map and cross-plot.

Save and put into the repository as the knowledge to analyze the same case and compare with the other an individual Marketing practitioner's expert (MPE-n).

IOP Conf. Series: Materials Science and Engineering **352** (2018) 012059 doi:10.1088/1757-899X/352/1/012059

6. Conclusion

Kelly's Repertory Grid known as WebGrid is an appropriate tool for tacit knowledge acquisition of Marketing practitioner's experts. It is also used for eliciting simple practitioner theory. The examination of the underlying assumptions and objective of framework and method in tacit knowledge acquisition of Marketing practitioner's experts serves a useful critical function by identifying the problem and solving problem rapidly. The understanding of the purposed method employs TKE and MPE capturing and acquiring tacit knowledge of Marketing practitioner's experts. The framework used in the process of acquisition tacit knowledge in the Marketing domain, and it also can be implemented in other work domains.

References

[1] Venkitachalam K, Busch P 2012 Tacit knowledge: review and possible research directions *Journal of Knowledge Management* **16** 357-372

[2] Walsham G 2001 Knowledge management: The benefits and limitations of computer systems *European Management Journal, Elsevier Science Ltd* **19** 599-608

[3] Bhardwaj M, Monin J 2006 Tacit to explicit: an interplay shaping organization knowledge *Journal of Knowledge Management* **10** 72-85

[4] Nonaka I, Takeuchi H 1995 *The Knowledge-Creating Company* (Oxford University Press)

[5] Polanyi M 1966 *Tacit Dimension* (New York: Doubleday & Co)

[6] Jiang Y, Liu T 2010 Research on knowledge creation mechanism based on Hypercycle Theory *International Conference on Management Science & Engineering (17th)* (Melbourne, Australia) 1490-1496

[7] Kaj U K, Pekka P, Hannu V 2003 Tacit knowledge acquisition and sharing in project work context *International Journal of Project Management* **21** 281-290

[8] Randy W, Harry S D 1996 *Knowledge Acquisition via Tracked Repertory Grids* Computer Science Department (The University of Alabama, Huntsville)

[9] Lars B 2005 *The Repertory Grid Technique: Making Tacit Knowledge Explicit* A.N. Other (Sense Publisher)

[10] Kelly G A 1995 *The psychology of personal constructs* (New York: Routledge)

[11] Awad E M, Ghaziri 2004 *Knowledge Management* (Prentice Hall)

[12] O'Dell C and Grayson C J 1997 *If We Only Knew What We Know: Identification and Transfer of Internal Best Practices* (Best Practices White Paper, American Productivity & Quality Center)

[13] Rhem A J 2011 *A Framework for Knowledge Acquisition* White Paper A J Rhem & Associtates, Inc

IOP Conf. Series: Materials Science and Engineering **352** (2018) 012060 doi:10.1088/1757-899X/352/1/012060

The configuration of residential area in urban structure of the palace in Siak Sri Indrapura - Riau

Muhammad Rijal
Lecturer, *Jurusan Arsitektur Fakultas Teknik Universitas Riau*,
Kampus Bina Widya, Jl. HR Subantas KM 12.5, Simpang Baru, Tampan, Pekanbaru, Riau, Indonesia

E-mail: mrijal.unri@yahoo.com

Abstract. This article is part of major research in describing the configuration of waterfront residential area in urban space structure of the palace and related to the Malay Kingdom in the waterside of the Strait of Malacca. This research aimed to identify the configuration of riverfront residential area in Siak Sri Indrapura City based on physical and non-physical aspects. The method used in this research was qualitative rationalistic referring to the components of urban design theory. The results of the research showed that the spatial configuration in Siak Sri Indrapura City is linear and related to the past events and socio-cultural and socio-economic interaction of the society.

1. Introduction

Siak Sri Indrapura City is a riverfront city formed since the existence of Siak Kingdom in the 18th century [1]. The city located in the eastern area of Sumatra Island is a port city formed by the process of life activities on the bank of Siak River (Figure 1). Riverside and riverfront cities usually have close relationship with the history of the formation and the existence of the cities or the discovery of a place to live and reside [2]. By looking at the history of the development of cities in the world, then this is understandable. Even some cities are still in close contact with water as in Venice and Naples [3]. Siak Sri Indrapura City has been formed by the material and spiritual culture of various ethnic, social, economic, and governmental systems in the past [4]. The passage of time that has transformed the economic system, government, technological developments, and mosaics formed shows indications and tendencies of development and growth without direction. These indications began to be seen since the city was used as the capital of Siak Regency, the implications for the expansion of territory by the Government of Riau Province. Siak Regency is an expansion of Bengkalis Regency, by the Decree of the Governor of Riau Province No. 253/U/1999 dated May 26, 1999 which decides Siak Sri Indrapura City as the district capital. To anticipate these indications, it is necessary to understand the characters of Siak Sri Indrapura City as a riverfront city, because the understanding of the characters of a region can be used as one of the considerations in creating the environment that residents do not feel strange to the environment [5].

Siak Sri Indrapura city as a riverfront city shows indications and tendencies of development and growth without direction because of the concept of structuring and development that is not clear yet. It will further result in the loss of the characteristic of riverfront city spatial patterns as one of the formers of city identity. This research aimed to describe the configuration of waterfront residential area by examining the characteristics of riverfront city spatial patterns in Siak Sri Indrapura City based on the physical aspects formed through the physical elements of the city as the accumulation of arrangement of urban design elements and based on non-physical aspect formed from non-physical elements, i.e., the social order formed from socio-cultural and socio-economic conditions.

Figure 1. Siak Sri Indrapura City located in the eastern area of Sumatra Island is a port city formed by the process of life activities on the bank of Siak River

2. Literature Review

The character of a city is needed to provide an understanding of the identity of a city [6] according to its potential. In this case, the character is a physical and non-physical manifestation, providing an image and identity of the city [7]. The understanding of the value of a place is the understanding of the uniqueness of a place in particular when compared to other places [8] [9]. Characteristic of riverfront city spatial patterns is the appearance of urban environment that has a dynamic mass development pattern in accordance with the character of the riverbank, has a linkage of activities and orientation with river water environment [10], as a result of historical products of decision making by many parties within a certain time and the development of the social life of the community [11].

The elements of a characteristic of riverfront city spatial patterns are classified into two, namely, the physical elements of the city as the visually formed physical conditions and the non-physical elements as the unmeasured elements in which the city is a social order of society that influences the formation of urban spatial patterns. The elements forming the character of the physical space of the city used as the basis of analysis in the study area are: land use, mass and form of building, circulation and parking, open space, pedestrian path, supporting activities, symbols, and preservation [12] as the accumulation of urban design elements [13]. The element forming the character of the non-physical space of the city used as the basis of analysis in the study area is social order [14] that is formed through the socio-economic and socio-cultural [15] conditions of Siak Sri Indrapura City community that form the character of the riverfront city spatial patterns.

3. Research Method

Based on the research objective of studying the characteristics of riverfront city spatial patterns in Siak Sri Indrapura City by linking physical and non-physical development of riverfront city spatial pattern, so to achieve the objective of this research, the research methodology used was through rationalistic approach with the qualitative paradigm. The design of rationalistic research comes from the theoretical framework constructed by interpreting the results of the previous research, known theories, the ideas of the experts and constructed into something that contains some problems that need to be examined further. This rationalistic qualitative research methodology comes from a holistic approach in the form of grand concepts, which is interpreted into a substantive theory. The object was studied without separating the context in particular focus/accentuation, and the results of the research were re-positioned on grand concepts [16]. To conduct this research, the components to be examined were determined first, then interpreted into three main groups of urban design theory, namely, the theory of figure-ground, linkage, and place. The components under study based on the physical aspects by the accumulation of urban design elements were: land use, mass and form of building, circulation, and parking, open space, pedestrian path, supporting activities, symbols, and preservation. As for the non-physical aspects based on the social order of Siak community were: socio-cultural and socio-economic conditions.

4. Results And Discussion

4.1 Characters of Physical Spatial Patterns

The character of the riverfront city physical spatial patterns formed in Siak Sri Indrapura City is essentially linear because of the dominant natural element, i.e., Siak River. Siak River which is directly adjacent to Siak Sri Indrapura City plays an important role in forming the residential orientation since the river is a medium of life for Siak community and a vital means in the process of forming the city. In the land area, there is a grid pattern constructed by massive mass construction and open space (Figure 2).

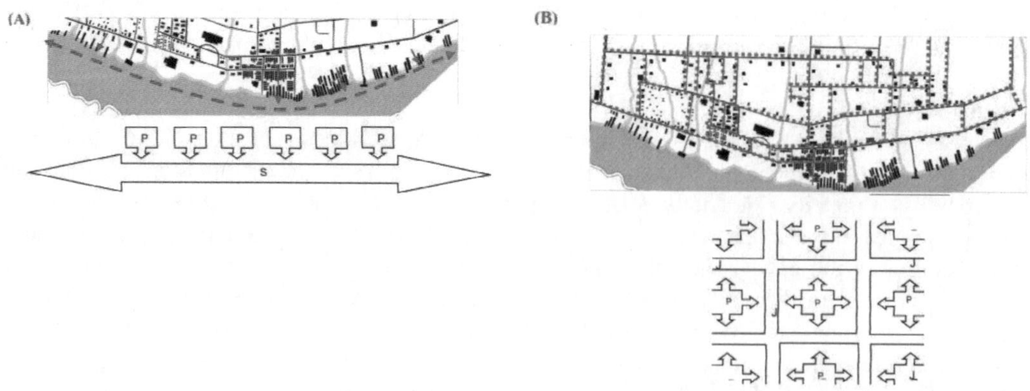

Figure 2. Riverfront residential area is the beginning of the growth of settlements formed in Siak Sri Indrapura City. The pattern of the settlement land formed on the riverbank is linear. (A). While land settlement area is a residential area formed by the expansion of settlement on the riverbank (the spread of settlement development in Siak Sri Indrapura City which tends to land due to erosion in the river side area). The pattern of the settlement land formed in the land area is a grid. (B).

The patterns of physical space formed in Siak Sri Indrapura City based on the physical elements of character formation of riverfront city spatial patterns:

4.1.1 Land Use

The formation of space based on the placement of land and river shows that the transition area between land area and water area is a transition area between land and water life, so that visually the relationship shows the focusing thus creating a central pattern where the transition area as a riverbank area becomes the center/core/nucleus of space for Siak Sri Indrapura City. So, the pattern of space created is centered towards the riverbank. From these spaces, the layout pattern of this riverfront city is divided into several patterns as follow:

- Grouping pattern. This pattern is found in the transition area between the land area and the water area in the form of market areas and settlements grouped at the inlet of Siak River.
- Spreading pattern. This pattern is found in the land area in the form of residential areas spread with the form of individual buildings.
- Elongated pattern. This pattern is found in the transition area between the land area and the water area in the form of market areas and settlements directly adjacent to the water area, elongated along the bank of the Siak River.

Urban space formed by the inlets, Siak River, and the connecting paths in the form of roads and bridges in a grid pattern. This can be seen from the imaginary paths formed from the secondary river in the form of the inlets elongated from the north-south, perpendicular to the primary river in the form of Siak River and the connecting paths in the form of roads and the bridges connecting the areas formed from the distance between the inlets elongated in linear from the east-west. The patterns of settlement land formed on the land area and transition area are:

- The pattern of settlement land formed on the riverbank in a linear pattern. This linear pattern is formed from a settlement orientation (solid) along the riverbank that tends to lead to the Siak River (void).
- The pattern of settlement land formed in the land area in a grid pattern. This grid pattern is formed from the settlement orientation (solid) in the land area that tends to lead to the road (void) in a grid pattern.

From the identification of the spaces formed based on grouping of similar activities or several types in this riverfront city, from the figure-ground analysis, the spatial pattern formed is a grid. The configuration between solid and void elements contained in each activity space forms the patterns as follow:

- The spatial pattern in the market areas formed by the trading activities forms a grid pattern formed by the urban solid in the form of building blocks in line with the orientation facing to the road of market area in the form of urban void supported by the border of the market area in the form of inlet and Siak River as urban void.
- The spatial pattern in Siak Palace area formed by the activities of government services forms a grid pattern formed by urban solid in the form of single building block with the orientation facing to the road of palace area in the form of urban void supported by border area of Siak Palace in the form of inlet as urban void open linear system and the center of the area in the form of open field located in front of Siak Palace in square shape.
- The spatial pattern in the mosque area formed by the activities of worship forms a linear pattern formed by the urban solid in the form of a single building block with the orientation facing the road of mosque area in the form of urban void supported by the border of mosque area in the form of inlet and Siak River as urban void open linear system.
- The spatial pattern in the river area formed by the waters activities forms a linear pattern formed by solid urban in the form of a single building block with the orientation facing to the Siak River in the form of urban void spread in linear.

4.1.2 Mass and Form of Building

From the explanation that a solid urban formed from a single block, the block defines the sides and block of the field, the spatial patterns formed from the mass configuration of buildings in Siak Sri Indrapura City are:

- The linear pattern formed by the settlement arrangement developed along the bank of the Siak River and the settlements located along Jalan Sultan Ismail. The orientation of the building faces the river and Jalan Sultan Ismail.
- The grid pattern formed by the arrangement of the row of buildings and intersection of circulation pathways on land area. The dominant single building form on the land area and market area that the building mass faces the road, except the place of worship of Muslims, i.e., the mosque that faces the west.

While the shapes of buildings in Siak Sri Indrapura City explain that the spatial patterns are in the form of:

- The buildings formed on a riverbank area are stilt houses, where the buildings are supported by poles planted in the riverside with the orientation toward the area of Siak River in the form of a single building and a row of buildings that form a linear pattern along the side of the Siak River.
- The buildings formed in the land area are the buildings directly connected with the ground, stilt houses with the foundation in the form of piles and the orientation of the buildings toward the road in the form of a single building and a row of buildings that form a grid pattern in accordance with the pattern of the roads formed in Siak Sri Indrapura City.

4.1.3 Circulation and Parking

From the pattern of achievement through the transportation system in Siak Sri Indrapura City, the circulation patterns formed are:

- The circulation pattern in river water area in Siak Sri Indrapura City formed by Siak River that serves as a means of river transportation used for mobile activities and the transfer of people and goods. The existence of river transportation facilities influences regional economic activities in the development and establishment of spatial structure in Siak Sri Indrapura City. The circulation patterns that occur in the Siak River are in the form of circulation of water transportation means (boat, ship) that move along the bank of the Siak River and crossings to Mempura Village, across from Siak Sri Indrapura City.
- The circulation pattern formed in the land area in Siak Sri Indrapura City seen on the road transportation path formed in the city area in a grid pattern. The circulation pattern formed in Siak City is using the road pattern formed by land transportation means such as pedicab, bicycle, motorcycle, private car, and pedestrian along the pedestrian path formed.
- The parking system formed in Siak Sri Indrapura City is in the form of parking system that existed in the land area formed on the roadside and the water parking system formed in the riverbank area in the form of docks.

4.1.4 Open Space

Open space as a character forming an element of riverfront city spatial pattern in Siak Sri Indrapura City in the area scale of Siak River itself as the largest open space of Siak Sri Indrapura City. The pattern formed of open space is based on the mobile activities that form linear pattern by the condition that makes up the river. The open spaces created in the land area are in the form of open space patterns created through:

- Corridor is a path category that serves as a circulation path that connects a function with other functions or more. This form of space is defined by the existing roads in Siak Sri Indrapura City and the elements of the existing building walls on both sides.

- Cluster: this distinctive shape is seen in open space located in front of Siak Palace in a square-shaped. The specific function of this open space gives the impression of strong nodes in the area. This space is formed because of the arrangement of the building mass around the open space.
- Inter-building space: the open space formed by a clearance between one building and other buildings on the settlement in Siak Sri Indrapura City.

4.1.5 Pedestrian Path

This pedestrian path pattern is inseparable from the role of the circulation path as a connector besides as a transportation infrastructure such as:

- Activities Connector. This type of connector is formed because of the activities of the Siak community that occur at a certain time in the form of activities that occur in high intensity of mobile activities.
- Compositional Connector. This type is formed by the pattern of the house that provides the form of a compositional connection. Linear circulation evenly causes the network. The development of the surrounding area influences this form of connection to accommodate the increasingly varied community activities. This connector aspect is an important element that connects the main circulation path to the filler area.
- Areas Connector. This type is formed as the main path between the areas, i.e., on the road that is typically linear parallel to Siak River segments. This areas connector supports a wider connection of areas about the areas outside Siak Sri Indrapura City

4.1.6 Supporting Activities

The spatial pattern formed based on supporting activities in Siak Sri Indrapura City from the result of the figure-ground analysis in the form of:

- The trading activities are taking place in the buildings in the market area that have the orientation toward the road following the pattern formed by the road so that the spatial pattern formed following the pattern of the road is in the form of a grid pattern.
- The port area in the form of a dock area by utilizing the functional spaces of riverfront city such as the docks formed along the bank of the Siak River. The port area is formed on the bank of Siak River as the implementation of water transportation activities for the community of Siak Sri Indrapura City in conducting trading activities to the outside or enter the riverfront city. The activities take place linearly along the journey from the port to the market area.

4.1.7 Symbols

The study of data showed that the symbols contained in Siak Sri Indrapura City are in the form of symbols contained in:

- River area: symbols and signs can be seen in the form of a barrier formed in the Siak River as the indicator of river depth that can be bypassed by water transportation path. The barrier is also the mangrove that grows along the Siak River as the boundary between land and river.
- Riverside area: symbols and signs can be found in the form of a port jutted into the river and the parking area for the water transportation means anchored.
- Land area: the symbols and signs formed are in the form of traffic signs and billboards/signboards.

4.1.8 Preservation

Analyzed from a solid-void urban pattern, Siak Palace is a landmark of a riverfront city that has a historical value with a unique shape and the size which has a larger scale compared to the surrounding buildings. Monumental impression with the presence of an urban void in front of the palace is in the form of an open field that provides visibility of a monumental visual of this Siak Palace. While the Siak River, viewed from its existence as a physical element, the existence of this urban void is felt firmly with the construction of a dam along the riverbank.

4.2 The Character of Non-Physical Spatial Patterns

While non-physical spatial patterns formed in Siak Sri Indrapura City are formed from two dominant ethnic groups in building social order in the form of social and cultural construction built by the Chinese community as an immigrant community who has been living in Siak Sri Indrapura City for a long time and the Malay community known as local population (Figure 3). Both communities have different patterns in constructing the social space, where the dominant Chinese community in the market area tends to form a firm grid spatial pattern, while the Malay community tends to form irregular spreading pattern in Siak Sri Indrapura City and does not have a firm space limit. This is due to the two concepts of differences between the two ethnic groups in interpreting urban space. Chinese community tends to have the concept as an urban community while Malay community tends to have the concept of traditional community.

Figure 3. The distribution of ethnic groups formed in Siak Sri Indrapura City affects the placement of religious spaces arranged by the presence of ethnic majority living. 1) the distribution of majority group of Malay as a local community. 2) the distribution of Chinese ethnic group in the market area. 3) the center of Muslim Malay ethnic group's religious space, namely, Syahbuddin Grand Mosque. 4) Muslim Malay ethnic group's religious space in the form of small mosques scattered in the land area. 5) the center of Chinese ethnic group's religious space in the form of a temple in the market area as the settlement of Chinese ethnic group majority. 6) religious space for Christian group in the form of churches.

4.3 The Configuration of Waterfront Residential Area

The configuration of settlement space in Siak Sri Indrapura City is linear. This is caused by the geomorphological condition of the area formed it and the more dominant water activities in the context of socio-cultural and socio-economic life of its people. This is in line with the opinion of Hornby that the pattern of settlement space on the waterside area of Malacca Strait tends to be linearly associated with the water activities forming it [17]. Also, past events also influence the configuration of settlement space in Siak Sri Indrapura City associated with the existence of the palace as a relic of Siak Sri Indrapura Kingdom. This palace is related to the history of the Islamic Malay Kingdom in the waterside of Malacca Strait. Efers state that the palace city as the relic of Malay Kingdom tends to separate the economic space and the administrative space of the kingdom [18]. However, in Siak Sri Indrapura City, there is a difference where the economic space merges with administrative and religious spaces and is only separated by an inlet.

In understanding the growth of waterfront city in the waterside of Malacca Strait, it is inseparable from the existence of the palace as the relic of the Malay Kingdom in Malacca. Likewise, in Siak Sri Indrapura City, the palace area that blends with open space located on the bank of Siak River shows that the palace is a city center, as a symbol of power in the city space. The existence of the palace gives an impact in the order of urban space that initially tends to be linear, creating the order of space as if in the form of centering. The grid and linear patterns blended with the centering pattern is a powerful concept in understanding the growth of Siak Sri Indrapura City. Thus, the aspect of historical value can be a guide in understanding the development of cities in the waterside of Malacca Strait associated with the Malay Kingdom as the palace city.

The socio-cultural and socio-economic interaction formed in Siak Sri Indrapura City are related to Malay and Chinese ethnic groups in filling the life activities of Riverside. This can be found in the cities formed in the waterside of Malacca Strait associated with the past government system that formed it. The migration in the early formation of Malay city and the existence of the European nation also influenced the configuration of waterfront space settlements associated with the social order of society that formed it.

5. Conclusion

Based on the research results, the character of the urban spatial pattern formed in Siak Sri Indrapura City is linear because of the dominant natural element in the form of Siak River on the riverbank. Siak River which is directly adjacent to Siak Sri Indrapura City plays a very important role in shaping the settlement orientation because the river is a medium of life and a vital means for the Siak community in the process of forming the city. In the land area, a grid pattern is found, built by mass construction and open space. While the character of the non-physical spatial pattern is formed from two groups of different social order in building social and cultural construction in Siak Sri Indrapura City, namely, a group of Chinese community as the immigrants who have been living in this riverfront city for a long time and the Malay community as the local community. Chinese community who dominate in the market area tends to form spatial grid pattern and has a firm limit on the region. While Malay community tends to form a spreading and irregular linear pattern with unclear and not firm space in forming the spatial pattern in Siak Sri Indrapura City. This is due to two different concepts in interpreting urban space. Chinese people tend to have character as an urban community while Malay people tend to have character as a traditional community.

Acknowledgments

LPDP BUDI-DN, Republic of Indonesia

References

[1] Norma D 1999 *Selintas Sejarah Kerajaan Siak Sri Indrapura dan Peninggalannya* (Pekanbaru: Departemen Pendidikan Dan Kebudayaan) (In Indonesia)
[2] Rijal M 2002 *Karakteristik Pola Ruang Kota Siak Sri Indrapura Sebagai Kota Pinggiran Sungai* (Universitas Diponegoro: Thesis) (In Indoonesia)
[3] Kostof S 1991 *The City Shaped: Urban Pattern And Meanings Through History* (London: Thames And Hudson Ltd)
[4] William M 1999 *Sejarah Sumatra* (Bandung: Remaja Rosda Karya) (In Indonesia)
[5] Trancik R 1986 *Finding Lost Space* (New York: Van Nostrand Reinhold Company)
[6] Lynch K 1961 *Good City Form* (Massachussets: The MIT Press)
[7] Budihardjo E 1997 *Lingkungan Binaan dan Tata Ruang Kota* (Yogyakarta: Andi) (In Indonesia)
[8] Lynch K 1960 *City Sense And City Design* (Massachussets: The MIT Press)
[9] Schulz C N 1980 *Genius Loci: Towards A Phenomenology Of Architecture* (London: Academic Editions)

[10] Breen A And Rigby D R 1994 *Waterfronts: Cities Reclaim Their Edge* (New York: McGraww-Hill)
[11] Henri C And Hasan M A 1999 *Panggung Sejarah: Persembahan Kepada Prof. Denys Lombard* (Jakarta: Yayasan Obor Indonesia) (In Indonesia)
[12] Shirvani H 1985 *The Urban Design Process* (New York: Van Nostrand Reinhold Company)
[13] Zahnd M 1999 *Perancangan Kota Secara Terpadu: Teori Perancangan Kota dan Penerapannya* (Yogyakarta: Soegijapranata University Press) (In Indonesia)
[14] Rapoport A 1977 *Human Aspects Of Urban Form: Towards A Man-Environment Approach To Urban Form And Design* (England: Pergamon Press Oxford)
[15] Koentjaraningrat 1999 *Manusia Dan Kebudayaan Di Indonesia* (Jakarta: Jambatan) (In Indonesia)
[16] Muhadjir N 1992 *Metodologi Penelitian Kualitatif* (Yogyakarta: Rake Sarasin) (In Indonesia)
[17] Hornby F W And Jones M 1991 *Settlement Geography* (Melbourne: Cambridge University Press)
[18] Evers D & Korff R 2002 *Urbanisme di Asia Tenggara: Makna dan Kekuasaan dalam Ruang-ruang Sosial* (Jakarta: Yayasan Obor Indonesia) (In Indonesia)

Parametric estimation for reinforced concrete relief shelter for Aceh cases

Atthaillah, Eri Saputra, and Muhammad Iqbal
Department of Architecture Malikussaleh University, Jalan Sultanah Nahrasiyah, Lhokseumawe, Indonesia 24351

E-mail: atthaillah@unimal.ac.id

Abstract. This paper was a work in progress (WIP) to discover a rapid parametric framework for post-disaster permanent shelter's materials estimation. The intended shelters were reinforced concrete construction with bricks as its wall. Inevitably, in post-disaster cases, design variations were needed to help suited victims condition. It seemed impossible to satisfy a beneficiary with a satisfactory design utilizing the conventional method. This study offered a parametric framework to overcome slow construction-materials estimation issue against design variations. Further, this work integrated parametric tool, which was Grasshopper to establish algorithms that simultaneously model, visualize, calculate and write the calculated data to a spreadsheet in a real-time. Some customized Grasshopper components were created using GHPython scripting for a more optimized algorithm. The result from this study was a partial framework that successfully performed modeling, visualization, calculation and writing the calculated data simultaneously. It meant design alterations did not escalate time needed for modeling, visualization, and material estimation. Further, the future development of the parametric framework will be made open source.

1. Introduction

The development of the parametric method was found very limited in low-tech society, particularly, in developing world and countries vulnerable to disaster. The minimum amount of literature found discussing adoption and implementation of the parametric method in low-tech society was a reason for the situation. Furthermore, The availability of infrastructure which supported the existence of digital technology was considered a major challenge in an area susceptible to disaster. Some earlier examples implemented digital technology combined with conventional one, in a low-tech environment, were the case of Solomon Island's earthquake and tsunami back in 2007 by Emergency Architect Australia shown in the work of Yeung and Harkins and Benros, Granadeiro, Duarte, & Knight in Haiti's earthquake for transitional relief shelters [1] [2]. Even though limited implementation discovered, parametric method, at the same time, offered a wide array of potential as well as challenges for exploration in a low-tech environment.

This research attempted to explore a rapid estimation framework for building materials using the algorithm with a parametric method. Exploration of this study applied to every simple single story reinforced concrete shelter, house or building with brick infills as its wall. This work invented a system that corresponded to design revisions without compromising time due to recalculation for building materials. The intended flexibility was required in a post-disaster reconstruction where time and accuracy need to be fulfilled simultaneously. The reinforced concrete shelter with brick infill for

IOP Conf. Series: Materials Science and Engineering 352 (2018) 012061 doi:10.1088/1757-899X/352/1/012061

its wall, for Acehnese, was more than just a shelter, yet it was also a dignity to whom it belongs to. For this reason, the researchers believed that the existence of this particular construction would be available for a prolonged time. As a consequence, result from this research benefited every actor in reinforced concrete construction for an extended amount of time.

As a response to the earlier mentioned situation, it was urgently required to invent an effective, efficient and simultaneous framework. A simultaneous framework was required by a designer or an architect to accurately inform and satisfy clients regarding the questions of the material required for construction, even at a preliminary stage. These were another reasons for the immediate needs of this research to be conducted.

2. Limitation

As has been noted previously, this research mainly based on observation toward permanent relief shelters available in the Province of Aceh, Indonesia. At this stage, this study was still a work in progress (WIP) which only intended to perform the calculation for the above-ground structure elements, bricks wall and excluding roof part, doors and windows. Due to some technical problems, rendering and molding materials were omitted at this stage. Furthermore, this study only performed the calculation for building materials required and omitted fee for builders and workers.

3. Definition of Terminologies

The simple reinforced concrete construction determined within this research was a two ways grid structure which used beams and columns made of reinforced concrete and its wall from bricks. A beam was a rigid structure component which beared transversal forces into vertical element called columns. Meanwhile, a column was a rigid component which took axial loads from the beam then transferred it to foundation system of construction [3]. Furthermore, Ching and Adams defined beam as a rigid bearing element which was designed to bear and transfer load into columns [4]. Within a reinforced concrete construction, a beam positioned above foundation was called *Sloof*. The most top beam was called *Rengbalk*. These terminologies were applicable in Indonesia wide construction industry. Beams and columns that were intended made out of reinforced concrete material. Walls at our case study was a brick wall which rendered/plastered both on its inside and outside.

In our algorithm, structural components were considered to be firstly erected followed by non-structural components. Besides, the method of erecting brick walls at first hand followed by structural columns was considered obsolete due to corrosive of the reinforced steel bars as a result of prolonged exposure of structural steel bars to weathering [5]. Furthermore, the method failed to comply with proper concrete covering for beams and columns as regulated. Therefore, it escalated the risk of corrosive of steel bars. It weakened the structural system of construction [6].

On the other side, regarding the method, there were terminologies such as parametric and algorithm that were utilized in this study. Parametric was defined as a method to solve design problems using algorithms as its workflow [7]. Furthermore, Khabazi explained algorithm was compilation or group of rules and instructions as a certain procedure to calculate, processes data and perform particular works that were already defined [8]. Within this study, we used Grasshopper as an interface for constructing algorithms. There were two objects needed to be understood within Grasshopper. Firstly, a parameter, a Grasshopper object stored or enclosed information or data. Secondly, a component, a Grasshopper object that executed commands or known as an operational object [9].

4. Trend in Literature

Sener and Torus utilized parametric technology, particularly Max Script under 3DSMAX platform, to establish attractive post-disaster refugees camps using container [10]. The work focused on creating the camps that were dynamically arranged for the unknown number of casualties. Another advantage offered was the ability to randomly arrange a camp. Thus, it did not look monotonous as happened in the most of the humanitarian cases. This feature was considered able to enhance the psychological performance of its users. Moreover, advantage used containers that it was easily reused and removed into a different location [10].

Implementation of parametric technology in a real humanitarian available in work Yeung and Harkins at post-earthquake and tsunami relief in Solomon Island [1]. They developed a system that facilitated input data from a spreadsheet which then connected to Grasshopper 3D interface. Data from the spreadsheet as a requirement from the field then translated to produce a latrine design for a public school. Furthermore, all cost estimation was integrated into the system which finally used off-site laser cutting technology to produce building components. In this case, they used wood construction for the building. Another special feature developed within their work was innovative working drawing which was presented as a simple graphics. It was a simplification of each building components which then used to the assembled building as simple as playing with a child's toy. This was significant for transferring building information effectively to everyone in the field, particularly for a humanitarian case.

Meanwhile, at Haiti earthquake back in 2010, Benros, Granadeiro, Duarte, & Knight utilized parametric technology with C# on Revit platform to explore design possibilities based on interpretation of local vernacular architecture [2]. The algorithm was able to generate design multiple variants. This was significant to improve psychological quality due to the capability to suit particular needs of post-disaster users. Also, the possibilities to integrate mass-customization techniques were discussed for Haiti case [2].

Further, Goncalves produced shelter typology design from modular building components. The most intriguing feature of study by Goncalves was the concept of *Do It Yourself* (DIY) [12]. DIY ability was a crucial factor for reconstruction in post-disaster case. The feature was considered to contribute to improvement in.the physical and psychological situation of post-disaster users. The feeling of belonging to the building also improved utilizing DIY concept.

The most current integration of parametric technology in the humanitarian case was shown in the work of Daher, Kubicki, and Halin. At the research, exploration was directed to use parametric potentials to help, direct and decision-making for a shelter and camp designs. At the study, they produced a shelter and camp prototype out of parametric design system based on International Red Cross requirements [11]. At work, the material calculation was not its focus.

From the discussions, there was focused on cost estimation shown in the work of Yeung and Harkins in Solomon Island. At the research, cost estimation was only limited to wood construction. For the reasons, the writer considered there was a need for immediate research for calculation of construction materials specifically to reinforced concrete shelter, this was particularly due to its wide availability at humanitarian aids, such as in 2004 Aceh post-earthquake and tsunami relieves. Therefore, this research focused on material estimation for reinforced concrete construction. The material estimation that was produced in the form that was widely understood by people from different backgrounds.

5. Methods

The method of calculation used an algorithm to perform the tasks. The intended algorithm was a visual algorithm in Grasshopper interface. Grasshopper was plug-in works under Rhinoceros platform. Advantage used the visual algorithm that it helped designer comprehending more easily even without a prior programming background. The steps within this research, firstly, prepared a design directly on Rhinoceros platform in the form of center lines. Secondly, organized and put together algorithms on Grasshopper 3D interface. Thirdly, wrote of algorithms for modeling, visualization and volume calculation on Grasshopper. Finally, organized algorithms for writing the result of the calculation into a spreadsheet. We used plug-ins under Grasshopper called ykTools to write the data into Microsoft Excel. Finally, authors used GHPython scripting to customize some modeling, calculation and data organization components. Those components were intended to simplify and optimize the algorithms.

The algorithm was able to simultaneously visualize, calculate and write the result into spreadsheet even as design changes, in a real-time. The changed could be room organizations, opening sizes, or structural element sizes. Flexibility, high accuracy and effectiveness to correspond to design alterations were the innovative way that helped to stimulate and to improve the culture in the practice of cost or material estimation. Thus, it promoted a highly efficient working culture for optimum result.

6. Validation

Calculation for each building elements based on Standard Nasional Indonesia (SNI), which was Indonesian Standard, of reinforced concrete construction. Those SNIs were as followed:

- SNI 6897 2008, for wall works [13];
- SNI 7394 2008, for concrete and molding's works [14];

For the weight of steel bars per cubic meter will be calculated based on the following formula:

$$The\ weight\ of\ Steel\ Bars\ (Kg) = V\ x\ BJ \quad (1)$$

Whereas**:**

V: Volume of steel bars
BJ: Weight of steel (kg) per cubic meter (7850 Kg/m^3)

The volume of a steel bar was the volume of the cylinder due the profile of steel bar used in the construction was the rounded shape. Using the formula (1) above, the height of steel bars were the accumulative length of some steel consisted of a structural component in a cubic meter. Meanwhile, the radius was half of the size of the steel bar used. Next, the weight obtained was converted into some steel bars by multiplying the total volume of a structural component such as columns. The result then divided the weight of steel bar in length as available in the market. The steel bars were tabulated for steel bar considering 12 meters length of each steel bar.

Next, the need for cement was converted into a unit of Zak with the assumption each Zak contained 40 kg of cement. So with brick that was tabulated into some brick. Sand and aggregate were in cubic meters. Generally, within low-tech environment, when the need for materials were given based on unit mentioned on SNI, most people failed to understand it, such as, the need for steel bars using kilograms as it units within SNI, meanwhile, steel bars available in the market was steel with the length of 12 or 10 meters each bar. This misunderstanding compromised toward the plotted budget. All of this validated data and formula was written in GHPython scripting to perform an optimum calculation.

7. Result and Discussion

The Grasshopper algorithms were structured in five main parts to model, visualize, calculate, organize data and write data into a spreadsheet. It needed one input of multiple curves, which were centerlines. The rest were parameters to control shelter's elements size, position, and elevation. All off controlling parameters were put at the most left side of algorithms for easier controlling of shelter modeling. These controlling parameters were set to the standard of shelter construction; therefore, by default, it did not need to change the input parameters of input centerline curves.

IOP Conf. Series: Materials Science and Engineering **352** (2018) 012061 doi:10.1088/1757-899X/352/1/012061

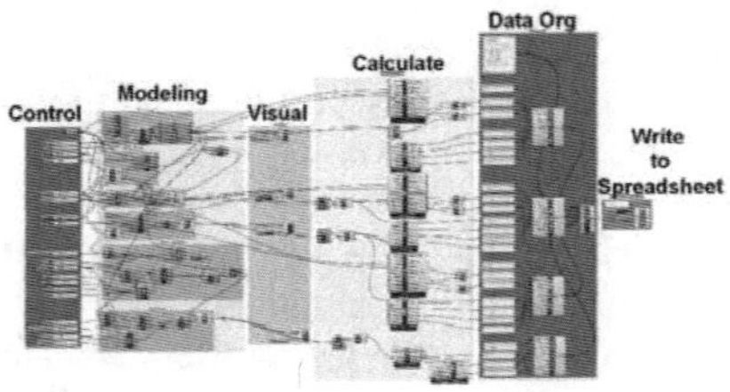

Figure 1. The rapid estimation algorithms

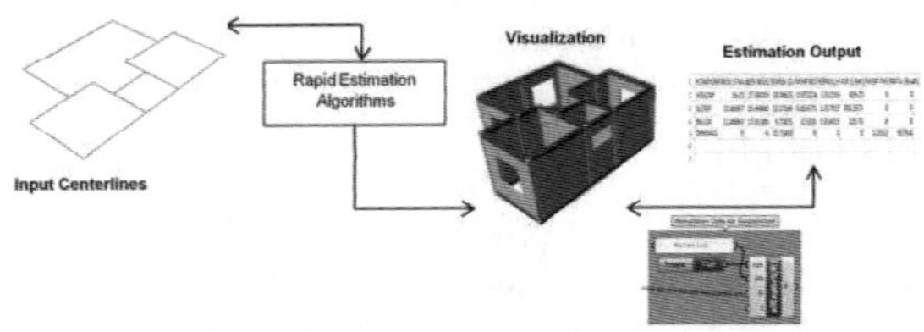

Figure 2. The framework flow

The algorithms took one curve input parameter from Rhinoceros, right clicked on the curve input in Grasshopper then selected the intended centerlines within Rhinoceros. Next, the algorithms performed simultaneous tasks of modeling to writing the estimation data into a spreadsheet. The curve input parameter was able to handle multiple centerline curves. By limiting one input, it had saved a significant amount of time for modeling, visualization and estimated the materials. To write the estimation data to a spreadsheet, it needed Microsoft Excel opened. Through Boolean toggle available within the algorithms, by double-clicking, the estimation data streamed automatically into Microsoft Excel.

Figure 1 demonstrates which one curve input all of the building components were automatically modeled based on reinforced concrete construction system after passing algorithms. For example, columns were automatically modeled by the framework based on the 12m^2 rule of the wall surface; it needed reinforcement. The logic for columns creation, firstly, the component took input curves then generated points for columns positioning. Furthermore, the algorithms evaluated the length of each curve and multiplied it by the height of columns. The height of columns was achieved from slider input of Grasshopper. If multiplication length of each curve with a height of columns was equal or larger than 12m^2, a new column was automatically added. Further, walls, bottom, and top beams were modeled based on input curves with slider controlled of component sizes, which represent shelter layout. Further, columns were used as trimming objects for walls and the beams, thus, it represented a real model of a built shelter. Meanwhile, the opening holes were limited to four holes for windows and holes for ventilation. If necessary, these windows and ventilation's holes could be easily added by plug in a new Grasshopper slider parameter into the corresponding components. In this way, flexibility aspect for adding and reducing openings was easily maintained. At this stage, with one curve parameter input, the automation rules based on reinforced concrete construction method was easily modeled to further used for estimation data.

Furthermore, the material calculation took the reference from the 3D model of the shelter generated through algorithms. Every shelter's component was calculated its volume and pass it through calculation components within algorithms. For example, to obtain some steel for beams, the volume data of the beams were pass through customized steel bars estimation component. However, for wall component, it needed to convert the volume into square meter since the calculation of wall based on an area of a wall surface. All of the data was automatically updated even as design changed in the number of time. Once the centerline passed through the algorithms everything become simultaneous; the modeling, visualization, and estimation.

KOMPONEN	BESI UTAMA (Batang)	BESI BEGEL (Batang)	SEMEN (Zak)	PASIR BETON (m3)	KERIKIL (m3)	AIR (Liter)	PASIR PASANG (m3)	BATA (Buah)
KOLOM	16.25	27.083333	18.08625	0.972214	1.512333	419.25	0	0
SLOOF	11.666667	19.444444	12.173437	0.654375	1.017917	282.1875	0	0
BALOK	11.666667	17.013889	9.73875	0.5235	0.814333	225.75	0	0
DINDING	0	0	31.71459	0	0	0	3.5312	6179.6

KOMPONEN	BESI UTAMA (Batang)	BESI BEGEL (Batang)	SEMEN (Zak)	PASIR BETON (m3)	KERIKIL (m3)	AIR (Liter)	PASIR PASANG (m3)	BATA (Buah)
KOLOM	21.66667	36.11111	24.115	1.296286	2.016444	559	0	0
SLOOF	13.53333	22.55556	14.12119	0.759075	1.180783	327.3375	0	0
BALOK	13.53333	19.73611	11.29695	0.60726	0.944627	261.87	0	0
DINDING	0	0	37.34763	0	0	0	4.1584	7277.2

Figure 3. Estimation of action

The design alterations did not compromise time needed for shelter material estimation. This had improved time efficiency, which in turn, it enhanced rapidity criterion as require in post-disaster cases. As shown in figure 3, each shelter's components were visually color coded for easy validation of component being calculated. This important in two folds, firstly, it helped authors organizing algorithms for proper calculation components. Secondly, it helped clients or users identifying components calculated for them and gave an overall overview of shelter design in 3D view. In this way, once users or beneficiaries objected to the design it could be altered in a real-time.

Also, the conversion of units as shown in figure 3 demonstrated of material information had extended the understandable of calculated information for people with non-technical backgrounds. This was important to assure precise information received by clients, which in post-disaster cases usually involved a wide range of people from different backgrounds. This had potential to enhance communication among people involved in a relief shelter. Based on the current result, this study had proved one potential of parametric technology for low-tech or area susceptible to a disaster such as Aceh. Within this method, it allowed architecture to explore for a more robust output of design dynamically to suit clients' needs, particularly in a post-disaster environment. Furthermore, it showed that parametric approach had demonstrated that design alteration did not escalate time for modeling, visualizing and calculating of preferred shelter design. This meant an improvement of rapidity criterion for shelter relieves.

Furthermore, some of the rapid estimation WIP customized GHPython components were a wall, beams, columns, all estimation components, and data organizer components. As explained earlier, the idea behind these customized components was for efficiency and optimization of algorithms. Thus, it reduced human error when constructing algorithms.

8. Future Development

All components that were produced within this study was made open source in its final release. In the future, the framework offered within this study was developed to model and calculate a complete construction of shelter including substructure, interior, opening, and roof. It was our vision to develop this parametric framework to perform complete bill of quantity (BoQ) for reinforced concrete construction. Also, authors envisioned altering this framework for a more complex building system, such as multi-story construction. Furthermore, we envisioned integrating optimization tool using Galapagos within our framework. The optimization benefited in two folds; firstly, it provided design alternative automatically with budget constraint. Secondly, authors integrated environmental analysis tool such as Ladybug and Honeybee for shelter generation and used Galapagos for optimization. Therefore, shelter design left a minimal negative environmental footprint.

9. Conclusion

The current state of the parametric framework within this study was still underway for its perfection. There was a partial part of construction estimated in the current state. However, a significant potential had been revealed from this study for performing rapid estimation of relief shelters. Firstly, the framework was able to shortcut modeling by limiting one input of curves parameter, to collect centerlines, to perform 3D visual of shelter designs while simultaneously calculating and writing data into a spreadsheet. Secondly, the framework provided the benefit of design alteration with no time extension for modeling, visualizing and estimating of construction materials. Thirdly, calculated data provided was understandable by a wide range of people. Lastly, color-coded visualization offered a better understanding of a design to a client and helped validation of the calculated component easier. Consequently, all of these contributed to a better and faster system of calculating culture of shelter materials. Potentials available within the framework of this study, also, it possibly is implemented on every reinforced concrete construction, of course, with the perfection of the framework. For example, integrated this framework into government procurement system of reinforced concrete building projects. For another type of construction, there were lessons learned as shown earlier in this study or conducted further research implementing parametric framework into current or future procurement system of government projects.

References

[1] Yeung W K and Harkins J 2010 Digital architecture for humanitarian design in post-disaster reconstruction *International Journal of Architectural Computing* **9** 17-31

[2] Benros D, Granadeiro V, Duarte J P, and Knight T 2011 Automated design and delivery of relief housing: the case of post-earthquake Haiti *Proceeding of the 14th International Conference on Computer Aided Architectural Design, Liege* 247-264

[3] Schodek D L 1995 *Struktur* (Bandung: PT Eresco) (in Indonesian)

[4] Ching F D and Adams C 2008 *Ilustrasi Konstruksi Bangunan, 3rd ed* (Jakarta: Erlangga) (in Indonesian)
[5] Zhou Y, Briceno S, Asplund B and Mangkusubroto K *2007 Handbook on Good Building Design and Construction Aceh and Nias Islands* (UNDP Regional Centre, Bangkok)
[6] Kertopati P I, Larsen O and Tedjo S 2007 P*edoman Membangun Rumah Sederhana Tahan Gempa* (Palang Merah Indonesia & International Federation of Red Cross and Red Crescent Societies Yogya-Jateng Earthquake Response, Yogyakarta) (in Indonesian)
[7] Atthaillah 2014 Arsitektur parametrik dengan Rhinoceros dan Grasshopper: Kajian workflow dari desain, fabrikasi hingga hitungan kebutuhan material *Arsitekno* **3** 10-23 (in Indonesian)
[8] Khabazi Z 2012 Generative Algorithm using Grasshopper Morphogenesism http://www.grasshopper3d.com/page/tutorials-1
[9] Akos G and Parsons R 2014 The Grasshopper Primer Third Edition http://modelab.is/grasshopper-primer/ (Accessed 2 April 2015)
[10] Sener S M and Torus B 2009 Container Post Disaster Shelter-CPoDS *Proceedings of eCAADe 27* 599-604
[11] Daher E, Kubicki S and Halin G 2015 A parametric process for shelters and refugees' camps design *Proceedings of the 33rd International eCAADe, Vienna* 541-548
[12] Goncalves A M d C F A 2014 A Grammar for Shelter: An exploration of ruled based designs n prefabricated and modular shelters *Proceeding of the 32nd International eCAADe, Newcastle* 327-336
[13] Anonymous SNI 6897:2008 Tata cara perhitungan harga satuan pekerjaan dinding (Badan Standardisasi Nasional, Jakarta) (in Indonesian)
[14] Anonymous SNI 7394:2008 Tata cara perhitungan harga satuan pekerjaan beton (Badan Standardisasi Nasional, Jakarta) (in Indonesian)

IOP Conf. Series: Materials Science and Engineering **352** (2018) 012062 doi:10.1088/1757-899X/352/1/012062

The historical and religious approach towards city park design in Banda Aceh, Indonesia *Case Study: Krueng Neng Park (Taman Krueng Neng)*

D Safriana[1], E Wulandari[2]
[1]Department of Architecture, Faculty of Science and Technology, State Islamic University Ar-Raniry, Darussalam, Banda Aceh 23111
[2]Department of Architecture and Planning, Engineering Faculty, Syiah Kuala University, Darussalam, Banda Aceh 23111

Email: dcsafriana@gmail.com, elysawulandari@yahoo.com

Abstract. Banda Aceh has a long history as a coastal city; it had ever been a cosmopolitan maritime city based on Islamic sharia. One of the physical traces of this history is located at the Krueng Neng River, a site which became a Turkish military training area to support the existence of the kingdom of Aceh in the 17th-19th century. Currently, the development of the Banda Aceh city has the goal of making the city as an Islamic tourism destination. One of the ways to achieve this goal is by the arrangement of Islamic open space in historical environments. Therefore, this paper intends to examine historical and religious approaches in the concept of open space arrangement in Krueng Neng, Lamjame Village, Banda Aceh. This is important with regards to the Banda Aceh's designation as one of the eight heritage cities in Indonesia, as Banda Aceh's city plan will be developed according to the goals of the heritage program. The study model is in the form of design research that will develop an activity program and pattern of the spatial arrangement based on history and religion, and supported by location-based approach (field condition) incorporating both geographical and socio-cultural contexts. The result of the study is a park design based on Islamic garden principles and incorporating historical aspects from 17th century Turkey in the form of replicas of a Turkish ship and cannon 'Lada Sicupak'. In summary, one of the options to enhance the goal of Islamic tourism in Banda Aceh is arranging a local park based on its historical aspects and applying Islamic garden principles.

1. Introduction

The city of Banda Aceh is a coastal city on a bay, located at the northern tip of the island of Sumatra. The natural environment exhibits three temperate climates that tend to be hot with strong winds during the west wind season and relatively moderate rainfall [1]. Therefore, the character of the natural environment makes the weather of the city of Banda Aceh hot. During the tsunami, most of the coastal areas were destroyed, also relatively open, and becomes dry. Therefore, it needs to undergo greening as the penetration of negative condition above. The development of city parks in the coastal area is one of the possible solutions to the problem of geographical conditions in the city of Banda Aceh.

According to the Banda Aceh City Spatial Planning (RTWK Kota Banda Aceh Year 2009-2029), one type of green open space that will be developed in Banda Aceh is city parks. One of the new city parks built by the Banda Aceh city government is Krueng Neng Park, with an area of 1.85 ha.

The city parks in Banda Aceh were developed to meet the standards of 30% urban green open space [1]. However, the parks were all designed with the same character without applying specific themes and had no unique features, which can attract visitors. It is important for the government to provide as many unique tourist destinations as possible, especially as Banda Aceh has become a tsunami tourism destination and halal tourism destinations.

Thus, to provide an example of a unique park design based on the historical and religious characteristic, this paper takes Krueng Neng Park which is one of the existing parks in the northwest suburbs of Banda Aceh, located on the riverbank as far as 500 m from Ulee Lheu Beach. The potential of this location is unique because it contains much history of the glory of the city of Banda Aceh in the past [2]. The park was a center of Turkish military activity while assisting the Kingdom of Aceh in building a cosmopolitan maritime city [3, 4]. In addition, Taman Krueng Neng area is also one of the central locations of city's religious activity, which is conducted every week in the mosque of the park. If the values of history and religious activity are incorporated as the themes in the design of Krueng Neng park, its role as a tourist destination will add significant value to Banda Aceh in the context of a tourism city. Therefore, it needs to display a replica object to revive historical memories, particularly of Aceh, but also of Indonesia in general. This is also a way to develop the character of the nation, as has been proclaimed by the government [5]. These religious values can also be elevated in the design of Islamic garden landscaping, so visitors will appreciate the greatness of God as the creator of nature.

2. Park design based on historical and Islamic principles

There are several aspects to be considered in designing an Islamic park intended to evoke the past.
Below I elaborate the ways Islamic garden is designed. Firstly, I discuss what is city park and its Islamic principles, followed by the discussion of Krueng Neng's historical aspects which are associated with the Turkish presence in Banda Aceh.

2.1 City Parks

"City Park according to the Regulation of Minister of Public Works No. 5 of 2008 is open space that serves social and aesthetics as a means of recreational activities, education or other activities at the city level" [6]. City parks have at least three functions: landscape, environmental and aesthetics. Also, there are three important elements in the landscape, i.e., hard elements, soft elements and supporting elements. In the design of landscape is known two major parts of landscape material, namely soft materials and hard materials [7].

2.2 Gardens According to Al Quran and Hadith

In the Qur'an the garden of heaven is described as having flowing rivers and fruits growing in it [8]. Types of plants contained in the garden are various fruits, cucumbers, nuts, onions, garlic, dates, grapes, pomegranates, olives, and fig. The main color mentioned in the Qur'an is green.

Traditionally, an Islamic garden is a cool place of rest and reflection and a reminder of paradise. The Qur'an has many references to gardens, and the garden is used as an earthly analog for life in paradise, which is promised to believers who carry out the command of God. The general themes of a traditional Islamic garden are water and shade, not surprisingly since Islam originated and spread in a hot and dry climate. Thoughts and ideas about Islamic gardens are derived from the beliefs of Muslims and their dreams of heaven. In some verses of the Qur'an, it is explained that believing Muslims will gain an appreciation on the Day of Judgment. In the Qur'an it is also said that believers will attain a very beautiful and magnificent paradise that is described as a paradise garden. Muslims believe that paradise gardens have pure water with flowing springs, fruit trees, and places sheltered under shady trees, as is written in the following verse [9].

إِنَّ ٱلْمُتَّقِينَ فِي جَنَّٰتٍ وَعُيُونٍ ٤٥

"The righteous are indeed placed in the Garden of Heaven, with springs sprinkled on it." [9]

2.3 Islamic Parks as Landscape

As a Landscaped gardens, which are designed with specific purposes, have to apply ideology and principles based on the Islamic religion and culture of Muslim society, and thus uses relatively special design elements [10]. The themes of traditional Islamic gardens are water and shade. Unlike English gardens, which are often created for recreation, Islamic gardens are meant for rest and reflection. Therefore, Islamic parks usually have seats for contemplation [11]. In Islamic parks, there are several vital aspects, i.e., water, shadows, color, and sound.

According to the 'harfiyah', there are four keys of Islamic values: peace, safety, submission, and cleanliness. Islamic values in architecture consist of zoning, space, hijab and privacy, centrally oriented, Islamic symbols and ornamental art [12]. The gardens are open spaces near a certain area of buildings in which planted a variety of plants to refresh the inner and outer space and to bring happiness. Due to the location of the Krueng Neng city park is in Banda Aceh which promotes Islamic sharia and has a long history, the park will be designed based on religious and historical concepts. It will incorporate the concept of viewing objects while walking so that visitors can experience the past as a reflection of the glory of Banda Aceh.

2.4 History of the Aceh-Turkey Relationship

The Sultanate of Aceh Darussalam was recorded to have had a close relationship with the Ottoman Empire. Turkey had provided military assistance in the form of ships, artillery, and troops, also other weapons that Aceh needed to attack the Portuguese. The evidence can be seen in two of the relics of the Sultanate of Aceh Darussalam, namely the Turkish ship and 'Meriam Lada Secupak' as proof of the history between Aceh and Turkey [13]. Therefore, to bring back the memory of the glorious Acehnese kingdom, the Turkish ship and cannon will be placed in Krueng Neng Park, which is located in a former Turkish military camp.

Figure 1. The cannon is known as 'Lada Sicupak' [19]

Figure 2. Ottoman fleet in the Indian Ocean in the 16th century [20]

IOP Conf. Series: Materials Science and Engineering **352** (2018) 012062 doi:10.1088/1757-899X/352/1/012062

2.5 A Semiotic Method for the replica object's representation of Turkey-Aceh history in Aceh.

In everyday life, people experience the place and space. Both place and space have their signs and will remain in people's memories [14]. Past experiences can be presented using semiotic methods [15]. The semiotic method is divided into two parts (dichotomy): signifier and signified. Signifiers, defined as physical forms, can be experienced through the form of architectural works, while signs can be seen as meaning revealed through the concepts, functions, and values contained in the work of architecture. The Saussure semiotics is the relation between the signifier and the marker by convention, commonly called the signification. Semiotics signification is a sign system that studies the relation of sign elements in a system based on certain rules or conventions. These social conventions are needed to be able to interpret the signs [16]. By using this theory, the author tries to bring nuance to the concept of Aceh glory in the 17th century, which is inseparable from the role of Turkey. Therefore, it is necessary to present a replica of a Turkish ship and a 'Lada Sicupak' canon as a signifier.

3. Methods

The study method includes two parts, namely:

i. Study of the aspects affecting the design problem (problem seeking), i.e., historical aspects, socio-cultural and religious aspects, and aspects of the natural environment.
ii. Study the aspects that guide the concept of design (problem-solving) related to the themes of environmental arrangement and the choice of specific historical and religious landscape objects in the context of socially/culturally, economically and environmentally sustainable development in urban park design.

The study was conducted descriptively and qualitatively, also presented the characteristics of signs (semiotics) of architectural objects that evoke historical meaning. The concept of design was developed using field surveys, especially in the context of the socio-cultural character of the community, and the spatial policy of the city of Banda Aceh such as disaster mitigation.

4. Results and Discussion

4.1 Existing conditions of Krueng Neng City Park

The design of the Krueng Neng park site was completed in 2015, however, the greenery is absent. The design of the park layout seems empty, which is very typical with other parks in the city of Banda Aceh, and also lacks of uniqueness. Therefore, it needs replica objects to develop a unique character. These replica objects will cause people to recall the history of the friendship between Aceh and Turkey as well as it would provide uniqueness for the place.

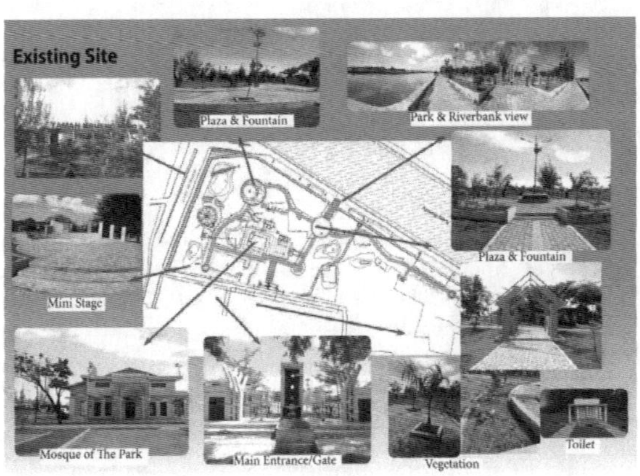

Figure 3. Existing condition of Krueng Neng Park

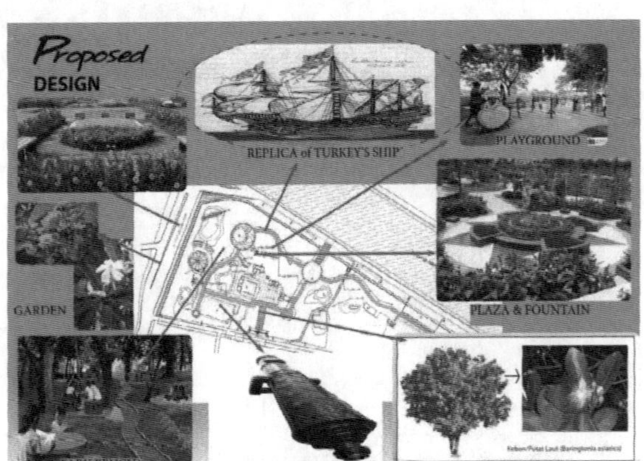

Figure 4. Proposed Design of Krueng Neng Park

4.2 Conceptual Design

4.2.1 *Placement of the Replica Objects.*

Based on the above picture, the space for greening area seems unplanned and remains empty. To improving the design of the potential space can be done by utilizing the empty space through the addition of recreational objects in the form of a replica of the 17th century. The historical values to be elevated in the design of this park are the placement of a replica of a Turkish ship and the placement of a replica of a 'Lada Sicupak' cannon. These two objects are the memorable objects to evoke memories of the past (interview with bu Azmah). The placement of the replica objects will occur in the north-center of the site:

- The ship replica will be placed on the north side of the Krueng Neng river.
- The replica of the cannon will be placed near the main gate of the park (the middle south) so that visitors can see the object immediately upon entering the garden yard.

The placement of the two objects is based on the principle of kinetic vision and visitors' experience of spacing. Medium objects are displayed in a semiotic manner that highlights the replicas and evokes the atmosphere of the past. The concept of kinetic vision in the design is divided into three scales (Figure 5):

1. Macro scale: the object is viewed from outside of the park and can induce awe and cause curiosity. On a large scale, the object is a focal point.
2. Meso scale: visitors can feel the experience of casting towards the object. The focus will be on the circulation space and garden around the ship.
3. Micro scale: visitors can experience the interior of the ship and can see the exterior view: interior and exterior details on board/deck will be a focal point.

View from the distance, monumental impression of the glorious image of the Aceh 17th century kingdom as a maritime country	View inside the site, the circulation room leads the visitors to the shipbuilding, depicting the religious atmosphere of Aceh through green media and water	The view from the ship: the interior and exterior atmosphere illustrates the toughness of the nation through the display space.

Figure 5. Illustration of Replica's placement (Turkish Ship)

IOP Conf. Series: Materials Science and Engineering **352** (2018) 012062 doi:10.1088/1757-899X/352/1/012062

4.2.2. Conceptual Design of Islamic Gardens.

There are four important elements in designing an Islamic garden:

- Water

  The water element is very important in the design of Islamic gardens because the water can be as soothing and calming as the atmosphere of heaven, which is described in the Qur'an (Al Quran 1:25) [17]. It can also lower temperatures during hot weather [18]. The arrangement of this park will include a water fountain placed in the middle of the plaza to make visitors feel comfortable.
- Shadows

  Shadow effects are also needed in the design of Islamic parks. In the design of this garden, the shadow effect is generated from tree vegetation, a pergola, and a gazebo. The shadow is defined as shade or shelter for a place of 'tafakkur' (contemplation).
- Color

  The color in the arrangement of Islamic gardens is also very important. In the design of this garden, the color will be produced from colorful flowers. In addition to providing aesthetic value, these colored flowers will affect the psychological state of visitors at the time of their visit.
- Sound

  Sound elements in the design of this garden consist of the sound of water gurgling, which will make visitors feel comfortable and calm, in addition to can being relaxing. The other sounds will come from the calls of endemic birds placed in several locations in this park.

5. Conclusions

The Krueng Neng Park area is one of the green areas in Banda Aceh that has the potential to become a destination for local and national tourists, as well as for the international community. However, to give the spirit of design in this park needs a more comprehensive design concept using a historical and religious approach. The placement of historical objects such as replicas of ships and Turkish cannons will bring visitors back to the 17th century, the glorious age of the Sultanate of Aceh, and will evoke historical memories to preserve the glory of Aceh in the past. In addition, this park will be designed following the principles of Islamic garden design and will be able to support religious activities carried out regularly every week, thus reinforcing the character of Aceh's religious community. It is expected that the concepts of this design through this research can help the government develop Krueng Neng Park as a tourist attraction in the city of Banda Aceh.

The approach to the history of the triumph of the Sultanate of Aceh aims to make younger generations remember the identity of the city, which had ever been as a great and glorious kingdom in the city of Banda Aceh, and to evoke the spirit of its glory.

This research recommends that the local community is important. This community will play the role of tour guide and will be the main actor in implementing this activity to increase promotion and economic development.

References

[1] Departemen Pekerjaan Umum 2009-2029 *RTRW Kota Banda Aceh* (in Indonesian)

[2] Wulandari E and Safriana D 2017 *Konsep Penataan Kota Banda Aceh sebagai Kota Wisata Tsunami* vol 1 (Bandung: Arcade) p 1

[3] Lombard D 1991 *Kerajaan Aceh, Jaman Sultan Iskandar Muda (1607-1636)* (Jakarta: Balai Pustaka)

[4] Wulandari E, Soetomo S, J Syahbana J A and Asnawi M 2017 The ecology character of Banda Aceh City in the 17th century *Journal of Islamic Architecture*

[5] Widodo J and Kalla J 2014 *Nawacita: Jalan Perubahan untuk Indonesia yang Berdaulat, Mandiri dan Berkepribadian* Visi Misi dan Program Aksi (in Indonesian)

[6] Regulation of Minister of Public Works No. 5 of 2008

IOP Conf. Series: Materials Science and Engineering **352** (2018) 012062 doi:10.1088/1757-899X/352/1/012062

[7] Hakim R and Utomo H 2008 *Komponen Perancangan Arsitektur Lansekap Prinsip-Unsur dan Aplikasi* (Jakarta: PT Bumi Aksara)
[8] Quran, Surah Al-Baqarah; 25
[9] Quran, Surah Al-Hijl: 45
[10] Clark E 2004 *The Art of The Islamic Garden* (Malborough: The Crowood Press) p 23
[11] Petruccioli A 1998 *Rethinking the Islamic Garden* Article of ResearchGate
[12] Widyastuti N, Aly A, and Ronald A 2014 *Islamic Values Expression in Surakarta Houses* (Surakarta: UMS)
[13] Feener M, Daly P and Reid A 2011 *Mapping the Acehnese Past* vol 268
[14] Hinds J and Paul P 2008 Engaging with the natural environment: The role of affective connection and identity *Journal of Environmental Psychology* **28** 109–120
[15] Christomy T and Yuwono U 2004 *Semiotika Budaya* (Pusat Penelitian Kemasyarakat dan Budaya Direktorat Riset dan Pengabdian Masyarakat Universitas Indonesia, Depok) (in Indonesian)
[16] Noth W 1992 *Handbook of Semiotica Indiana University Press* Bloomington and Indianapolis
[17] Quran, Surah Al-Baqarah; 25
[18] Fireza D 2007 *Desain Taman Islami* (Jakarta: PT Mizan Publika)
[19] Museum Bronbeek No.27. Photo Fiona Kerlogu ottomansoutheastasia.org
[20] Ottoman fleet http://www.istanbulhaber.com.tr

IOP Conf. Series: Materials Science and Engineering **352** (2018) 012063 doi:10.1088/1757-899X/352/1/012063

Architectural heritage in post-disaster society: a tool for resilience in Banda Aceh after the 2004 tsunami disaster

Cut Dewi[1,2] and Era Nopera Rauzi[1]
[1] Department of Architecture and Planning, Faculty of Engineering, University of Syiah Kuala, Darussalam, Banda Aceh, 23111, Indonesia
[2] Centre for Cultural Studies (PPISB) of University of Syiah Kuala, Darussalam, Banda Aceh, 23111, Indonesia

E-mail: cutdewi@unsyiah.ac.id

Abstract, This paper discusses the role of architectural heritage as a tool for resilience in a community after a surpassing disaster. It argues that architectural heritage is not merely a passive victim needing to be rescued; rather it is also an active agent in providing resilience for survivors. It is evidence in the ways it acts as a signifier of collective memories and place identities, and a place to seek refuge in emergency time and to decide central decision during the reconstruction process. This paper explores several theories related to architectural heritage in post-disaster context and juxtaposes them in a case study of Banda Aceh after the 2004 Tsunami Disaster. The paper is based on a six-month anthropological fieldwork in 2012 in Banda Aceh after the Tsunami Disaster. During the fieldwork, 166 respondents were interviewed to gain extensive insight into the ways architecture might play a role in post-disaster reconstruction.

1. Banda Aceh and its Disaster and Cultural Background

Banda Aceh is the capital city of Aceh province, which is located in the northern part of Sumatra. As one of most affected cities by the largest natural disasters in the 21st century, during the 2004 Indian Ocean Tsunami and earthquake, its population reduced from 239,146 people to 177,881 after the Tsunami, then rise again in 2011 reaching 228,562 [1]. During the disaster, In Aceh, the loss of human life was somewhere between 130,000 to 170,000 deaths, about 250,000 houses were destroyed [2], and over 500,000 people were made homeless and displaced [3].

Due to its position, the city is very prone to disaster especially earthquake, flooding, and in 2004 it was exposed to a tsunami which is predicted has happened for several times in the past [29]. Its position at the mouth of a river has resulted in a sequence of major and minor floods [see for example 30]. In the past several earthquakes have been recorded by the travelers such as Beaulieu visited Aceh around 1620-30 witnessed flooding and also the recurrence of earthquake three to four times a year; in which of the large earthquake occurred on 7 March 1621 that caused panic [30]. In addition to these two disasters, he also noted a large fire, which burnt approximately 260 houses, on 4 June 1621, just a couple months after the earthquake [30]. It is, however, hard to find written records and archaeological proof of the existence of ancient tsunamis. A recent geophysical research conducted by Aron J. Meltzner et al. [29], which documents the history of earthquakes and tsunamis in Aceh between 1390 and 1455. From the travelers' notes, Acehnese has demonstrated resilience by the quick reconstruction after disasters using traditional techniques and materials.

In response to the 2004 Tsunami Disaster, in 2005, Indonesian government established the Aceh and Nias Reconstruction Board (BRR) to manage the aid and funding which is recognized as one of the largest humanitarian aid projects in history for a-four-year period [4]. BRR dealt with huge not only infrastructure loss, but also human and psychological loss. The survivors suffered the loss of infrastructure that supported their life such as roads, bridges, electricity, and telecommunication, agricultural and fisheries sector, and others. The loss of the built environment, a familiar place to where they tie up their memories has left trauma. In addition to this, survivors lost their beloved people and experienced a huge trauma too as many of them were hospitalized for injuries and some lost their jobs, homes and other crucial elements of their livelihood. Therefore, no wonder then this disaster has attracted a great deal of attention from many countries and donors across the world. International emergency aid and donors, including some vital United Nation's organizations, have been involved since the beginning of the recovery process.

The Tsunami Disaster has become a turning point for Acehnese to move away from thirty-year conflict to peaceful lives. Aceh has been under the violence since 1873 when the Dutch war occurred and thirty years of conflict between Aceh Free Movement (Gerakan Aceh Merdeka/GAM) and the Indonesian government. Therefore, the conflict memories and disaster memories have been mixed with the memories of the "glory" of the past Islamic kingdom, its central place in society and the role of Islamic Sharia. These socio-cultural aspects have contributed to the ways Acehnese survived surpassing disaster and uses architectural heritage in the process of reconstruction and resilience. Before discussing my case study, below I explore several problems of understanding of architectural heritage in post-disaster context.

2. Architectural Heritage: Its Problems in Post-disaster Context

As I have argued everywhere, at least two major problems, which lead to a lack of understanding of the role of architectural heritage in post-disaster context. The problems are arguably caused by two main factors. Firstly, The Eurocentric Authorised Heritage Discourse (AHD), a term coined by Smith [7], which emerged in in Europe around the 19th Century, has influenced the ways global heritage organizations understand heritage. The AHD reflects the heritage understanding of European elites and understands heritage as tangible. Most initial founders of modern heritage conservation were European elites such as Ruskin, Moris, and le-Duc who worked as architects and art critics. Inextricably, the architectural view has influenced the way heritage has been defined as tangible, monumental and aesthetic things and at the same time architectural heritage has been shaped and reshaped by this AHD discourse [5,7]. Therefore, architectural heritage has been strongly associated with aesthetic and monumental buildings whose values are strongly linked exclusively to their form [5]. Architectural forms: façade, plan, ornamentations, amongst other elements shape the meaning of architecture, like other artworks [8]. In fact, architecture has two important aspects: forms and uses or functions. The later has been overlooked in the discourse. The relationship between tangibility and intangibility (between architectural form and its function) has gained little attention [9,10], and this interrelationship requires in-depth exploration. In post-disaster reconstruction, the destruction of architectural heritage has been only mapped and understood as the destruction of physical destruction without emotional consequences.

Secondly, little attention has been given to cultural issues in post-disaster contexts, so that this creates gaps in the development literature and debate over the importance of cultural issues for peoples' resilience [11]. This only to attract attention after a series of major disasters, including the 2004 tsunami and earthquake in Aceh, and those of 2011 in Japan, and Hurricane Katrina, in the US in 2005. The attention is shown by the discussion of cultural aspects of post-disaster reconstruction [see for example 11-14]. This means cultural issues, including heritage problems and reconstruction after a tremendous disaster is secondary. The cultural issues have been given in the later stages of reconstruction or if it is given; it is separated from emergency, technical and other more urgent issues such as health, infrastructure, and housing. The issues of food supply, shelter, and health, will, of course, be a priority in post-disaster conditions. The aid distribution, especially during emergency stages, stands a better chance of being successfully distributed if the cultural context of its beneficiaries is understood and their cultural resilience also contributes to the ability of people to recover from such extreme situations [15]. Therefore, the role of culture should also be recognized as important in the early phases of recovery. For example, "cash for work", a scheme launched during the early stage of post-disaster reconstruction in Aceh, has been criticised for changing community's values on communal works like *gotong-royong* and disregarding the local tradition. In addition, there is separation between cultural practices, such as traditional ceremonies, artisanship, and other intangible cultural and building conservation. In fact, some cultural practices occur and take place in heritage buildings and to build heritage building requires artisanship and craftsmanship. This also means there is a tendency to separate tangible from intangible responses, which negatively impact on post-disaster society.

3. Architectural Heritage: Resilience and Reconstruction in Banda Aceh

The issue of resilience has become central in post-destruction contexts [16]. This can be observed in the increasing the discussion and expose of cultural resilience following disaster and war [14]. For example, in the case of Japanese cities under threats of continuous disaster, they show that traditional principles and building techniques have contributed to their resilient and this traditional knowledge has enabled them to recover from the disasters and, even, had private initiative reconstruction [17]. Another example is resilience developed outside the disaster-hit area. Read [19] shows the ways survivors develop their resilience outside the affected area and the ability to return to a destroyed site. In this sense, the resilience is evidence in the ways people show their ability to rebuild their lives while coping with memories of a painful past, the loss of beloved people, places, buildings, and other familiar environments [20-22]. In this regard, resilience is the ways people recover from extreme situations, aftershocks and back to a 'new normal', in which activities are retrieved, the city and its infrastructure are rebuilt, and social, cultural and economic life function again [14]. In literature of relation to resilience and architecture, two interrelated elements of resilience have emerged: rebuilding the city [14] and establishing public memory discourse [20] which is usually associated with the creation of physical public space for commemorations, such as memorials and monuments [14,20,23], and the discussion in public domain such as newspaper, magazines and the like; it is in a more abstract nature which Freud would named a "talking cure" [22].

To understand the ways architectural heritage play a role in providing resilience, I have interviewed 166 respondents at three iconic sites of Banda Aceh: the Masjid Baiturrahman, the Tsunami Museum, and Masjid Peulanggahan (Teungku Dianjong), using architectural anthropology as a method and then analyzed the data using NVivo®. Architectural anthropology enables an ethnographic approach emphasizing in-depth engagement to fully understanding of a particular setting of the subject being researched, to have access to a wide audience of its human validity, yet its explanatory theory that can't be applied to many settings like grounded theory [24]. The full immersion of the researcher in a particular context through a reliance on unstructured data is required and a focus on single case or small number of cases, and data analysis that emphasizes the meanings and functions of human actions [24]. I make no claim that the interviewees are representative of the society in Banda Aceh.

I asked several main questions such as: what is overall reasons for visiting the sites, what do you experience when visiting the sites, what do you remember when visiting the sites, what does the building mean to Banda Aceh, what is the building of importance for Banda Aceh before and after the tsunami, and finally in what ways does Banda Aceh remain the same or change after the tsunami, and to what extent these changes affect people attachment to places?

There are at least two ways Architectural heritage provide resilience to the survivors of the 2004 Tsunami Disaster in Banda Aceh. Firstly, architectural heritage, due to its iconic nature, has played important role in relocating survivors and aid workers in emergency situation. The iconic architectural heritage through its physical elements which have high visibility features and ubiquitous presence, spatially guides survivors. As an iconic landmark, it is popular [25], so that it is familiar to the observers. It becomes a familiar object that helps people orient themselves after the disaster and guide visitors and embrace returning citizens on their first days in Banda Aceh. In addition, in describing changes and losing places after the disaster, people use the physical aspects of iconic architectural heritage as tools. Three things create the familiarity of the form of architectural heritage: its tangible monumental (massive) form [25], complex and unique form, and central location [26]. At least, after the 2004 earthquake and Tsunami and the reconstruction that followed, heritage places got the highest recognition 91%, from interviewees as places that were familiar and unchanged in the city [27]. Here example from my interviews expressed the roles of architectural heritage as place marker.

> The Masjid Baiturrahman is a place marker, especially for people outside Banda Aceh such as Jakarta. They refer the masjid as a place landmark to locate other places nearby. For example, when they asked about my house, whether or not her house next to the mosque. (Female_Baiturrahman 37_Bachelor_Housewife_Acehnese)

> I still recognize Masjid Teungku Dianjong, although its wall has been changed from timber into bricks. The roof is still the same….
> (Male_Baiturrahman 32_Highschool_Selfemployed_Acehnese)

Table 1: Summary of Data

Item	Masjid Baiturrahman	Tsunami Museum	Masjid Peulanggahan	Outside Sites	Total Interview
Sex					
Male	19	22	7	25	73
Female	50	22	10	11	93
Total	69	44	17	36	166
Age					
<17					
18-24	22	10	3	25	60
25-34	17	14		9	40
35-44	8	11	8	1	28
45-54	12	3	3		18
55-64	8	4	2	1	15
over 65	2	2	1		5
Place of Residence					
Banda Aceh	51	25	16	30	122
Within Aceh	14	8		2	24
Within Indonesia	4	5	1	2	12
Outside Indonesia		6		2	8
Duration in Banda Aceh					
1-7 days	15	17	1	1	34
8 days-11 months	2			2	4
1 year - 8 years	15	11	2	18	46
≥8 years	37	16	14	15	82

Secondly, architectural heritage has supported reconstruction process in the way it provides a place for seeking refuge and decision-making. Heritage has the potential for providing people with resources for resilience. As argued by Rico [28], heritage is one of the sources of resilience due to its ability to preserve traditional knowledge, so that it is not a passive victim which is at risk and should be protected. Architectural heritage like survived mosques gained added value due to their survival rates during the tsunami, which also helped create resilience in the community as well. In addition, Daly and Rahmayati [16] argue that the familiarity created by heritage and social and cultural aspects of recovery are important for the reconstruction process. For them, survivors, the landscape has two roles: abstract places embodying their long-term processes of cultural understanding and material places to contain activities contributing to resilience. Therefore, changes in the built environment like mosques and *meunasah* have contributed to the loss of familiar places [16]. Religious buildings have served as shelter, meeting point, etc. in the face of calamity [18]. In Aceh, the mosque and *meunasah*[1] have served as meeting place for discussing reconstruction process [16]. Along with the importance of architectural heritage, Samuels [17] argues that the non-physical environment like neighborhoods has provided tsunami survivors resilience too. Everyday activities have contributed to rebuilding processes along with physical reconstruction. In addition, the survival architectural heritage has served as places to seek refuge during the 2004 Tsunami and following earthquakes. Moreover, architectural heritage functioning as mosques has greater chance to be used as evacuation place [6]. Here some examples from my interview:

> During the earthquake and tsunami in 2004 we firstly gathered at Pasar Aceh (Aceh Market) and then we moved to the Masjid Baiturrahman to find safer place. I got home I might die. The Masjid saved my life. A lot of people seek refuge in the Masjid.....
> (Female_Baiturrahman 19_Highschool_Unemployed_Acehnese)

> ...this masjid is one of the survived masjids during the Tsunami and a lot of people seek refuge here. Even during another earthquake after the 2004, people also run to the Masjid to save their lifes. This means that the masjid is in their hearts and minds as an evacuation center.
> (Female_Baiturrahman 18_Highschool_Universitystudent_Acehnese)

> This Masjid (the Masjid Baiturrahman) is a comfortable place to pray. What I remember about Banda Aceh is about this masjid. This masjid is *keuramat* (sacred) because the water couldn't came inside the masjid; not like those luxurious houses were destroyed. This is the Allah (God) will.
> (Female_Baiturrahman 34_Elementary_Housewife_Acehnese)

4. So What?: Implication and Further Research

The most important implication of this research is we should rethink the ways modern disaster response, and reconstruction treats architectural heritage during emergency and reconstruction process. Aid workers and donors should take into account architectural heritage as an active agent too, not merely passive victim. They should consider to empower the survived architectural heritage and give a proper treat during the emergency stage, so that local people or survivors have enough room to reconstruct their architectural heritage in the ways they traditionally do.

Therefore, the immediate action to restore the function of architectural heritage, especially those related to every day (communal) activities such as mosques, *meunasah*, and others. It is central for the Acehnese after the tsunami. At least, let people use the available place in the ways they traditionally do. The following reconstruction or restoration of the damaged architectural heritage is also should be undertaken in the ways people familiarly do. Since the ways experts in heritage studies define architectural heritage reconstruction after the tremendous disaster is potential for diminishing people resilience. For example, the notion to restore architectural heritage in authentic form, so that it might delay its reconstruction and it requires a lot of resources in contrast to what people want. People, especially in Banda Aceh, value the authenticity of uses or function; rather than material [27].

---

[1] A communal place which is located in village level. It serves as a prayer and gathering place as well as a place for education and communal activities are conducted.

References

[1] Badan Pusat Statistik Aceh 2011 *Jumlah Penduduk Provinsi Aceh Menurut Kabupaten/Kota 2000-2011* http://aceh.bps.go.id/?r=data/eprint&id=62 (Retrieved 24 December 2012)
[2] Kenny S 2010 *Reconstruction through Participatory Practice?* In Clarke M, Fanany I and Kenny S *Post-Disaster Reconstruction: Lessons Learn from Aceh.* (London: Earthscan, Ltd)
[3] Ananta A 2007 *The Population and Conflict* In Ananta A and Onn L P *Aceh: A New Dawn* (Singapore: Institute of Southeast Asian Studies)
[4] Badan Rekonstruksi dan Rehabilitasi Aceh dan Nias/ BRR 2009 *Seri Buku BRR*
[5] Orbasli A 2008 *Architectural Conservation* (UK, USA and Australia: Blackwell Publishing)
[6] McCaughey J W, Mundir I, Daly P, Mahdi S and Patt A 2017 trust and distrust of tsunami vertical evacuation buildings: Extending protection motivation theory to examine choices under social influence *International Journal of Disaster Risk Reduction* **24** 462-473
[7] Smith L 2006 *Uses of Heritage* (New York: Routledge)
[8] Whyte W 2006 How Do Buildings Mean? Some Issues of Interpretation in the History of Architecture *History and Theory* **45**153-177
[9] Leathart J 1940 *Style in architecture (*London: Nelson)
[10] Ching F D K 2007 *Architecture: Form, Space, and Order* (New Jersey and Canada: John Wiley and Sons, Inc)
[11] Barakat S 2007 Post War Reconstruction and Recovery of Cultural Heritage : Critical Lessons from the last Fifteen Years in Cultural Heritage in Post War Recovery. In N. S. Price *Cultural Heritage in Postwar Recovery* **6** (Rome, taly: ICCROM Conservation Studies)
[12] Al-Nammari F M. and Lindell M K 2009 Earthquake Recovery of Historic Buildings: Exploring Cost and Time Needs. *Disasters* **33** 457-481
[13] Lakoff A 2010 *Disaster and the politics of intervention* (Columbia University Press)
[14] Vale L J and Campanella T J 2005 *The resilient city: how modern cities recover from disaster (*New York: Oxford University Press)
[15] Ascherson N 2007 Cultural destruction by war, and its impact on group identities. In N. Stanley-Price *Cultural Heritage in Postwar Recovery* **6** (Rome, Italy: ICCROM Conservation Studies)
[16] Daly P and Rahmayati Y 2012 cultural heritage and community recovery in Post-Tsunami Aceh In *From the Ground Up: Perspective on Post-Tsunami and Post-Conflict Aceh* edited by Daly P, Feener R M and Reid A 57-78 (Singapore: ISEAS Publishing)
[17] Samuels A 2010 Remaking neighbourhoods in Banda Aceh: Post-tsunami reconstruction of everyday life In Clarke M, Fanany I and Kenny S *Post-Disaster Reconstruction: Lessons Learn from Aceh* (London: Earthscan, Ltd.)
[18] Sugimoto S and Sagayaraj A 2011 *Social frame, religious networks, miracles: experiences from tsunami disaster management in south India* (Kentucky, the United of America: The University Press of Kentucky)
[19] Read P 1996 *Returning to Nothing: the Meaning of Lost Places* (Cambridge: Cambridge University Press)
[20] Huyssen A 2003 *Present Pasts: Urban Palimpsets and the Politics of Memory* (Stanford, California: Stanford University Press)
[22] Geertsma A 2011 Redefining Trauma Post 9/11: Freud's Talking Cure and Foer's Extremly Loud and Incredibly Close *Aspeers* **4**
[23] Fredericks V 2011 Remembering Katyn: Mourning, Memory, and National Identity. *Memory Connection* **1** 197-210
[24] Groat L and Wang D 2002 *Architectural Research Methods* (Canada: John Wiley and Sons, Inc)
[25] Sklair L 2011 Iconic Architecture and Urban, National, and Global Identities In D. E. Davis

and N. L. d. Duren *Cities and Sovereignty: Identity Politics in Urban Spaces* (India, USA: Indiana University Press)

[26] Appleyard D 1976 *Planning a pluralist city : conflicting realities in Ciudad* (Cambridge, Mass: MIT Press)

[27] Dewi C 2017 Rethinking architectural heritage conservation in post-disaster context *International Journal of Heritage Studies* **23** 587-600

[28] Rico T 2014 The limits of a heritage at risk framework: The construction of Post-disaster cultural heritage in Banda Aceh, Indonesia *Journal of Social Archaeology* **14** 157-176

[29] Meltzner A J, Sieh K, Chiang H W, Shen C C, Suwargadi B W, Natawidjaja D H and Galetzka J 2010 Coral evidence for earthquake recurrence and an A. D. 1390–1455 cluster at the south end of the 2004 Aceh–Andaman rupture *Journal of Geophysical Research* **115** 1-46

[30] Lombard D 1991 *Kerajaan Aceh : Jaman Sultan Iskandar Muda (1607-1636)* (Jakarta: Balai Pustaka)

Bring back history alive through transformation of old building into museum

Norashikin Abdul Karim, Siti Norlizaiha Harun[1], Salwa Ayob[2]
[1]Faculty of Architecture, Planning and Surveying, Universiti Teknologi MARA, Perak Branch, 32610 Seri Iskandar, Perak, Malaysia
[2]Faculty of Art and Design, Universiti Teknologi MARA, Perak Branch, 32610 Seri Iskandar, Perak, Malaysia

E-mail: shikin6463@gmail.com

Abstract. When looking at the old building, it has aroused the curiosity of the glorious history of architecture and interior space in the past, such as the furniture, features, fittings, furnishings and interior layout. To save our old and heritage building from decay and loss of the heritage value, one of the Malaysian government efforts is to restore it to become a new function as a museum. There are some criteria should be considered for the transformation of an old and heritage buildings into a museum, there are; integrity of history; original structure and materials; new space and new function; showcase display and interpretation; and visitors' perception. This paper will highlight the literature review on consideration factors in heritage museum restoration. This writing paper is also part of continuous research aimed at developing the criteria for assessing the old and heritage building as a museum. The proposed criteria could serve as a basis for heritage organizations to prepare guidelines to manage the transformation of an old and heritage building into a museum in Malaysia. Indirectly, it contributes the ultimate goal to give best heritage experience for the tourist in a museum tour.

1. Introduction

An old building is a historic building that was built at any time in the past [1]. It can also be termed as a heritage building because of its high historical value to a community. But not all old buildings can be labeled as heritage buildings. Old buildings also defined as a historic building. Briefly, a historic building is one that can give a sense of wonder and to know more about the people, culture, architecture, aesthetic, historical, documentary, archaeological, economic, social and even political, spiritual or symbolic values and the first impact is always emotional, for it is a symbol of our cultural identity and continuity is a part of our heritage [8]. While, Heritage building may be perceived as having some aspects that are considered important to the community. It also called a heritage-listed building had been evaluated as having aspects that it is considered necessary to preserve it [10]. Almost all of the old buildings, are deteriorating and need to be restored. Restoring it into a museum is a one way to preserve the old building and it can disclose continuous care of the buildings. It is important to have this concern because, without it, the historic buildings will never be preserved or conserved properly. Also, it is important to retain existing and possible rare examples of buildings, architecture or craftsmanship or historic interior scheme which would otherwise disappear [1].

Why can the museum be a tool to bring back history alive of historic buildings? From a traditional understanding of what form a museum was defined in functional terms. Hence the intent of museums was realized as concrete and tangible, paralleling the substance of the 'material evidence' [19] which historically has been the main focus of museums [4]. A museum also as an interpretive medium communicates focusing on a big group of the public with the intent of transfer information, ideas and reaction relate the material evidence of man. The main purpose to set up a museum was revealing and unravel the history of the building and building components can be as a collection, thus confirming public trust in the museum as caretaker of the societal record [5].

2. The transformation of historic buildings into museum in Malaysia

Most of the historic building in Malaysia is scattered around in the country consists of various types of buildings and functions. The chart below (Figure 1) presents the historic building that can be divided into 16 building categories such as fort/military, churches/mosques, palaces, clock tower, prison, government office, institutional and commercial, residential, schools, railway stations, hotels & guest houses, monuments, mausoleum, dam, warehouse/storage, and public building. Of these 16 categories, some of the buildings have changed their function and some have remained their original function. There are 5 categories of historic buildings have changed their function into a museum such as 4-fort/military, 5-palaces, 7-government office, 11-institutions and commercial and 3-residential, the total of numbers are 31 buildings and most of the buildings are gazetted.

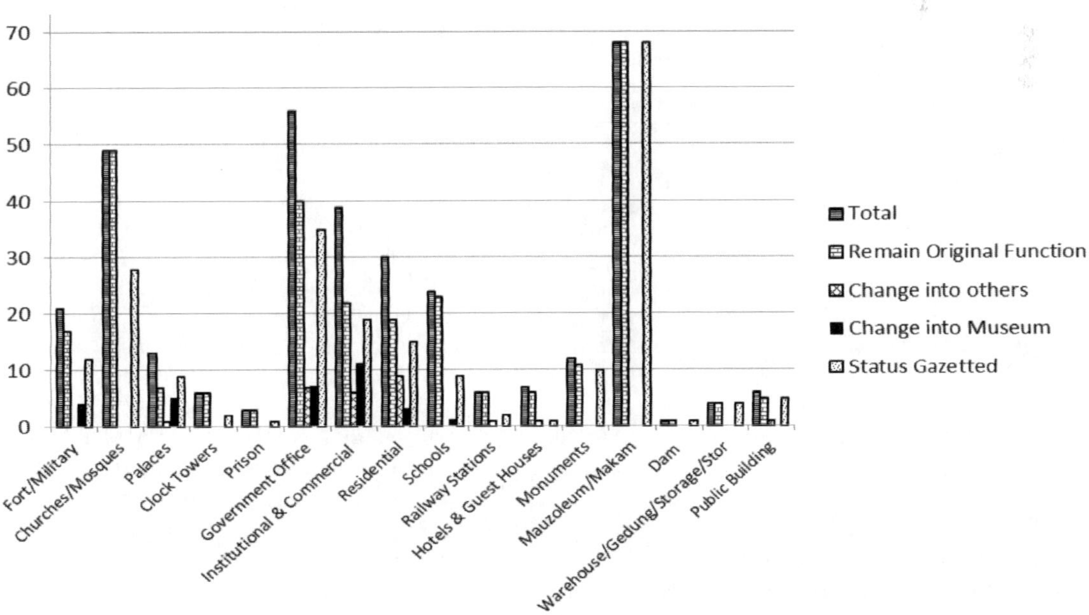

Figure 1. The categories and change of used of historic building in Malaysia

IOP Conf. Series: Materials Science and Engineering **352** (2018) 012064 doi:10.1088/1757-899X/352/1/012064

3. Transformation criteria of historic building into a museum

3.1. The meaning of building transformation in the understanding of historic building conservation

The term 'transformation' in the understanding of historic building conservation, has carried their own meaning. The meaning transformation here referred to the building has fewer changes [8] and it has been a change in the term of 'conjectural reconstruction' or 're-creation' [3]. These philosophies represent an acceptable action and seek to create an earlier appearance when the building transform their function. It can be concluded that the building transformation is a process to change the old building appearance according to the new function of the building like to be a museum. It requires an idea and the history of the building to recreate the same scheme as the original scheme of the building as an exhibit material in the museum. Indirectly, this transformation can restore and revive the historic interior scheme to visitors. There are some criteria should be considered for the transformation of historic buildings into a museum, including the integrity of history; original structure and materials; new space and new function; showcase display and interpretation; and visitors' perception (Figure 2).

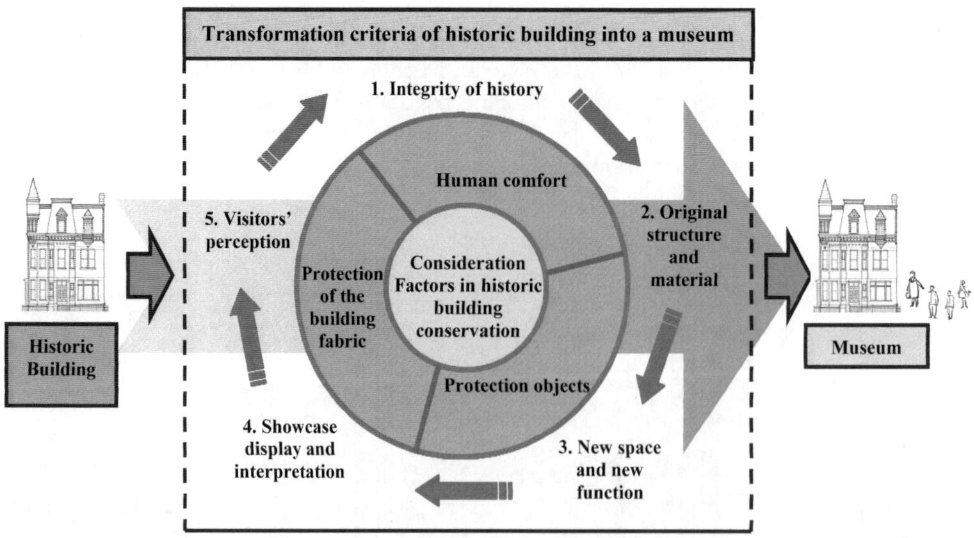

Figure 2. The transformation criteria and consideration factors for historic building conservation

3.2. Integrity of history

The most important criteria for setting up a museum in a historic building is honesty in reviving history. To apply integrity of history in the museum building, it is crucial to understand the importance of the building through a combination of survey on the ground and a desktop study to expose the building's history and to understand deeply what in the interior is authentic [7]. However, any historical study of such a building should include the client who commissioned it, together with his objectives which led to the commissioning of the project and an assessment of the success of its realization; the study should also deal with the political, social and economic aspects of the period in which the structure was built and should give the chronological sequence of events in the life of the building. And also the names and characters of the actual creators should be recorded, if known, and the aesthetic principles and concepts of composition and proportion relating to the building should be analyzed [8].

3.3. Original Structure and Materials
To transform the historic building into a museum, the original structure and presentation of material should be concerned because these are the potency medium to bring back the building history alive. These criteria are an important part of an exhibitor material which most visitors excited want to see. Because of that, the aim of exhibiting the conservation is to create and contribute preservation-friendly exhibits that attracting and to be informed to the public. And an effective museum exhibit can achieve its educational purpose, be aesthetically appealing, and meticulously safeguard the components on building display [16]. The structural and material condition must also be studied: the different phase of construction of the building complex, later interventions, any internal or external peculiarities and the environmental context of the surroundings of the building are all relevant matters. The original structure which is common in the historic building and should be respected are beams, arches, vault, domes, trusses and frame, walls, piers and columns. And he considers there are two main structural materials found in historic buildings which is timber and mansory [8]. Every existence material and original structure built up in the historic building should be strictly observed and the below table (Table 1) is some guidance need to be considered [3] ;

Table 1. The International Charters Guidelines

International Charters	Guidelines for works on site
Athen Conference (1931), Venice Charter (1964), Burra Charter (1979) & Appleton Charter (1983)	Should be highlighted in the choice of material and the techniques, even the respectfulness against traditional practice.
New Zealand Charter (1992) and code of ethics (1993)	The use of modern material substitutes is appropriate only when; they provide a significant advantage which can be identified; their material used has a solid scientific basis; and the material physical has been carried by a body of involved.
Thessaloniki Chater (1992)	The new material should compatible with the expression, appearance, texture and form of the original.
Declaration of Tlaxcala (1982)	Should meet the requirement of both the physical and geographical conditions and the manner of life of the population.
Venice Charter (1964), Budapest Resol's (1972), Washington Charter (1992), Burra Charter (1979), Desch. Decl.(1982), Washington Charter (1987), Appleton Charter (1983) & Thessaloniki Chater (1992).	When assessing the aesthetic integrity and coherence of the whole, the factors should be taken into account are; the relations of mass and colour and ; the traditional background, the stability of its structure and its correlation with its environment; form, scale, colour, texture and materials; tone of shape, appearance, segment, filled with ornament , and overall formation; existing and original patterns of movement and layout; and lastly the (vernacular) plan, volume and shell.

3.4. New Space and New Function
To proposed new space and new function into a museum, it should start with exhibition layout and understand the exhibition environment [16]. However, the most important concern is in the context of

interpreting new scheme of space and function. Participation and obtained knowledge always present the best context in the study of the scheme; and each scheme needs understandeness and difference that will require particular research to allow it to be allocated firmly within its setting of time and sense. Compatible work may glance at another contemporary scheme by the identical client. Perhaps, it involves a comprehensive study and assimilation of contemporary sense such as colour theory, archaeological and antiquarian obsession, or scrutinize about style-setters [9]. Interpreting research is used to recreate method to generate scheme for existing space in historic building to be used as a new function for the museum exhibition. Below (Table 2) are some of anxiety that can undermine the final look of a reinstated scheme [9] ;

Table 2. The typical concerns on reinstated scheme for historic building converted into a new function

Main concern	Advices
Function & Room Usage	The usage of room, historic room layout and original surface finishes should be investigated.
Replacement of furnishing or material	Be conditional on what already exists in the room, furniture, upholstery, carpets, wallpapers, painted decoration and pictures may be needed. All will need to be sourced with an understanding of authentic making and also of what is technically practicable.
Existing furnishings, fixtures and decoration	Avoiding the technical complexities of the different conservation disciplines, and the various admissible ways of cleaning and making good, conserved object ultimately have to be satisfied within a decorative scheme.
Artificial lighting	The reinstated scheme required a constant environment, and artificial lighting may provide to reinforce an authentic sense, or affect it, both in the fittings impart and in the type, color and quality of the light fabricate.
Gaps in evidence	Well-dispute cases need to be made to fulfill gaps or doubts, as the evidence base for a room's reinstatement is seldom finish. The last choices become easier once the project team has an understanding of the situations and feel that shaped the original scheme.
Clarifying a scheme	There are many interpretation approaches that may explain a room's display. These options may require developing, and a display be fine-tuned, in answer to changing technologies. The main criterion is that no approach affects the practical conservation of the architecture or collection and that all significant hardware and display material is fully reversible.

3.5. Showcase Display and Interpretation
Interpretation of the object display is the physical presentation of an interior itself, there are further levels of interpretation to assist the visitor experience involved in the story of how the room was first created, and how it came to be recreated. Exhibitions, either in a minimal form, a discreet display within the recreated room or larger, separate exhibitions covering wide aspects, can incorporate to the understanding of a recreated scheme within a building [9]. The museum showcase display should always be open to identifying and attracting a new audience. Factors such as special interests within the museum's constituency change, and as the factors change, so does the visitors and potential visitor. The museum staff must be aware of the changes and make the exhibition and programming decisions about developing

new audiences [5]. However, the most important link is that of the type of successful visitor interpretation aims to create, and interpretation is attempting to produce mindful visitors; which is visitors who are vigorous, excited and always questioning. On the other hand, has been discovered to be related to reduce visitors' attention, which can be seen as an indicator of mindlessness. It is possible that mindfulness and mindlessness are valuable concepts in thoughtful how visitors react to the interpretation of object display in an exhibition [14].

3.6. Visitors' Perception

Visitors that going to museums or other historical location are especially searching an image of the past which is recognized as very far [13] rather than an authentic historical exposure [18]. The sensory perception of authenticity in a museum is multiplex because both authentic and inauthentic exposure coincides in equal parts in being part of the visitor experience [11]. There are 3 different levels to be taken into consideration when involving with museums and authenticity such as artifacts, edifices and encounters [15]. However, objects displayed in museums and the information provided by these objects are in general recognized by the visitors to be authentic. As a result, the museums must appear themselves as places where visitors can acquire knowledge, to discover, acquaint something, and enhance their own culture, rather than as spaces created in a destination in order to attract visitors' attention [2]. What do people expect to find at the museum? The main visitors wish to see the genuine object the 'real thing' [5]. Visitor observes show that the most dominant types of interactive exhibits are those in which visitors react to an informed guide, a museum demonstrator or an interpreter, perhaps in period costume, this person-to-person interaction can be among the most effective experiences that a museum can bid, especially if staff or volunteers are trained to obtain questions and visitors will be involved in their presentations [12].

3.7. Consideration factors in historic building restoration

The consideration factors in historic building restoration are, it is often the case that the contrast with needs of these three requires a level of compromise such as human comfort; safekeeping objects, including internal finishes and furnishings; and protection of the building fabric [6] (Figure 2). The history and archaeology of the building must be carefully studied, so that the architect is aware of the sequence in which the fabric was put together. Another important part is also should consider the original material, archaeological evidence, original design and authentic documents [8]. For the most historic building if it does not use the honesty of local materials it should exchange them to new uses for a modern material as long as a building proves to be in harmony with itself, its surroundings and its users [17].

4. Discussion and conclusions

Transforming a historic building into a museum is a challenge in creating an interior scheme and require guidance. Because, it involves both not only recreating but preserving the historic scheme. Research satisfies the responsibility to record, acquaint and preserve for historic buildings, collections, schemes and tastes, and helps to unlock the depths of meaning and value behind the display of interiors, it not only exposes artifacts and craftsmanship, but also open windows on history, customs, people and interacting the social life [9].Many benefits can be obtained by restoring the historical scheme inside the historic building such as visitors can re-experience the true historical scheme. This is the moment that every visitor has waiting to know and they will feel excited approaching historic buildings. Therefore, the guidelines on recreating and preserving historic scheme were needed when the historic building transforms into a museum.

References

[1] Ahmad, A. G. (1997). *British Colonial Architecture in Malaysia 1800-1930* (Kuala Lumpur: Museums Association of Malaysia)
[2] Brida J G, Disegna M and Scuderi R 2014 The visitors' perception of authenticity at the museums: Archaeology versus modern art *Current Issues in Tourism* **17** 518-538
[3] Bell D 1997 The Historic Scotland Guide To International Conservation Charters (Edinburgh: *Technical Conservation Research and Education Division*)
[4] Corsane G 2005 *Heritage, museums and galleries : an introductory reader / edited by Gerard Corsane Heritage* (London and New York: Routledge)
[5] David D and Gary E 1996 *The Handbook for Museums (Heritage: Care-Preservation-Management) 1st Edition* (London and New York: Routledge)
[6] David D 2012 Environment in the historic interior. In M. F. White *Interior Finishes & Fittings for Historic building Conservation* (United Kingdom and USA: Blackwell Publishing Ltd) p 187
[7] Forsyth M 2012 Conservation of the historic interior. In M. F. White, *Interior Finishes & Fittings For Historic Building Conservation* (United Kingdom and USA: Wiley-Blackwell Publishing Ltd) p 4
[8] Feilden B M 2002 *Conservation of Historic Buildings* (Amsterdam, Boston, Heidelberg, London, New York, Oxford, Paris, San Diego, San Francisco, Singapore, Sydney and Tokyo: Architectural Press.)
[9] Finlay J 2012 Recreating historic schemes of interior decoration. In M. F. White *Interior Finishes & Fittings for Historic Building Conservation* (United Kingdom: Balckwell Publishing Ltd) p 173-186
[10] Hay Gaol Museum (2004) Just because it's old: museums and galleries in heritage buildings. *Museum and Galleries Foundation* 1–26
[11] Hall CM 2007 Response to yoeman et al: the fakery of 'the authentic tourist' *Tourism Management* **28** 1139–1140
[12] Lord B and Lord G D 2002 The purpose of Museum Exhibitions. In B. L. Lord, *The Manual of Museum Exhibitions* (Walnut Greek, Lanham, New York and Oxford: Altamira Press) p 11-25
[13] Lowenthal D 2005 *The Past is a Foreign Country* (New York: Cambridge University Press)
[14] Moscardo G 1996 Mindful visitors: Heritage and tourism *Annals of Tourism Research* ***23*** 376–397
[15] Pine II B J and Gilmore J H 2007 Museums and authenticity *Museum News* http://www.strategichorizons.com/documents/MuseumNewsMay07 Museums%26Authenticity.pdf (Accessed 8 september 2012)
[16] Raphael T J 2005 Guidleine-Preventive conservation and the exhibition process; development of exhibit guidelines and standard for conservation. *Journal of the American Institute for Conservation* **44** 245–257
[17] Robertson M 2007 Conservation and authenticity In Forsyth M *Understanding Historic Building Conservation* (Oxford and Australia: Blackwell Publishing) p 31
[18] Schouten F 2007 Cultural tourism: between authenticity and globalization *In Cultural : Global and Local Perspectives* Richards G (The Haworth Press Inc) 25-37
[19] Weil S 1983 The Multiple crises in our museums' *in Beauty and the Beast: On Museums, Art and the Law and the Market* (Washington: Smithsonian Institute)

IOP Conf. Series: Materials Science and Engineering **352** (2018) 012065 doi:10.1088/1757-899X/352/1/012065

The study of selective property of college student's learning space

Mizuki Nagai, Yuji Matsumoto, Ryusuke Naka
Design and Architecture, Kyoto Institute of Technology, Japan

E-mail: mizu.1913jks@gmail.com

Abstract. These days, college students study not only at places designed for learning such as libraries in colleges, but also cafes in downtown while the number of facilities for learning run by colleges is increasing. Then I have researched facilities in college and those in downtown to find selective properties of college students' learning space. First, I found by questionnaire survey that students chose ''3rd place'' such as cafes and fast food shops, second to their houses and libraries in college. Next, I found ''psychological factor'' were also affected their choice. Furthermore, they studied different subjects at different places. In experiments, I researched how effectively they studied each subject at every place. The results show that I find that places you like and places where learning efficiency is good are different. They learned the least effective at ''3d place'' regardless of what they learned. The result of how long they kept high-level intellectual activity at each place shows that they could work on the study with more motivation at their favorite place and 3rd place. On the other hand, at the 2nd place, they could study rather effectively, but could not keep concentration and motivation for a long time. In this way, college students have 2 patterns of choosing learning space.

1. Background and purpose

In recent years, "shared space of universities that includes learning services managed by universities supporting knowledge creation" such as learning commons is spreading. It is expected that further improvement of learning · commons will be advanced further in the next 3 years. However, learning places are spreading not only to places that have learning functions like schools but also to third places such as cafes and fast food stores. In this research, I focus on "learning efficiency" for each learning content and "intelligent activity degree" in time transition. I conduct research on learning places inside and outside the campus and examine the spatial selection characteristics of university students.

1.1. Previous Study on Third Place

Third place is the place defined by urban sociologist Ray Oldenburg in 1999 and he insists on the importance of informal people gathering place in urban life [1]. In addition, it is the third place that is located between the two of the individuals' homes, workplaces, and connects them and keeps balance [2]. Such a way of thinking is also spreading in college students and forms a third place in town. Among studies on college students' third place, research on Okumoto [3], which clarifies in which part and where places are likely to be formed. Kawakami [4] has research that puts the focus on college students' homes and favorite places other than university and clarifies the actual condition of the third place. One example of third place, in particular, is a cafe. There are many papers on whereabouts in the cafe space, such as research on cafe space that emphasizes communication by [2]. On the other

IOP Conf. Series: Materials Science and Engineering 352 (2018) 012065 doi:10.1088/1757-899X/352/1/012065

hand, the conventional cafe was a place intended for eating, drinking, and conversation, but pointed out that users are increasing for personal work such as work and study. As mentioned above, the use of cafes aimed at work and study, etc. is progressing, but few studies focus on the third place as a student's learning space.

1.2. Previous studies of college students' learning places

Ishizuka [5] analyzed the actual situation of use of learning commons and space composition, and gained insight to create future learning space. Megaki [6] observed the usage situation on a regular basis and described the possibility of how to use Learning Commons in the future. In the study of Learning Commons, most are research on planning and space. Endo [7] "Research on Third Place as a College Student's Learning Place" pointed out that the relationship between the decision on the decision to use third place and the relationship between third place's location and decision making is not clear.

Therefore, he conducted a questionnaire survey and an interview survey with a specific human group capable of being captured as an analysis target, captured the actual state of learning carried out nomadically, and clarified the following things.

① Process of determining learning place and its location trend
② Characteristics of consciousness of college students viewed from the reasons of nomad learning
③ Value of Third Place as a learning place in the city

As mentioned above, research has been conducted for each place, but as a learning place for college students, there are no studies that compare the selection characteristics of places, including the on-campus facilities and the off-campus facilities.

2. Experimental/Methods

2.1. Questionnaire survey

Previous studies showed that the movement of establishing Learning Commons is in the whole university and that college students use third places as learning space. Therefore, a survey was conducted for university students in Japan to grasp the actual conditions of Japanese college students' learning places. The questionnaire survey outline is shown below (Table 1)

Table 1. Outline of Questionnaire survey

Purpose	Understand the actual conditions of college students' learning places				
Object	College students in Japan (86 valid responses)				
Period	September 20, 2016 - September 30, 2016				
Contents	① Choose places to use for studying from the following (Multiple answers possible)				
	A Learning Commons	B canteen	C campus café	D laboratory	E classroom
	F library	G house	H cafe outside campus	I train	J Other
	② In each place selected in ①				
	the frequency of use	Reason for selection		Usage time zone	
	Tools to use		Average learning time per one time		

In the following sentences, "first place" is "house", "second place" is "library", "third place" is "outside campus or fast food store". In addition, "First Place" is written as "1st P", "Second Place" as "2nd P", "Third Place" as "3rd P".

IOP Conf. Series: Materials Science and Engineering **352** (2018) 012065 doi:10.1088/1757-899X/352/1/012065

2.2. Interview survey

According to the questionnaire survey, it turned out that students are using 3rdP as learning space after home and library. Questionnaire survey also reveals that college students are selecting and learning multiple places, and think that there is learning content in one of the factors that select place. Therefore, we aim to conduct the interview survey for the purpose of the following 2.

① Understand the actual usage of 3rdP as a learning space in more detail.
② Reveal the location selectivity for each student's main learning content

First, only subjects using 3rdP as a learning space answer questions about 3rdP.

Next, the contents of learning mainly extracted by the student in the preliminary survey, "memorizing (memorization of words etc.)", "calculation (calculation of mathematics and chemistry etc.)", "thinking (Report, etc.) ", And "creative (Idea drawing, etc.)", learning places and reasons considered to be appropriate for each study subject are asked to be answered by all subjects.

The interview survey outline is shown below (Table 2).

Table 2. Outline of interview survey

Purpose	① Understand the actual usage of 3rdP as a learning space in more detail. ② Reveal the location selectivity for each student's main learning content
Object	Kyoto Institute of Technology undergraduate student (32 valid responses)
Period	October 25, 2016 - October 31, 2016
Contents	Ask 3rdP users to ask the actual situation of use and ask all subjects to answer each place and reason why they think they can concentrate on the content of the study.

2.3. Experiment

According to the interview survey, it became clear that the places considered able to concentrate on each learning content are different, and it was found that the type of learning content is one of the factors that are selecting the place. Psychological factors such as motivation and mood are also considered to have some influence on student's place selection.

Therefore, it aims to analyze "relation between learning efficiency by space and learning content" and "relationship between space and intellectual activity" to clarify selection characteristics of college students' learning place.

Ask the subject to answer the preference of study place of "learning overall", "memorizing", " Calculation ", "thinking", and "creative". Have the subjects take the four kinds of learning at 1stP, 2ndP, and 3rdP, to ask a subjective evaluation of spatial impression evaluation and learning efficiency.

We use the VAS (Visual Analog Scale) method to know the subject's psychological change. Placing words expressing a pair state such as "very sleepy", "very waking" at both ends of the 100 mm horizontal line segment, and the state closest to the state of the subject at the time point is placed on a line perpendicular to the vertical line. In this manner, the psychological quantity at the time of measurement is measured. Let the left end be 0 and the right end be 100. From there, the distance to the point of intersection between the vertical line drawn by the subject and the horizontal line segment is measured in millimeters, and the value is evaluated in 101 steps as the score of the psychological quantity of the subject. The outline of the experiment survey is shown below (Table 3).

Table 3. Outline of Experimental survey

Purpose	① Relationship between learning efficiency for space and learning content ② Relationship between space and intellectual activity Analyze above to clarify selection characteristics of college student's study space
Object	Kyoto Institute of Technology undergraduate students(59 valid responses)
Period	December 1, 2016 - December 27, 2016

	"Questionnaire about the preference of study place" to answer.
Contents	subjects take the four kinds of learning (Table 4) at three places, 1stP, 2ndP, and 3rdP, to ask a subjective evaluation of spatial impression evaluation and learning efficiency.

Table 4. Learning contents

Memorizing(total 4 minutes)	Memorizing 20 daily words
Calculation(total 7 minutes)	Junior high school level equation (sentence title and calculation problem)
Thinking(total 15 minutes)	English translation of English class 3 grade
Creative(total 3 minutes)	Write out associative words from specified words

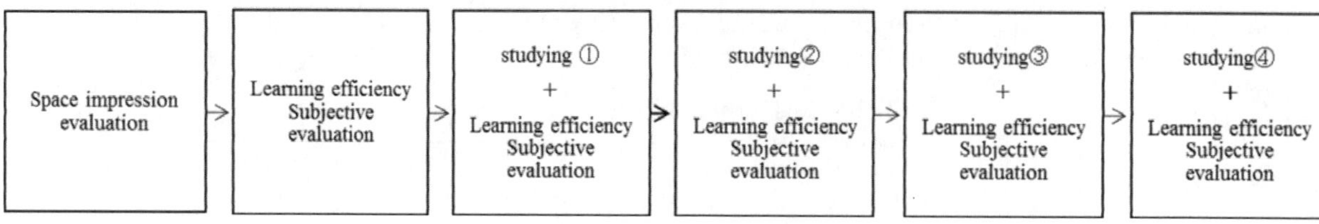

Figure 1. Experiment order

3. Results and Discussion

3.1. Questionnaire survey
Students were using cafe outside the campus (34%) and learning commons (24%) after the house (65%) and the library (57%). In addition, they select and learn an average of 2.6 kinds of places. In the library, studying mainly about university lectures, while at home and cafe, in addition to that, highly studying privacy such as reading and qualification study was done. Therefore, one of the reasons for selecting a place when they study is the type of study. Students study "studying for the test" most and are doing in various places. Therefore, it was assumed that there are places considered suitable for each learning content.

3.2. Interview survey
The reason why the student selects 3rdP is that in 1stP, there is temptation such as manga and television, but people of various age group and occupation use 3rdP, and it is not too quiet like 2ndP. It was also found that "psychological factors" were affected.

From Table 5, it is found that the places where each learning content is likely to concentrate are different, and learning contents are one of the factors of location selection. 1stP is selected for “memorizing” and 2ndP is selected for "calculation" and in "creative system". 3rdP is the best choice for reasons of receiving ideas from external sources.

Table 5. Where students can concentrate on learning content

	1stP(ranking)	2ndP(ranking)	3rdP(ranking)
Memorizing	17(1)	11(2)	3(3)
Calculation	10(2)	17(3)	4(3)
Thinking	12(1)	11(2)	8(3)
Creative	10(2)	4(3)	17(1)

3.3. Experiment

3.3.1. Relationship between space and learning efficiency.
Table 6 shows "favorite places" and "average points" by learning content. 3rdP is selected as "favorite place" as well and as "place likely to concentrate" for only "creative". (Table 3). Based on the results of the hearing 3.2, I think that students are using 3rdP as a place to change their moods and learn at ease, instead of using them for learning they want to concentrate. Also from Table 6, "favorite place" and "place where learning efficiency is good" were different.

Table 6. Where students can concentrate on learning content

	A place of your choice by learning content			Average point by learning content		
	1stP(ranking)	2ndP(ranking)	3rdP(ranking)	1stP(ranking)	2ndP(ranking)	3rdP(ranking)
Memorizing	30(1)	21(2)	8(3)	14.05(3)	15.41(1)	14.36(2)
Calculation	15(2)	31(1)	13(3)	13.22(1)	13.14(2)	12.39(3)
Thinking	23(2)	24(1)	12(2)	3.23(2)	3.35(1)	3.09(3)
Creative	22(2)	8(3)	29(1)	21.03(1)	17.34(2)	15.85(3)

In more detail, factor analysis was conducted to find out where learning efficiency is good. The correlation between the obtained four factors and the learning efficiency was examined. The first factor named "cozy factor", the second-factor "efficiency factor", the third factor "quiet factor" and the fourth-factor "closing factor".

Table 7. "Averages of spatial factor scores" and "Correlation between factor and learning efficiency"

	Average value of spatial factor score			Correlation between factor and learning efficiency			
	1stP	2ndP	3rdP	Memorizing	Calculation	Thinking	Creative
Cozy factor	0.812	-0.324	-0.164	-0.001	-0.055	-0.021	.240**
Efficiency factor	-0.51	0.445	-0.379	.177*	-0.14	-0.028	-0.08
Quiet factor	0.386	0.352	-1.09	.164*	0.11	0.114	0.052
Closing factor	0.337	0.052	-0.441	-0.068	-0.031	-0.02	0.033

*. Correlation coefficient is significant at 5% level (both sides) **. Correlation coefficient is significant at 1% level (both sides)

From Table 6, 1stP was chosen as a favorite place in "memorizing", but learning efficiency was the best at 2ndP.As shown in Table 7, a weak correlation was found between "learning efficiency" and "efficiency factor" of "memorizing ", so it seems that it is appropriate to do it in a "quiet place likely to concentrate". In "creative ", I learned that learning efficiency is the best in 1stP.Students thought that they often get ideas from outside "things" and "people" at 3rdP, but the result is that ideas are easy to emerge in places like "comfortable" like 1stP. From the above, it can be said that it is better to use a place suitable for each learning content rather than a favorite place to improve learning efficiency.

3.3.2. Relationship between space and intellectual activity.
When comparing intellectual activity degrees by location over time, the "feeling of loss of motivation" (Figure 2) always showed a low value which means state with a relatively high motivation)in the favorite place. "General vitality" (Figure 3) has a relatively high value at a favorite place and 3rdP compared with other places. It turned out that it was possible to learn with high motivation there. On the other hand, at 2ndP, "feeling of loss of motivation" (Figure 2) declined and "general vitality" (Figure 3) increased until after the second learning, but both later deteriorated sharply. 2ndP has a high score of "efficiency factor" (Table 7), it is a place where learning efficiency is relatively good, but

concentration ability and motivation do not last long, it seems that it is suitable for learning that limits time.

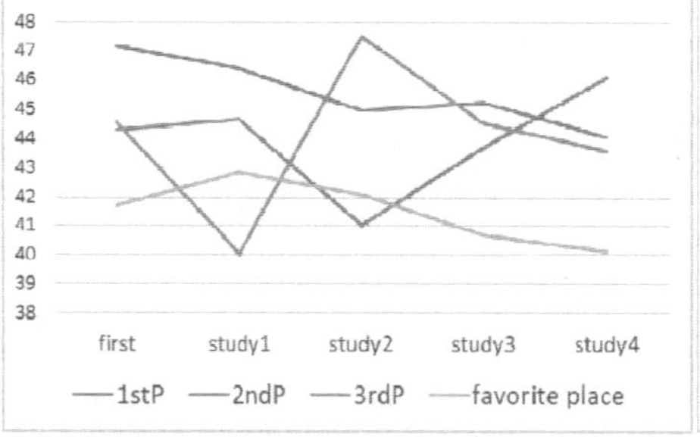

Figure 2. Loss of motivation

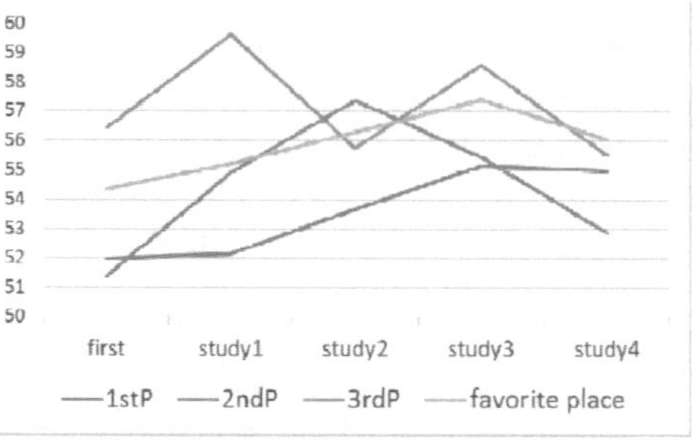

Figure 3. General vitality

4. Conclusion

Students selected various places as learning places for each learning content. Many students use 3rdP, but 3rdP was found to be poor in learning efficiency and not used for improving learning efficiency in every learning content. I learned that students could work on learning with relatively high motivation at 3rdP and their favorite places. From the above, it can be said that there are 2 patterns as a motivation for students to select a learning space. The first is a pattern that selects a place that emphasizes "learning efficiency" such as "2ndP". The second is a pattern that emphasizes "motivation" and selects "favorite place" or "3rdP". In the former, it is considered good for learning that requires concentration and last short time. In the latter case, it is considered good for learning that does not set the clearly end.

References

[1] Oldenburg R 1999 The Great Good Place *MARLOWE & COMPANY NEW YORK*

[2] Isamu I, Mitsuo T and Ken M 2005 A Study on the Cafe Space as seen from the Third Place Concept: Through Survey of the Cafe Space in the Central Area of Kyoto City *Summary of the Architectural Institute of Japan* 515-516

[3] Okumoto Y, Yasushi Y H and Hayashida H 2010 Consideration on 'Places' College Students Build in the Streets: A Case Study of Wakayama University, Kagoshima University, Kanazawa University *The Japan Architectural Institute Research Presentation* 197-200

[4] Kawakami N and Sakai T 2014 Study on Third Place of University Students: Targeting Regions around Kyushu University Campus *Journal of the Japan Institute of Architecture of Kyushu* Section **53** 277-280

[5] Yuko I, Mizuno and Yanagisawa 2014 A case study in active learning and self-study space: A study on advanced learning space in university Part 3 *Summary of the conference of the Japanese Society of Architecture Conference Pap*ers 431-432

[6] Megaki R 2008 Possibility of learning commons: Ochanomizu University's Challenge to an Attractive Learning Space *Information Science and Technology* **58** 341-346,

[7] Yuta E, Haruhiko G and Takashi Y 2014 Studies on Third Place as a Learning Place for College Students *Public Interest Corporation Japan City Planning Association Urban Planning Proceedings*, **49** 1083-1088

IOP Conf. Series: Materials Science and Engineering **352** (2018) 012066 doi:10.1088/1757-899X/352/1/012066

A review of spatial comfort in shophouse in humid tropics

Laina Hilma Sari[1,3], Siti Zulfa Yuzni[1], Muhammad Haiqal[1], Evalina Z[2]
[1] Department of Architecture and Planning, Faculty of Engineering, University of Syiah Kuala, Darussalam, Banda Aceh 23111, Indonesia
[2] Department of Regional and Spatial Planning, University of Syiah Kuala, Darussalam, Banda Aceh 23111, Indonesia
[3]Center for Thermal, Vibration and Applied Acoustic Studies of Unsyiah, Banda Aceh 23111, Indonesia

E-mail: laina.h.sari@gmail.com

Abstract. Shophouse is the most building type built along the road with various commercial functions. Due to the high rate of land price and the high percentage of building basic coefficient, therefore the shophouse is commonly built in maximum. This characteristic creates a lack of apertures through the buildings due to being trapped by surrounding shophouses. Shophouses are also built in heavyweight construction, i.e., concrete and brickwork. Due to such limitations, this study aims at assessing the spatial comfort of the shophouse, i.e., thermal and lighting. It is quite visible that quality. This study was carried out through the field measurement. Taking Banda Aceh climate as the case study in humid tropics, this study found that the high inside air temperature mostly runs throughout the upper floor during the day, and the ground floor during the night; and the small value of daylight occurs almost on the entire rooms during the day. This paper further proposes the possibility to improve the comfort quality in the shophouse which is conducted through the literature survey.

1. Introduction

Shophouse is the most building type built along the road with various mix-use functions. Due to the high rate of land price and the high percentage of building basic coefficient, therefore the shophouse is commonly built in maximum size. This characteristic creates a lack of apertures through the buildings due to being trapped by surrounding shophouses. Shophouses are also built in heavyweight construction, i.e., concrete and brickwork [1,2]. In humid tropics area such as Indonesia, thermal comfort is an essential need to be provided along with the building construction. Most of the shophouses use air- conditioner (AC) to provide cool air; therefore, it shows that there is lack of thermal comfort. Due to the maximum use of land, shophouses are commonly built with limited openings and ventilation; therefore, the electrical light is also widely used during the day.

As in the other city, the main roads in Banda Aceh are majorly surrounded by shophouses. The town is further developed with some new roads which are also followed with various new shophouses. The characteristics of the shops as previously described are built densely with heavyweight construction. The facade is designed with high surface temperature material [3]. With some problems firstly discussed the use of energy in providing cool air and light into the shophouse will be continuously excessive if there is not an intervention to reduce the energy use. In studying this phenomenon, this study reviews two different shophouses in some dense areas in Banda Aceh. The aim of this study is to assess the quality of thermal and lighting comfort in the shophouses which was evaluated through field survey, and to find the solution through literature review.

2. Spatial Comfort

Spatial comfort from the perspective of architectural science includes thermal comfort, adequacy Light and room acoustic [4,5]. In tropics for home cases room, acoustic is not a matter. Therefore thermal and lighting are mainly reviewed. According to ASHRAE Thermal comfort in the building is a condition that expresses satisfaction with the surrounding environment [6]. Factors influencing it are air temperature (^{0}C), mean radiant temperature (^{0}C), air velocity (m/s), relative humidity (%), human activity level (Metabolic Rate) and the type of clothing worn (Clo). Comfort temperature all over the globe is different which is depending on the average daily outdoor temperature (Td). Karyono collected some thermal comfort studies in Indonesia and from those found a regression equation of predicted comfort temperature (PCT) for Indonesia on the mean daily outdoor temperature as follows [7]:

$$PCT = 0.749Td + 5.953 \quad (1)$$

PCT is the predicted comfort temperature, and Td is the average daily outdoor temperature. The coefficient determination of $R^2 = 0.38$ ($r = 0.61$), and the regression is significant at a 95% confidence level. For Banda Aceh where the average daily outdoor temperature (Td) is 27.5^{0}C; while the maximum, and minimum value are 31.8^{0}C; and 23.2 ^{0}C respectively, therefore the comfort temperature range following formula 1 is as shown in table 1.

Table 1. Predicted Comfort Temperature (PCT) of Banda Aceh [8]

	Max (^{0}C)	Min (^{0}C)	Avg (^{0}C)
Td	31.8	23.2	27.5
Tco	29.7	23.4	26.6

To bring thermal comfort in the tropics with warm moist character, it needs to remove the excessive heat through adequate ventilation. Stilt building is required in this area to capture the wind that moves faster above the average height of the building. Buildings are made of lightweight construction with apertures facing north and south [4]. Another factor of spatial comfort is lighting. Lighting is a process, a way, the act gives light. Light is a prerequisite for human vision, especially in recognizing the environment and run activities. The object that we see is the reflection of light from the object. Therefore, how we view and respond to our surroundings depends on the type of lighting used. There are fundamental differences between the light and illumination. Lighting emphasizes the properties of radiation that must be considered in designing interior. The application of good lighting cannot be separated from the optimal use of natural light and artificial efficient. While the light makes the room bright, it also brings the heat which can contribute discomfort. Therefore unwisely designing large apertures will refer to energy waste. On the other hand, poor lighting can make it difficult to respond around, whereas excessive exposure can lead to glare (glare) so that users are not comfortable during the day. It is necessary to present the building which is convenient for the user visually, at the same time it also needs a proper consideration for the light coming without bringing excessive heat effects. In this study, the illumination (lux) of the daylight is measured. Illumination is the intensity, as perceived by the human eye, of light that hits or passes through a surface. The recommended standard of illumination for shophouse with ordinary activities, i.e., homes, warehouse and routine office work is 120- 500 lux [10].

3. Research Method

This study is a quantitative work which was developed through field measurement and literature review. Due to the limitation of the tool, the field measurement was limitedly conducted only to air temperature and relative humidity recorded for 1-3 days; air velocity and illuminance which were recorded for 1 hour alone during the day. The inside measured data were compared with the outside data. The measuring tool utilized is thermal data logger, wind speed meter, and lux meter. This study assessed two type of shophouse which has the same pattern namely built-in row. House-type A is located in suburban while the house- type B is in urban which is closed the CBD (Central Busines District) area of Banda Aceh. The two house is built of brick and concrete.

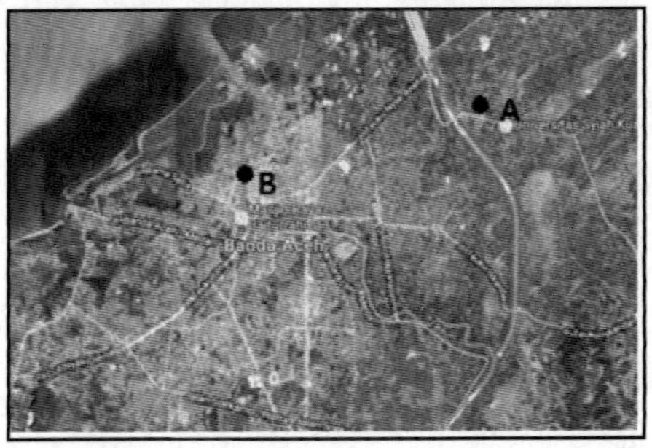

Figure 1. Location of the two houses measured in the study

Type A is the shophouse that is utilized as an office on the ground floor and as the house on the second floor (figure 2). It is located in Darussalam where the location is slightly dense. Type 2 is the house built by NGO in 2005 for tsunami survivors which resemble shophouse. This two-story house is built in a row attached to each other. Some people utilize the ground floor as business space and the second floor as the bedroom, while most of the people reside the entire area as homes (figure 2). The house is designed differently from the first type where the roof is pitch roof. The living room has high ceiling intended as a void connecting the upper ceiling to the bedroom on the upper floor. This house is located in Peunayong which is a highly-density living area. These two house types were measured at different time. Thus, the analysis assessed the inside thermal parameters of each house related to the outside thermal parameters during its measurement.

Figure 2. a. Shop house-type A **b**. Shophouse-type B

4. Result and Discussion

The spatial comfort data collected from the survey are shown in table 2. It shows that the two shophouses have the same pattern of air temperature that the inside air temperature during the night is significantly higher than the inside air temperature. This can be understood that the brickwork and concrete wall act as the thermal mass which releases the heat during the absence of the sun into the interior. The condition is worse due to the building attached to each other in a long row. The closed apertures during the evening and night also promote such discomfort because of the absence of adequate airspeed. However, table 2 shows that shophouse-type A has higher inside air (Ta), relative humidity (RH), and airspeed compared with the value of shophouse-type B. The condition may emerge due to the opaque shape and flat roof of shophouse-type A aggravated by limited apertures which are only accessed from the front and the rear. No roof shading the apertures also creates the discomfort. Shophouse-type B with more openings on the top wall as well as the front and the rear wall creates more paths for circulating air. The gable roof also promotes less inside air temperature compared with the flat roof. Illuminance was also recorded which was measured within 11 am to 3 pm during the measuring period. Shophouse-type A gained 53 lux of the average illuminance which does not meet the requirement which is at least 120 lux should be in homes space. While shophouse-type B can receive 137,6 lux which meets the illuminance standard for homes space.

Table 2. Thermal data of the two houses collected from the survey

Thermal parameter	House- type A			House- type B		
	Inside	Outside	difference	Inside	Outside	difference
Ta- avg (^{0}C)	33,25	28,5	4	31,5	29	2,5
Ta- max (^{0}C)	35	33	2	34	33	1
Ta- min (^{0}C)	31,5	24	6	29	25	4
RH – avg (%)	49,52	64,95	-15,43	54,67	63,1	-8,43
RH- max (%)	44,92	50,22	-5,3	47,49	50,22	-2,73
RH- min (%)	54,67	84,73	-30,06	63,1	79,81	-16,71
Av (m/s)	0,4	2	-1,6	1,7	3,21	-1,51
Illuminance (lux)	53	Bright day	N/A	137,6	Bright day	N/A

Although the thermal performance of shophouse-type B is slightly better compared with shophouse-type A, both shophouse types perform higher air temperature correlated with the predicted comfort temperature (table 1). It, therefore, needs some other strategies to create better spatial comfort in the shophouse.

5. Design Strategies

To find the solutions in providing sufficient and comfortable thermal and lighting parameters some relevant studies are discussed. The reviews are grouped into three characteristics, namely courtyard design; aspect ratio of the road and the verandah; and optimal roof design and vertical greeneries.

5.1. Courtyard design

Khan Ly in his study on shophouse in Hanoi found that tube house which is similar to shop house type commonly built; the courtyard has been approached by the ancient people to obtain the comfort [11]. Figure 3 shows that tube house or long narrow building has been separated by the courtyard which can improve the quality of spatial comfort, i.e., cooling and lighting.

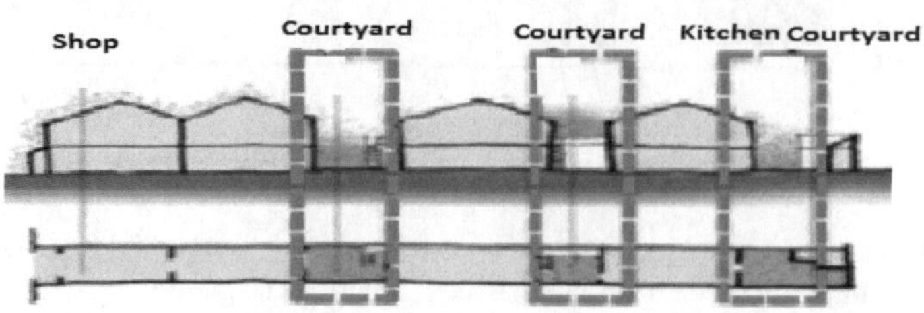

Figure 3. Courtyard in ancient tube house in Hanoi [11]

Even though Hanoi has the warm-humid sub-tropical climate, this design is significantly suitable to be applied in shophouse design in tropics such as Banda Aceh and extensively across Indonesia. The courtyard will let the air move throughout the house plan. The high inside air temperature will be solved through the shade from the vegetation planted in the yard and the shade from the next rooms.

The comfortable daylight also can be extensively achieved through the shaded openings along the rooms.

In Malacca, Malaysia such study on traditional Chinese Shophouses (CSHs) was conducted by Kubota et al. [12], They investigated indoor thermal conditions in traditional Chinese shophouses (CSHs) in Malacca, Malaysia, using field measurements and focuses on the cooling effects of courtyards. The results indicate that the indoor air temperature in the living rooms of CSHs was approximately 5-6°C lower than the outdoor temperature during the day primarily due to structural cooling effects with night ventilation, whereas the indoor air temperature at night was similar to the outdoor temperature. If the thermal adaptations of the occupants were considered, then the thermal conditions in the living rooms were acceptable for most of the day. The results indicate that the front courtyards functioned as a cooling source for the surrounding spaces in the CSHs. Kubota also found through his comparative study between the traditional Malay house and traditional masonry Chinese shophouse that the small courtyards which are commonly built in Chinese shophouse were effective to cool the high thermal mass structures through nocturnal ventilative and radiative cooling [1].

5.2. Aspect ratio of the road and the verandah
To gain thermal comfort in the shophouse can also be obtained through the aspect ratio of the road and the verandah. The narrower road can lead to the better thermal environment. In a case of the height of the beside building is 15 m, the road with 12 m-to-15 m's width (or the aspect ratio at 1 to 1.25) will conduct a comfortable microclimate. A more extensive verandah will also form a more comfortable microclimate [13].

5.3. Optimal roof design and vertical greeneries
Designing roof either through color, material or the form will affect indoor thermal performance. Emmanuel through his study found that white roof or roof insulation combination can reduce indoor air temperature up to 3^0C [14]. The white roof is also an efficient roof in reducing the effect of the urban island [15]. Building houses with apertures should consider a higher ceiling to cool down the inside air temperature. A higher ceiling can reduce the energy use since it has smaller vertical temperature gradients once the zone is occupied. A higher ceiling will allow greater ventilation is, therefore, allowing the room to meet thermal comfort level [16]. Pitch roof as evidenced in the case study works effectively in reducing the heat gain. It can decrease the inside air temperature down to 1^0C [8]. In Singapore, the shophouses are built with jack pitch-roofs for natural light and ventilation for cooling down the microclimate (Figure 4.B) [17].

A. B.

Figure 4. A. White roof for reflecting solar radiation [18]. B. Shophouses in Singapore with jack pitch-roofs [17]

Roof materials can be approached through Green roof which is also broadly known as the easiest way to reduce the air temperature. Green roof (the plants and the soil) can absorb 60% of the total solar radiation, and only 27% is reflected [19]. The heat and light absorption works for the greeneries life on the roof which means automatically less heat radiated into the interior. Further, optimal use of greenery around the building such as vertical garden also works efficiently in providing internal thermal comfort. Vertical greeneries also has the ability in producing fresh social space through the rise buildings [20].

5. Conclusion

Shophouse is the most building type built along the road with various mix-use functions configured in the limited land. Through field survey such limitations in humid tropic zones evident to emerge spatial discomforts such as high inside air temperature and relative humidity, low air velocity and poor illuminance. Some literature reviews provide some design strategies to cool down the microclimate. The use of courtyard as inherited in Chinese shophouse design is effective as a cooling source for the surrounding spaces in the shophouse as well as the void for searching the daylight. The aspect ratio of the road and the verandah also influence the spatial comfort in shophouses. The narrower road with the aspect ratio at 1 to 1.25 can lead to the better thermal environment. Roof design also works to create good spatial comfort. Whitening and greening the roof are the strategies to reduce the heat radiated into the interior. Vertical greeneries is also an excellent choice to cool down the microclimate of the shophouse which tends to be modified and expanded vertically. This study recommends further studies to analyze the cooling strategies in the shophouse design extensively by simulating them with thermal analysis software or through experiments.

References

[1] Kubota T and Toe D H C 2015 Application of passive cooling techniques in vernacular houses to modern urban houses: A case study of Malaysia *Social and Behavioral Sciences* **179** 29-39

[2] Sari L H 2015 An ecological assessment of housing in Aceh Province Indonesia *Proceeding International Conference On Architecture, Technology And Urban Infrastructure* **15** 1-12

[3] Irfandi 2015 dampak penggunaan elemen arsitektural pada koridor jalan terhadap terjadinya *urban heat island Prosiding Seminar Nasional Dengan Tema Permukiman Dan Kota Lestari, Jurusan Arsitektur Fakultas Teknik Universitas Syiah Kuala* Isbn: 978-602-74128-0-4,

[4] Szokolay S V 2004 *Introduction To Architectural Science, The Basis Of Sustainable Design*, (Oxford, Uk: Elsevier, Architectural Press)

[5] Lecher N 2012 *Heating Cooling and Lighting, Metode desain untuk Arsitektur* (PT Raja Grafindo Persada)

[6] CIBSE A 2006 *Environmental Design, The Chartered Institution of Building Services Engineers* (London: Guideline)

[7] Karyono 2001 Penelitian Kenyamanan Termis Di Jakarta Sebagai Acuan Suhu Nyaman Manusia Indonesia *Dimensi Teknik Arsitektur* **29** 24 – 33 (In Indonesian)

[8] Sari L H, Irwansyah M, Izziah and Meutia E 2017 An Environmental Assessment of Vernacular Housing in Banda Aceh, Indonesia *Tesa Arsitektur Journal of Architectural Discourses* **15**

[10] SNI 03-6197-2000, *Konservasi Energi Pada Sistem Pencahayaan* (In Indonesian)

[11] Khanh L 2013 *Study On The Thermal Environment And Residents' Behavior In The Courtyard Of Traditional Tube Houses In Hanoi, Vietnam*. (Available from: http://www.envr.tsukuba.ac.jp/~jds/pdf/2013Seminar%20presentation/Le%20Khanh%20Ly.pdf)

[12] Kubota T, Toe D H C and Ossen D R 2014 Field investigation of indoor thermal environments in traditional chinese shophouses with courtyards in Malacca *Journal of Asian Architecture and Building Engineering* **13** 247-254

[13] Yin S and Xiao Y 2016 Scale Study of Traditional Shophouse Street in South of China Based on Outdoor Thermal Comfort *The 4th International Conference on Counter measures to Urban Heat Island (UHI), Procedia Engineering* **169** 232–239

[14] Emmanuel R 2010 Linking the 'in' and 'out:' new comfort goals for the rapidly urbanising equatorial tropical megacities in a changing climate *Proceedings of Conference: Adapting to Change: New Thinking on Comfort* (Cumberland Lodge, Windsor and UK: 9-11 April)

[15] Life Science 2010 *White Roofs Could Reduce Urban Heating* (online, Available from: http://www.livescience.com/8069-white-roofs-reduce-urban-heating.html)

[16] Hashimoto Y and Yoneda H 2009 Numerical Study on The Influence of A Ceiling Height for Displacement Ventilation *Eleventh International IBPSA Conference Glasgow* (Scotland July

27-30 online, Available from: http://www.ibpsa.org/proceedings/BS2009/BS09_1045_1052.pdf)

[17] Guan T C 2011 *Learning From The Singapore Shophouse: Towards A Sustainable Tropical Architecture*

[18] https://thecornersoftheworld.wordpress.com/tag/north/

[19] Ekaterini E and Dimitris A 1998 The contribution of a planted roof to the thermal protection of buildings in Greece *Energy and Buildings* **27** 29-36

[20] Nugroho A M 2013 *An evaluation of building passive cooling achieved through application on vertical garden Proceeding of The 14th International Conference on Sustainable Environment and Architecture (SENVAR)* (Banda Aceh)

IOP Conf. Series: Materials Science and Engineering **352** (2018) 012067 doi:10.1088/1757-899X/352/1/012067

Model of slums rejuvenation in Telaga Tujuh village: the case of Langsa city, Aceh, Indonesia

Mirza Irwansyah[1], Irin Caisarina[1], Dini Solehati[2]
[1]Department of Architecture and Planning, Faculty of Engineering, Syiah Kuala University, Banda Aceh 23111, Indonesia
[2]Master of Civil Engineering Student, Department of Civil Engineering, Syiah Kuala University, Banda Aceh 23111, Indonesia

E-mail: mirza.hasan@unsyiah.ac.id

Abstract. Telaga Tujuh village is the only island inhabited compared to other islands in Langsa City, Aceh. Most of the houses are on stilts with very limited infrastructure such as lack of road facilities, local drainage, drinking water, wastewater, and garbage disposals. In determining the model of the slum settlements arrangement of Telaga Tujuh Village, there are some things to know that the characteristics of slums themselves and the causes of slum settlement. The aim of this study is to determine model of slum settlement arrangement that is suitable to be applied in the location. The method used is qualitative with sampling technique and qualitative analysis. To obtain the primary data used observation method, questionnaires, and interview. Secondary data obtained from agencies related to slum settlement arrangement. Based on characteristic analysis found that 365 residential buildings are irregular with the percentage of 100%, 365 residential buildings do not have safe drinking water supply, 365 residential buildings do not have waste water management. From the analysis shows that the appropriate model to be applied to Telaga Tujuh village is the rejuvenation model with the land consolidation system through the re-arrangement divided by two, 60% for the existing residential development and 40% for commercial development.

1. Introduction

Along with the growth of population in urban areas, the needs of housing, provision of facilities and infrastructure increased, either an improvement or new construction. Settlements and infrastructure need both in terms of housing or the environment of the neighborhoods that are affordable and livable yet fully can be provided by the community itself or the Government. Thus, the carrying capacity of environmental facilities and infrastructure of existing settlements began to decline and eventually contribute to the occurrence of the slums. One of the slums located in Aceh is in Langsa City, located in Telaga Tujuh village. The village has a total area of settlement 5.94 ha with a population of 1,574 people comprising of 420 families. Based on area and population, level of average population density is 152 people/ha. The number of built houses in Telaga Tujuh village totaled 365 buildings. The city government of Langsa has implemented programs related to the improvement of the environment, including the improvement of facilities and infrastructure of settlements, environmental sanitation and garbages, housing sanitation acceleration program, and provision of clean water, but was unable to reduce the slums. This is shown from a drinkable water supply is only 5%, the lowest order in Langsa city, sanitation for only 3%, ranks last in the city, low levels of public health is demonstrated from the number of cases of the disease environment often inflicted on the community i.e. cases of diarrhea.

Based on these conditions, then it is important to know the model of the arrangement of slums that are suitable to be applied on-site research.

2. Literature Review

2.1. The Slums

The slums are the settlement that is not habitable because of the irregular buildings, building density level is high, and the quality of buildings and facilities and infrastructure are not eligible. While slum housing is a housing that decreased the quality of function as occupancy (Act of the Republic of Indonesia number 1 Year 2011 about Housing and Settlement Area). The slums are part of the environment outside protected areas, urban and rural environment, functioning as residential or residential neighborhoods and places of activity that support livelihoods and livelihoods.

2.2. Characteristics of Slums

The characteristics of slums are:

1. The condition of houses in the slums were forced under the standards, an average of 6 m^2/person;
2. These settlements for the physical benefit of principal that is close to the workplace and home prices are also cheaper both buying or renting.

Besides that, there are other opinions that mention the characteristics which are based on the slums, namely:

1. The slums inhabited by a dense population and huddle rooms, due to both natural population and high migration from the village;
2. Slums are inhabited by citizens who are on low incomes or producing sub-systems, living below the poverty line;
3. Housing in the neighborhood of low quality or fall into the category of home emergency (sustainable housing condition), i.e. the building houses made from the traditional materials, such as bamboo, wood, reeds, board and other rapidly destroyed materials;
4. Conditions of hygiene and sanitation are low;
5. Lack of urban services such as clean water, toilet facilities, sewage and garbage disposal systems and fire protection;
6. Unplanned growth so that physical appearance is irregular and neglected;
7. Socially isolated from other layers of society;
8. The settlements are generally located around the center of the city and are often unclear as well as the legal status of the occupied land.

2.3. Model of Rejuvenation

According to Law number 1 Year 2011, the rejuvenation model is an effort to dismantle part or all of the housing and settlement environment and then in the same place constructed new housing and settlement infrastructure and settlements that are more suitable in accordance with the regional spatial planning [1, 2]. The main purpose of this activity is to increase the value of optimal land use in accordance with the potential land. In addition, it is expected to provide the consequences of the technical form of structuring as well as land consolidation, land re-adjustment, and land sharing [2]. According to Trianto, one of the strategies in rejuvenating is consolidation of land [2]. Land consolidation is the incorporation of several parcels of land (owned by different owners) and removes the ownership limit into a large site that can be used for a planned development project. This scheme is typically conducted in the old neighborhoods with low-density in the center of the city, and transforming it into part of a new, high-density city with smaller units of houses, more efficient designs, better public facilities and infrastructure [3].

The principle of land consolidation can be done in several ways including [2]:

1. Rights to land before and after consolidation are not changed to be higher or lower.

2. Land pooling by the government. The land collection is a land adjustment technique whereby the entire rebuilding process must be carried out by the government and allocated in some areas for the old occupants.
3. Land consolidation (reblocking) in the slums. Land consolidation can be done by leveling the entire building with the land, and rebuilding from the beginning a denser and more efficient settlement in the utilization of its space.
4. Involve the active participation of landowners.
5. Land given back to the owner has a higher value than before the land consolidation.

3. Research Location

Pusong. The village of Telaga Tujuh is a densely populated island and is the only inhabited compared to other islands in Langsa City. Geographically Village of Telaga Tujuh is located between 04 ° 46'77.05 "North Latitude and 97 ° 96'57.04" East Longitude. The width of Village of Telaga Tujuh is 5.94 ha with a population of 1,584 people and has 420 Family Heads (KK) spread over 5 (five) Dusun (hamlet), Dusun Aman, Dusun Damai, Dusun Rukun, Dusun Sejahtera, and Dusun Sentosa. The village of Telaga Tujuh has borders as follows:

1. The West is bordered by Village of Kuala Langsa;
2. The East is bordered by Village of Manyak Pahed (Aceh Tamiang);
3. The North is bordered by the Strait of Malacca; and
4. The South is bordered by Village of Langsa Lama.

The village of Telaga Tujuh has a height of 2 m above sea level. The distance from Kuala Langsa to Village of Telaga Tujuh 6 miles or 9.6 km with 45 minutes travel time using boat or boat with cost per person Rp. 5,000.

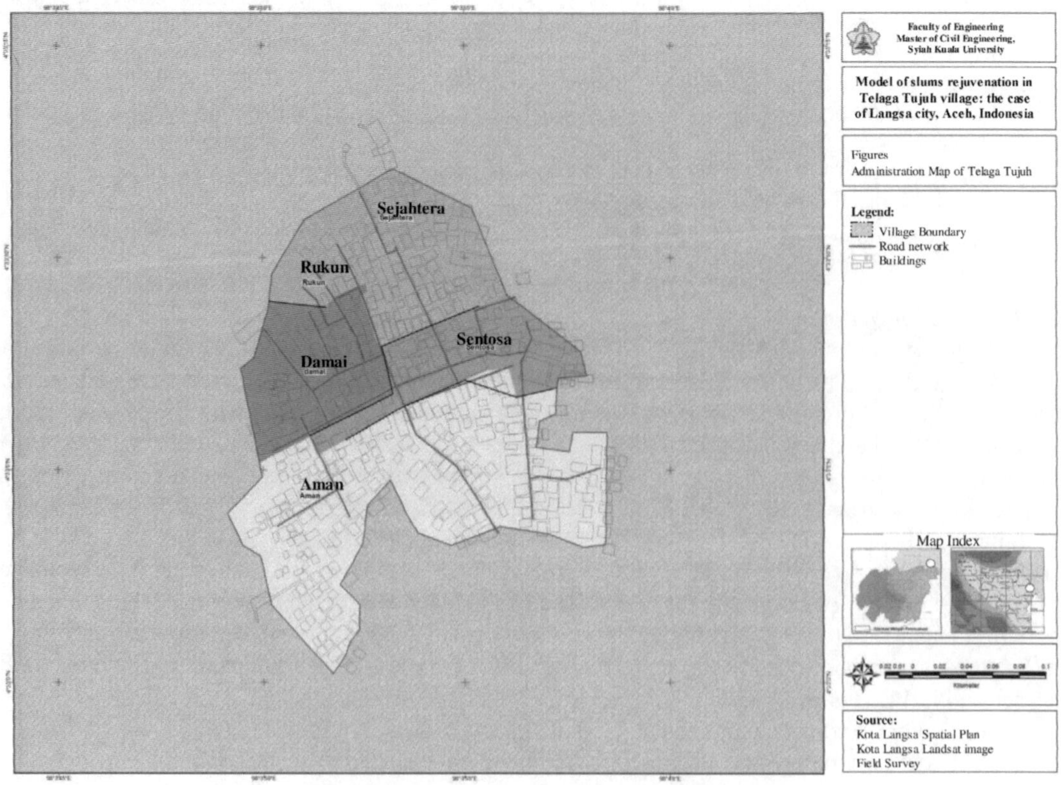

Figure 1. Map of slums area of Village of Telaga Tujuh

4. Methodology

4.1 Data Collection Method
Data collection methods in this study divided by 2 (two) namely the primary data collection and secondary data collection.

4.1.1 Primary Data Collection
Primary data collection is done by observation, questionnaire, and interview. Observation is done by visual observation, visual recording, and measurement. Questionnaires were conducted by absorption of aspirations to obtain information on the problems and potential of the area at this time and to explore the perception of the inhabitants of the region on the condition of their living environment. Interviews conducted in the form of semi-structured by asking questions to the policyholders associated with the slum settlement.

4.1.2 Secondary Data Collection
Secondary data collection was conducted through surveys to several government agencies that became the source of data, namely: Regional Development Planning Board (BAPPEDA) of Langsa City, Public Works (PU) of Langsa City, Central Bureau of Statistics (BPS) of Langsa City, Forestry Office of Langsa City, PDAM of Langsa City, Tourism Office of Langsa City.

4.2 Data Processing Method
If the data collection has been done, then the data that has been collected must be processed. The data processing procedure that will be done in this research is as follows:

1. Editing, i.e. sorting back the completeness and correctness of data needed;
2. Tabulation, i.e. by grouping data to simplify the process of analysis;
3. Classification, the data selected based on the needs of the analysis to be done.

After the data is processed and classified, then presented in certain forms such as tables, diagrams, graphs, and others to facilitate the reading and understanding.

4.3 Data Analysis
After doing the data processing, continue the data in the analysis. The analysis used is qualitative descriptive analysis for the characteristics of slum settlement of Village of Telaga Tujuh, and determine the model of slum rejuvenation of Village of Telaga Tujuh, Langsa City, Aceh.

4.4 The number of samples
This research uses data probability sampling technique in the form of stratified random sampling. Prior to sampling, the population has been calculated first. The total population used is the total of residential buildings of slums of Village of Telaga Tujuh as many as 356 units. Research sample using Isaac and Michael formula [4]:

$$s = \frac{\lambda^2 . N . P . Q}{d^2 (N-1) + \lambda^2 . P Q} \quad (1)$$

Whereas:
s : Sample
λ^2 : Z_{tabel} value with significance level of 5% (z=1,96)
N : Population
P : Proportion in population (0,5)
Q : 1-P (1-0,5=0,5)
D : Significance level 5% = 0,05
The sample size is 188 units using *stratified random sampling.*

5. Results and Discussion

5.1. Characteristics of Slum areas of Village of Telaga Tujuh

5.1.1. Condition of Residential Buildings

Ideal building with a distance of 3 m, based on the existing condition, the residential building is 0-1 m. According to Prayitno [5], the high density of buildings is more than 100 buildings/ha with a population density of 200 souls/km, this is experienced by Village of Telaga Tujuh where the density of buildings of more than 150 buildings/ha and population density 200 soul/km. According to PERMENPERA Number 22 Year 2008 regarding Minimum Service Standards in the Housing Sector of Provincial Region and District or City Region, the house can be categorized as habitable occupancy if the per capita floor area is greater than 7.2 m^2. Meanwhile, on the existing condition of floor area per capita less than 7.2 m^2 amounted to 280 units. So that 80% of residential buildings of Village of Telaga Tujuh unfit for habitation.

Figure 2. Residential Building Density Condition of Slum of Village of Telaga Tujuh

5.1.2. Local Road Condition

The local road network is a major component in a residential area. An access can be made only through the existence of the local road. A matter of urgent which importance to the existence of local roads to be passed by four-wheeled vehicles. This is related to the interests of fire-fighting vehicles. Likewise, irregular road patterns cause inconvenience to road users. Although with the condition of Village of Telaga Tujuh 1-1.5 m, it is still possible to pass by bicycle and pedestrians. However, because many misused such as drying salted fish, drying clothes, and used by traders until this road is not comfortable or safe to pass. From the analysis result, the length of road with width of $\geq$ 1.5 m, hardened surface and not damaged only 78 meter from total road network of 2,350 meter causing road network quality decreased, this is not in accordance with General Directorate of Cipta Karya (2013) standard width minimum pavement for 2-way road 4 meters.

Table 1. Quality of Local Road of Slums Area of Village of Telaga Tujuh

Quality of Road Surface	Length (m)	Percentage (%)	Area of Settlement (ha)	Percentage (%)	Road Infrastructure	Photo
Local road concrete (good condition) good quality	105	21	1,76	22	There are no street lights and no drainage	

Quality of Road Surface	Length (m)	Percentage (%)	Area of Settlement (ha)	Percentage (%)	Road Infrastructure	Photo
Road without pavement (dirt road or wood) bad quality	395	79	6,24	78	There are no street lights and no drainage	
Total	**500**	**100**	**8**	**100**		

5.1.3. Local Drainage Condition

Drainage is an important part of the spatial structure in the settlement. Based on the analysis of the inundation area in the 5 ha settlement, where Village of Telaga Tujuh does not have drainage with 0% percentage. This is not in accordance with the mandate of Law No. 1 of 2011 on Housing and Settlement Area, where the drainage of the environment is not connected with the city system, must be prepared local recharge or retention ponds. The drainage condition of the slum neighborhood of Village of Telaga Tujuh 100% is not served by drainage from 8 ha of slums. This is based on the parameters of the assessment of slum conditions that refer to PERMEN PUPR. No. 2 of 2016. Besides drainage can also affect the bad environmental conditions as happened in Village of Telaga Tujuh.

Figure 3. Condition of Local Drainage of Slums Area of Telaga Tujuh (Observation Result, 20 Mei- 4 Juni 2017)

5.1.4. Condition of Drinking Water Supply

Slums are identical to inadequate infrastructure, including clean water or drinking water infrastructure. Based on observations in the field, 365 residential buildings in the study area did not have drinking water facilities for drinking, bathing, and washing with a non-piped system with 0% percentage. The drinking water used by the community obtained from PDAM is in the form of Water Tank. Drinking water from PDAM Langsa City is brought to Village of Telaga Tujuh once every 7 days with good quality water taps no smell, no taste and no color just water quality from PDAM not able to fulfill the requirement of everyday citizen needs because of limited distribution and water price quite expensive. Residents buy water from PDAM for Rp. 20,000 to Rp. 25,000 per 20 liters. Regarding the habit of the residents during the rainy season that provides empty places to accommodate when it rains, the water will be used as bath water and washing. Based on the above analysis, there are 365 residential buildings unfilled drinking water 60 liters /day for 1,574 people with 100% percentage caused by the community only rely on clean water supply from PDAM, based on this matter not in accordance with PERMEN PUPR Number 2 Year 2016, minimum requirement is 60 liters/person/day with minimum service capacity of 20l/person/day. The need for drinking water can be fulfilled with the Drinking Water Supply System with the piped network (SPAM) and the Non-piped Water Supply System (SPAM BJP).

IOP Conf. Series: Materials Science and Engineering 352 (2018) 012067 doi:10.1088/1757-899X/352/1/012067

Figure 4. Clean Water Condition of Slum Area of Gampong Telaga Tujuh

5.1.5. Sanitation Management Conditions

Based on the results of the analysis it is found that the slums of Village of Telaga Tujuh are using individual waste management system. In addition, the sanitary waste disposal system is conducted on-site, where the wastewater is not collected in a channel on the network that will lead to wastewater management or receiving water body, but to the land part of the house or directly discharged to the coast. Such wastewater management systems are poor because the waste that directly sinks into the soil can pollute groundwater quality and also damage the coast. The waste management system in the slums of Village of Telaga Tujuh is very poor as all households dispose of household waste individually and on-site. From the analysis result, 365 residential buildings do not have a good wastewater management system with 100% of the slum area of Village of Telaga Tujuh slum, this is not in accordance with the mandate of Law no. 1 of 2011 that settlements are habitable settlements that have a centralized or local wastewater management system.

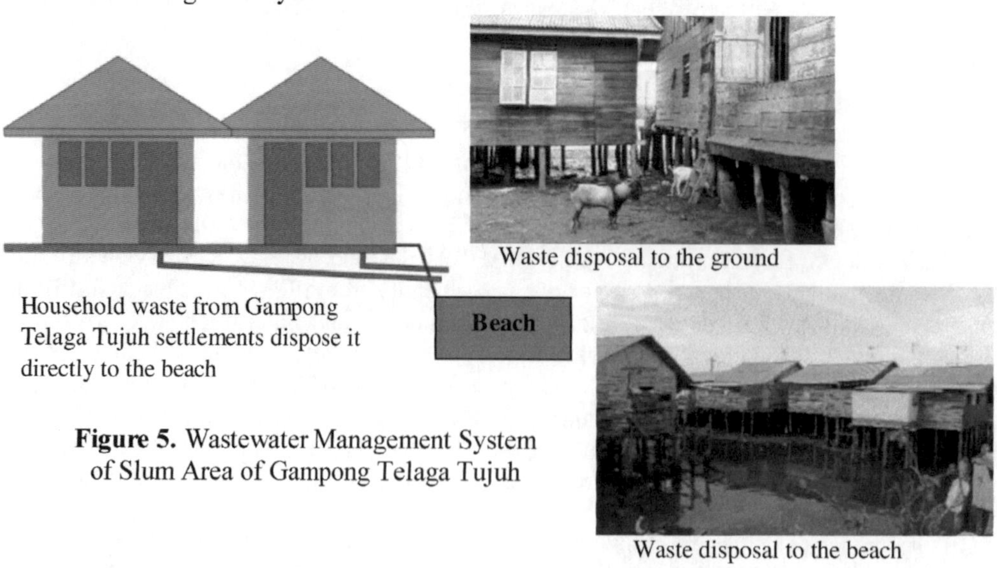

Figure 5. Wastewater Management System of Slum Area of Gampong Telaga Tujuh

5.1.6. Condition of Waste Management

Based on analysis result, 365 residential buildings do not have good waste management with 100% percentage do not have technical condition of facilities and infrastructure. This is evidenced by

IOP Conf. Series: Materials Science and Engineering **352** (2018) 012067 doi:10.1088/1757-899X/352/1/012067

household waste placed on certain places side or back of the house that has a yard or vacant land. There are 365 residential buildings with no garbage bin or TPS or TPST, and no garbage transportation by officers or government of Langsa City and no 100% processing system. The distance between Langsa City to the slums.

Figure 6. Waste Management of Slum Areas

5.1.7. Condition of Fire Protection
Based on PERMEN PUPR Number 2 Year 2016, slums area have narrow roads and alleys that make it difficult for citizens to access emergency services. Fires are a frequent disaster in the slums. Based on the analysis results obtained the existence of water supply from the sea, meanwhile from the analysis that there are no means of supporting fire protection such as fire protection facilities related to the provision of Fire Extinguishers (APAR), pump cars, thrust pumps, motor pumps and ladder cars. This is because the supporting facilities for firefighting operations from the Langsa City government have not been able to balance the need for fire prevention. Public awareness of fire hazards is also very low where the fire prevention system is one of the priorities that must be owned by people in slums.

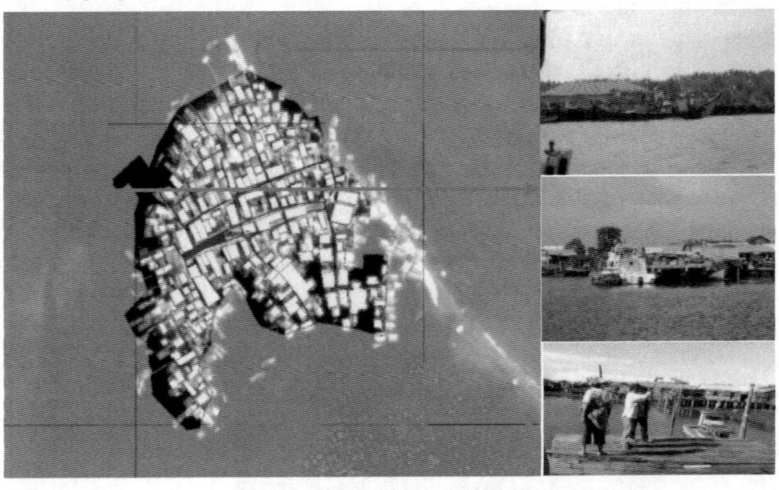

Figure 7. Location of Water Resources as Fire Protection Infrastructure of Gampong Telaga Tujuh (Observation Result, 20 Mei- 4 Juni 2017)

5.1.8. The Condition of Land Potential
The results of the analysis obtained from observations and interviews with the group of Program Improvement of the Quality of Settlement Area (P2KKP) and Geuchik that Village of Telaga Tujuh has potential as fishing port spread in 6 (six) locations. The village of Telaga Tujuh also has potential as mangrove ecotourism because it is adjacent to mangrove island so that it can improve the regional economy. The potency of land owned by Village of Telaga Tujuh with the high category. Slums that have a high potential land model suitable for structuring are rejuvenated models with a land

IOP Conf. Series: Materials Science and Engineering **352** (2018) 012067 doi:10.1088/1757-899X/352/1/012067

consolidation system such as Sulestianson [6] states that slum settlement arrangements with a consolidated model of land rejuvenation should have the potential to be developed into functional areas that are more strategic than residential.

The distance of Gampong Telaga Tujuh to Mangrove Island 50 meters

The area of mangrove is approximately 500 ha and has potential as mangrove ecotourism for Gampong Telaga Tujuh so it can minimize negative impact to local environment and culture and can increase community economic income and conservation value. The mangrove island of Kota Langsa, including the largest mangrove forest in Indonesia (Kota Langsa City Forestry Agency, 2017)

Figure 8. Mangrove Ecotourism Potency

5.2. Arrangement Model of Slum Settlement of Village of Telaga Tujuh

The result of analysis related to 8 (eight) characteristics of slum settlement in the form of residential building condition, local road condition, drainage condition, condition of drinking water supply, wastewater management condition, solid waste management condition, fire protection condition, land potential condition as base in determining model arrangement of slum settlements of Village of Telaga Tujuh. Based on the slum characteristic analysis that Village of Telaga Tujuh has high density, poor facilities and infrastructure, low sustainability level, and has the potential of land that can be developed to improve the regional economy, it is still possible to become a residential area. This makes it possible to apply rejuvenation models with land consolidation systems, as Sulestianson [6] states, the terms of land consolidation for unpatronized settlement layouts, with diverse uses, and the potential to be developed into functional areas more strategic than shelter. Land use in land consolidation is divided into 2 (two), 60% is used for resettlement for existing densification landholders, and 40% of the land is used as commercial land to cover the cost of housing, facilities, and infrastructure. This is as Prayitno [5] says, on 60% land, alternatives are created for two schemes, namely the development of a series house or flats where to accommodate 100% of the population currently occupying the existing land.

Figure 9. Arrangement Model of Rejuvenation Settlements of Village of Telaga Tujuh

5.3. Concept Model of Fisherman Village Rejuvenation

5.3.1. Concept of Residential Buildings
The concept of residential development in slums should consider safety, comfort, health, and convenience for building users, and have permits. Several stages of the concept carried out in achieving decent residential development:

1. The mass of the building extends eastward and northwestward or elongated in the direction of the axis of the road;
2. The standard used for building floor plan is 60%: 40%, which are 60% for residential building and 40% for commercial area;
3. The minimum requirement of building area 9m2 / person, and the standard number of residents on average 5 persons/families;
4. The shape of a house is a permanent building, the building undergoes a shift shape. Form of the house based on relationship or characteristics between buildings:
   a. A Single house, a stand-alone house and a plot, separated from the house adjacent to it;
   b. The house series, the house in pairs (coincide) usually one roof in a plot is usually a maximum of 6 rows;
   c. Flats, terraced houses built vertically in structurally functional sections.

Figure 10. Concept of Residential Buildings

5.3.2. Concept of Local Road
The concept of the development of roads in the settlement neighborhood of fishing village is as follows:

a) Opens the access area to connect with the city collector path
b) Development of the main line of the area surrounding the area
c) Development of access to central environmental units
d) Improving the quality of local roads in order to pass through three-wheeled motor vehicles.
e) Development the path of circulation of environment that is safe, comfortable and friendly for people with disabilities and the elderly
f) Development of area orientation with the development of marking elements as a benchmark of settlement in the of form of gate area
g) Development of local roads using paving block and concrete and absorption wells to accommodate rainwater flow to maintain road quality, road width 3 m.

Figure 11. Concept of Local Road

5.3.3. Concept of Drainage Management
The drainage handling concept in the Fisherman's Village Area is generally a concept of sub-catchment pattern arrangement of areas where the benefits of sub-catchment distribution smaller are the reduction of the burden on the parent channel serving the sub-catchment especially when on the channel has a lot of solid surfaces. Several stages of concept were done in achieving optimal drainage development include:

1. Development of drainage area and create new networks in the inundation area so as to support the creation of a healthy and comfortable settlement environment;
2. Development and improvement of drainage channels with minimal dimensions for local drainage L=40 cm, for main local drainage L=50 cm with river stone paving and minimum dimensions for drainage on primary circulation L=100 cm with river stone pavement;
3. Coastalization in the form of dredging, straightening, slicing of the narrow coastal part and the making of a wall of beach wall reinforcement on the vulnerable coastal wall;
4. Control of the beach not to become a garbage disposal by the community that it can cause siltation of the beach;
5. Control of the buildings around the beach and the orientation of the building is not back to the beach;
6. Greening along coastal areas bounded by road network and green open space on the beach border;
7. Production of absorption wells in every home construction, and local road.

Figure 12. Concept of Drainage Management

5.3.4. Concept of Drinking Water Supply
The concept of developing clean water infrastructure is enhanced by its capacity through the development of retention ponds as a source of water and flood control. Several stages of the concept that is carried out in achieving the optimal supply of drinking water include:

1. Pool retention as a source of water and flood control;
2. Clean Water Treatment Plant (IPAB) of sub-regional scale;
3. Primary drinking water network area;
4. Secondary sub-regional networks.

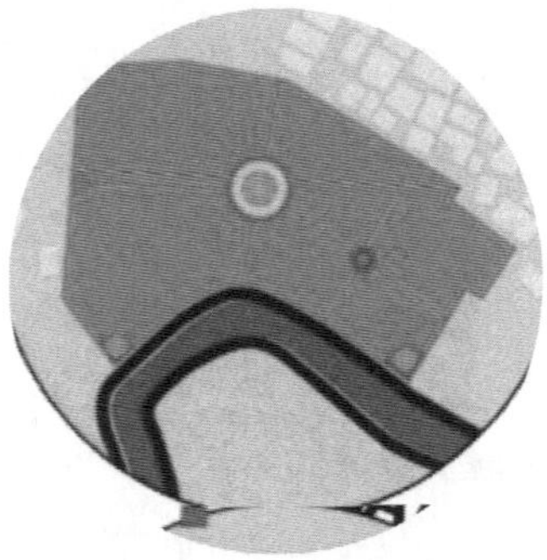

Figure 13. Concept of Drinking Water Supply

5.3.5. Concept of Wastewater Management

The concept of wastewater management of the area is directed to an integrated system of water treatment so that the quality of the discharged water is handled first and then flows to the coast in relatively good quality condition. This is done to reduce water pollution by ecoli bacteria as well as for handling rainwater directed to the provision of groundwater reserves through the procurement of absorption wells which located on the main local road.

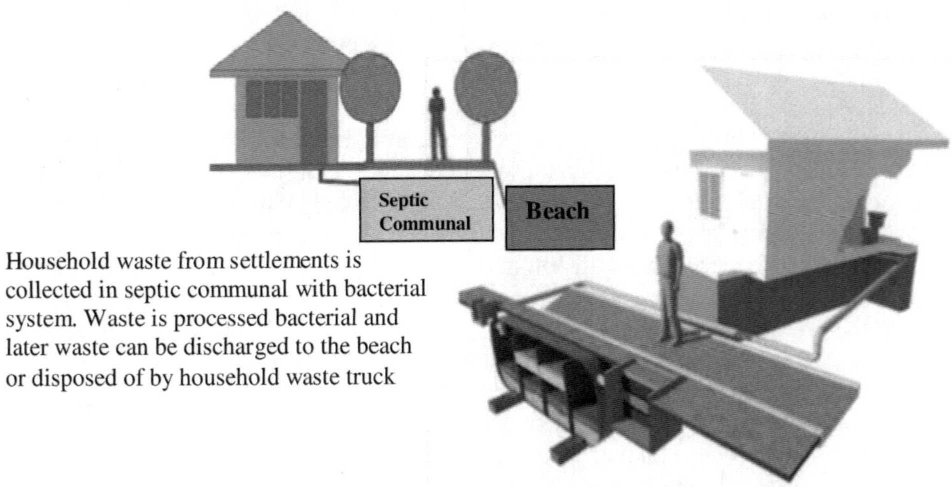

Figure 14. Concept of Wastewater Management

5.3.6. Concept of Waste Management

The concept of development of garbage in fisherman's village is spread in every settlement along the coast and the sub-regional center garbage in RW or Dusun. Development of access for trash carts on neighborhood streets and for trucks transporting trash to trashtrack.

Several stages of the concepts carried out in achieving optimal waste disposal include:

1. Provision of iron containers at Trashtrack in RW or sub-village;
2. The construction of road ramp with a maximum slope of 3% with ratio (1:5) and road width of at least 3 (three) meters so that the road is quite sloping and can be passed by garbage cart, and garbage in this location can be transported out to TPS, or waste management is done by composting method by placing the composting points at some RTH points;
3. On the coast every 10 m will be placed household-scale trash can, so it is expected that residents can use the trash and do not throw garbage to the beach;
4. Conducted management 3R (reduce, reused and recycled) for example by sorting and composting organic waste.

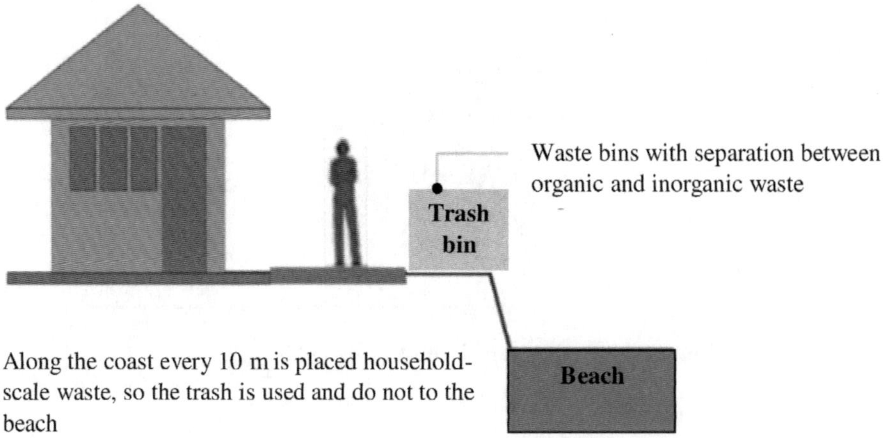

Figure 15. Concept of waste management

5.3.7. Concept of Fire Protection
Development of the concept of fire protection arrangement is as follows:
1. Provision of fire hydrants in each settlement environment;
2. Set the distance between buildings and the use of building materials.

5.3.8. Concept of Mangrove Tourism Potential
Mangrove forest tourism is a form of travel activities by traveling around the area of mangrove forest and enjoys all the uniqueness contained in it. Mangrove forest as one of the tourist area has a function as a place of travel for local and foreign tourists, with facilities provided by the manager. Mangrove area with a diversity of types of mangrove plant species and its animals can provide a unique tour activity and provide insight and experience with a different atmosphere. Types of tourism or tourism potentials offered by mangrove forest as one of the tourist attractions such as bird watching, fishing, cross country tracking, enjoying the scenery and the diversity of mangroves characteristic of mangrove tourism [7]. The concept of mangrove ecotourism cannot be separated from:
1. The value of education;
2. Local community participation and economic value;
3. Conservation and environmental management efforts;
4. Minimize environmental impacts and impacts.

Figure 16. Concept of Mangrove Tourism

6. Conclusion
Based on the results of research and analysis and discussion that has been done can be concluded as follows:

A. The results of the analysis of the characteristics of slum settlement of Village of Telaga Tujuh are identified as 365 residential buildings having high density with 100% percentage of irregular buildings and unsuitable for occupancy, 20% of poor local roads with dirt pavement, 365 buildings do not have local drainage with 100% , 365 residential buildings are not met the need to drink, bathe, and wash at least 60 liters/person/day with 100% of safe drinking water unavailable, 365 residential buildings do not have waste water management with 100% percent unserved, 365 buildings do not have good waste management, and identified 50% of fishery land potential in 6 locations and 1 location of tourism.

B. The village of Telaga Tujuh is an area that has the potential to be developed as a function of the settlement, and commercial. The location with fish dock and mangrove forest facilities, makes this area has a high-value dan it is worth doing a rejuvenation area in that area.

C. Model of rejuvenation with land consolidation system by building a house series are:
   a. The rehabilitation of the Village of Telaga Tujuh rebuilding house series was established on 8 ha of land with 36 type houses. While the size of the building space which is the demand of the space of case studies and interviews with related parties can be used as a reference in the design strategy, in addition to the standard space for ideal or predefined house occupancy.
   b. Development of settlement is done by dividing 2 (two), 60% for existing building as much as 365 buildings, and 40% for the commercial area which will be developed as fishery and mangrove ecotourism.

Acknowledgements
This research was supported by Faculty of Engineering of Syiah Kuala University.

References

[1] Law of the Republic of Indonesia Number 1 Year 2011 on Housing and Settlement Area. State Gazette of the Republic of Indonesia Year 2011 Number 5188 (Jakarta: Secretariat of State, from http://ehousing.perumahan.pu.go.id. (accessed on 28 November 2016)
[2] Trianto A 2016 Pattern on handling slum settlements Cimahi City *Master Program Thesis* (Bandung Institute of Technology)
[3] UN-HABITAT (United Nations Human Settlement Program) 2008 Housing for the Poor in Asian Cities, Bangkok http://unhabitat.org. (accessed on December 4, 2016)
[4] Sugiyono 2016 *Quantitative Research Methods, Qualitative and R & D* (Alfabeta Publishers; Bandung)
[5] Prayitno B 2016 *Scheme of innovation for handling slum areas* (Gadjah Mada University Press; Yogyakarta)
[6] Sulestianson E 2014 Handling slum settlements with characteristic approach and slum factor case study: slums in Taman Sari and Braga Urban Villages *Journal of Regional and City Planning B Sappk* **3** 261-270
[7] Pranatha I M A, Arthahana I W, and Utmani N W F 2015 Mangrove tourism landscape planning at Akame Restaurant, Benoa, *Journal of Landscape Architecture* (Faculty of Agriculture, Udayana University, Bali) **1**